M. Altmaier, V. Montoya, L. Duro, A. Valls (Eds.)

Proceedings of the Second Workshop of the HORIZON 2020 CEBAMA Project

Karlsruhe Institute of Technology
KIT SCIENTIFIC REPORTS 7752

Proceedings of the Second Workshop of the HORIZON 2020 CEBAMA Project

Edited by
M. Altmaier, V. Montoya, L. Duro, A. Valls

KIT Scientific Publishing

Report-Nr. KIT-SR 7752

Impressum

KIT Scientific Publishing

Karlsruher Institut für Technologie (KIT)
KIT Scientific Publishing
Straße am Forum 2
D-76131 Karlsruhe

KIT Scientific Publishing is a registered trademark
of Karlsruhe Institute of Technology.
Reprint using the book cover is not allowed.

www.ksp.kit.edu

Print on Demand 2019 – Gedruckt auf FSC-zertifiziertem Papier

ISSN 1869-9669
ISBN 978-3-7315-0825-0
DOI 10.5445/KSP/1000084618

FOREWORD

The present document contains the proceedings of the Second Annual Workshop (AWS) of the EURATOM H2020 Collaborative Project CEBAMA (Cement-based materials, properties, evolution, barrier functions). The electronic version of these proceedings is available on the webpage of the project (*http://www.cebama.eu/*) and at KIT Scientific Publishing (*www.ksp.kit.edu*). The project started in June 2015 and has a duration of four years. The project is implemented by a consortium with 27 Beneficiaries, from 9 EURATOM Signatory States, Japan and Switzerland. National Waste Management Organizations contribute to the running project by participation in the End-User Group, by co-funding Beneficiaries, and provide for knowledge and information transfer.

These proceedings serve for several purposes. The key purpose is to document and make available the progress of the CEBAMA project to a broad scientific community. For this purpose, a considerable part of the project activity is reported by the proceedings, together with the scientific-technical contributions containing details of the work. The progress shown corresponds to the first two years of life of the activities, and we are very proud to see that all groups have started their research and that good results have already been obtained. The project facilitates training mobility measures for students, and the workshop also served to put in place and communicate the procedures for application of these grants. Additional purposes of the proceedings are to ensure on-going documentation of the project outcome, promote systematic scientific-technical development throughout the project and to allow thorough review of the project progress.

All Scientific and Technical papers (S&T) included in the proceedings have been reviewed by the workpackage leaders and the EUG (End-User-Group) of CEBAMA. The EUG is specifically set up within the project representing the interest of the national waste management and/or national regulatory organizations that may use the results of the project for their Safety Cases. For this reason, this EUG had a strong involvement on guiding the priorities of the activities of the project and it is a very useful tool to guarantee that the research is done in a way that allows for future knowledge transfer.

The proceedings give only very brief information about the project structure and the different activities around the project. This type of information is available in detail under *http://www.cebama.eu/*.

The editors of the proceedings thank all contributors to the project, especially those submitting Scientific and Technical contributions, and the workpackage leaders who provided the summary of the different workpackages for publication in these proceedings. Special thanks are given to the reviewers, the members of the EUG, whose effort and hard work reflected their commitment and dedication to the project and contribute to a high quality of the research performed within CEBAMA.

TABLE OF CONTENTS

THE PROJECT

Cebama is a Collaborative Project funded by the European Commission under the Horizon 2020 Reasearch and Training Programme of the European Atomic Energy Community (EURATOM) (H2020-NFRP-2014/2015), section B - Contribute to the Development of Solutions for the Management of Ultimate Radioactive Waste, Topic NFRP 6 - 2014: Supporting the implementation of the first-of-the-kind geological repositories).

The Collaborative Project Cebama addresses key issues of relevance for long term safety and key scientific questions related to the use of cement-based materials in nuclear waste disposal applications. These materials are key components in the barrier system of repositories, independent on the actual host rocks. They are used as waste forms, liners and structural components as well as sealing materials in a broad variety of applications. Waste forms and their behaviour as well as the technical feasibility and long-term performance of repository components are key topics detailed in the Strategic Research Agenda (SRA) of the Implementing Geological Disposal Technology Platform (IGD-TP) and cover studies related with: (a) the release of radionuclides, (b) long-term behaviour of seals and plugs, (c) evolution of cement-based seals, (d) interaction of cement with clays and (e) optimisation aspects.

The overall objective of Cebama is to support the implementation of geological disposal by improving significantly the knowledge base for the Safety Case for European repository concepts. The research planned in Cebama is largely independent of specific disposal concepts and addresses different types of host rocks in addition to bentonite. Cebama is not focusing on one specific cement material, but aims to study a variety of important cement-based materials in order to provide insight on general processes and phenomena.

The ambition of Cebama is the development of comprehensive models for predicting the alteration processes and their impact on transport characteristics such as porosity, permeability and diffusion parameters of different cement-based materials in contact with the engineered and natural barriers of repositories in crystalline and argillaceous host rocks. Dedicated studies on radionuclide retention processes on relevant hydrated cement phases, concrete and alteration products are also part of this advanced approach.

The project started on 1st June 2015 and will last 4 years. The project is implemented by a consortium with 27 Beneficiaries, from 9 EURATOM Signatory States, Japan and Switzerland. National Waste Management Organizations (WMO) contribute to the running project by participation in the End-User Group, by co-funding Beneficiaries, and provide for knowledge and information transfer.

Indispensable for CEBAMA is the documentation of the scientific and technical state of knowledge and the dissemination of the generated knowledge not only to the WMO but also to the general scientific community and other interested parties. This dissemination strategy will be implemented by presentations at international conferences and publications in peer-reviewed journals. Interaction with social stakeholders is of relevance for facilitating global understanding for the need of research in nuclear waste management, and CEBAMA aims at helping to bridge this communication process. The aspect of training and education of the next generation of waste management professionals is specifically addressed in CEBAMA.

2nd ANNUAL WORKSHOP

The 2nd Annual Workshop of the Collaborative Project CEBAMA was held in Espoo (Finland), 16th – 19th May 2017. The workshop was hosted by VTT at the GTK (Finnish Geological Survey) building. There were 66 attendees, including beneficiaries, the End-User-Group and project external organizations. The workshop was organized in four days being the last one an optional day to visit the Posiva site at Olkiluoto. The general agenda of the workshop is shown in Figure 1.

Preliminary Agenda *(16th–19th May 2017)*

Tuesday 16th May 2017

11:30-13:00	1st ExCom meeting *(restricted to ExCom members)*
13:00-14:00	Welcome + Presentation by the Coordination Team
14:00-16:00	Individual WP sessions
16:00-17:00	Smaller Task group session
17:00-18:30	Finland Profile: Posiva & Finland's Geological Survey. Guided visit to the Geological Museum.

-- Workshop DINNER --

Wednesday 17th May 2017

08:30-10.00	Associated Groups & Safety Case presentations
10.00-12:00	**WP1** plenary session
	LUNCH
13:00-14.30	**WP3** plenary session
14:30-17:00	Socio-political **Stakeholder** panel discussion (public event)
	POSTER SESSION (Finger food)
18:00-19:00	2nd ExCom meeting with EUG *(restricted to ExCom members)*

Thursday 18th May 2017

09:00-11.00	**WP2** plenary session
11:00-12:30	**EUG** feedback + **General Assembly** **Workshop Adjourns**
	LUNCH
Optional sessions:	
13.30-14.30	**VTT Underground Research Lab tour**
13:30-16:30	**PhD and young researchers** session + VTT lab tour

Friday 19th May 2017

07.00-18.00	**Posiva site excursion:** Olkiluoto Visitor Centre, Low & intermediate level repository

Figure 1: *General agenda of the 2nd Annual Workshop of Cebama.*

The objectives of the 2nd Annual Workshop are presented below.

- Ensuring that all project partners are aware of the project objectives, work program and reporting obligations
- Promoting joint research activities within Cebama
- Providing for detailed agreements on the work program, activities and training measures
- Communicating the status of the different project activities, work and state of progress between all project partners including non-technical stakeholders
- Discussing the status of the work programme and decide on next steps, including deviations from original work planning, if required
- Preparing for organization of public Workshop Proceedings for effective documentation and communication of the project achievements

In contrast to the 1st Annual Workshop, the 2nd workshop not only included the ExCom/EUG meetings, WP sessions and General Assembly but it also held new sessions such as a PhD session or a Stakeholder Panel. Figure 2 details all activities / sessions that took place during the 2nd Annual Workshop. A brief description of each activity is presented below.

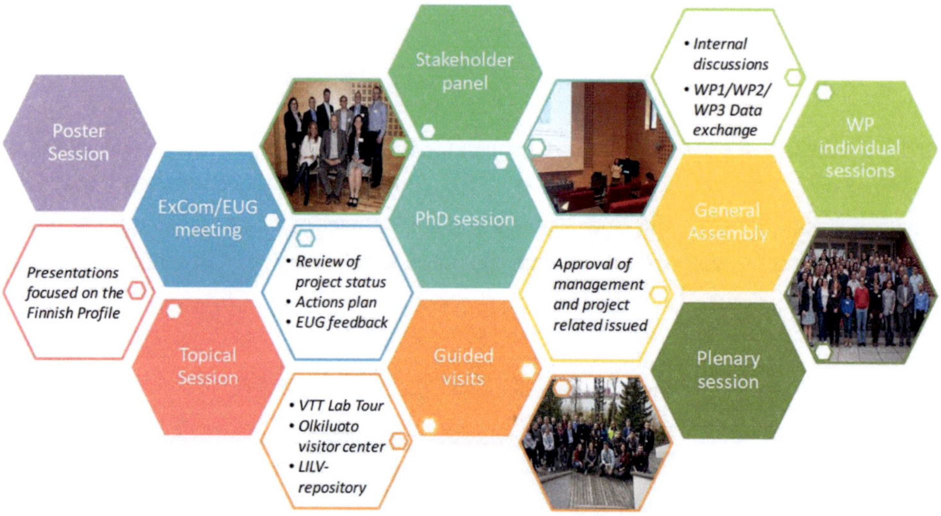

Figure 2: *Activities and sessions held at the 2nd Annual Workshop of Cebama.*

<u>ExCom and EUG meetings</u> took place the first and second day of the workshop focused on the project management and discussion with EUG on the project progress.

<u>*Individual WP sessions*</u> were devoted to internal discussions on the progress of each WP and to agree on the data exchange between work-packages in order to achieve the milestones scheduled for the next months.

<u>*The topical session*</u> was focused on the Finnish Profile. It consists of different presentations and <u>*guided visits*</u>, which were optional. The following talks were given in this session.

- Siting process for HLW disposal in Finland. *T. Ruskeeniemi (GTK),*
- Overview of Posiva's programme for final disposal of spent nuclear fuel. *T. Jalonen and M. Vuorio (Posiva),*
- Use of scientific/technical knowledge and applicability to the safety case. *B. Pastina (Posiva).*

Regarding the guided visits, most of the workshop attendees visited both the GTK museum and the VTT underground Lab Tour. Both visits were considered valuable additions to the Workshop.

Friday took place the visit to the Olkiluoto Visitor Centre and the Low & Intermediate level repository. Anne Niemi, visit manager of TVO, guided the group during the tour, starting with a presentation on TVO Company. Two more presentations were given before visiting the Low & Intermediate level repository:

- Posiva overview of cementitous material R&D (Marja Vuorio, Posiva),
- TVO overview of cementitious material safety (Timo Kukkola, TVO).

Friday afternoon was devoted to visit the underground repository for low-intermediate level waste (LILW-repository), the ONKALO Research Gallery and the scientific exhibition "Electricity from Uranium" (Olkiluoto Visitor Centre). Good feedback was received by the visitors.

<u>*Plenary sessions of WP1, WP2 and WP3*</u> (2nd and 3rd day) consisted on a brief overview of the work progress given by the workpackage leader followed by a series of presentations given by partners. These presentations were devoted on showing partners' work and the obtained results. The talks included in each session are provided below.

WP1 Session

- WP1 Overview - 2nd year progress. *F. Claret (BRGM), E. Holt (VTT), U. Maeder (UniBern)*
- Evolution of porosity in cementitious materials during early stage of alkali-activation: a spin echo small angle neutron scattering and SEM/EDS investigation. *A. Sabau, D. Bykov, W. Bouwman, C. Duif, J.-L. Kloosterman (TU-Delft)*
- Comparison of experimental and modelled pore solutions of low-pH Ordinary Portland Cement based mix designs. *T. Vehmas, M. Leivo, E. Holt (VTT)*
- Preliminary experimental results on the changes in microstructure, mineralogy and transport properties of Boom clay - concrete interface. *F. Claret, S. Gaboreau, N. Maes, Q. Tri Phung (SCK-CEN, BRGM)*
- Effects of the hydrogeochemistry of the FEBEX *in-situ* test on the aging of the concrete plug. *J. Cuevas, R. Fernández, A.I. Ruiz, M.C. Alonso, J.L. García-Calvo, J. Turrero, E. Torres, A. Garralón, P. Gómez, L. Sánchez (UAM, CSIC, CIEMAT)*
- Experimental investigation of hydro-mechanical behavior of claystone-concrete interface. *Z. Liu, J. Shao (LML)*
- Interaction between cement and Czech bentonite under temperature load and in *in-situ* conditions: Results after first testing period. *R. Červinka, R. Vašíček (UJV, CTU)*
- WP1 benchmark comparison of analysis to-date. *M. Leivo (VTT)*

WP2 Session

- Status of WP2. *B. Grambow (Subatech/Armines)*
- Study of radium uptake by cementitious materials relevant for LILW disposal in the Czech Republic. *B. Drtinová, J. Kittnerová, D. Vopálka (CTU)*
- Preliminary results from the study of radionuclide retention in cementitious systems and single mineral phases. *M. Isaacs, M. Felipe-Sotelo, E. Rastrick, D. Read (SURREY)*
- Effect of redox conditions on sulfur and selenium binding in AFm phases. *L. Nedyalkova, B. Lothenbach, J. Tits, E. Wieland, U. Mäder (PSI/Empa)*
- C-14 sorption on CEM I : effect of HCP degradation on C-14 uptake. *C. Bucur, I. Florea, R. Dobrin, A. Dinu (RATEN) – talk given by B. Grambow (Subatech/Armines)*
- Structural investigations on the uptake of long-lived safety relevant radionuclides by cementitious materials. *S. Lange, M. Isaacs, M. Klinkenberg, D. Read, D. Bosbach, G. Deissmann (JUELICH)*
- Radionuclide through-diffusion experiments in unsaturated carbonated and non-carbonated hardened cement paste using the osmotic technique. *S. Rasamimanana, K. Perrigaud, B. Grambow, C. Landesman (Subatech/Armines)*
- Update on beryllium activities at KIT-INE within WP2: solubility, hydrolysis and uptake by cement. *X. Gaona, N. Cevirim, M. Böttle, E. Yalcintas, N. Ait Mouheb, V. Montoya, T. Rabung, M. Altmaier (KIT-INE)*

- The use of a flow-through reactor to study AFm anion exchange. *N. Marty, S. Grangeon, F. Claret (BRGM)*
- Building blocks on molybdenum retention processes in cement systems. First experimental results from solubility and sorption onto pure cement phases. *M. López-García, J. Olmeda, M. Grivé (Amphos21)*

WP3 Session

- Status of WP3 and Modelling task. *A. Idiart (Amphos21)*
- Long-term non-isothermal reactive transport model of the interactions of concrete with the compacted bentonite and the clay host rock in a HLW repository in clay: Discretization errors and feed-back effect of the changes in porosity caused by dissolution/precipitation reactions. *J. Samper, A. Mon, J. Fernández, L. Montenegro, A. Naves (UDC)*
- Modelling of interactions between hydrated OPC and Czech bentonite under *in-situ* conditions and by heating. *T. Rosendorf, D. Vopálka, R. Červinka (CTU)*
- Compositional parameters for solid solution CSH and the applicability to thermodynamic modelling. *T. Vehmas, A. Itälä (VTT)*
- Rapid development of a reactive transport code with FENICS and Reaktoro. *L. Hax Damiani (PSI)*
- Final results of modelling chemo-mechanical couplings of calcium leaching experiments. *A. Idiart, E. Coene, M. Laviña (Amphos21)*
- H-M-C coupling analysis considering several scenarios of long-term alteration in cement-bentonite system: Part 2: Permeability. *S. Ito, S. Tachibana, A. Iizuka, H. Owada, D. Hayashi (Kobe Univ., RWMC)*
- Reactive transport model in the low pH cement / bentonite interface and effects on radionuclide migration. *V. Montoya, N. Ait Mouheb, T. Schäfer, V. Metz (KIT-INE)*

✎ *PhD session* was organized the 3rd day of the Workshop aimed at promoting training of the students and a forum for networking. The audience was restricted to students, supervisors, ExCom and EUG.

The session was chaired by two of the students (S. Lange (JUELICH) and R. Vasconcelos (USDF)). Nine students gave a presentation on their investigations within CEBAMA (see the list below) and discussed on their results.

- Characterization and sorption properties of low pH cements. *N. Ait Mouheb (KIT-INE)*
- Characterisation of UK cement backfill material and preliminary groundwater leaching experiments. *R. Vasconcelos (USDF)*
- Short-Term Interaction of Concretes with Ground-Waters: Percolation Tests. *A. Fernández Pérez (CSIC)*
- Nanostructural modification of Ordinary Portland Cement. *T. Vehemas (VTT)*
- Preliminary Results From the Study of Radionuclide Retention in Cementitious Systems and Single Mineral Phases. *M. Isaacs (SURREY)*

- Rapid development of a reactive transport code with FEniCS and Reaktoro. *L. Hax Damiani (PSI)*
- Diffusion of radionuclides through cement based materials: from laboratory experiments towards evaluation and modelling. *T. Rosendorf (UJV-CTU)*
- Chemical evolution of bentonite by modelling in repository conditions. *A. Itälä (VTT)*
- H-M-C coupling analysis considering several scenarios of long-term alteration in cement-bentonite system. *S. Ito (Kobe Uni.)*

☙ Stakeholder Panel was organized, for the first time in the framework of CEBAMA, in conjunction with its 2nd annual workshop. The panel sessions are intended to provide an area of interaction and discussion between the attendants to the workshop (representing mainly beneficiaries and End-User-Group), the invited panellists and external participants (e.g., other stakeholders than the panellists, students, etc.). These events are open and free of charge.

The 1st stakeholder panel was composed by five high level Finnish panellists from the Government, Electricity Generators, Safety Authority, Eurajoki municipality, Waste Managers and Social Experts have been organized. The panel was chaired by Erika Holt (VTT). As it was an important session within the workshop and it is a significant milestone/objective of the project, a special section is included in this Proceedings extending the information related with this event (see next chapter).

☙ Poster session was held at the end of the second day of the workshop at the same venue. Beneficiaries had the opportunity of presenting their results. The list of posters is provided below and all of them are included in the section *POSTERS*.

- Experiments on Interface Processes at the Cement/Callovo-Oxfordian Claystone Interface and the Impact on Physical Properties; Mechanical results from Callovo-Oxfordian Claystone. *R. Cuss, A. Wiseall, J. Harrington, J. Talandier, X. Bourbon (BGS, Andra)*
- Use of Neutron and X-ray Tomography based visualization of water uptake in cementitious materials for OPERA disposal. *Y. Yigittop, Z. Zhou, D. Bykov, A. Sabau, L. van Eijck, J.-L. Kloosterman (TUDelf)*
- Implementation of crystallization and precipitation mechanisms in pore-scale models based on the Lattice-Boltzmann method. *S. Rohmen, A. Idiart, G. Deissmann, D. Bosbach (JUELICH, AMPHOS21)*
- Thermal alteration of C-S-H phase in cementitious materials. *D. Hayashi, M. Ida, H. Owada, N. Fujii, K. Negishi (RWMC, Taiheiyo Consultant)*
- RWMC's contribution for Cebama in Period 1. *H. Owada, D. Hayashi, N. Fujii, H. Satoh (RWMC, Mitsubishi Material Corp.)*
- H-M-C coupling analysis considering several scenarios of long-term alteration in cement-bentonite system. *S. Tachibana, S. Ito, A. Iizuka, H. Owada, D. Hayashi (Kobe Univ., RWMC)*
- Characterisation of UK cement backfill material and preliminary groundwater leaching experiments. *R.G.W. Vasconcelos, N.C. Hyatt, J.L. Provis, C.L. Corkhill (USDF)*
- Bentonite interaction with saline high-pH solutions. *T. Heikola1, S. Kumpulainen, L. Kiviranta (VTT, Saanio & Riekkola Oy)*

- Quantitative visualization of heterogeneous transport processes at the host rock - cement interface. *J. Kulenkampff, M. Gründig, S. Gruhne, H. Lippold, J. Lippmann-Pipke, K. Jantschik, H.C. Moog (HZDR, BGR, GRS)*
- Integrated routing to study the concrete-bentonite interface to converge in the temporal and spatial upscaling. *J. Cuevas, A.I. Ruiz, R. Fernández, A. Ortega, M. Angulo, D. González Santamaría, M.C. Alonso, J.L. Garcia Calvo, M.J. Turrero, E. Torres, A. Garralón, P. Gómez, L. Sánchez (UAM, CSIC, CIEMAT)*
- Rapid development of a reactive transport code with Fenics and Reaktoro. *L. Hax Damiani, P. Krejci, S.V. Churakov, G. Kosakowski (PSI, UNIBERN)*
- Low-pH cement carbonation rate and impact on microstructure gas transport and shrinkage. *E. Kangni-Foli (IRSN)*
- Chemical and mineralogical changes at the interface between cementitious materials and groundwater. *E. Rastrick, M. Isaacs, D. Read (SURREY, NPL)*

At the end of the workshop, Seif Ben Hadj Hassine (EUG chairman) gave some statements on behalf of the EUG about the progress of the project. Following, the general assembly closed the workshop.

Structure of the proceedings

The present proceedings are divided into the following sections:

1) **Stakeholder Panel:** Description of the panel and summary of the main outcome.
2) **WP Overview:** An overview of the work conducted in the frame of the technical Work Packages.
3) **S+T contributions:** Scientific and technical manuscripts. All these contributions submitted by the beneficiaries were reviewed by the EUG members.
4) **Posters:** Posters presented during the poster session of the workshop.

STAKEHOLDER PANEL

During the 1st Annual Workshop, all consortium members agreed on organizing different panel discussions with stakeholders during the project, in combination with the annual workshops, so that the use of resources from the project is optimised. Panel sessions are intended to provide an area of interaction and discussion between the attendants to the workshop (representing mainly beneficiaries and End-User-Group), the invited panellists and external participants (e.g., other stakeholders than the panellists, students, etc.). These events are open and free of charge.

The 1st stakeholder panel was composed by six high level Finnish panellists (see Table 1) from the Government, Power Utility, Safety Authority, Eurajoki municipality, Waste Managers and Social Scientists. The panel was exceptional, considering that Finland is the first country licensed to construct a deep geological repository for nuclear waste, and that the panellists represented the main actors in the societal acceptance of the process of licensing this one-of-a-kind repository. The panel was chaired by Erika Holt, from VTT, the Finnish partner member of the ExCom of the project.

Table 1: *List of panellists.*

Sector	Panellist	Organization/Institution
Government	Liisa Heikinheimo (LH)	Deputy Director General, Energy Department Finnish government – Ministry of Economic Affairs and Employment
Power Utility	Sami Hautakangas (SH)	Head of Spent Fuel and Disposal, Fortum
Safety Authority	Jaakko Leino (JL)	Head of Nuclear Waste Safety Assessment, Finland's Radiation and Nuclear Safety Authority (STUK)
Eurajoki Municipality	Vesa Jalonen (VJ)	President, Municipal Council, Eurajoki
Communication expert	Timo Seppälä (TS)	Senior Advisor, Nuclear Waste Services Unit, Saanio & Riekkola Consulting Engineers
Social Sciences	Matti Kojo (MK)	Post-Doctoral Researcher, Faculty of Management, University of Tampere

The panel was attended by the Workshop project participants as well as 14 external participants to the Cebama project from VTT, University of Helsinki, University of Tampere, Saanio & Riekkola Oy and Geological Survey of Finland.

The 2 hour panel was organized in three parts: (i) introduction and background of the panellists, (ii) questions and answers and (iii) final conclusions and statements from each panellist. After the welcome to all attendees, each panellist gave a brief introduction on their background and experience related to the stakeholder engagement. A brief summary of their profiles is included below.

Liisa HEIKINHEIMO, Deputy Director General, Energy Department Finnish government – Ministry of Economic Affairs and Employment

Dr. Liisa Heikinheimo has over 20 years of experience in the nuclear technology and nuclear waste management field. She earned her MSc in materials science at Helsinki University of Technology (today

Aalto University) and Doctoral degree from the Technical University of Eindhoven (NL). Since their founding, she has served on the planning board as well as the research and steering groups of Finland's national waste management fund. She has served on several project and programme board memberships, such as in Euratom and Tekes co-operation. She was a Posiva R&D supervisor from 2009-2017. She started her career at VTT in 1985, where she was later the Technology Manager for VTT's area of "Materials for Power Engineering" (2004 - 09). She then worked at TVO nuclear power company (2009 - 2017) as head of R&D covering the responsibility of waste management R&D and administration. Earlier in 2017 she began working for the Finnish government, Ministry of Economic Affairs and Employment directing the nuclear energy section.

Sami HAUTAKANGAS, Head of Spent Fuel and Disposal, Fortum Oy

As the Head of Spent Fuel and Disposal Services at Fortum Corporation, Dr. Sami Hautakangas works closely with nuclear decommissioning and waste management. During his career in Fortum he has had various roles in nuclear waste management activities, which are related to Fortum's fully owned Loviisa nuclear power plant and other Fortum's partly owned NPPs. He has worked in Posiva and SKB Board of Directors and participated actively to the public discussions related to the waste management programs nationally and internationally. In addition, he has been closely contributing to the national nuclear safety and waste R&D programs. He holds a PhD in Physics.

Jaakko LEINO, Head of Nuclear Waste Safety Assessment, Finland's Radiation and Nuclear Safety Authority (STUK)

Jaakko Leino has been the Head of the Nuclear Waste Safety Assessment Section for the past 4 years at STUK's department of Nuclear Waste and Material Regulation. His section is responsible for oversight of post-closure safety and review of post-closure safety assessment of nuclear waste disposal facilities. The section also reviews repository design, performance of the engineered barrier system and material issues. He was responsible for the review of post-closure safety case in Posiva's construction license application and thus very familiar with the KBS-3 concept. He has a Master of Science in material chemistry and metallurgy, specialized in corrosion science. He has worked at STUK since 2010, as an inspector responsible for regulatory oversight and review of engineered barrier system for waste disposal concepts.

Vesa JALONEN, President, Municipal Council, Eurajoki

Vesa Jalonen is the chairperson of the municipal government of Eurajoki, the regional area on the west coast of Finland where Posiva and TVO power companies are located. He is also principal of the Eurajoki school. He has been involved in various aspects of stakeholder dialogue as a community representative and familiar with waste management issues.

Timo SEPPALA, Senior Expert at Saanio et Riekkola,

Timo Seppälä is an expert in nuclear waste management and communications. With 17 years of experience as a communication manager in Posiva Oy, he is familiar with communication of nuclear waste disposal with different audiences. Over the years he has gained substantial international and multicultural experience working in different projects with numerous international organizations like OECD, IAEA and the European Commission dealing with nuclear waste management issues. He has an International MBA (2015) in Communications and Leadership, as well as an MSc in Environmental Sciences (limnology) in 1986 from Helsinki University. Recently he has served as in expert for international peer review on the site selection for the HLW repository in Japan on an OECD/NEA assignment.

Matti KOJO, Post-Doctoral Researcher, Faculty of Management, University of Tampere

Matti Kojo has several years of experience in the field of studying governance with regard to nuclear power and nuclear waste and carbon capture and storage. His research topics include e.g. public and stakeholder participation and compensation issues. He has participated as a researcher in many Finnish nuclear waste research programmes (JYT2001; KYT2014; KYT2018) and international research projects (EURATOM FP7 InSOTEC and IPPA; FP6 ARGONA) since 1997. Matti defended his doctoral thesis in political science on "The public engagement turn of nuclear waste policy" in 2014 at the University of Tampere. One of the projects Matti is currently working in is 'Governing Safety in Finnish and Swedish Nuclear Waste Regimes' (SAFER) funded by the Finnish Research Programme on Nuclear Waste Management (KYT2018).

After their introductions, E. Holt proceeded with questions to the panellists. Questions focused on different topics: 1) Communication with Stakeholders, 2) Dissemination and Interaction, 3) Use of Technical Information, 4) Assisting Less Advanced Programs, 5) CEBAMA project specific (cementitious & clay materials, radionuclides, modelling…) and 6) Other.

Three types of questions were asked to panellists:

- Questions already prepared before the panel by E.Holt and agreed between the ExCom members of Cebama,
- Questions submitted by Workshop attendees during the Workshop. A document was pre-distributed the first day in order to compile questions for the panellists,
- Open questions from the audience (questions raised during the panel discussions).

The main issues and conclusions of the discussion are summarised below.

The panel opened with an analysis of trust **of the Finnish people on nuclear electricity generation** and whether incidents such as the one occurred in Fukushima in 2011 had an impact on this trust.

SH commented that the Loviisa power plant marks the 40th anniversary of nuclear generation in Finland, as it was commissioned in 1977. Later, plans for the construction of Olkiluoto were developed and Loviisa 2 was commissioned. The long history of nuclear generation without incidents contributes to maintaining the trust of the Finnish citizens. Operators, regulators and in general all stakeholders in Finland find nuclear energy reliable. The government has provided, from the very beginning, very simple and stable rules to run nuclear power. FORTUM and TVO, which are the license holders of the NPPs in Finland, are also solely responsible for nuclear waste management as stated by the legislation. The support to nuclear in Finland has not changed much, with about 40% against and 40% for, and this has not changed dramatically over the years.

TS commented that the history of Finland having been involved in two big wars, has generated a very pragmatic society, which contributes to the shaping of the trust in the government. These statements were backed up by MK who showed in his presentation where the residents of Eurajoki and Pyhäjoki get their information on final disposal of nuclear waste and that the residents of the host municipalities in Finland has trust in regulators and the scientific community and perhaps therefore many of them believe that nuclear power is safe.

The next question addressed within the panel was how **to communicate with non-technical stakeholders**.

VJ, from his experience in the municipality, pointed out that there is a big gap between "common people" and the technical audience of the room. The general public is interested in how companies work, which is mainly based on confidence. Nuclear companies in Finland have made a great effort in communicating information to show that they are trustworthy. The public do not want technical details, they want to receive information in a clear and understandable manner.

TS, explained there are two main approaches when conveying communication to the general public: either you put communicators in front of the public or you train technical persons to be able to communicate scientific issues to people. The last approach is complicated because not all experts are interested and also because not all of those interested have the ability to develop communication skills.

MK pointed out that a continuous dialogue must be kept. The process must start with the gathering of the concerns of the public.

TS commented on the fact that designing, building and operating a permanent repository is a long-term project and this means much more than communication. Several factors affect the multi-decade process: i) The framework for communication must be sound; ii) The process must be transparent; iii) The information must be accessible; iv) The roles of the authorities and the different stakeholders must be clear and v) The regulators and the implementers must be independent.

VJ also commented on the relevance of preparing a summary understandable by the people, ensuring that they are going to be informed if something goes wrong and that the companies in charge know what to do in different situations.

Several questions were also raised by the audience regarding different subjects, which are summarised below.

Trust developed in Finland and groups of opposition.

The success of the Finnish process is based on the trust of the population. Several discussions were held on this regard. The conclusions were: there has been a low level of opposition due to a lack of argumentation for an alternative solution to deep geological disposal (DGD); Finnish society is shaped by electricity dependency where there is a high level of imports and nuclear is a good source to ensure domestic energy supply; long-term and good quality job of the implementers and regulators in involving and informing society in an understandable, clear and responsible process of acceptance.

Detailed Q&A on this subject:

> *B. Grambow (Subatech, France):* In many countries there are opposition groups that may lead to a loss of trust. How is this managed in Finland?
>
> The answers addressed minimising mistakes. VJ commented that there is a higher level of trust in the municipality as there are advantages for the Community due to the presence of the nuclear power plants, such as economic benefits in terms of occupation and other advantages.
>
> *M. Cuñado (ENRESA, Spain):* What is the attitude of young generations, the future decision makers, towards waste management? Which sort of questions do they have?
>
> VJ answered that in the municipality they are accustomed to having nuclear facilities and some of them work for the nuclear power plant. For this reason, their attitude is more positive than of older people.

L. Duro (Amphos 21, Spain): It looks strange that Greenpeace is the stakeholder receiving the lowest level of trust from the public. Is this only in Finland or also in other countries?

MK explained that the pressure of NGOs is weaker in Finland than in other European countries and one of the reasons is that they have less resources, so they are less visible.

TS explained that one of the reasons for the lower confidence is the lack of solid argumentation: if you oppose to deep geological disposal, you are supposed to give an alternative. In Finland, politicians and decision makers agree that there is no better alternative, and this helps.

LH explains that there are opposing views not towards waste management but to nuclear energy in general. However, there is a general criteria for the use of nuclear energy in the Finnish Nuclear Energy Act, that use of nuclear energy should promote the overall good of the society. This criteria is used in the licensing decisions at the Government level. The Finnish mentality accepts well these decisions and does not open the discussion much once this has been earlier agreed and decided.

Erika Holt commented that it had been discussed when planning this event whether to invite the opposition to the panel but it was difficult to identify who would be a party representing opposition in Finland.

It was explained that the opposition is very minor in Finland and that there are not many resources available for them.

M. Altmaier (KIT-INE, Germany). We are a European Project and the opinions in other European countries are very different from those in Finland. How do other opinions influence that in Finland?

LH stated that the situation is very country specific in Finland, for instance 20% of the electricity is imported since there are currently no other sufficient self-resources. Therefore, the acceptance of new nuclear power plants is easier, as it is thought to be a straight way to increase domestic energy resources. Other countries with greater energy options may have a different view of nuclear energy. Nevertheless, due to the proximity and similarity, any issue regarding safety which may arise in Sweden, would be taken very seriously also in Finland.

E. Neeft (COVRA, Netherlands): Why in the opinion of VJ, other countries are not as successful in the process as Finland?

VJ: Local acceptance is very important, and we are a small municipality easier to deal with.

SH: The process and facilities bring welfare to the local municipality but it is also important at a national, not only at a municipal level. One must admit that nuclear has a positive influence for the entire nation, so that no opposition arises.

Use of Technical information.

Scientific and technical information is mainly used by the implementers. The regulator conducts an independent review and the communication to the public must consider the characteristics of the audience.

Detailed Q&A on this subject:

Q from the audience. Is technical information produced by the projects such as ours (CEBAMA) used by the regulators? How?

JL explained that the investigation and the use of the results is the responsibility of the implementers. It is not the role of the regulator to do research but to follow the advances of the projects.

Erika Holt asked how the regulators organize the review process of the work submitted by the implementer.

JL explained that - internal people review the work but that STUK also has framework contracts with review groups of external experts, mostly international.

Erika Holt asked for some examples of good outcomes from very technical projects such as CEBAMA.

JL explained that most of the EU projects deal with small parts of the Safety Case, and that the Safety Case is not meant to be communicated but to be a tool of communication between the implementer and the regulator. CEBAMA as a Project does not have direct interfaces with the public but is used to address questions that have arisen by the regulator to the implementer, where better technical understanding is needed.

Erika Holt pointed out that CEBAMA has a group of stakeholders composed of end users of the results, i.e., implementers, (EUG from End Users Group) and that there is a specific outcome (deliverable report) intended to contribute to the Safety Case implementation.

Lara Duro explained that one of the intentions of the Project is to identify what is the contribution of CEBAMA to the Safety Case. In general, we have many results from research projects but it is very seldom clear which is the use of the results. For this reason, the EUG is involved in the Project from the beginning, they will review the project plans at intermediate steps, review results and see how the outcomes can contribute to the Safety Case. The outcome will be summarized in a deliverable.

Erika asked whether there are examples on "how not to do things" from scientific projects.

TS explained that the communication approach has changed a lot in the last years. He stated that the types of communication used in the 1980s did not consider the knowledge of the audience, the local conditions, etc., and this must not be repeated. The context is essential.

How to go forward. Lessons to other countries.

It was discussed that Finland can help other countries following the same process for implementation of repositories. Regarding communication, the conclusion was that implementation and stakeholder acceptance are very dependent on local considerations, so that Finland's know-how cannot be directly exported. However, the framework for the communication process to work properly is defined by some requirements: transparency of the process for site selection; clear division of duties and responsibilities between the different players and actors; maintaining trust, clear legal framework and information being easily accessible.

Detailed Q&A on this subject:

LH stated that the most important is to have a good communication with the stakeholders as end users. Research must follow independency and challenge the in-house programmes. It helps advance and shows if questions can be solved in different ways. She reminded the audience that Finland did not have the acceptance from the beginning, everything is a process.

SH said that the SNF repository by Posiva has been producing results for 30 years, what has meant a big investment for the Posiva owners. This is an example of an in-house programme and agreed with LH that independent programmes are needed to give different insights to the topic. In Finland there are national research programmes (KYT) with are independent from the nuclear utilities.

B. Grambow (Subatech). Finland is one of the most advanced countries. Now the European Joint Programme has the intention of ensuring that knowledge is transferred to other programmes. How do you perceive that initiate?

SH recognised that harmonising procedures for spent fuel management have been on-going at an European level. However, in this specific process it can be difficult to find common procedures, since it depends very much on how other countries are approaching the spent fuel management, as the predefined technical concepts can be very different. Obviously, Finland would be happy to help if the concept is based on KBS3..

LH recalled that these are long-term processes and agrees with SH in that this is concept dependent. However, she is of the opinion that lessons on how to make up a Safety Case can be very useful for other countries.

TS said that communication has a cultural dimension and this cannot be easily exported. He emphasized different aspects that put in place the framework needed for communication to succeed: transparency of the process for site selection; clear division of duties and responsibilities between the different players and actors; clear legal framework and information easily accessible.

VJ complemented that local people need to see there will be good work, welfare and benefits for the municipality resulting from the repository.

SH summarised TS's words with "the rules must be very clear". He added that this has not been probably the case in some other European countries. Newcomers and newbuilding countries must consider waste management from the very beginning of the construction process.

MK added several elements: expertise, communication, engagement, ethics.

N. Marcos (Saanio & Rikkola) pointed out the relevance of education of the population. She noted that the education of people in Finland is very high since the 1920s and that this makes a difference. She recalls the case of other countries with lower educational levels where one cannot expect the same type of reactions and cannot communicate in the same way. One problem is how to adapt the information to different education levels.

MK and TS thanked N. Marcos comment and shared a final remark: adapt your communication to your audience.

Erika Holt closes the panel after thanking all panellists and the audience for their active participation.

WP OVERVIEW

Overview of Work Package 1

Experiments on interface processes and the impact on physical properties

WP leaders:
Erika Holt (VTT)
Francis Claret (BRGM)
Urs Mäder (UNIBERN)

The 2nd CEBAMA workshop (Espoo, Finland, May 16-19, 2017) provided the fourth occasion for discussing Work Package 1, after the project kick-off meeting (Brussels, July 2015), a meeting in London (November 2015) and the 1st Annual Project Workshop in Barcelona (June, 2016). This 2nd Annual Workshop had participants from 18 of the 19 partner organizations of WP1. Some partners had multiple persons attending, including participation of new junior staff or students getting integrated into the project.

This Workshop marks approximately the mid-point of the CEBAMA Project, and accordingly there was much more focus on presentation of data and results, but it also involved technical reporting on work progress. A specific session for WP1 was scheduled early in the meeting to provide a brief overview of all projects on 2-3 slides, and to address a number of central issues, such as collaborations and data "freeze" / data transfer, primarily with WP3 (interpretation and modelling). This format allowed for fewer (7) but more detailed presentations of select projects during a two-hour WP1 plenary session. A PhD session was organized as a new feature where participants from all work packages had the opportunity to present and to exchange experiences in a less formal setting. A poster session, with motivating drinks and finger food, provided an additional floor for WP1 project presentations and discussions across work packages.

M. Leivo (VTT) presented results of characterisation of the WP1 bench mark material, a low-pH ternary mix including silica fume and blast furnace slag, such as used for massive concrete plug applications. This mix was prepared by VTT at the time of the 1st Workshop and was distributed to interested partners. It is an initiative to bundle efforts and foster collaborative work within WP1, but also with data transfer to WP3. The proceedings include the short summary presentations of all projects, the more detailed plenary presentations, as well as the posters and student presentations.

Collaboration amongst partner projects had clearly increased, including stages by students/staff at research facilities (in part supported by CEBAMA mobility measures). Several projects generated quality-assured data that are ready to be used by WP3 (modelling), and WP3 had taken the lead in producing a data freeze document to this end. Dissemination of results is well under way, with presentations at the "Mechanisms and Modeling of Waste/Cement Interactions" workshop in Murten, Switzerland (May 2016) with a couple of paper contributions now published in the conference volume (Physics and Chemistry of the Erath, 99, 2017). Upcoming venues with WP1 participation include the Clay Conference 2017 (7th International Conference on Clays in Natural and Engineered Barriers for Radioactive Waste Confinement, http://www.clayconferencedavos2017.com/congress.html) and NUWCEM, October 2018 in Avignon (3rd International Symposium on Cement-Based Materials for Nuclear Wastes, http://www.snetp.eu/events/event/nuwcem-2018/). None of the projects claimed significant deviations from their research plans.

The first Deliverable report D1.01 provided detailed scientific work summary plans from each partner, serving as a basis for future reference. The next four Deliverables, D1.02 on systems to be studied, D1.03 as a state-of-the-art report on WP1 topics to be used as external reference by the general public, D1.04 on experimental method boundary conditions to be used in WP1 studies, and D1.05 on Experimental materials to be used in the WP1 program were all available before or shortly after the last workshop time (June 2016).

Overview of Work Package 2

Radionuclide retention in high pH concrete

WP *leader:*
Bernd Grambow (ARMINES)

The work on radionuclide retention processes in high pH concrete environments is progressing well. The aim is to provide insight on general processes and phenomena and their couplings in overall interaction mechanism, which can then be transferred to different disposal situations and water access scenarios in a high pH repository environment with cementitious materials. It also assesses the impact of chemical alterations (e.g., high pH concrete ageing, carbonation, m/V ratio, temperature, transition from oxidizing to reducing conditions, aging time) on radionuclide retention.

The studies conducted in the reporting period including synthesis of model phases containing various radionuclides in various oxidation states, leaching and solubility tests, diffusion tests, sorption experiments and co-precipitation studies, all under well controlled redox conditions and modelling of experimental results. Characterization of the cements and equilibrated waters has been carried out by a large number of techniques and a series of batch sorption experiments has been performed identifying systems of interest. The solubility, through-diffusion and additional batch sorption experiments have been started.

Structural investigations on the sorption and uptake kinetics of long-lived radionuclides (^{226}Ra and ^{99}Tc, I^-, MoO_4^{-2}) by cementitious materials have shown strong uptake of radium by calcite and by CSH phases as well as incorporation of iodide and molybdate into AFm-SO_4, but no significant uptake of Tc(VII). Within the iodine systems, iodate was shown to adsorb very quickly to a range of the solids tested including CEM I, AFt and AFm. Concerning the incorporation of Se and S in AFm phases, the X-ray spectra revealed the water content and an intercalation of the Se and S anions in the AFm interlayers for the S(IV)-AFm.

Anion exchange (MoO_4^{-2}/Cl^-) in AFm phases were studied using flow-through experiments. Modelling of the experiments allowed the determination of Mo/Cl and OH/Cl exchange constants and the estimation of anionic-exchange capacity (AEC) of the studied material. Powellite ($CaMoO_4$) was also successfully synthesized and the determination of its solubility product was in very good agreement with the ones reported in the literature.

Inorganic C-14 interaction on non-degraded and degraded hardened cement pastes has been studied both by sorption and diffusion tests. Solid/liquid distribution ratios were 10 times higher for the non-degraded HCP than for the most degraded state. Through-diffusion experiments were conducted with HTO and C-14 both in unsaturated carbonated and non-carbonated hardened cement paste using the osmotic technique. A specific experimental set-up using an osmotic technique has been developed in order to determine the diffusion parameters (porosity and diffusion coefficient) in partially saturated conditions.

Overview of Work Package 3

Interpretation & modelling

WP leader:
Andrés Idiart (AMPHOS21)

The objective of Work Package 3 (WP3) is to contribute to the interpretation of experimental results generated in CEBAMA. Numerical modelling is used to assess the impact of several physical and chemical processes on transport properties of cementitious systems as well as their interface with clayey materials. In addition, research within WP3 is intended to improve the validity of existing models to predict changes in transport properties of cementitious systems and to contribute to our ability to extrapolate models of system-level to modelling for Safety Case application.

An essential basis for WP3 is the outcome of the experiments and their characterization performed within the CEBAMA project. Every WP3 partner has already established collaborations with the experimental Work Packages to model specific sets of experiments. During the 2nd Annual Workshop, new collaborations have been defined. WP3 activities started right after the kick-off of the project. During the first year, WP3 partners used existing experimental data, relevant to their objectives within the project, in order to test and verify their modelling approaches and main developments during the first stages of the project. Deliverable *D3.04: Description of and results from the modelling of external lab and/or field experiments* describes the work performed in WP3 during the first year.

During the 2nd Annual Workshop in Helsinki, the procedure to conduct the 1st Data Freezing was discussed and defined. Based on those discussions, a template to collect and compile experimental data about the laboratory tests was sent to WP1 partners after the workshop. Another important topic addressed in this workshop was the setup of a common modelling task between WP3 partners. After the 1st Annual Workshop, deliverable *D3.03: Consolidated plan for Upscaling Modelling Task* was submitted. This deliverable describes the objectives and scope of the modelling task. During the 2nd Annual Workshop, the details of the modelling task were discussed and agreed upon and participants to this modelling task were clearly identified. A total of 8 WP3 partners are involved in the Modelling Task. It was also agreed that two subsequent meetings specific for the modelling task will be held after the workshop in Barcelona (13th September 2017, during the Migration Conference), and Davos (27th September 2017, during the 7th International Conference on Clays in Natural and Engineered Barriers for Radioactive Waste Confinement) to define the details of the model implementation in each modelling tool.

Seven (7) WP3 partners presented their progress at the 2nd Annual Workshop:

- UDC: Javier Samper. Long-term non-isothermal reactive transport model of the interactions of concrete with compacted bentonite and clay host rock in a HLW repository: discretization errors and feedback effect of the changes in porosity.
- CTU: Dušan Vopálka, Tomáš Rosendorf. Modelling of interactions between hydrated OPC and Czech bentonite under in-situ conditions and by heating.

- VTT: Tapio Vehmas. Compositional parameters for solid solution CSH and the applicability to thermodynamic modelling.
- PSI: Leonardo Hax. Rapid development of a reactive transport code with Fenics and Reaktoro.
- Amphos 21: Andrés Idiart. Final Results of Modelling Chemo-Mechanical Couplings of Calcium Leaching Experiments.
- RWMC: Hitoshi Owada, Daisuke Hayashi. H-M-C coupling analysis considering several scenarios of long-term alteration in cement-bentonite system: Part 2: Permeability.
- KIT: Vanessa Montoya. Reactive transport model in the low pH cement / bentonite interface and effects on radionuclide migration.

In addition, nine (9) scientific and technical contributions from WP3 have been presented in this 2nd Annual Workshop Proceedings. The models presented mainly focus on (1) cement/concrete matrices and cement / clay geochemical interactions; (2) reactive transport models, also coupled to hydro-mechanical models; (3) mostly at the laboratory length scale, but also at pore-scale (JUELICH) and macro-scale (UDC, RWMC), with time scales covering the duration of lab tests, but also long-term predictions. A brief overview of these contributions is given below.

KIT developed and implemented a 1D model to simulate the on-going experiments at KIT of the interaction between a low-pH cement and bentonite porewater. The model considered a low-pH cement paste (pH ~11.0) synthesized in the CEBAMA project. The simulations predict the formation of a degraded area in the cement paste of ~2 mm after one month of alteration, carbonation of cement, and an increase in total porosity.

Amphos 21 developed a multi-scale chemo-mechanical model for cementitious systems to study accelerated leaching experiments performed at Chalmers University. The model, implemented in iCP, simulates the alteration of transport and mechanical properties of cement paste and concrete as a function of the degree of calcium leaching from the samples. In general, comparison with existing experiments regarding the effect of calcium leaching on physical and chemical properties of tested samples show a good agreement. The measured effect of calcium leaching on increased diffusivity of cement paste is also captured by the model.

BRGM performed a series of spectral induced polarization measurements on low-pH concrete and cement paste to monitor the evolution of their mineralogy and microstructure. They found that concrete and cement paste are very resistive and exhibit high phase shifts between injected current and measured voltage that are not due to experimental artefacts. They attributed this behaviour to the nanoporous nature and highly charged C-S-H. Their findings will serve reactive transport models for predicting the long-term performance of concretes. Modelling by combining an electrostatic surface complexation model with an induced polarization model based on membrane polarization is on-going.

JUELICH continues with the development of a pore-scale reactive transport modelling code, named iPP. The code, based on the Lattice-Boltzmann method, has been extended to support the simulation of heterogeneous precipitation processes leading to the formation of grain overgrowth via distance field transformation. Their contribution shows how different kinds of precipitation processes can be simulated with this technique, using different transfer functions. A mathematical formalism is developed which allows controlling the locality of precipitation, although different types of transfer functions are currently under development. The simulated precipitation patterns will be compared to results obtained by other simulation codes and experimental data produced within WP1 in CEBAMA, addressing in particular the experiments performed by USFD and SURREY.

RWMC has continued their work on two-dimensional Hydro-Mechanical-Chemical (H-M-C) coupling analyses of long-term interaction between concrete and bentonite. Simulations were performed to estimate changes in the mechanical and hydraulic properties of the buffer material. They present results of a series of H-M-C coupled models focusing on the changes in stress distribution and permeability of the bentonite buffer as a result of interaction with concrete. They also introduce a new constitutive model describing the impact of volume change of clay minerals induced by the interaction with concrete on the mechanical behaviour of bentonite.

UDC presents a set of reactive transport models of a generic radioactive waste repository in clay according to the Spanish Reference Concept. Their focus in this contribution is on the assessment of discretization errors and the effect of porosity feedback on the long-term reactive transport modelling of the interaction between concrete and clayey materials. The model consists of a non-isothermal multicomponent reactive transport setup of the long-term (1 My) interactions of concrete with compacted bentonite and the clay host rock. Their results underline the importance of a correct discretization of the system for accurate predictions.

CTU presents their on-going modelling work of the interactions between hydrated OPC and Czech bentonite under in situ conditions and explicitly considering the effect of low and high temperature. The interaction between hardened OPC and bentonite was modelled using a simplified description of solid phases and considering a Czech groundwater composition. They found that the predominant mechanisms are the dissolution of quartz, illite, kaolinite, portlandite, and under certain conditions also ettringite.

VTT presents their research about an advanced model for the representation of C-S-H gels as a solid solution. Their goal is to define reliable modelling parameters for C-S-H gels. Ion activity product and Gibbs formation energies were calculated from various solution compositions. Their results enable the determination of solubility products and formation energies with varying silicate chain lengths and CaO/SiO_2-ratios. C-S-H gels were modelled as a solid solution and the results compared to the experimentally measured pH, calcium and silicon concentrations, showing a good agreement.

PSI present their most recent developments of a new flexible reactive transport simulator based on the FEniCS software combined with the Reaktoro geochemical code. Their approach combines the advantages of Gibbs Energy Minimization and Law of Mass Action methods. They present two examples of validation of their tool, consisting of a literature benchmark and modelling of an experiment that shows the diffusion of electrochemical coupled species.

S + T CONTRIBUTIONS

Characterization and sorption properties of low pH cements

Naila Ait Mouheb[1*], Vanessa Montoya[1], Dieter Schild[1], Eva Soballa[1], Christian Adam[1], Frank Geyer[1], Thorsten Schäfer[1]

[1] KIT, Institut für Nukleare Entsorgung (DE)

* Corresponding author: naila.ait-mouheb@kit.edu

Abstract

Characterization of three low pH cement pastes, including the description of their sorption properties for tritiated water (HTO), $^{36}Cl^-$ and $^{129}I^-$ is described in this work. SEM-EDX and NMR analyses show that after 90 days of hydration, the main hydrated phases are C-S-H and C-A-S-H gels with a Ca:Si ratio between 0.8 - 1.0 and a Al:Si ratio of 0.05. TG-DSC and XRD indicate the presence of calcite in the mixtures where limestone filler has been used. Additional techniques were used to identify minor hydrated solid phases like ettringite (i.e., XRD and solid state NMR). Porosity and pore size distribution was characterized by MIP observing that the size of the pores in the hydrated cement phases varies from the micro to the nanoscale. Uptake studies of HTO, $^{36}Cl^-$ and $^{129}I^-$ from batch sorption experiments indicate very weak sorption ($K_d < 0.40 \pm 0.13$ L/kg) for the 3 selected radionuclides. The uptake process of $^{36}Cl^-$ and $^{129}I^-$ is probably associated with surface processes in the C-S-H and C-A-S-H phases with competition for sorption sites, between them. In the case of HTO, isotopic exchange with the interlayer water of the C-S-H and the C-A-S-H seems to be the main uptake process.

Introduction

The deep geological repository concept of nuclear waste is based on the confinement of the radioactive waste over a long period of time by multiple barriers. Many of the concepts developed internationally use concrete and clay as confinement barriers (ANDRA, 2005; ENRESA, 1995). The different barriers are not in geochemical equilibrium and during the prolonged period of post disposal, cementitious material will undergo alterations, possibly changing the chemical and physical properties of this barrier. One important process that could reduce the durability of a concrete barrier is the leaching / degradation of the solid especially at the cement / clay interface due to the contact of clay pore water with the cementitious material. In order to minimize the interaction between the "classical" cement materials and the bentonite porewater (pH ~ 7.5) low pH cements were developed within the nuclear waste disposal context in the late 90's.

The main characteristic of these low pH cements is the absence of portlandite as hydrated solid phase, which reduces the pore water pH to ~ 11 compared to the "classical" cements (pH ≥ 12). Thus, to obtain 'low pH' cement, the clinker content in the binder is substituted for supplementary materials like silica fume or fly ashes (Coumes et al., 2006). From a chemical point of view the Ca/Si ratio composition in the hydrated low pH cement is less than 1.0 and C-S-H phases are the main solids present.

In the framework of the CEBAMA project, KIT-INE is studying the interaction between low pH cement materials and bentonite pore water, assessing the impact of mineralogical and microstructural modifications on transport properties and its effect on the mobility of radionuclides. To evaluate time dependent changes on transport properties and mineralogical composition in the geochemical perturbed system, it is crucial to have a detailed physicochemical characterization of the cementitious starting materials under equilibrium conditions. The objective of this contribution is to complete the characterization of the cement samples started in the first year of the project (Ait Mouheb et al., 2017) using X-ray diffraction (XRD), thermogravimetric - differential thermal analysis (TG-DTA), ^{29}Si and ^{27}Al magic angle spinning nuclear magnetic resonance spectroscopy (^{29}Si and ^{27}Al MAS NMR), scanning electron microscopy - energy dispersive X-ray spectroscopy (SEM-EDX) and mercury intrusion porosimetry (MIP). In addition, sorption properties of the low pH cements to HTO, $^{36}Cl^-$ and $^{129}I^-$ at tracer concentrations have been investigated.

Materials

Cement Samples preparation

Three different low pH cement pastes (MIX 3C, 3D and 3E) consisting either of binary or ternary binder composition with sulphate resistant Portland cement (CEM I 52.5N SR, from Lafarge), silica fume (microsilica ELKEM 940U) and limestone filler ($CaCO_3$) have been prepared and described in Ait Mouheb (2017) (see Table 1). An inorganic superplasticizer (SioxX®) was used to achieve good workability in the MIX 3C.

The mixtures have been hydrated with de-ionised water using a water/binder ratio (w/b) = 0.6 and the samples have been cured in a chamber at 98% relative humidity (RH) and 21 ± 2°C during 90 days.

Table 1: *Composition (w.w.%) of the solid mixtures proportions used in this work.*

	MIX 3C (%)	MIX 3D (%)	MIX 3E (%)
OPC (CEM I 52.5 N)	39	40	50
Silica fume (ELKEM 940U)	39	40	50
Limestone filler	19	20	-
Superplasticizer (SioxX)	3	-	-

Low pH artificial cement water (ACW)

A low pH artificial cement pore water (ACW) was prepared by dissolving: 645.5 mg KOH, 639.9 mg Na_2SO_4, 363.7 mg $Na_2Si_3O_7$, 617.8 mg $CaCl_2$:$2H_2O$ and 295.9 mg $Ca(OH)_2$ in 2 L Milli-Q water. Then, 1.2 mg $CaCO_3$ and 295.9 mg $Ca(OH)_2$ were added. The solution was shaken for one week inside of a glove box filled with Ar. The pH was adjusted with 180 mL of H_2SO_4 (0.04 M) at pH = 11.0. Finally, the ACW is centrifuged (Heraeus Megafuge 2.0R) for 30 min. Later, the supernatant was filtrated by PVDF membrane with a pore size of 0.22 µm. The composition of the synthetized water is specified in Table 2.

Table 2: *Chemical composition of the artificial cement water (pH = 11.01) analysed by ICP-OES and ion chromatography.*

Element	Concentration (M)	Element	Concentration (M)
Na	$5.87 \cdot 10^{-3}$	**CO_3(*)**	$2.76 \cdot 10^{-6}$
K	$3.76 \cdot 10^{-3}$	**S(VI)**	$5.75 \cdot 10^{-3}$
Ca	$3.56 \cdot 10^{-3}$	**Cl**	$3.84 \cdot 10^{-3}$
Si	$1.46 \cdot 10^{-3}$		

(*) Total concentration

HTO, ^{36}Cl- and ^{129}I- stock solutions

A stock solution containing HTO and ^{36}Cl was prepared by adding 450 µL of HTO solution ($1.01\cdot10^7$ Bq/mL) and 538.9 µL of ^{36}Cl solution ($7.42\cdot10^5$ Bq/mL) in 2 L of low pH artificial cement water. The concentration of the radiotracers were [HTO] = $1.86\cdot10^{-9}$ M and [^{36}Cl] = $4.55\cdot10^{-6}$ M. A separate stock solution containing ^{129}I was prepared by adding 270.3 µL of ^{129}I ($3.7\cdot10^5$ Bq/mL) in 2 L of low pH artificial cement water ([^{129}I] = $5.93\cdot10^{-5}$ M).

Methods

Analytic techniques used for cement characterization

Different techniques (SEM-EDX, XRD, TG-DSC) were used to characterize the three mixtures as described in Ait Mouheb et al. (2017). Additionally, ^{29}Si and ^{27}Al magic angle spinning nuclear magnetic resonance spectroscopy (^{29}Si and ^{27}Al MAS NMR) have been used in this contribution. The ^{29}Si and ^{27}Al MAS NMR spectra were acquired on a Bruker Avance III 400 wide-bore spectrometer. The instrument operates at 104.28 MHz for ^{27}Al and 79.50 MHz for ^{29}Si. The samples were placed in a disposable Kel-F container and then inserted into a standard ZrO_2 rotor. All experiments were carried out at room temperature with a magic angle spinning frequency of 15 kHz.

Spectra were recorded using standard Bruker pulse sequences. For iron-free samples, the FID (free induction decay) was recorded with 32 k data points and zero filling to 64 k was applied. For ^{27}Al, after a 90° excitation pulse, 512 scans with a relaxation delay of 500 ms were acquired. For ^{29}Si, the excitation pulse angle was 30° and at least 4 k scans with a delay of 15 - 20 s were used. The paramagnetism of iron-containing samples required different experimental conditions. Only 8k - 16k data points were sampled. For ^{27}Al, 8 k scans with a delay of 250 ms were performed; for ^{29}Si, 64 k scans were accumulated and the delay reduced to 500 ms. For the processing, no zero-filling was applied, and multiplication with Gauss functions was used (5 - 10 Hz line broadening).

The observed ^{29}Si resonances were analysed using the Q_n(mAl) classification, where a Si tetrahedron is connected to n Si tetrahedral with n varying from 0 to 4 and m is the number of neighbouring AlO_4 tetrahedrons.

Kinetic batch sorption experiments

The sorption studies of HTO, ^{36}Cl and ^{129}I were carried out by duplicate in a glove box under a controlled Ar atmosphere (O_2 and CO_2 concentrations < 5 ppm). 2.67 grams of each cement sample (mix 3C, 3D and 3E) were added in different polystyrene tubes. Then, 20 mL of the prepared stock solution containing HTO and $^{36}Cl^-$ were added obtaining a suspension with solid-liquid ratio of 133 g/L. The same procedure was used to do the sorption experiments with $^{129}I^-$.

The suspensions were mixed continuously for 1 day to 30 days using roller mixers. After the selected time, mixing was stopped and the suspension is filtrated using Polyvinylidene fluoride filters (PVDF) with a pore size of 0.45 µm inside of the glovebox. Subsequently, the supernatant was measured by liquid scintillation spectrometry (LSC) in order to determine the activities of each radionuclide. Results are expressed as distribution ratio R_d (Eq. 1) which corresponds to the ratio S/C, where S is the amount of tracer adsorbed onto the unit mass of the solid phase, and C is the amount of tracer per unit volume of solution.

$$R_d = \frac{Quantity\ of\ a\ radionuclide\ sorbed\ per\ unit\ mass\ cement}{Equilibrium\ concentration\ of\ the\ radionuclide\ in\ pore\ water}\ [\frac{L}{kg}] \qquad \textbf{\textit{Eq. 1}}$$

Results and discussion

Characterization of the cement samples

<u>*X-Ray diffraction after 90 days of hydration*</u>

According to Ait Mouhed et al. (2017) and based on XRD analysis the three synthesized mixtures were highly hydrated after 90 days of hydration not observing any phase of the clinker in the difractogram. The only clear identifiable crystalline solid phase was calcite for the samples 3C and 3D. C-S-H phases could not be detected in any of the samples probably due to its low crystallinity.

The Rietveld refinement using the Topas Academic v4.2 software was applied to quantify the crystalline and amorphous phase composition of the three mixtures. The amorphous phases were hypothesized to be C-S-H phases and are present as major phases in all the samples: 82% for the samples 3C and 3D and 92% for 3E (see Figure 1). Additionally, calcite is present as 12% in sample 3C and 3D and in less of 2% in 3E. Small amounts (< 6%) of crystalline C-S-H phases (Tobermorite type) was quantified as well (see Figure 3).

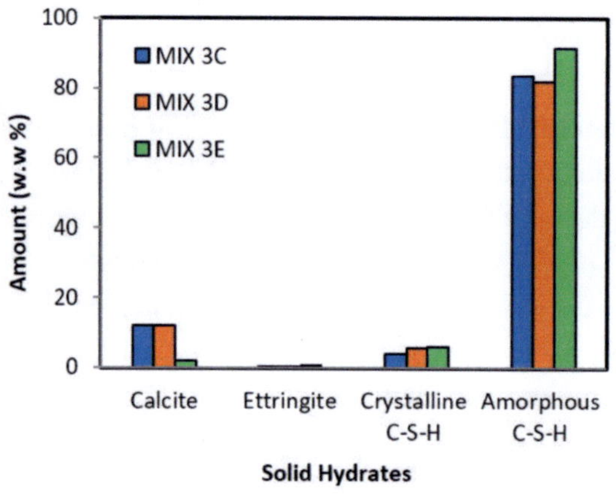

Figure 1: *Quantitative Rietveld phase analysis of the three samples: MIX 3C (in blue), MIX 3D (in orange) and MIX 3E (in green).*

<u>*TG - DSC after 90 days of hydration*</u>

Thermogravimetric analysis (TG) can give quantitative information of the different phases present in the samples by measuring the temperature dependent mass loss (wt.%). In addition, differential scanning calorimetry (DSC) measures the heat difference related to enthalpy changes caused by loss of water or CO_2 or recrystallization reactions. In Figure 2, the mass loss (%) and the differential scanning calorimetry (DSC) data of the three samples are presented. The interpretation of the TG analyses (Ait Mouheb et al., 2017) revealed C-S-H, ettringite and calcite phases. The mass loss at 820°C, not discussed in the previous contribution, is attributed to the decomposition of C-S-H to wollastonite ($CaSiO_3$) by dehydroxylation (Beaudoin et al., 1990).

Additionally, thermogravimetric analysis has been applied to measure the water bound in cement samples corresponding to the interlayer water in C-S-H hydrate, structural water in ettringite, and adsorbed water, except water in micropores or in larger pores. The amount of water bound in the 3 samples was calculated according to the method described in Lothenbach et al. (2012) which consists of measuring the total mass loss up to 500°C of the 3 samples, after exchange with isopropanol and drying the samples at 40°C. An average value of 28.17 ± 1.70 wt.% has been calculated for the 3 studied mixtures.

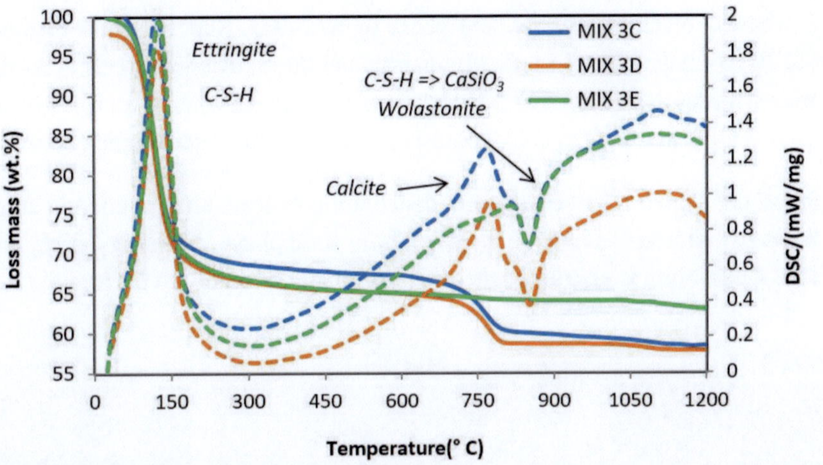

Figure 2: *Mass loss (wt.%) recorded by TG analysis (solid line) and differential scanning calorimetry (DSC) (dotted line) for MIX 3C, 3D and 3E (after 90 days of hydration).*

4.1.3 SEM-EDX analyses after 90 days of hydration

The three pastes were analyzed by SEM-EDX to observe the morphology and chemical composition of the different hydrated phases (Ait Mouheb et al., 2017).

In order to determine the different phases and the Ca/Si ratios of the C-S-H phase elemental maps of the surface were conducted by SEM-EDX. Sixty different spots have been investigated to determine the atomic ratio of Ca, Al, Si, C, Fe, S, Na, Mg and K. In Figure 3, the data analyses for the 3 samples is depicted by the plot of Al/Ca versus Si/Ca atomic ratios analyzed at several points of the mix 3C, 3D and 3E samples. According to Ait Mouheb et al. (2017) and Figure 3, the analyses show that mix 3C, 3D and 3E are composed essentially by C-S-H phases with Ca/Si ratios in the range of 0.6 to 0.8.

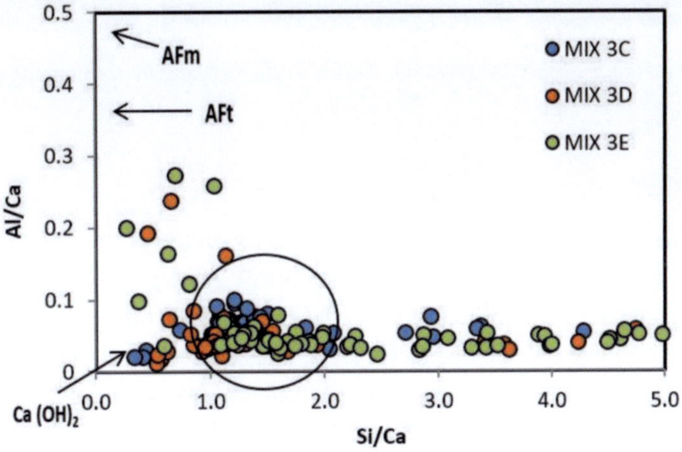

Figure 3: *Al/Ca versus Si/Ca atomic ratios for different microanalyses spots of the mixture 3C, 3D and 3E after 90 days of hydration.*

Finally, to complete the interpretation of the data and to show the heterogeneity of the sample, elemental maps of calcium (yellow), silicon (red) and aluminium (blue) and the measurement of the Ca/Si ratio by SEM-EDX on all samples is depicted in Figure 4. The atom percentage increases with the intensity of the colour in the mix 3C, 3D and 3E (maximum silica content 35%, 29% and 59%, calcium 26%, 22% and 24%, aluminum 13% - 5% and

10%) respectively, whereas black areas represent pores or fractures. Superposition of the calcium, silica and aluminum content as RGB image allows to clearly distinguish three different zones (see Figure 4). The orange color in the RGB images represents areas of C-S-H phase predominance, whereas purple - blue colors are probably attributed to non-reacted OPC and the red color represents areas of non-reacted silica fume.

Figure 4 also shows the spatial resolved element distribution of iron, aluminium and sulphate. Superposition of these elements indicate that iron is mainly present as a pure solid phase, probably mainly as ferrite or hydroxide not detected by XRD. Aluminium is also associated with sulphate, probably as ettringite.

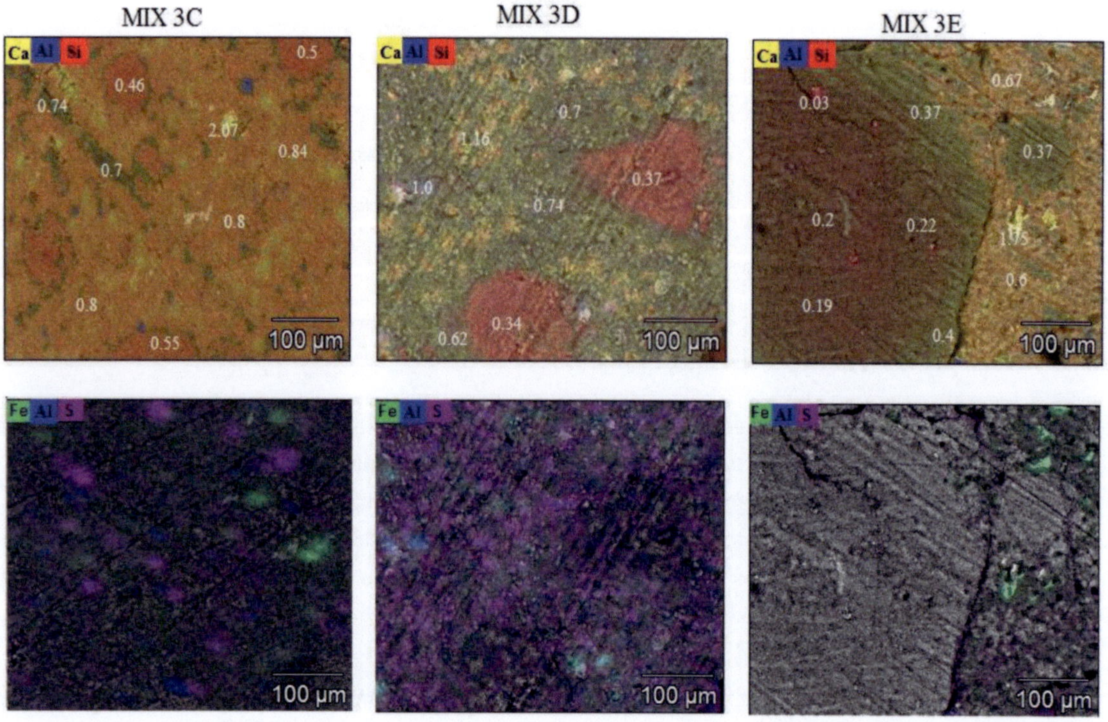

Figure 4: *SEM-EDX elemental maps of calcium (yellow), silicon (red)and aluminium (blue) together with local Ca/Si ratios of samples 3C, 3D and 3E.*

<u>*Solid state NMR spectroscopy*</u>

The main advantage of solid state NMR spectroscopy is its ability to detect amorphous phases and the possibility to provide structural information. The ^{29}Si MAS NMR spectra of 3 cement samples are presented in Figure 5 with assignment of the bands. A broad signal at the chemical shift range between -79 and -85 ppm is the main feature of the spectra and can be assigned to the Si present in the C-S-H phases. The overlap of the peaks does not permit an unambiguous deconvolution between different spectra arising from Si nuclei in different surroundings but it is known that the spectra of C-S-H phases consists of at least three resonances called Q^1, Q^2(1Al) and Q^2 (Lothenbach et al., 2012 and 2014; Richardson, 1999) that could be identified in Figure 5. Q^1 represents a silicate tetrahedron at the end of the C-S-H chain appearing at -79.1 ppm, Q^2(1Al) represents a SiO_4 chain unit connected to one SiO_4 and one AlO_4 tetrahedron (δ = -82.9 ppm) and Q^2 represents a silicate tetrahedron at the middle of the C-S-H chain (δ = -84.5 ppm, see Figure 5).

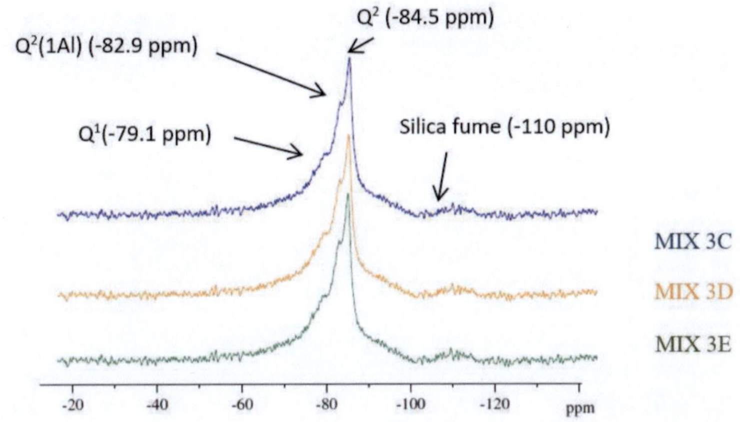

Figure 5: *^{29}Si MAS NMR spectra of hydrated cement (MIX 3C, 3D and 3E) with assignment the chemical shifts.*

The small broad peak at -110 ppm is characteristic for the silica fume which has not reacted during the hydration of the samples (Lothenbach et al., 2012) indicating that only a small amount of Si will be present in this form.

The ^{27}Al-NMR spectra recorded at the three low alkali hardened cement pastes are closely resemble each other presenting all of them two different signals at 58.8 and 13.1 ppm (Figure 6). Al in a tetrahedral coordination environment is generally observed between 50 and 80 ppm, while octahedral Al resonance signals appear between -20 and 20 ppm (Skibsted and Hall, 2008). Figure 6 shows that Al is mainly in an octahedral coordination, and there are only relatively minor amounts of tetrahedral coordinated aluminium. The observed resonances of tetrahedrally ^{27}Al resonances are associated with the aluminum in the nonbridging position of the C-S-H phases and the octahedral ^{27}Al with the presence of ettringite (L'Hôpital et al., 2016).

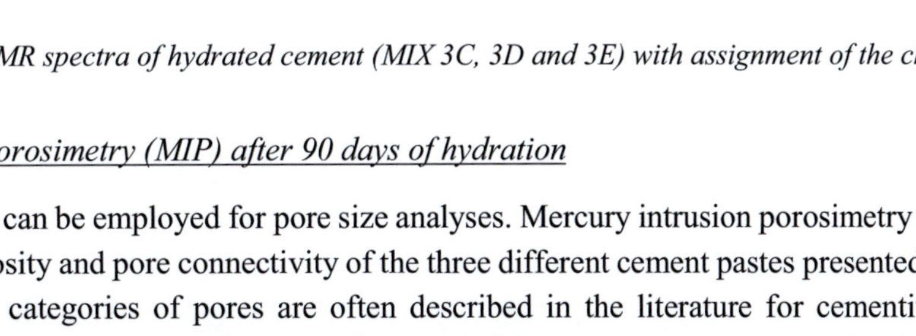

Figure 6: *^{27}Al MAS NMR spectra of hydrated cement (MIX 3C, 3D and 3E) with assignment of the chemical shifts.*

Mercury intrusion porosimetry (MIP) after 90 days of hydration

Various methods can be employed for pore size analyses. Mercury intrusion porosimetry was initially selected for studying the porosity and pore connectivity of the three different cement pastes presented in Ait Mouheb et al. (2017). Three main categories of pores are often described in the literature for cementitious phases: macro, capillary and gel pores (Bajja et al., 2017). As observed in Figure 7, no macro porosity is detected at the samples (> 0.05 μm) and the size of the pores in the hydrated cement phases varies from the micro to the nanoscale. The presence of superplasticizer (mix 3C) eliminates the nano-porosity present in the C-S-H phases (see Figure 7) as observed in Khatib and Mangat (1999).

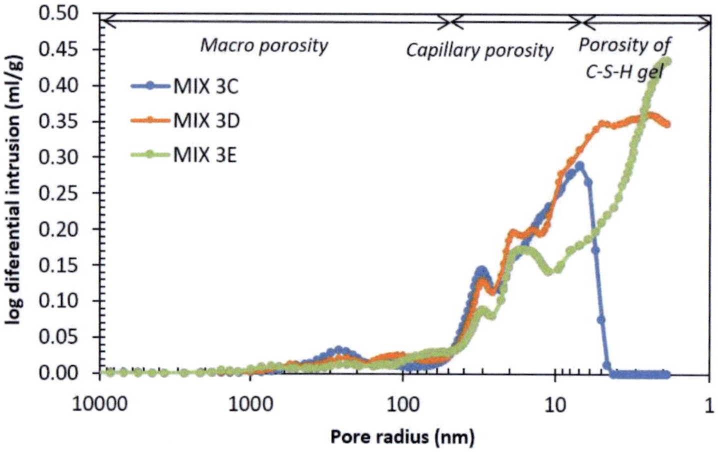

Figure 7: *Pore access at 90 days age of hydrated cement (MIX 3C, 3D and 3E).*

Batch sorption experiments of radiotracers

Sorption values for HTO and sorption mechanism

The results of HTO batch sorption kinetic experiments are shown in Figure 8. The determined R_d value of 0.40 ± 0.13 L/kg and its associated error have been calculated as an average value from all the data obtained from the kinetic sorption experiments. The R_d value for HTO in the three different low pH cement pastes is very weak and independent of the time and the chemistry of the solid indicating that the uptake process presents a similar fast mechanism. In the literature, no R_d values are determined for HTO in low pH cements and lack of sorption data in sulphate-resistant Portland cement (only two sets of experiments) is described in Ochs et al. (2016).

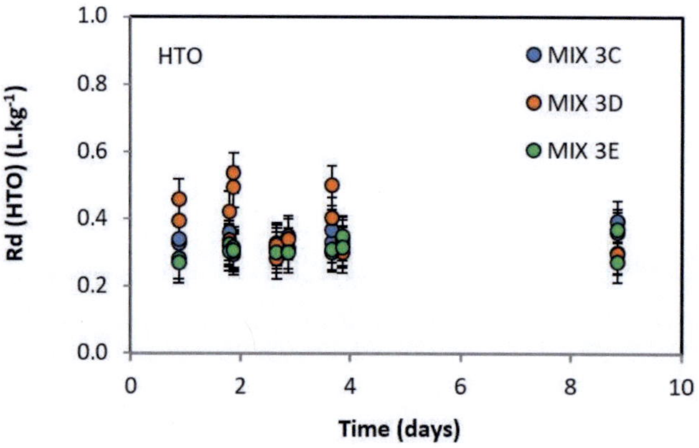

Figure 8: *Sorption of HTO on low pH cement pastes (MIX 3C, 3D and 3E) as a function of time. S/L ratio = 0.133 kg/L. HTO initial concentration = $1.86 \cdot 10^{-9}$ M. Background electrolyte is the one at equilibrium with the low pH cement.*

Comparison of the obtained results with the ones obtained in the literature with other cement pastes (Tits et al., 2003) show that tritium sorption is low in all cases ($R_d < 0.8$ L/kg). Isotope exchange of HTO with the water bound to the cement has been postulated as possible uptake mechanism (Tits et al., 2003; Ochs et al., 2016). Additionally, Wieland and Van Loon (2002) revealed that higher water / cement ratios can results in higher R_d values for HTO.

In order to demonstrate the postulated mechanism and to derive a comprehensive sorption model, the R_d value has been estimated assuming that HTO and H_2O behave identically and that 100% of the bound water is accessible.

The mass of water bound in three cement samples have been determined experimentally by TG-DSC following the procedure described in Lothenbach et al. (2012) (see previous section *TG - DSC after 90 days of hydratation*). Therefore, the expected R_d will be 0.29 ± 0.03 L/kg which is in agreement with the measured value of 0.3 ± 0.1 L/kg.

Sorption values of $^{36}Cl^-$ and sorption mechanism

The results of $^{36}Cl^-$ batch sorption kinetic experiment are shown in Figure 9. The determined R_d value of 0.22 ± 0.15 L/kg and its associated error have been calculated as an average value from all the data obtained from the kinetic experiments. The R_d value for $^{36}Cl^-$ in the three different low pH cement pastes is very weak and independent of time and chemistry of the solid indicating that the uptake process presents a very similar fast mechanism. In the literature there are not R_d values determined for $^{36}Cl^-$ in low pH cements. However, Ochs et al. (2016) summarized the reviewed R_d values for chloride in different cementitious materials (Nirex reference vault backfill, hardened cement paste, white and grey Portland cement, ordinary Portland cement, sulphate-resisting Portland cement and C-S-H phases) as a function of pH (11.8 - 13.5) for different states of cement degradation. Examination of available data by the authors indicate that the most influential factor on chloride sorption values is the aqueous total chloride concentration, including stable chloride present in cement solid phases and leached into the pores, while pH becomes important only after the effect of total chloride concentration has been accounted for. All available studies (Ochs et al., 2016), indicate a decrease of chloride sorption with increasing total (stable chloride and ^{36}Cl) aqueous concentration. Then, a different chloride behaviour within two concentration ranges has been identified, postulating two different possible uptake mechanism based on experiments performed at pH > 11.8 which are still not well understood (Ochs et al., 2016):

- At Cl^- concentrations lower than few millimolal, $R_d > 10^1$ L/kg is described which might indicative of surface processes (electrostatic interaction and/or anion exchange).
- At Cl^- concentrations above few millimolal, $R_d < 10^1$ L/kg are observed which probably indicate the formation of Friedel´s salt or solid solution with cement phases such as AFm.

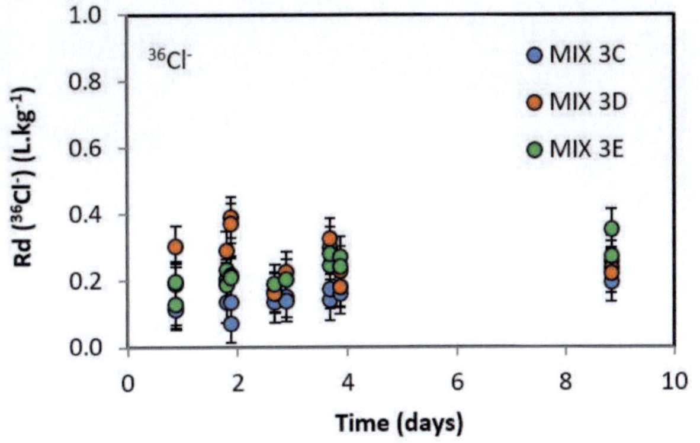

Figure 9: *Sorption of ^{36}Cl on low pH cement pastes (MIX 3C, 3D and 3E) as a function of time. S/L ratio = 0.133 kg/L. Initial total chloride concentration ($^{36}Cl^- = 4.55 \cdot 10^{-6}$ M + $Cl^- = 3.84 \cdot 10^{-3}$ M). Background electrolyte is the one at equilibrium with the low pH cement.*

In the present study, the total concentration of chloride (stable + $^{36}Cl^-$) is equal to $3.8 \cdot 10^{-3}$ M which would indicate the possible formation of solid solutions with Friedel´s salt or ettringite. In order to demonstrate the postulated mechanism and to derive a comprehensive sorption model, the possibility of formation of other phases in the sorption experiments was tested by performing XRD and TG analysis of the solid sample after 30 days of sorption studies. Formation of other solid phases was not detected by these techniques. At this moment, with the available information, it has been postulated that the uptake process could be associated with surface processes in the C-S-H and C-A-S-H phases identified in the three samples instead of precipitation of secondary phases (Friedel´s salt was not present in the initial cement pastes). In that case zeta potential on cement surfaces could play a significant role for chloride uptake. Available data indicate that sorption of Cl decreases if pH decreases in the range between 12 and 12.5 and are interpreted as being due to the decrease in the zeta potential on cement surfaces. According to Ochs et al. (2016), the zeta potential decreases from pH 12.5 (20 mV) to pH 11 (-10mV) which indicates that the R_d value for Cl would follows the same trend. This mechanism would explain the weak R_d value determined in this work.

The uptake of $^{129}I^-$ and sorption mechanism

The results of $^{129}I^-$ batch sorption kinetic experiment is shown in Figure 10. The determined R_d value of 0.13 ± 0.13 L/kg and its associated error have been calculated as an average value from the data obtained from the kinetic experiments until 10 days. The R_d value for $^{129}I^-$ in the three different low pH cement pastes is very weak, observing almost not sorption. Sorption experiments performed at 25 days indicate a small increase in the sorption which could indicate a slow change of the sorption mechanism which would need to be proved in the future. In the literature (Ochs et al., 2016) there are R_d values determined for $^{129}I^-$ in different cementitious materials (Ordinary Portland cement, concrete, C-S-H phases, AFm and AFt phases) and under different experimental conditions (time, S:L ratio, pH, initial aqueous concentration). Comparison of the obtained results in this work with the ones obtained in the literature using other cementitious materials and calcite indicate that iodine sorption strongly depends on the pH, the Ca:Si ration of the C-S-H phases, the initial I^- aqueous concentration, time and the concentration of other competing anions like Cl^- and sulphate (Ochs et al., 2016). The sorption mechanism of iodine in the 3 studied cements pastes is expected to be similar to the one described for chloride. For this reason, the uptake process for iodine in this study is probably associated with surface processes at the C-S-H and C-A-S-H phases where chlorine competes for sorption sites with chloride (present in higher concentrations).

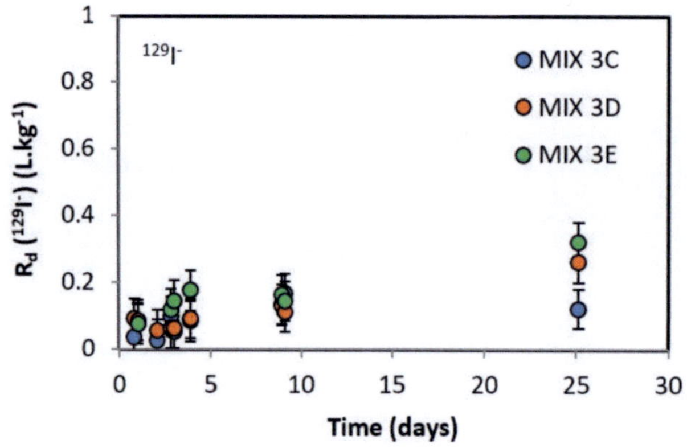

Figure 10: *Sorption of ^{129}I on low pH cement pastes (MIX 3C, 3D and 3E) as a function of time. Total iodine concentration (I^-+ ^{129}I) = $5.93 \cdot 10^{-5}$ M.*

Conclusions and Future work

The chemical properties of the 3 low pH cement pastes prepared in this work with Ordinary Portland cement (OPC) and silica fume (40 - 50%) are clearly different from the ones typically found with only OPC. The main hydrated phases formed after 90 days of hydration are the calcium silicate hydrates (C-S-H) with a low Ca/Si ratio which varies between 0.6 and 0.8 and C-A-S-H phases where a partial substitution of the Si for Al have been produced. The ^{29}Si and ^{27}Al MAS NMR analyses demonstrated the incorporation of aluminium as tetrahedrally coordinated Al(IV) in the bridging position of the C-S-H phases. Unreacted silica fume has also been identified and calcite is present in two of the samples where limestone filler was added. Neither precipitation of strätlingite, katoite nor Friedel´s salt were identified. Additionally, as seen by MIP, the presence of superplasticizer clearly decreases the porosity and the smallest connected pore size detectable of the cement pastes.

Uptake studies of HTO, $^{36}Cl^-$ and $^{129}I^-$ from batch sorption experiments indicate very weak sorption ($K_d < 0.40 \pm 0.13$ L/kg) for the 3 selected radionuclides. The uptake process of Cl and I is probably associated with surface processes in the C-S-H and C-A-S-H phases with competition for sorption sites, between them. In the case of HTO, isotopic exchange with the interlayer water of the C-S-H and the C-A-S-H seems to be the main uptake process.

Besides the completion of the cement sample characterization with additional techniques (especially the characterization of the pore structure) and the potential of the surface, degradation studies of the low pH cement paste by the bentonite pore water using imaging techniques are in progress.

Acknowledgement

The research leading to these results has received funding from the European Union's European Atomic Energy Community's (Euratom) Horizon 2020 Programme (NFRP-2014/2015) under grant agreement, 662147 – Cebama.

Additionally, we would like to acknowledge the colleagues from KIT-INE: Nicolas Finck for the XRD advise, Stefanie Hilpp and Sylvia Moisei-Rabung for the IC and ICP-OES analyses, Tanja Kisely for the TIC, Stephanie Heck, Martina Schlieker, Elke Bohnert and Claudia Joseph for the lab assistance and advise. We thank Francis Claret (BRGM), for his comments and suggestions which have contributed to improve this paper.

References

Ait Mouheb, N., Montoya, V, Borkel, C., Schäfer, T. (2017). Experimental studies on low pH cements / clay interface processes: characterization of low pH cements. Proceedings of the 1st CEBAMA Annual Workshop. KIT Scientific Publishing, KIT-SR 7734.

ANDRA (2005). Andra Research on the Geological Disposal of High-level Long-lived Radioactive Waste - Results and Perspectives. June, 2005.

Bajja, Z., Dridi, W., Darquennes, A., Bennacer, R., Le Bescop, P., Rahim, M. (2017). Influence of slurried silica fume on microstructure and tritiated water diffusivity of cement pastes. Constr. Build. Mater., 132, 85-93.

Beaudoin, J.J., Ramachandran, V.S., Feldman, R.F. (1990). Interaction of chloride and CSH. Cem. Concr. Res., 20, 875-883.

Coumes, C.C.D., Courtois, S., Nectoux, D., Leclercq, S., Bourbon, X. (2006). Formulating a low-alkalinity, high-resistance and low-heat concrete for radioactive waste repositories. Cem. Concr. Res., 36, 2152-2163.

ENRESA (1995). Almacenamiento geológico profundo de residuos radiactivos de alta actividad (AGP). Diseños conceptuales genéricos. Publicación Técnica ENRESA, 11/95.

Khatib, J.M. and Mangat, P.S. (1999). Influence of superplasticizer and curing on porosity and pore structure of cement paste. Cem. Concr. Compos., 21, 431-437.

L'Hôpital, E., Lothenbach, B., Kulik, D.A., Scrivener, K. (2016). Influence of calcium to silica ratio on aluminium uptake in calcium silicate hydrate. Cem. Concr. Res., 85, 111-121.

Lothenbach, B., Le Saout, G., Ben Haha, M., Figi, R., Wieland, E. (2012). Hydration of a low-alkali CEM III/B–SiO_2 cement (LAC). Cem. Concr. Res., 42, 410-423.

Lothenbach, B., Rentsch, D., Wieland, E. (2014). Hydration of a silica fume blended low-alkali shotcrete cement. Phys. Chem. Earth, Parts A/B/C, 70-71, 3-16.

Ochs, M., Mallants, D., Wang, L. (2016). Radionuclide and metal sorption on cement and concrete. Springer.

Richardson, I.G. (1999). The nature of C-S-H in hardened cements. Cem. Concr. Res., 29, 1131-1147.

Skibsted, J. and Hall, C. (2008). Characterization of cement minerals, cements and their reaction products at the atomic and nano scale. Cem. Concr. Res., 38, 205-225.

Experiments on interface processes at the cement / Callovo-Oxfordian claystone interface and the impact on physical properties; mechanical results from the Callovo-Oxfordian claystone

Robert Cuss[1*], Andrew Wiseall[1], Jon Harrington[1], Jean Talandier[2], Xavier Bourbon[2]

[1] BGS, British Geological Survey (UK)
[2] Andra, Agence nationale pour la gestion des déchets radioactifs (FR)

* Corresponding author: rjcu@bgs.ac.uk

Abstract

The evolution of flow and strength properties of the concrete/Callovo-Oxfordian Claystone (COx) interface is of importance to the long-term performance assessment of a repository. A series of shear / flow experiments using the bespoke Direct Shear Rig will examine the evolution of flow and strength as the cement ages. Two lithofacies of COx will be tested, the so-called repository and high carbonate varieties, both cored from the Meuse / Haute-Marne underground research laboratory. To date, all tests prior to concrete casting have been conducted, bringing a series of samples of smooth and rough surfaces up to full saturation.

Introduction

The weakest part of any gallery or deposition hole seal is likely to be at the interface between the sealing components and the host rock. A single seal completion may comprise a number of elements reflecting different design criteria in order to address specific engineering challenges associated with changes in geochemistry and stress. The interaction of these components with the host rock, their evolution in terms of strength / bonding, cation exchange behaviour and interfacial permeability, and the sensitivity of these properties to an evolving geochemical and physical environment will be key factors in determining the long-term seal performance. The current study aims to examine the temporal evolution of the host-rock/low-alkali cement interface in the French repository concept to changes in geochemistry, mineralogy and stress to assess their impact on the development of hydraulic permeability and strength (shear strength).

Since the previous Workshop Proceedings (Cuss et al., 2016), experimental emphasis has been placed on preparing the Callovo-Oxfordian claystone samples ready for casting of concrete to create the interfaces of interest. This has included re-hydration of all samples and shearing of intact samples in order to create realistic, rough, surfaces to act as representative interfaces.

Experimental apparatus

Two Direct Shear Rigs were used. These comprise 5 main components.

1. Rigid frame that had been designed to deform as little as possible during the experiment;
2. Normal load system comprising a rigid loading frame, a hydraulic ram, and a normal load thrust block;

3. Shear force actuator designed to drive shear as slow as 14 microns a day at a constant rate (equivalent to 1 mm in 69 days);
4. Pore pressure system that can deliver either water or gas through the centre of the top sample directly to the fracture surface; and
5. A state-of-the-art custom designed data acquisition system using National Instruments LabVIEW™ software facilitating the remote monitoring and control of all experimental parameters.

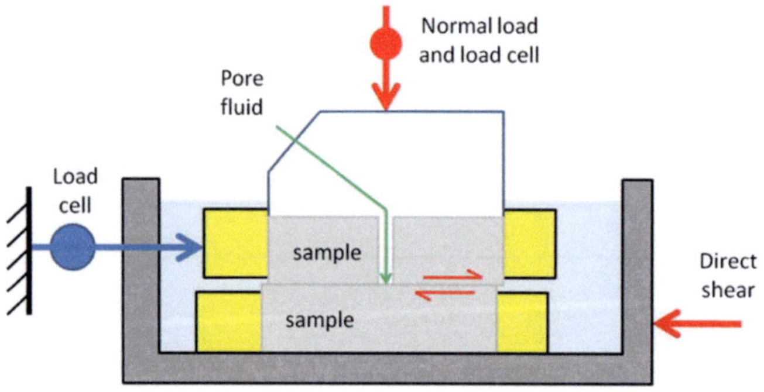

Figure 1: *Schematic of the Direct Shear Rig.*

The rig was designed to achieve 10 MPa normal stress (36 kN force) on a cubic sample 60 mm x 60 mm. At this normal load, it is expected that shear force will not exceed 25 kN. Normal load is imposed by a hydraulic ram that is pressurised using an ISCO/Teledyne syringe pump. For the Cebama study the apparatus has been modified in order to use cylindrical samples of 60 mm diameter. This was deemed necessary as it makes sample preparation more straightforward. This also acts to increase the rig capacity to 12.75 MPa normal load.

Test protocol

Well preserved samples of Callovo-Oxfordian Claystone (COx) were provided by Andra from freshly cored boreholes at the Meuse / Haute-Marne URL at Bure, approximately 300 km east of Paris, France. Each core barrel was sub-sampled into 65 mm long sections using a diamond tipped saw, from which a 60 mm diameter and 53 mm height sample was prepared using a lathe. At all times during sample preparation the "exposure" to air time was minimized and the samples stored in vacuum-packed plastic as much as is practicable. Off-cut material was weighed and basic geotechnical properties determined.

Samples of COx were placed in the experimental apparatus and were initially re-hydrated to bring up to full saturation under a normal stress of 6.5 MPa, representative of the effective stress at repository depths. The samples were hydrated using a synthetic pore fluid under a pore pressure of 0.25 MPa. Following hydration, some samples were sheared for 2.5 days at a rate of 2 mm per day. Data recorded included pore-fluid flow, shear displacement, shear stress, normal stress, and vertical displacement. This gave; initial flow rate, shear yield stress, shear modulus (G), shear peak strength, shear residual strength, dilation / contraction during shear, and change in flow during shear. Each fracture surface was scanned using a NextEngine 3D laser scanner, giving: roughness average, root mean square roughness, peak-to-valley height, kurtosis, skewness, texture direction, texture aspect ratio, and texture direction index. All samples were vacuum packed for medium-term storage.

Example shear test: Test Cebama_01

Test Cebama_01 was started on 21st November 2016. Figure 2 shows the starting test sample. The cylindrical sample was machine lathed and had no visible defects. Once weighed and measured the cling-film was wrapped around the sample and sample holders were bolted on. Both the upper and lower sample holders positioned flush with the top / bottom of the sample.

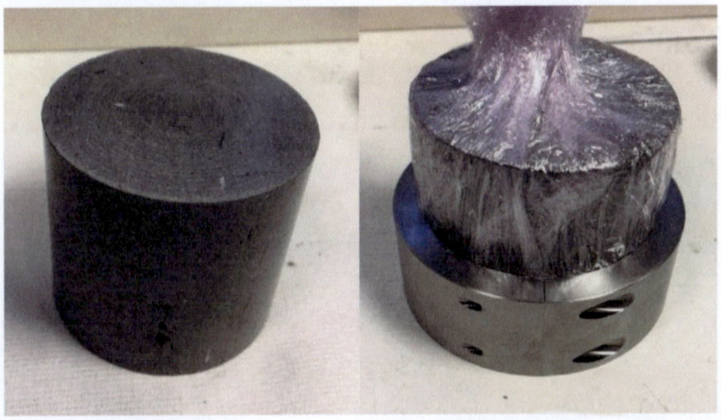

Figure 2: *Sample Cebama_01 prior to shearing.*

The initial stage of testing aimed to raise the saturation of the COx back to full saturation using a synthetic pore water representative of *in-situ* water chemistry (1.95 g/L NaCl, 0.13 g/L $NaHCO_3$, 0.035 g/L KCl, 0.63 g/L $CaSO_4 \cdot 2H_2O$, 1.02 g/L $MgSO_4 \cdot 7H_2O$, 0.08 $CaCl_2 \cdot 2H_2O$, 0.7 g/L Na_2SO_4). The shear arrangement only allowed fluid flow, fluid pressure, and sample height to be monitored. The initial flowrate of water into the sample was higher than the imposed maximum of 150 µL/h, a limit gained from previous experimental experience to ensure that erosion of the COx was not initiated (Cuss et al., 2017). As seen in Figure 3, hydration was continued for more than 900 hours. Periodically through this stage, the injection pump was switched to a constant pressure of 250 kPa in order to ascertain the sealing of the sample in the sample holders. However, failure of the seal between the two halves of the sample holders meant that pressure could not be held. Therefore, full hydration could not be established by the stabilisation of flow into the sample, and as a result the cessation of swelling was used as a proxy for full hydration. As can be seen in Figure 3, this was achieved within 200 hours and was followed by a small degree of compaction. Approximately 50 µm dilation was observed (approximately 0.1% strain), suggesting that the starting saturation of the test material was high.

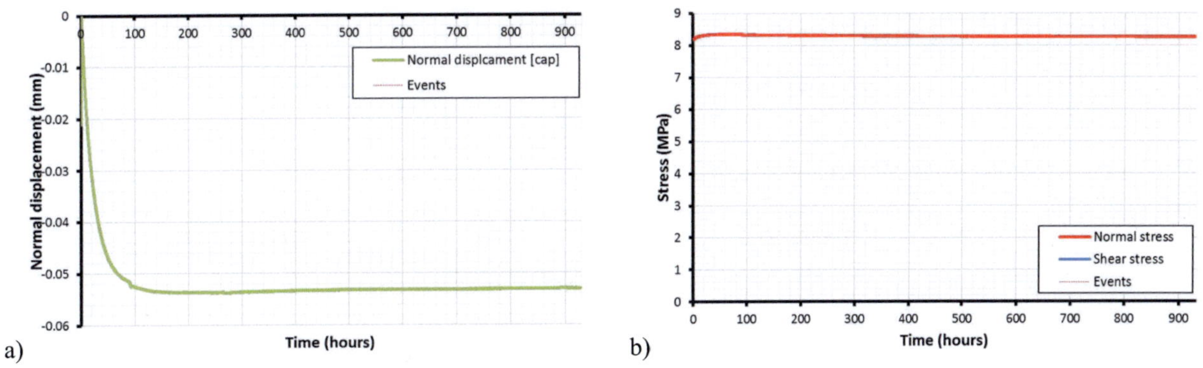

Figure 3: *Results for test Cebama_01 during the hydration phase. a) normal displacement (negative = dilation); b) normal and shear stress.*

Following 933 hours of hydration, sample shearing was started, the results are shown in Figure 4. As can be seen, the initial increase in shear stress was non-linear. This is commonly observed in shear tests using the Direct Shear Rig. The event seen at 943.3 hours is likely to be a small movement of one of the sample holders and is not suspected to be representative of sample failure due to a lack of corresponding sample contraction / dilation. Soon after this event, the stress-strain response can be seen to become linear, giving the shear modulus of COx (~ 340 MPa). Approaching 6 MPa shear stress, deviation from linear stress-strain behaviour is seen. This occurs at the yield shear stress and can be determined as the shear stress when the difference between the linear stress-strain region and the observed shear stress is 0.1 MPa, for Cebama_01 at 6.04 MPa. This is soon followed by peak stress of 6.19 MPa, followed by classic brittle deformation with a stress drop to 3.05 MPa and dilation of approximately 60 µm. At this point the stress starts to increase again, probably as a result of the mismatch of the two formed fracture surfaces and a second low-angle peak is reached, before a second stress decrease occurs. A secondary residual stress level was achieved, but as can be seen in the vertical displacement data, one fracture surface was now riding above the other. The test was terminated at approximately 9.5% strain. Table 1 summarises the data for test Cebama_01.

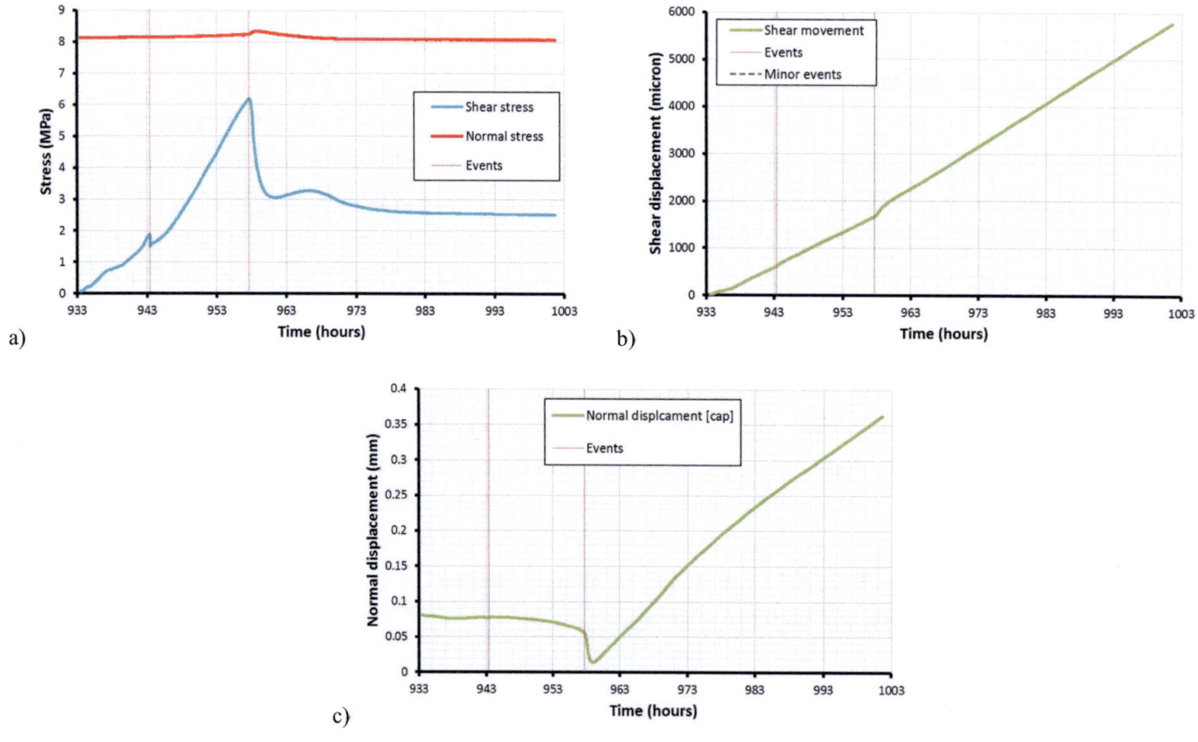

__Figure 4:__ Results for test Cebama_01 during the shear phase of the experiment.

__Table 1:__ Summary of test results for test Cebama_01.

Test	Cebama_01		Test	Cebama_01	
Rock type	COx		Displacement at peak	1.668	mm
Core barrel	EST 53639		Strain at peak	2.781	%
Sample diameter	59.986	mm	Residual shear stress	3.052	MPa
Sample height	52.873	mm	Residual norm[l] stress	8.270	MPa
Sample volume	149.423	cc	Displ[t] at residual	2.120	mm
Sample weight	360.307	g	Strain at residual	3.534	%

Test	Cebama_01		Test	Cebama_01	
Sample density	2.411	g/cc	Shear rate	90.740	µm/h
Rig used	DSR1		Shear rate set point	150	µl/h
Test start date	28/11/2016 15:21:18		Maximum displ[t]	5.767	mm
Shear start date	06/01/2017 13:01:04		Total strain	9.614	%
Shear finish date	09/01/2017 09:07:30		Average norm[l] stress	8.136	MPa
Shear duration	68.107	hours	Shear modulus	339.598	MPa
Total test duration	41.740	days	Intercept	-3.047	MPa
Starting yield stress	0.070	MPa	Start height	0.032	mm
Starting normal stress	8.133	MPa	Start date	28/11/2016 16:23:01	
Yield shear stress	6.042	MPa	Peak height	-0.021	mm
Yield normal stress	8.239	MPa	Peak date	09/12/2016 21:24:15	
Displacement at yield	1.624	mm	Total swelling	0.054	mm
Strain at yield	2.707	%	Swelling strain	0.102	%
Peak yield stress	6.185	MPa	Swelling duration	11.209	days
Normal stress at peak	8.251	MPa			

Figure 5a shows the fracture surface created during test Cebama_01. Some clear smearing and polishing was observed, as was the disaggregation of some clay as a fault gouge had started to form. The sample was laser scanned and vacuum packed for storage until concrete casting. Figure 5b shows the scan result for the test sample, with Table 2 summarising the surface textural data achieved for the lower sample of test Cebama_01.

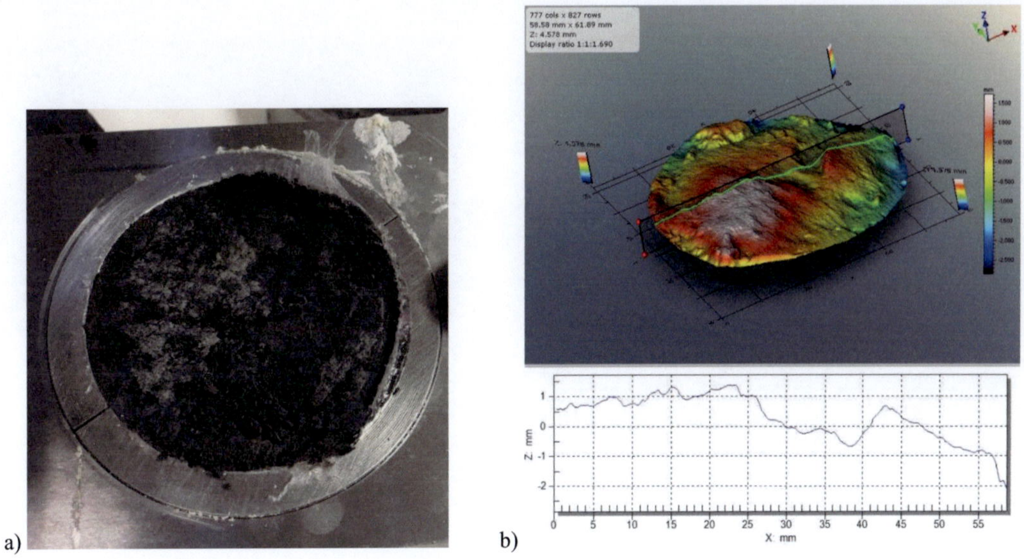

Figure 5: *Fracture surface created in test Cebama_01: a) photograph; b) scan results.*

Table 2: *Selected surface texture results for Cebama_01.*

Parameter	Value	Description
Sa	0.697 mm	Average roughness
Sq	0.879 mm	RMS roughness
St	4.578 mm	Peak to valley height

Parameter	Value	Description
SPaX	0.585 mm	Mean profile Pa x-axis
SPaY	0.465 mm	Mean profile Pa y-axis
SPtX	2.846 mm	Mean profile Pt x-axis
SPtY	2.37 mm	Mean profile Pt y-axis
Std	0.334°	Texture direction

Shear properties of COx

All sample hydration tests have been conducted, including 6 shear tests to create 12 rough interface surfaces and 8 hydration tests of planed (smooth) interface surfaces. These results are summarised in Figure 6. Two varieties of Callovo-Oxfordian have been tested; the repository depth rock (COx) and the overlying high carbonate variety (USC). As can be clearly seen in Figure 6a, USC has considerably greater shear strength than COx. In fact, the shear apparatus had to be modified in order to achieve shear failure of the sample at just about 20 MPa. Good repeatability was seen between the 3 repeat tests conducted on the two varieties, even though samples derived from more than one core barrel. Shear modulus (Figure 6b) was seen to be higher in USC compared with COx. It has to be noted that values are lower than previously reported values and the apparatus is currently undergoing a thorough calibration to determine whether the shear stress or displacement are being under / over represented. Figure 6c shows the results for starting, yield, peak, and residual shear stress for all 6 shear experiments. As shown, USC is much stronger than COx. In COx, there is little difference between the yield and peak stresses, whilst in USC there is considerable difference. In both varieties, the residual stress is less than half the peak stress condition.

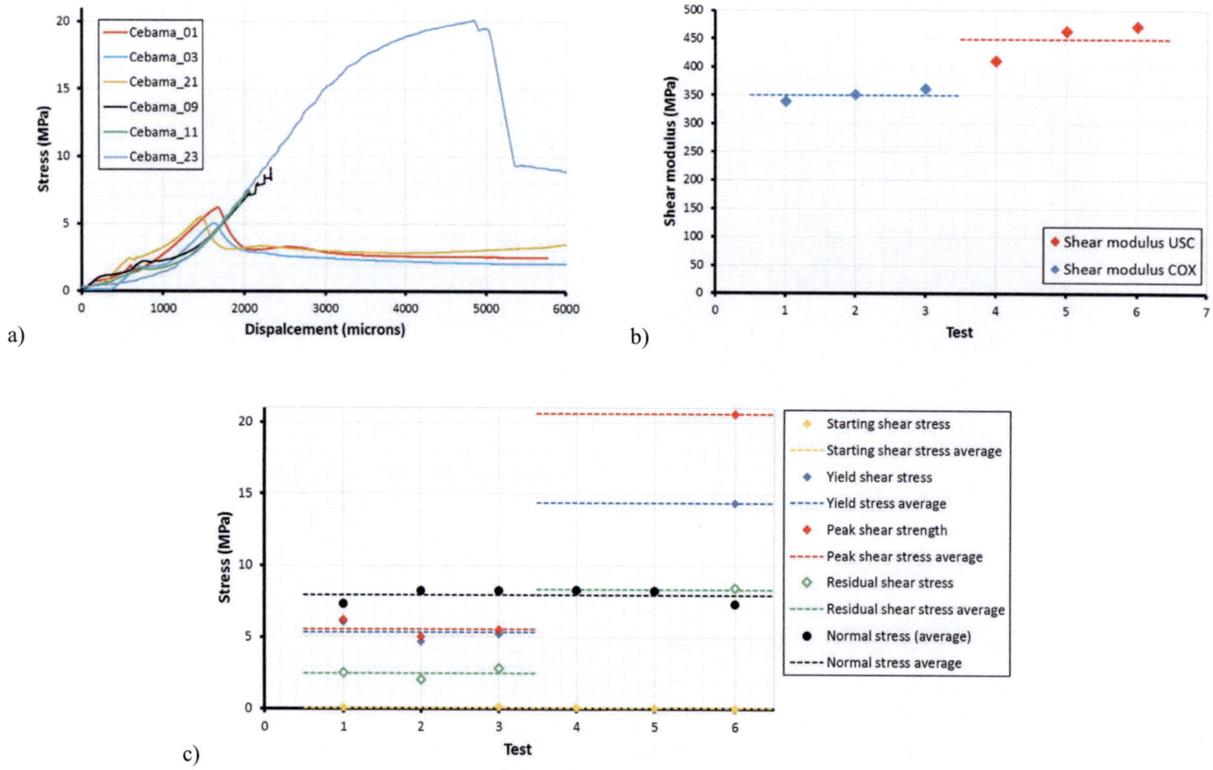

Figure 6: *Summary of shear data: a) Shear stress versus displacement for all shear tests conducted; b) Shear modulus; c) Shear stress results.*

Current status and future activities

All COx and USC samples have been sheared/prepared and re-hydrated, scanned, and stored in preparation for concrete casting to create the desired interfaces. Casting will occur in late May / early June and will be followed by a period of curing. The T_L variant of low pH cement will be prepared. The use of additional sample holders will allow concrete to be cast directly onto the COx giving a total sample height of 53 mm. A water layer will be used on top of the concrete to ensure that the concrete is fully hydrated. A mixture of samples with an irregular and planed COx/T_L interface will be produced, half with COx from the repository depth and half of the high carbonate variety. In addition, test samples of concrete / concrete (T_L/T_L) will also be produced as will thirty-six samples cast of 54 mm diameter and 108 mm length for uniaxial testing. Care will be taken to ensure that the COx does not go into free-swell and that water exchange at the interface will not result in any influence on the test results.

The test program calls for four tests soon after casting at Yr0; these tests will be repeated annually. The concrete/concrete tests will be tested at Yr1 and Yr2. Additional control tests will be conducted at Yr1 and Yr2. Uniaxial tests will be performed after 3, 6, 12, 20, 28, and 36 months after casting to determine how the strength properties change as the concrete ages.

Prior to shear testing of the COx/T_L interface a 4 mm-diameter bore will be drilled through the COx to the depth of the interface between the COx and concrete. Normal load of 8 MPa will be used with a target pore pressure of 1 MPa, which is dependent on the overall permeability of the COx/T_L interface. Once stable flow has been achieved, the sample will be sheared 5 mm at a rate of 0.25 mm/day. Flow along the interface and the shear properties will be determined. The resulting failure surfaces will be laser scanned to determine fracture roughness parameters. Fluorescent scanning will be conducted to identify flow paths of the injection water. SEM analysis will be used to determine chemical aging of the concrete, geochemical reactions between the COx and cement and identify flow paths.

The uniaxial tests on concrete will be performed using an MTS apparatus and will determine; Young's modulus (E), Poisson's ratio (ν), bulk modulus of compressibility (K), yield stress (σ_y), and the unconfined compressive strength (q_u). Repeat testing will show how these parameters vary as the concrete ages.

Acknowledgement

The research leading to these results has received funding from the European Union's Horizon 2020 Research and Training Programme of the European Atomic Energy Community (EURATOM) (H2020-NFRP-2014/2015) under grant agreement n° 662147 (CEBAMA).

The BGS authors publish with the permission of the Executive Director, British Geological Survey (NERC).

References

Cuss, R.J., Wiseall, A.C., Harrington, J.F., Talandier, J., Bourbon, X. (2016). Experiments on interface processes at the cement / Callovo-Oxfordian Claystone interface and the impact on physical properties; initial findings. Proceedings of the 1st CEBAMA Annual Workshop. KIT Scientific Publishing, KIT-SR 7734.

Cuss, R.J., Harrington, J.F., Sathar, S., Norris, S., Talandier, J. (2017). The role of the stress-path and importance of stress history on the flow of water along fractures and faults; an experimental study conducted on kaolinite gouge and Callovo-Oxfordian mudstone. Applied Clay Science, 150, 282-292.

Hydrogeochemical and isotope modifications of the FEBEX in-situ test: Clues to understand concrete evolution

María J. Turrero[1*], Elena Torres[1], Antonio Garralón[1], Paloma Gómez[1], Lorenzo Sánchez[1], Javier Peña[1], Belén Buil[1], Alicia Escribano[1], Juan Manuel Durán[1], Roberto Domínguez[1]

[1] CIEMAT, Research Centre for Energy, Environment and technology (ES)

* Corresponding author: mj.turrero@ciemat.es

Abstract

This contribution provides a compilation of the work done on the hydrogeochemical characterization of the *in-situ* FEBEX test, as part of the task 1.2 of CIEMAT in the WP1 of the CEBAMA project. The general aim of this task was to analyse aged concrete samples in contact with natural host rocks and bentonite in order to evaluate changes in physical transport parameters, mineralogy and microstructure that could affect transport characteristics.

To accomplish this task, the impact of the hydrogeochemical evolution of the FEBEX site on the concrete plug was studied from mm to cm scale.

This document gives an overview of that evolution covering the baseline conditions of the experiment and its evolution based on the analysis of conservative elements (e.g., Cl), exchangeable cations and stable isotopes ($\delta^{18}O$, $\delta^{2}H$ and ^{13}C). The results help to explain the more detailed mineralogical analyses and provide with the clues to contextualize and understand the evolution of the system. The work was made in close collaboration with CSIC and UAM groups.

Introduction

A general view on aspects related to the groundwater and pore water in the deep geological repository (DGR), including geochemical processes due to the interaction of granite groundwater and bentonite and those occurring by concrete-rock and concrete-bentonite interaction, was made in the first S&T contribution of the project (Turrero et al., 2016). As a natural system cannot be completely understood if a complete dataset on the composition of the baseline chemistry of groundwater and its evolution over time is not made, CIEMAT planned to address this task to understand the aging of the EBS in the FEBEX *in-situ* test in the context of hydrogeochemical processes of the EBS-granite system. The characteristics of the experiment were succinctly described in the first S&T contribution, but extended information as well as dismantling operations and state of the buffer material after 18 years of test progress can be found in Villar et al. (2016).

Results presented in this work come from the analysis of two types of samples:

(1) Granite groundwater sampled from radial and parallel boreholes around the FEBEX gallery. Previously to the installation of the experiment, nineteen boreholes were drilled radial to the gallery axis (1996). After the first dismantling (2005), two parallel boreholes were drilled at 20 (FU-1) and 60 cm (FU-2) from the wall of the gallery and packed-off to sample and check the evolution of the groundwater (Figure 1). The last sampling campaign was made at the end of 2015, after the finalisation of the dismantling operations. The granite deep groundwater sampled in radial boreholes around the FEBEX gallery is Na-Ca-HCO_3-type water, with a low electric conductivity (76 - 236 μS/cm) and pH neutral to alkaline. The evolution of granite groundwater will give an idea of the granite-EBS interaction at experiment scale.

(2) Aqueous leaching, cation exchange capacity and stable isotopes of some of the concrete-bentonite samples collected close to the granite wall and at different positions towards the inner part of the barrier (Figure 2). These analyses will provide insights on the geochemical processes at mm to cm scale.

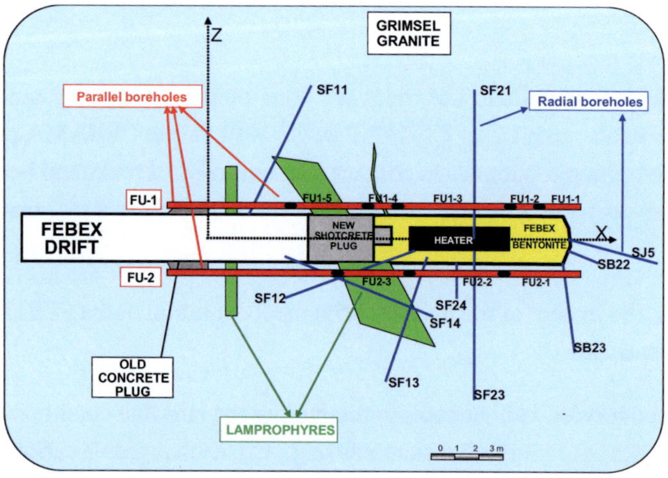

Figure 1: *Location of radial and parallel boreholes for groundwater sampling around the FEBEX gallery.*

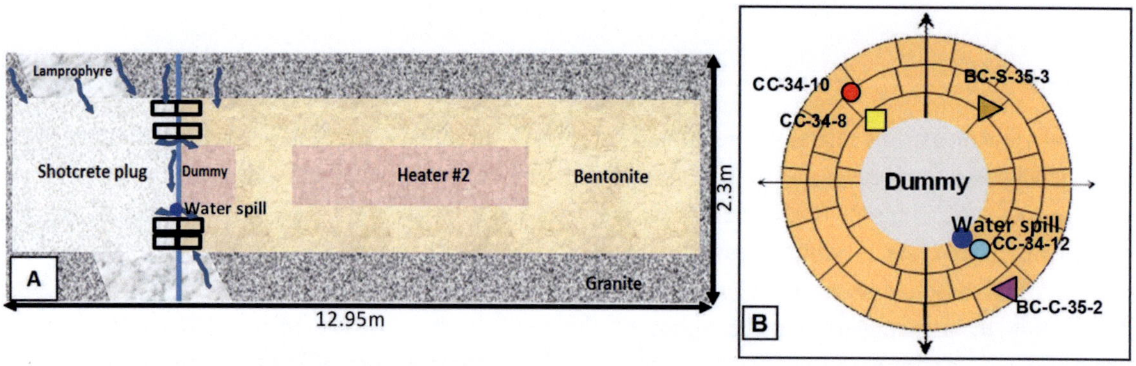

Figure 2: *A) Simplified picture showing sampling position (rectangles) and water fluxes under consideration. The configuration of the experiment shows the Heater#2 that was switched on up to dismantling in 2015, and the dummy, which is a substitute metal cylinder in place of the gap left by the Heater#1 once dismantled; B) Schematic situation of concrete and bentonite samples analysed for this work in the concrete plug / bentonite barrier interface.*

Granite groundwater and its interaction with the EBS

Granite deep water sampled in the radial boreholes exhibits a neutral to slightly alkaline pH, with an electrical conductivity average value of 101 ± 31 µS/cm. The interaction with granite and the minerals in the fracture fillings controls the chemical composition of Grimsel groundwater. Dissolution of carbonates and silicate minerals by groundwater recharge has resulted in the formation of Ca-Na-HCO_3 type water (Table 1). Ca-Na-HCO_3 type water was initially the dominant type of groundwater in all the packed-off sections of the two parallel boreholes as well. However, in sections FU1-3, FU1-4 and FU2-2, where boreholes intersect small lamprophyres and quartz dykes, groundwater chemical composition evolved from Ca-Na-HCO_3 to Na-Cl type water. In fact, five different types of water (Ca-Na-HCO_3, Ca-Na-SO_4, Na-Ca-SO_4, NH4-SO_4 and Na-Cl) were found as a function of two main factors: the distance to the bentonite barrier and the saline composition of the bentonite porewater (Garralón et al., *in prep.*).

Table 1: *Chemical composition of Grimsel granite water (mean of the chemical composition of radial boreholes), bentonite porewater and water spill (bentonite-concrete leachates, see Figure 2 for location).*

	pH	CE	meq/L									‰ SMOW	
		µS/cm	Cl	SO_4^{2-}	HCO_3^-	CO_3^{2-}	OH^-	Na	K	Ca	Mg	$\delta^{18}O$	δ^2H
Granite groundwater	8.1	104	0.1	0.2	0.7	--	--	0.5	0.0	0.6	0.0	-12.5	-90.1
Bentonite porewater (w = 23.8%)	7.4	13,230	112.7	26.3	16.1	--	--	91.3	0.4	25.4	32.2	-5.3	-50.0
Water spill (bentonite-concrete leachate)	10.2	10,850	23.0	18.0	4.9	55.2	5.8	114.5	13.5	3.1	0.3	-8.2	-69.1

FEBEX bentonite porewater is Na-Cl-type water with high magnesium and calcium contents. Its average EC value is 13,230 µS/cm and its pH is neutral (~ 7.4) (Fernández, 2004) (Table 1). Thus, the increase of salinity in the parallel boreholes can only be explained by the ion diffusion from bentonite towards granite. The quartz dykes and the small lamprophyres have acted as preferential flow paths that facilitate the solute mass transfer between the bentonite and the granite. The assumption of diffusive transport of these natural tracers was confirmed by 1D transport modelling of the Na^+ and Cl^- concentrations in the granite groundwater (Buil et al., 2010; Garralón et al., 2016). Those ions were considered as natural tracers of the mass solute transfer between bentonite and granite. Chloride concentration in bentonite porewater is five orders of magnitude higher (around 113 meq/L) than in granite groundwater (0.04 meq/L), whereas Na content is three orders of magnitude higher. The highest Cl^- and Na^+ concentrations occurred at 20 cm of the contact with the bentonite in section FU1-4, followed by FU1-3 and FU2-2 (at 60 cm to the contact with bentonite) (Figure 3). The fast decrease in Cl^- concentration in these sections took place after the dismantling of the bentonite barrier (source of Cl^- and Na^+). Simultaneously, an important increase of sulfate concentration in groundwater was detected.

The interaction of granite groundwater and/or bentonite porewater with other components of the experiment, for this work the concrete plug, can modify the chemical composition of the groundwater in the experiment as well. During the dismantling operations of the second heater (May'15), a water spill occurred close to sample C-C-34-12 (see Figure 2 for location). It was collected in a polypropylene bottle of 30 mL and analyzed. The chemical composition is shown in Table 1. Its analysis revealed that it was a Na-CO_3-type groundwater with high chloride, sulfate and potassium contents. Electrical conductivity, pH and chemical composition indicate that water collected from the spill is a mixture of granite groundwater and concrete and bentonite porewater. Effort in geochemical modeling to estimate the contribution of each component of the granite-bentonite-concrete system in a three-component mixing model is being made.

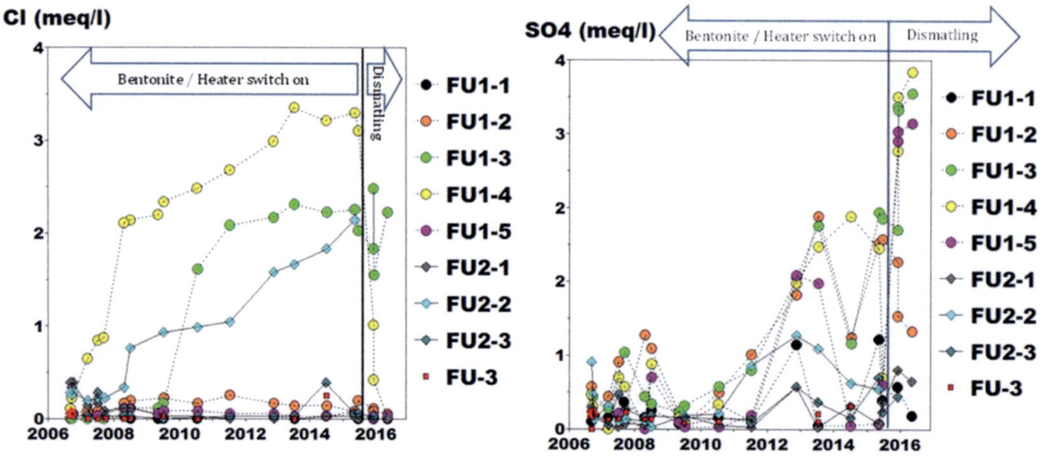

Figure 3: *Chloride and sulphate concentration (meq/L) evolution in water samples coming from parallel boreholes (Garralón et al., 2016).*

Concrete-rock and concrete-bentonite interaction

Aqueous leaching and exchangeable cations

Figure 4 (right) shows soluble Cl and Ca measured in concrete and bentonite samples from the vicinity of the concrete / bentonite interface (all the leaching tests were made at a S:L ratio of 1:8). According to the obtained data, a heterogeneous chemical behavior is observed in the interface, especially in the concrete side.

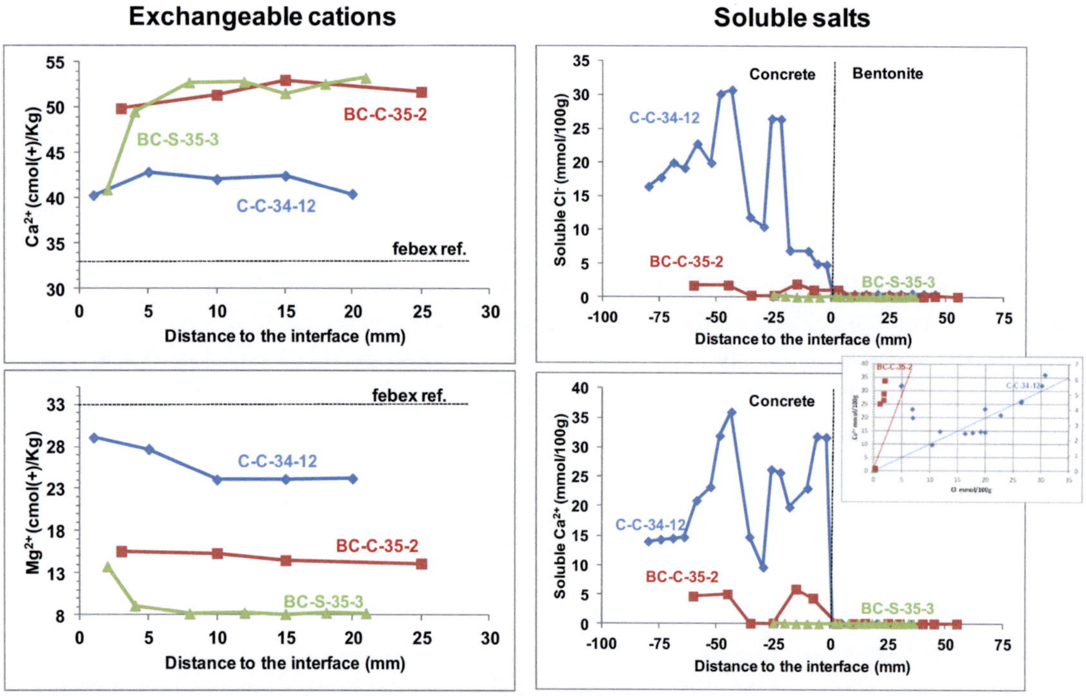

Figure 4: *(left) Exchangeable cations of three samples of bentonite at the interface with the concrete, and (right) soluble salts (S:L ratio of 1:8) (Cl^- and Ca^{2+}) in concrete-bentonite interface samples. There is a Ca:Cl 1:1 relation in the sample C-C-34-12 indicating the possibility of formation of Friedel's salt in the concrete.*

Depletion of soluble salts in bentonite points to solute diffusion between the bentonite barrier and some areas of the concrete plug. The bentonite samples closest to the granite seem to have undergone a more intense leaching that the samples in the inner part of the barrier. On the other hand, Ca:Cl 1:1 relation has been found in the concrete sample C-C-34-12. This could indicate the possibility of formation of Friedel's salt in the concrete. Alonso et al. (2016) found the formation of this salt in the concrete plug as far as 5 cm from the bentonite interface. Soluble sulfate usually increases close to the interface in the concrete side. So, the formation of secondary ettringite is favored. Alonso et al. (2016) and Cuevas et al. (2016) found secondary ettringite in the first millimeters from the interface. Soluble Mg was depleted at both sides of the interface. In fact, the chemical environment is not favorable to the presence of soluble Mg^{2+}. Chemical equilibrium diagram made with the software MEDUSA (Puigdomenech, 2015) shows that considering the conditions of the water spill analyzed near the sample C-C-34-12 (Table 1) magnesium should precipitate as silicates (Figure 5).

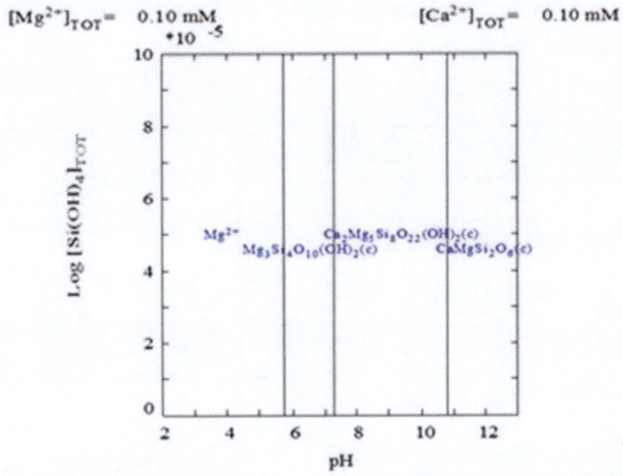

Figure 5: *Chemical equilibrium diagram showing stable forms of magnesium considering the chemistry of the spill water analysed in the vicinity of sample C-C-34-12 (Table 1).*

The exchange complex in the bentonite close to the concrete plug shows a considerable enrichment in Ca^{2+} and depletion in Mg^{2+} regarding the original values of the FEBEX (Figure 4, left), CEC value remaining almost constant and close to the original FEBEX bentonite. At the plug scale, Ca increase is higher close to the granite, decreasing to values closer to the reference (still around 20% higher) towards the inner part of the barrier. Leaching of the concrete is providing with calcium that enters in the exchange complex displacing magnesium.

Stable isotopes

Stable isotopes have proved particularly useful for understanding groundwater systems. Deuterium, oxygen-18 and carbon-13 are the most frequently used environmental isotopes. In natural scenarios, the combination of δ^2H, $\delta^{18}O$ and *d-excess* has been extensively used to trace water sources and mixing processes along a flow path (Dietzel et al., 2016). In the case of cementitious systems, the source of carbonates can be elucidated on the basis of the $\delta^{13}C_{VPDB}$ values. The information about the chemical composition and distribution of stable carbon and oxygen isotopes is a robust tool to elucidate the environmental conditions of carbonate formation and their subsequent transformation.

Figure 6 displays the plot of $\delta^{13}C$ and $\delta^{18}O$ values measured in concrete and bentonite samples along the x-axis of the gallery. $\delta^{13}C$ and $\delta^{18}O$ values range from -1.23 to -9.14‰ and -5.91 to -9.62‰ (V-PDB) in bentonite and concrete, respectively. In concrete samples, narrower ranges were observed: -2.32 to -8.0‰ for $\delta^{13}C$, and -5.90

to -9.56‰ for $\delta^{18}O$. In the proximity of the interfaces (between -20 and 20 cm), $\delta^{13}C$ values seem to shift towards more positive values. In the studied samples, isotopic signature of carbonates depends mainly on the isotopic composition of concrete and bentonite porewaters and granite groundwater.

The notable variations of $\delta^{13}C$ data observed in the plug/bentonite interface seem to be related to the neoformation of secondary calcite in both materials. $\delta^{18}O$ values were calculated for the solutions where secondary calcite was precipitated using the oxygen isotope fractionation factor (α) at 20ºC. This oxygen fractionation factor between calcite and solution is given by the following equation:

$$\alpha_{calcite-water} = \frac{(\delta^{18}O/\delta^{16}O)_{calcite}}{(\delta^{18}O/\delta^{16}O)_{solution}} \qquad \textit{Eq. 1}$$

Calculated isotopic values confirmed the different degree of exposition of each sample to bentonite porewater (Figure 7), granite groundwater and the concrete / bentonite leachates.

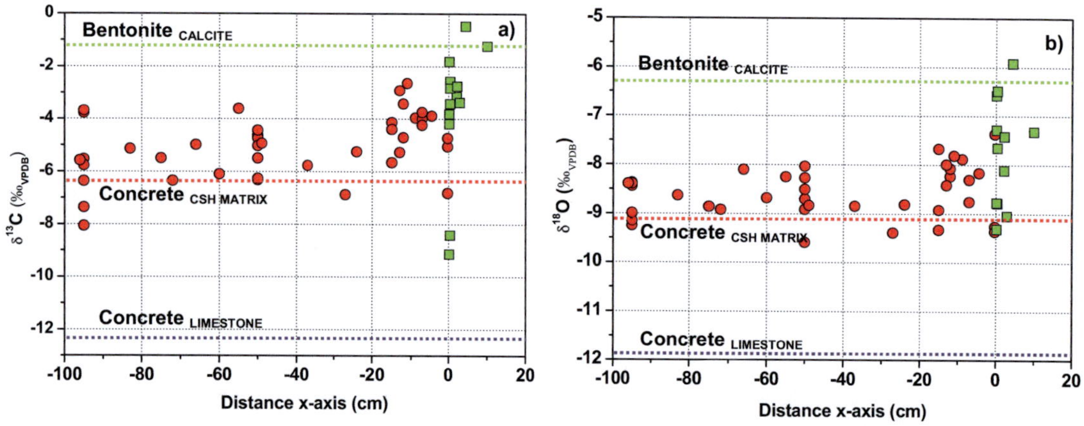

Figure 6: *$\delta^{13}C$ and $\delta^{18}O$ values plotted along the axis of the gallery. As reference values, isotopic values measured in raw FEBEX bentonite, cement binder and filler are plotted. Red circles represent concrete samples. Green squares correspond to bentonite samples.*

A positive correlation between $\delta^{13}C$ and $\delta^{18}O$ values was observed in both, concrete and bentonite (Figure 8). This covariation could a priori be interpreted as an indicator of alteration processes occurring in the materials during the experiment. Most of the isotopic signatures measured in concrete and bentonite samples fall within the data measured for the concrete C-S-H matrix and raw bentonite. Regarding the carbon sources in the binder, the most likely contributions to $\delta^{13}C$ are the limestone filler and the C-S-H matrix. However, these contributions would lead to a shift to more negative $\delta^{13}C$ values than values currently measured in the proximity of the concrete-bentonite interface. On the contrary, average $\delta^{13}C$ value measured in raw bentonite is -1.35‰. Thus, the precipitation of secondary calcite in concrete due to carbonate diffusion from bentonite should result in a shift into more positive values.

Although, additional data is necessary to determine the origin and distribution of carbonates at both sides of the interface, $\delta^{13}C$ values seem to be in agreement with the existence of a diffusion front of carbon species from bentonite towards the concrete / bentonite interface.

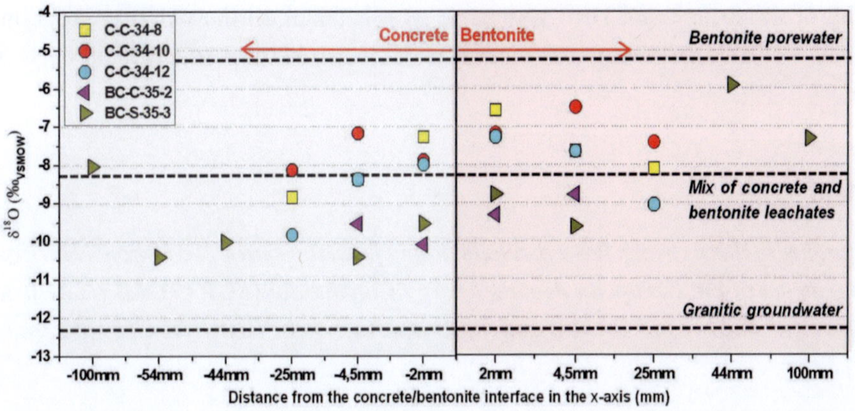

Figure 7: *$\delta^{18}O$ values calculated for formation solutions of secondary calcite at both sides of the concrete/bentonite interface using the oxygen isotopic fractionation factor (α).*

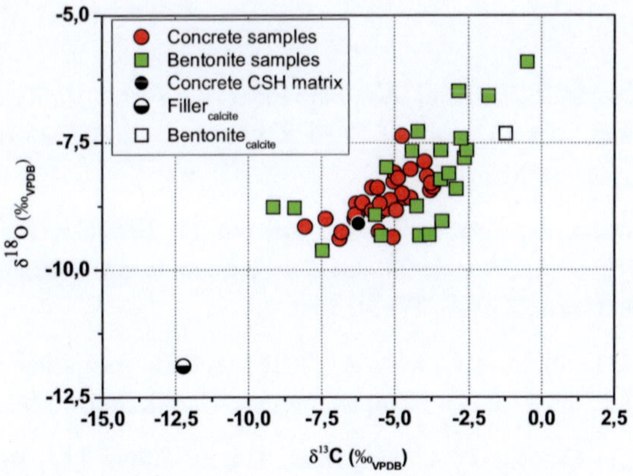

Figure 8: *Plot of $\delta^{13}C$ vs. $\delta^{18}O$ values measured in concrete and bentonite samples.*

Conclusions and Future work

Regarding the chemical composition of granite groundwater at the experimental scale, the most remarkable result obtained is the continuous diffusion of chloride from bentonite porewater towards granite through the small fractures isolated in the packed-off intervals of the boreholes, then the chemical composition evolving from Ca-Na-HCO_3 to Na-Cl type water. Considering the concrete-bentonite interface scale, the depletion of soluble salts (e.g., Cl and SO_4) in the bentonite and the enrichment in the concrete are also related to diffusion processes. Precipitation of secondary ettringite and formation of Friedel's salt in the concrete support this fact. On the other hand, isotopes indicate the existence of a diffusion front of carbon species from bentonite towards the concrete / bentonite interface. Data of soluble ions and exchangeable cations at the interface scale show therefore some key processes: (1) leaching of calcium from concrete and (2) enrichment in Ca^{2+} in the exchange complex of bentonite close to the concrete and depletion in Mg^{2+}. The characteristics of the pore solution at the interface favour the precipitation of Mg as Mg-silicate phases.

Further work will focus on advancing in the interpretation of mass transport processes from the analysis of groundwater (boreholes) and pore water (soluble salts and isotopes) at experiment and interface scale. Therefore, data on porosity and specific surface area are under way to evaluate any change due to the dissolution / precipitation processes observed.

The dismantling of the experiment HB6 was made as scheduled on the working program. Analysis of the different components, concrete, bentonite and water is under way and the results should be available for the next months.

Acknowledgement

The research leading to these results has received funding from the European Union's Horizon 2020 Reasearch and Training Programme of the European Atomic Energy Community (EURATOM) (H2020-NFRP-2014/2015) under grant agreement n° 662147 (CEBAMA) and the Annex XLII of the ENRESA-CIEMAT framework agreement (CAMBAR).

FEBEX-DP consortium provided concrete and bentonite samples for this study and made possible groundwater sampling (http://www.grimsel.com/gts-phase-vi/febex-dp/febex-dp-introduction).

References

Alonso, M.C., García-Calvo, J.L., Cuevas, J., Turrero, M.J., Fernández, R., Torres, E., Ruiz, A.I. (2017). Interaction processes at the concrete-bentonite interface after 13 years of FEBEX-Plug operation. Part I: Concrete alteration. Phys. Chem. Earth., 99, 38-48.

Buil, B., Gómez, P., Peña, J., Garralón, A., Turrero, M.J., Escribano, A., Sánchez, L., Durán, J.M. (2010). Modelling of bentonite-granite solutes transfer from and *in-situ* full-scale experiment to simulate a deep geological repository (Grimsel Test Site, Switzerland). App. Geochem., 25(12), 1797-1804.

Dietzel, M., Schön, F., Heinrichs, J., Diditius, A.P., Leis, A. (2016). Tracing formation and durability of calcite in a Punic-Roman cistern mortar (Pantelleria Island, Italy). Isotopes Environ. Health Stud., 52(1-2), 112-127.

Fernández, R., Torres, E., Ruiz, A.I., Cuevas, J., Alonso, M.C., García-Calvo, J.L., Rodríguez, E., Turrero, M.J. (2017). Interaction processes at the concrete-bentonite interface after 13 years of FEBEX-Plug operation. Part II: Bentonite contact. Phys. Chem. Earth, 99, 49-63.

Garralón, A., Gómez, P., Peña, J., Buil, B., Turrero, M.J., Torres, E., Sánchez, L. (*in prep*). Hydrogeochemical characterization of the groundwater in the FEBEX gallery. NAGRA Arbeitsbericht, NAB 16-014.

Turrero, M.J., Torres, E., Garralón, A., Gómez, P., Sánchez, L., Peña, J., Buil, B., Escribano, A., Durán, J.M., Domínguez, R. (2016) Interaction concrete/FEBEX bentonite: outlining experimental conditions and characterization approaches at laboratory and in-situ scale. Proceedings of the 1st CEBAMA Annual Workshop. KIT Scientific Publishing, KIT-SR 7734.

Villar, M.V., Iglesias, R.J., Abós, H., Martínez, V., Rosa, C., Manchón, M.A. (2016). FEBEX-DP on-site analyses report. NAGRA Arbeitsbericht, NAB 16-012.

Evolution of porosity in cementitious materials during early stage of alkali-activation: a Spin-Echo Small-Angle Neutron Scattering and SEM-EDS investigations

Andrea Sabău[1*], Zhou Zhou[1], Denis M. Bykov[1], Wim G. Bouwman[1], Chris P. Duif [1], Jan-Leen Kloosterman[1]

[1] Delft University of Technology, Faculty of Applied Science (NL)

* Corresponding author: A.Sabau@tudelft.nl

Abstract

The scope of this work is to understand degradation of concrete exposed to carbonation, chloride attack and external sulfate attack by means of Spin-Echo Small-Angle Neutron Scattering technique (SESANS). CEM I 52.5 N SR3/NA and CEM III 42.5 N LH/SR were exposed to Na_2SO_4 solutions, while CEM III 42.5 N LH/SR and a composite concrete containing ground blast furnace slag (GBFS) were exposed to NH_4Cl solutions. CEM I 52.5 N SR3/NA was subjected to carbonation. SEM-EDS and XRD techniques were used additionally to analyse degradation mechanisms of the concrete samples under different exposure conditions. Results showed that, degradation of each specimen varied, depending on the exposure conditions. CEM I 52.5 N SR3/NA subjected to sulfates showed important damages, as was highlighted using SESANS, SEM-EDS and XRD while CEM III 42.5 N LH/SR and GBFS showed similar behaviour to each other with less degradation. Samples immersed in chlorides suffered weakness due to Ca leaching compared to samples subjected to the sulfate attack. Carbonation needs longer time for investigations.

Introduction

As in most European countries, the Netherlands is investigating deep geological disposal as a final solution to dispose the radioactive waste. Within the Dutch research program OPERA (Onderzoeks Programma Eindberging Radioactief Afval) the feasibility of a repository in Boom clay formation is investigated. Cementitious materials with various functions (gallery lining, waste package concrete and backfill materials) are relevant for this programme. The geochemical degradation processes of concrete strongly depend on the composition of the infiltrated ground water coinciding with the formation of new precipitates. Cl^- and SO_4^{2-} are ions that are found in the chemical composition of the ground water and can be in contact with concrete structures. Various chemical degradation processes (chloride ingress, sulfate attack, carbonation) will alter the concrete and will change the pore water composition.

Studies showed that reaction between chlorides and monosulfates leads to the formation of Friedel's salt ($Ca_4Al_2Cl_2(OH)_{12}{\cdot}4H_2O$) (Glasser et al., 1999; Yamada et al., 2005). Friedel's salt can be formed in many circumstances. This compound is unstable and dissolves at 20°C generating a pH value of 12 for the solution. When free Cl^- ions from the chloride solutions (seawater, interstitial water) penetrate into the concrete some of them will be captured by the cement hydration products, especially by the aluminates. Another phase that is possible to be formed is the Kunzel's salt ($Ca_4Al_2(SO_4)_{0.5}Cl(OH)_{12}\cdot 6H_2O$), a chloro-sulfate Afm phase. The formation of Friedel's salt and Kunzel's salt may cause expansion and cracking of concrete.

Sulfate attack is a complex process, because chemical, physical, and mechanical processes contribute to the deterioration. In highly saline water (Dutch Boom Clay pore water has a higher salinity compared to pore water in Belgian Boom Clay the range of 0.53 - 0.83 M (Berhends et al., 2016), beside chlorides, sulfates are present in significant amount (Berhends et al., 2016). Another source of sulfates that can affect concrete and lead to the formation of new phases comes from pyrite oxidation, which is found in abundance in Dutch Boom Clay (OPERA-PU-SCK514, 2015).

Cementitious materials exposed to sulfate containing solutions will deteriorate forming precipitates such as gypsum, ettringite or thaumasite. Ettringite and gypsum are more likely to be formed closer to the top surface of concrete (Glasser et al., 2008). Gypsum can be formed in a layer close to the exposed surface in which sulfate reacts with Ca^{2+} leached from Portlandite or from decalcified CSH phases. Ettringite is formed in the reactions between aluminates from AFt or AFm phases and sulfates (Liu et al., 2014). Thaumasite is formed by the reaction of calcium silicate of the hydrated paste with sulfates and carbonates. Sulfate concentration is an important parameter in the formation of ettringite, gypsum or thaumasite. Alexander et al. (2012) mention limiting sulfate concentrations of Na_2SO_4 of 1 - 2 g/L leading either to formation of gypsum above this concentration or ettringite below it.

Carbonation initiates when CO_2 penetrates within cements and is a chemical process of reactions between dissolved carbon species and hydrated cement pastes. It is known also as a deterioration process because it makes the pH drop, leading to a higher corrosion of reinforced bars in concrete (Phung et al., 2015). Principal source of CO_2 is the atmosphere (316 ppm), but also micro-biological systems.

The objective of this work is to understand the influence of the anions Cl^- and SO_4^{2-} (ions that are in important amounts in Dutch Boom clay pore water) and carbonation on the porosity evolution of cementitious materials that can be used as backfill material and to reveal the differences in deterioration for concrete materials made using Portland cement and Blastfurnace Slag containing cement. Changes in porosity and variations in concentrations of leached elements as well as formation of crystalline phases were monitored during these investigations. Spin-Echo Small-Angle Neutron Scattering was used to determine structure of the pores in cementitious materials.

Materials and Methods

Materials

Experiments were carried out on three types of concrete under laboratory conditions. CEM I 52.5 N SR_3/NA is provided by IBR company, and is foam concrete based on Portland cement. CEM III 42.5 N LH/SR is a blast furnace slag cement provided by COVRA (Centrale Organisatie Voor Radioactief Afval). The concrete samples CEM I 52.5 N SR3/NA and CEM III 42.5 N LH/SR were produced using a w/c = 0.45. Another sample used in this study is a mortar provided by CUGLA company. This is a composite cement based on blast furnace slag (GBFS). All samples are in the form of cylinders with a diameter of 28 mm and a height of 100 mm. The samples were cut to 28 mm diameter and 1 mm height specimens using a saw machine and immersed in sulfate or chloride solutions, the final liquid to solid ratio being 5 g/L. These dimensions were chosen in accordance with the requirements for the SESANS measurements and were used without curing Mineralogical composition of non-degraded samples based on XRD analysis is given in Table 1 and SEM micrographs of samples CEM III 42.5 N LH/SR and GBFS mortar are shown in Figure 1.

***Table 1:** XRD analysis of specimens used in this study.*

Cement name	Mineral phases
CEM I 52.5 N SR3/NA	$CaCO_3$, $Ca(OH)_2$, SiO_2 and Gismondine ($CaAl_2Si_2O_8 \cdot 4H_2O$)
CEM III 42.5 N LH/SR	SiO_2, $CaCO_3$ and $CaSO_4 \cdot 2H_2O$, Ettringite
GBFS mortar	SiO_2, $CaCO_3$ and $CaSO_4 \cdot 2H_2O$

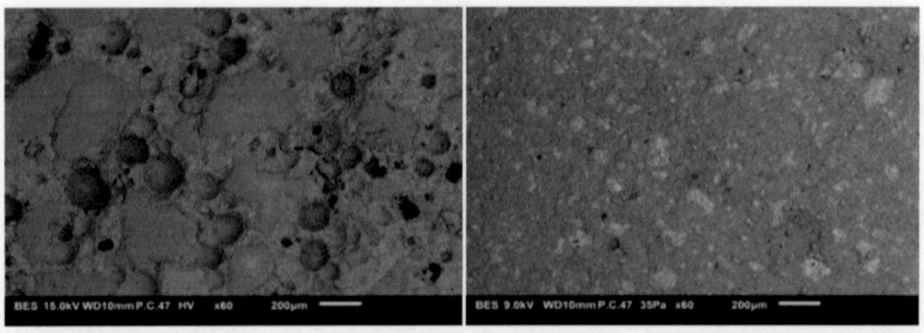

***Figure 1:** Backscattering electron micrographs of CEM III 42.5 N LH/SR (left) and GBFS (right).*

XRD analysis on CEM I 52.5 N SR3/NA revealed the presence of calcite, Portlandite, quartz and Gismondine. According to the SEM/EDS analysis, the CEM III 42.5 N LH/SR samples contained quartz, but also a multitude of pores with various dimensions (2 - 500 µm). GBFS presented only small pores (10 - 200 µm) on the surface and smaller regions of brighter intensity corresponding to SiO_2. On the top of analysed sample, calcite spots can be detected (especially for higher magnifications) and locally gypsum, confirming the XRD findings. The calcite layer formed on the surface can be formed also due to slow cutting process. The samples were used further in the experiments without any treatment.

Chemical degradation procedure

After cutting, the samples were immersed in solutions containing sulfates (33.8 g/L and 4.1 g/L Na_2SO_4) and chlorides (9.8 and 29.8 g/L NH_4Cl) with the scope to investigate the actions of leaching and degradation. The samples were stirred during the whole period of reaction (2 weeks and 2 months respectively), while the pH of solutions was checked periodically. The concentrations used in these studies are close to the concentrations existing in the pore water of the Dutch Boom Clay.

Degradation of concrete was performed using carbonation method. Samples were placed in a carbonation chamber for a period of two weeks, keeping constant relative humidity at 65% and CO_2 at 3.5%. At the end of each exposure time, samples were dried for at least 24 hours at 50 - 80°C, until a constant mass was obtained and placed independently in plastic bags and put in desiccator in order to preserve the samples and to prevent ulterior carbonation from the atmospheric CO_2.

In this study, the degraded and non-degraded samples will be compared. The non-degraded samples, consist in samples without being subjected to corrosive solutions, Na_2SO_4 or NH_4Cl and in this study will be mentioned as reference sample.

<u>SESANS</u>

Neutron beam techniques are considered progressively as non-destructive methods to investigate the dense materials at microscale level (50 nm to 20 µm). Small-Angle Neutron Scattering (SANS) was developed to afford examination of materials microstructures up to several hundreds of nanometers (Andersson et al., 2008). SANS instrument measures the neutrons intensities scattered into solid angle detector positioned far away from the sample. Spin-Echo Small-Angle Neutron Scattering (SESANS) was developed to have a much higher resolution compared to SANS in order to measure larger structures. SESANS studies were performed to determine the average pore size, shape of the pores and the volume fraction in nuclear graphite (Zhou et al., 2016). The SESANS approach relates the polarization P(z) of the neutron beam to the projection G(z) of the autocorrelation function (describes the density of the sample) γ(r) of the density distribution ρ(r) of the sample. The spin-echo length, z, is the real space distance over which correlations are measured in the sample. Through the transmission of polarization, SESANS measures G(z), the projection of the autocorrelation function of the density distribution (Andersson et al., 2008). Another important measure is the average number of times a neutron scatters when crossing a sample, which for a two-phase system is expressed as Eq. 1.

$$\Sigma_t = \lambda^2 t(\Delta\rho_0)^2\phi(1-\phi)\xi \qquad \textit{Eq. 1}$$

where t is the thickness of the sample, λ is the neutron wavelength, $\Delta\rho_0$ is the scattering length density (SLD) and 1-Φ is the subtraction of average density of the sample s and ξ the correlation length of the sample inhomogeneities. The raw data generated by SESANS measurements is the polarization as a function of the parameters mentioned above.

$$P(z) = e^{\{\Sigma t\,[G(z)-1]\}} \qquad \textit{Eq. 2}$$

Normalization with the empty beam polarization gives P(0) = 1. As the spin-echo length increases it exceeds the length scale of characteristic sample inhomogeneities, i.e., G(∞) = 0. This insight enables us to compute the fraction of neutrons that does not scatter when crossing the sample.

SANS was used to study dense polymer cements (Allen and Thomas, 2007; Rogante et al., 2013; Rogante et al., 2015), but no studies using SESANS have been performed. The main objective of this work is to investigate porosity changes in cementitious materials at microscale level using SESANS. In this study carbonation, chloride ingress and sulfate degradation were considered as main processes. In Figure 2 a standardized plot for a generic SESANS measurement is given.

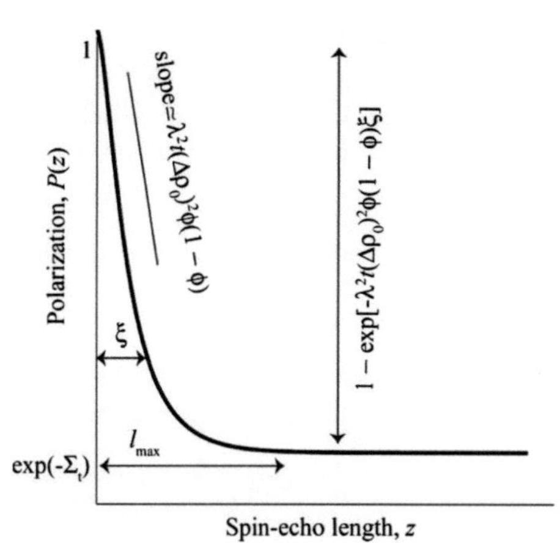

Figure 2: *Generalized graph of a SESANS measurement for dilute isotropic materials without any ordering where the polarization is plotted against the spin-echo length (Krouglov et al., 2003).*

A SESANS plot, P(z) versus z starts at 1, decays and saturates at spin echo length equal to the size that characterizes the sample; at this point no more density correlates are seen in the sample. At this point no more density correlations are observed in the sample or the sample pores are very large.

Since SESANS does not give information about elemental composition and mineralogy of the sample, it needs to be combined with complementary techniques. In this work, microstructural alterations resulting from exposure to solutions were subsequently analyzed using SEM-EDS and/or XRD to identify crystalline phases.

Results and Discussion

CEM I 52.5 N SR3/NA

CEM I 52.5 N SR3/NA was subjected to carbonation and sulfate attack. After carbonation the samples were taken as they are and post-analysis methods (SESANS, SEM-EDS and XRD) were used to interpret qualitatively the results. In Figure 3, spin-echo signals plotted as a function of spin echo length are presented for samples subjected to 33.8 g/L Na_2SO_4 (Figure 3, left) and carbonation (Figure 3, right).

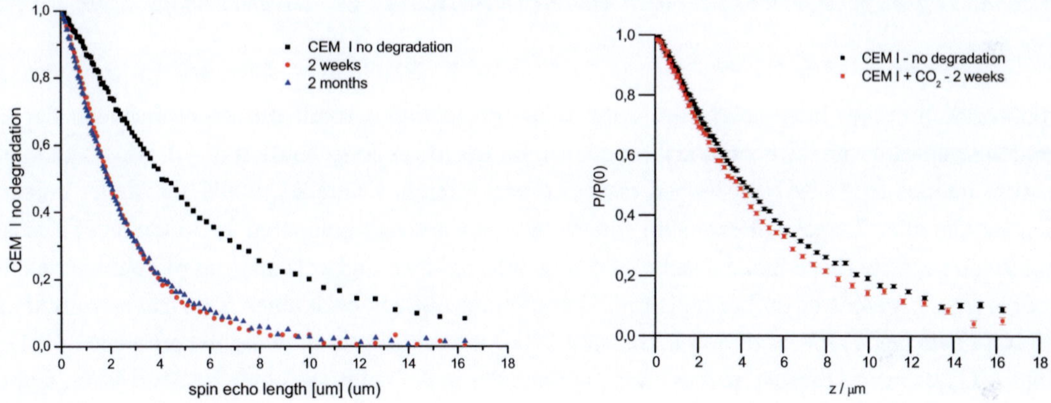

Figure 3: *SESANS signals of CEM I 52.5 N SR3/NA immersed in Na_2SO_4 for 2 weeks and 2 months (left) and carbonated for 2 weeks (right).*

The shape of the polarization decay is similar for both, carbonated samples and samples degraded with sulfates. In Figure 3, the degraded samples are compared with the reference ones. Polarization of carbonated sample (red) is slightly higher than the reference sample (black). which can be explained by an increase of porosity. Samples immersed in Na_2SO_4 showed clear changes in depolarization. Compared to reference samples (black curve), depolarization is observed for a both samples (blue and red curves), which means an increasing in the porosity. The increasing in the porosity of the degraded samples compared to the reference ones can be explained by a higher scattering of neutrons that is related to large pores (few hundreds of μm) or higher number of pores close to each other. In order to correlate the results obtained using SESANS, XRD was necessary to determine the mineral phases present after degradation. For samples immersed in Na_2SO_4 for 2 weeks, quartz, calcite, Portlandite and gismondine were present. After 2 months immersion in sulfate, the presence of $CaCO_3$, SiO_2 and tobermorite were observed.

The formation of $CaCO_3$ was not identified using XRD for the carbonated samples. By carbonation, the reduction of porosity is expected (Glasser et al., 2008) and the formation of a protective layer on the surface of the concrete. Crystallization pressure is the most probable cause of expansion of concrete immersed in sulfate solutions (Piastra, 2017).

CEM III 42.5 N LH/SR

CEM III 42.5 N LH/SR was subjected to both, sulfate and chloride solutions. In the case of sulfate, only small changes were noticed by SESANS visually during the exposure even after 2 months for both concentrations (4.1 and 33.8 g/L) as shown in Figure 4. A comparison between reference and degraded samples is shown.

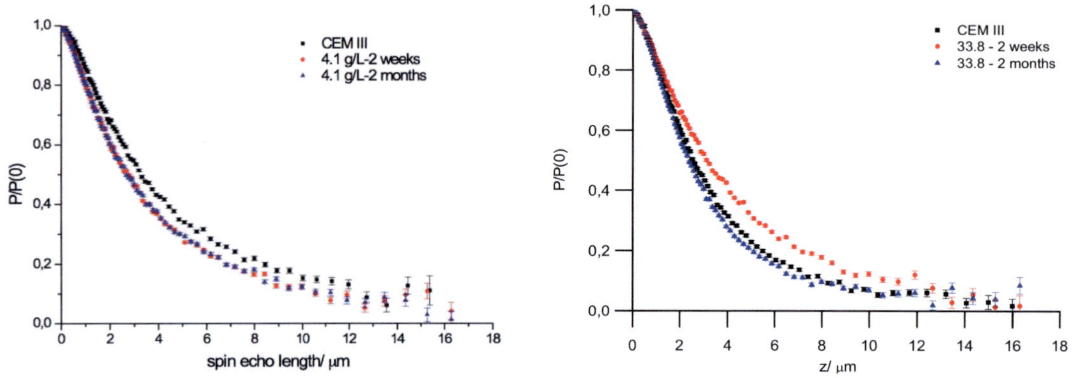

Figure 4: *SESANS signals of CEM III 42.5 N LH/SR immersed in Na_2SO_4 (4.1 g/L - left and 33.8 g/L - right) for 2 weeks and 2 months.*

The polarization curves have nearly the same shape presenting a small difference between degraded and non-degraded samples. From the depolarization curves, for the lower concentration of 4.1 g/L Na_2SO_4, almost no changes were noticed by SESANS between degraded and reference samples, while for higher concentrations (33.8 g/L), sample after 2 weeks of exposure showed a lower porosity compared to 2 months of exposure. The increasing in the porosity (for 2 months sample) can be inferred by a higher scattering of neutrons that is related to large pores (few hundreds of µm) or higher number of pores close to each other. This can be caused mainly by the expansion of ettringite content during immersion, XRD measurements showing the presence of SiO_2, $CaCO_3$ and ettringite. Gypsum was present neither on the surface nor in the voids as shown by SEM-EDS. Gypsum may be formed in highly concentrated sulfate solutions (higher than 24 g/L) (Gollop and Taylor, 1992), but was not the case under our experimental conditions. SO_4^{2-} concentration shows nearly no effect under these experimental conditions.

Additional CEM III 42.5 N LH/SR samples were exposed to highly concentrated NH_4Cl solutions (29.8 g/L) for 2 weeks and 2 months. NH_4Cl reacts with the hydrated products of cement, with transitional formation of $CaCl_2$ and NH_3 (Słomka-Słupik and Zybura, 2009). In these samples leaching played an important role, and the release of Ca (measured by ICP-OES) was greater than in the case of samples subjected to sulfate attack. Samples were weighted and changes mainly due to leaching of Ca^{2+}occurred while immersed in a low pH NH_4Cl solution. SESANS measurements were performed on these samples (Figure 5), nearly no changes being indicated between degraded and non-degraded samples.

The samples placed in NH_4Cl solution showed visual changes and also weakness and cracking after taking it out from the solution (Figure 6).

The XRD analysis was performed on these samples but no Friedel's salt formation was observed. Friedel's salt has a higher content in pastes made of cements with higher tricalcium aluminate content (Słomka-Słupik and Zybura, 2009). No crystals corresponding to Friedel's salt could be observed in specimens exposed for 2 months in NH_4Cl solutions (Figure 6). At the same time, high porosity is observed, by SEM. Friedel's salt dissolves at low pH in the case of longer action of NH_4Cl, which explains the absence of this compound.

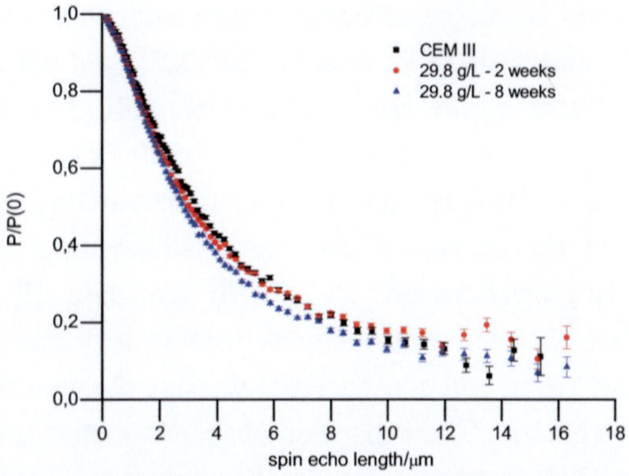

Figure 5: *SESANS signals of CEM III 42.5 N LH/SR immersed in 29.8 g/L NH_4Cl for 2 weeks and 2 months.*

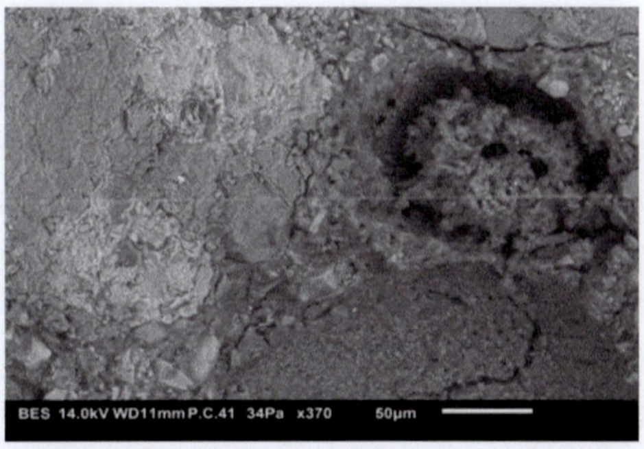

Figure 6: *Morphology of CEM III 42.5 N LH/SR sample after contact with NH_4Cl solution for 2 months, obtained with backscattered electrons.*

GBFS concrete

The mortar specimens were immersed in both, sulfates (33.8 g/L Na_2SO_4) and chlorides (29.8 g/L NH_4Cl) for the same exposure time (2 weeks and 2 months). The SESANS decay curves demonstrate different behavior between samples immersed in sulfates and chlorides (Figure 7). The shape of the depolarization decay is similar for both chloride corroded samples and samples degraded with sulfates.

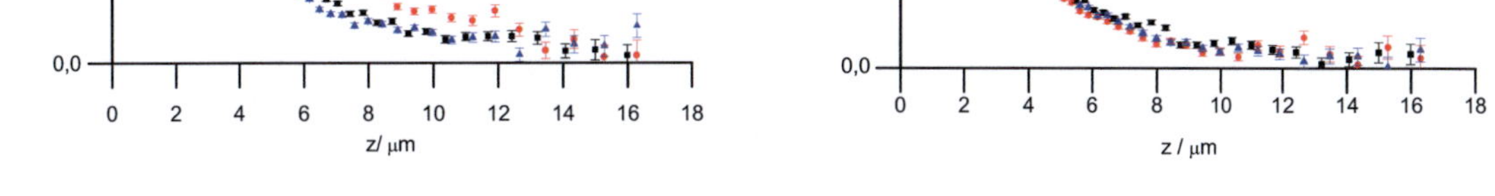

Figure 7: *SESANS signals of GBFS mortar immersed in 33.8 g/L Na_2SO_4 (left) and 29.8 g/L NH_4Cl (right) for 2 weeks and 2 months.*

XRD measurements were performed, highlighting beside quartz and calcite the presence of gypsum, that was maintained even after 2 months of contact with SO_4^{2-} solution, but XRD did not reveal the presence of Friedel's salt under investigated conditions. These results are in agreement with (Xiong et al., 2016) under the same experimental conditions but for longer exposure time (540 days). Friedel's salt is always formed in the presence of large amounts of chlorides in neutral exchange solutions when monosulfate is present among hydrated phases, but is dissolving after longer exposure time. In view of pore water composition, salinities up to 24 g/L have been reported from Rupel Clay in Zeeland region (Behrends et al., 2016). The main effect of chlorides is the destruction of the protective layer on the surface of steel reinforcement, leading to corrosion of the steel. The corrosion products contribute to stresses around rebar, and consequently damage the concrete cover. The nature of sulfate attack depends on the type of cations present. The expansion of foam concrete under very severe sodium sulfate environment may be more pronounced under $MgSO_4$ attacks (Kunther et al., 2013).

Conclusions and Future work

The samples degraded in Na_2SO_4 solutions showed visually small changes, while the samples which were in contact with NH_4Cl solutions were softened and most of them cracked. SESANS is an appropriate technique to investigate porosity at microscale level. The porosity of corroded materials increased at the surface areas that can be caused by Portlandite decomposition and decomposition of other amorphous hydrated phases, not detectable by XRD, such as C-S-H phases. Even at high concentration of sulfate ions, solutions with a mixture of cations, similar to those commonly found in ground waters, lead to much less deterioration than solution of sodium sulfate alone (OPERA-PU-SCK514, 2015). A comparison between the concrete specimens used in this study showed higher deterioration of CEM I in sulfate solution, while the deterioration of GBFS mortar and CEM III was similar. Carbonation experiments need a deeper investigation. Chloride attack revealed similar behavior in the case of all samples, no additional phases being identified in this study by the techniques used.

Acknowledgement

The research leading to these results has received funding from the European Union's Horizon 2020 Research and Training Programme of the European Atomic Energy Community (EURATOM) (H2020-NFRP-2014/2015) under grant agreement n° 662147 (CEBAMA).

References

Alexander, M., Bertron, A., De Belie, N. (2012). Performance of Cement-based Materials in Aggressive Aqueous Environments. Springer, 2012. State-of-the-Art Report, RILEM TC 211-PAE.

Allen, J.A. and Thomas, J.J. (2008). Analysis of C-S-H gel and cement paste by small-angle neutron scattering. Cement and Concrete Research, 37, 319-324.

Andersson, R., van Heijkamp, L.F., de Schepper, I.M., Bouwman, W.G. (2008). Analysis of spin-echo small-angle neutron scattering measurements. Journal of Applied Crystallography, 41, 868-885.

Behrends, T., van der Veen, I., Hoving, A., Griffioen, J. (2016). First assessment of the pore water composition of Rupel Clay in the Netherlands and the characterization of its reactive solids. Netherlands Journal of Geosciences, 95, 315-335.

Cohen, M.D. and Mather, B. (1991). Sulfate attack on concrete. Research needs. ACI Materials Journal, 88, 62-69.

Gollop, R. and Taylor, H.F.W. (1992). Microstructural and Microanalytical Studies of Sulfate Attack, I. Ordinary Portland Cement Paste. Cement Concrete Research, 22, 1027-1038.

Glasser, F., Marchand, J., Samson, E. (2008). Durability of concrete - Degradation phenomena involving detrimental chemical reactions. Cement and Concrete Research, 38, 226-246.

Kunther, W., Lothenbach, B., Scrivener, K.L. (2013). On the relevance of volume increase for the length changes of mortar bars in sulfate solutions. Cement and Concrete Research, 46, 23-29.

Liu, Z., Deng, D., De Schutterb, G. (2014). Does concrete suffer sulfate salt weathering? Construction and Building Materials, 66, 692-701.

Phung, T.Q., Maes, N., Jacques, D., Brunneel, E., Van Driessche, I., Ye, G., De Schutter, G. (2015). Effect of limestone fillers on microstructure and permeability due to carbonation of cement pastes under controlled CO_2 pressure conditions. Construction and Building Materials, 82, 376-390.

Piasta, W. (2017). Analysis of carbonate and sulphate attack on concrete structures. Engineering Failure Analysis, 79, 606-614.

Rogante, M., Domanskaya, I.K., Gerasimova, S.E., Vladimkrova, E. (2013). Feasibility Study for a Neutron-Based Investigation of Polymer Cement Concretes. Universal Journal of Applied Science, 1, 11-17.

Rogante, M., Domanskaya, I.K., Gerasimova, S.E., Len, A., Rosta, L., Székely, K.N., Vladimkrova, E. (2015). Nanoscale investigation of polymer cement concretes by Small Angle Neutron Scattering. Science and Engineering of Composite Materials, 24, 67-72.

Seetharam, S. and Jacques, D. (2015). Potential Degradation Processes of the Cementitious EBS Components, their Potential Implications on Safety Functions and Conceptual Models for Quantitative Assessment. OPERA-PU-SCK514.

Krouglov, T., de Schepper, M.I., Bouwman, G.W., Rekveldt, M.T. (2003). Real-space interpretation of spin-echo small-angle neutron scattering. Journal of Applied Crystallography, 36, 117-124.

Słomka-Słupik, B. and Zybura, A. (2009). Chloride ions diffusion in hydrated cement paste immersed in saturated NH_4Cl solution. Cement Wapno Beton, 29, 232-239.

Xiong, C., Jiang, L., Xu, Y., Chu, H., Jin, M., Zhang, Y. (2016). Deterioration of pastes exposed to leaching, external sulfate attack and the dual actions. Construction and Building Materials, 116, 52-62.

Zhou, Z., Bouwman, W.G., Schut, H., Desert, S., Jestin, J., Hartmann, S., Pappas, C. (2016). From nanopores to macropores: Fractal morphology of graphite. Carbon, 96, 541-547.

Thermal alteration of C-S-H phase in cementitious materials

Daisuke Hayashi[1*], Masaya Ida[1], Hitoshi Owada[1], Naoki Fujii[1], Kumi Negishi[2]

[1] RWMC, Radioactive Waste Management Funding and Research Center (JP)
[2] Taiheiyo Consultant (JP)

* Corresponding author: d-hayashi@rwmc.or.jp

Abstract

In cementitious material of the engineered barrier system (EBS) in geological disposal, calcium silicate hydrate (C-S-H) gel may change into crystalline C-S-H by thermal effect. Thereby bentonite alteration behaviour by cementitious material may change. Aged concrete which has been exposed to thermal effect and aged of 40 to 70 years was analysed for condition of C-S-H and its dissolution properties. In the aged concrete, C-S-H gel decreased C/S and changed to crystalline C-S-H (similar to tobermorite) by thermal effect. Due to these phenomena, Ca dissolution and pH value were decreased.

Introduction

In the concept of geological disposal of TRU waste in Japan, the temperature in the facility of the engineered barrier system (EBS) is kept below 80°C in order to mitigate the thermal alteration of cementitious material caused by heat generation of the waste (JAEA and FEPC, 2005). Previous research has suggested that crystalline calcium silicate hydrate (C-S-H) is generate at 80°C or less (Cong and Kirkpatrick, 1996; Glasser and Hong, 2003; Kalousek and Roy, 1957). Synthetic C-S-H gel changes into crystalline C-S-H (11 Å tobermorite) at 85°C (Atkins et al., 1991). Thus, C-S-H gel in cementitious material may change into crystalline C-S-H at 80°C or less and thereby bentonite alteration behaviour by cementitious material may change. Therefore, it is expected that change of physical properties (mechanical properties and porosity) of cementitious materials and pH degradation of groundwater in EBS. This study investigated thermal alteration of C-S-H gel in aged concrete which had been exposed to thermal effects. In addition, dissolution test of aged concrete was carried out.

Aged concrete specimens

Aged concrete specimens were collected from the foundations of rotary kilns for cement product at several abandoned cement factories. Specifications of the concrete specimens are shown in Table 1. Overview of the location where specimen ^{33}H and ^{33}N were collected is shown in Figure 1 and Figure 2. The heat of rotary kiln had propagated from above the foundations. The positions of the sample ^{33}H, ^{60}H, ^{59}H and ^{75}H are near the top of foundations. The position of specimen ^{33}N located at a part distant from top surface of the foundation. Specimen ^{33}H and ^{33}N were in same foundation. Specimen ^{33}H, ^{60}H, ^{59}H and ^{75}H were in different foundations. Specimen ^{60}H was collected by horizontal drilling same as ^{33}H and ^{33}N. Specimen ^{59}H and ^{75}H were collected by vertical drilling from the front. The rotary kilns had been abandoned at the time of drilling. Therefore, the temperature of the sample positions in the foundations was estimated. The temperature was estimated from heat conduction analysis using some actual measured temperature value at top surface of similar foundation of operating rotary kiln (see Figure 3).

Table 1: *Specifications of Aged Concrete.*

Specimen	Influence of Thermal effect	Year of construction	Age of samples at drilling (years)	Period of exposing to thermal effect (years)	Estimated temperature in drilling points (°C)
^{33}H	Yes	1933	73	50	70 - 100
^{33}N	No	1933	73	--	--
^{60}H	Yes	1960	47	30	70 - 100
^{59}H	Yes	1959	47	25	40 - 60
^{75}H	Yes	1975	40	30	around 50

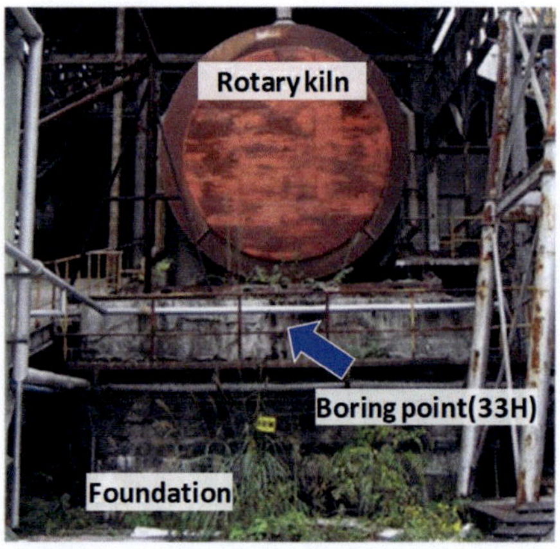

Figure 1: *Drilling point of specimen ^{33}H (foundation of cement rotary kiln).*

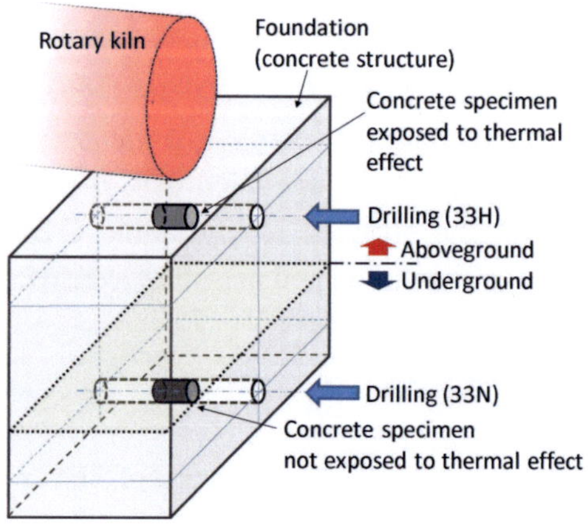

Figure 2: *Sampling point of specimen ^{33}H and ^{33}N (foundation of cement rotary kiln).*

Estimated mix proportion (Estimate), compressive strength and unit weight of aged concrete are shown in Table 2. Mix proportion was estimated by the test of F-18 (JCA, 1967). Each concrete foundations made from Japanese ordinary Portland cement (OPC), which is similar to European type I cement. The cross section of aged concrete are shown in Figure 3. According to Figure 3, maximum aggregate size of each aged concrete are estimated from 15 to 25 mm.

Table 2: *Mix proportion (Estimate) and unit weight of Aged Concrete.*

Specimen	Mix proportion (Estimate*) Unit Weight (kg/m^3)				Unit weight of concrete (kg/m^3)
	W	C	S+G	W/C (%)	
^{33}H	204	289	1,871	70.6	2,180
^{33}N	178	237	1,976	75.1	2,224
^{60}H	176	352	1,882	50.0	2,338
^{59}H	188	280	1,922	67.1	2,232
^{75}H	173	351	1,772	45.0	2,297

* Estimate from the test of F-18 (JCA, 1967)

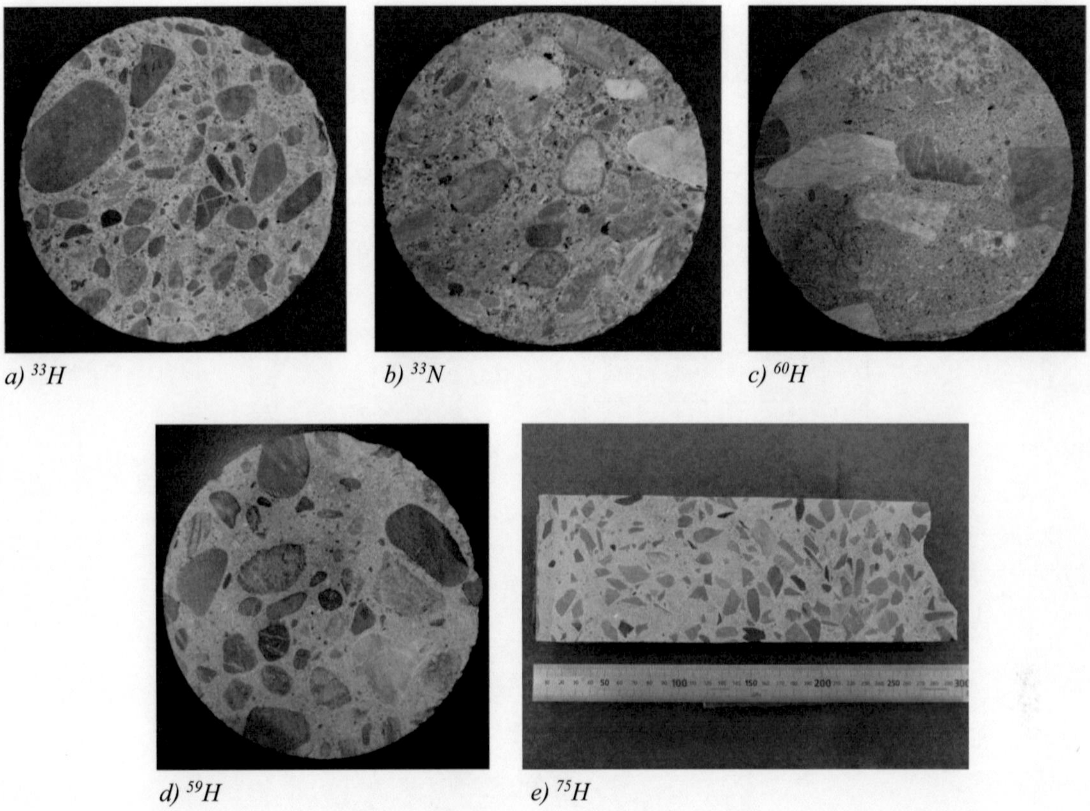

Figure 3: *Cross section of aged concrete (diameter of a - d are 100 mm).*

Experimental method of aged concrete

Concrete specimens were pulverized after removal of the aggregate and then the pulverized samples used for XRD analysis and measure of pH value. The minerals of the samples (^{33}H, ^{33}N, ^{60}H, ^{59}H, ^{75}H) were identified by XRD (X'Part PRO MPDP, PNAalytical), and then checked for crystalline C-S-H (such as tobermorite). Two concrete specimens, ^{33}H and ^{33}N, were immersed in ion-exchanged water at liquid-to-solid ratios (L/S) of 10, 200, 1,000 and 2,000. After the immersion (1 month and 2 years; however L/S = 1,000 was only 1 month), Ca concentration of the soaking water was analysed by ICP, and pH value of the soaking water was measured (by D-50, HORIBA).

The distribution of the Ca/Si molar ratio (C/S) of C-S-H in the cross-section of concrete specimens was obtained by an area analysis of Ca, Si using electron-probe microanalysis (EPMA: JXA-8100, JEOL). Analysis area of EPMA is 300 μm square. Pixel size is 0.5 μm.

Experimental result of aged concrete

Results of XRD analysis of aged concrete are shown in Figure 4. A mineral similar to tobermorite was identified in the samples that had been exposed to a thermal effect: ^{33}H, ^{60}H and ^{75}H. On the other hand, the mineral was not identified in the samples that had not been exposed to a thermal effect: ^{33}N and ^{59}H.

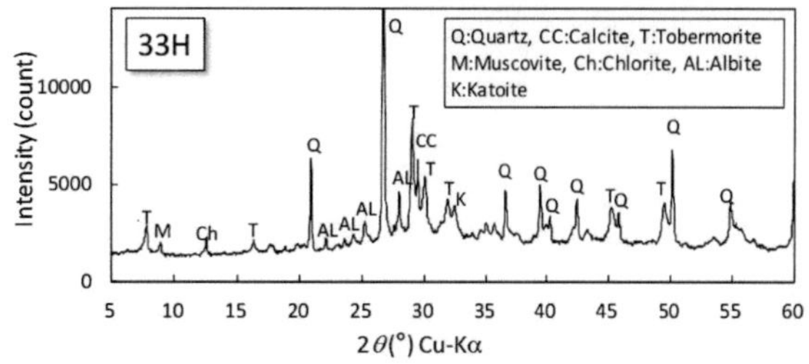

a) ^{33}H

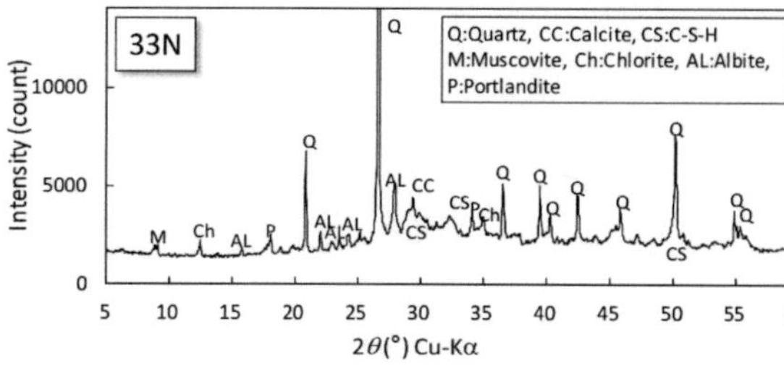

b) ^{33}N

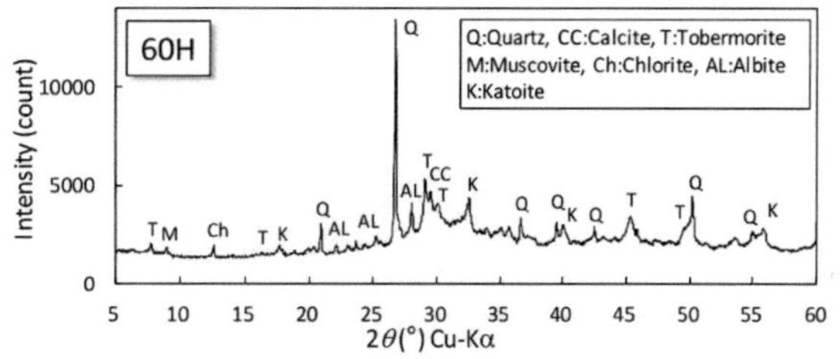

c) ^{60}H

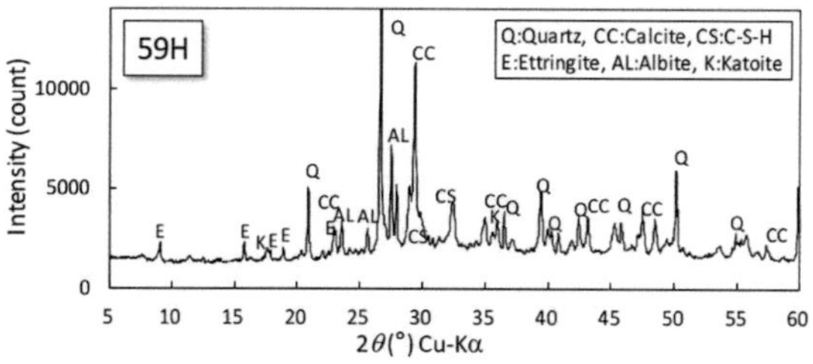

d) ^{59}H

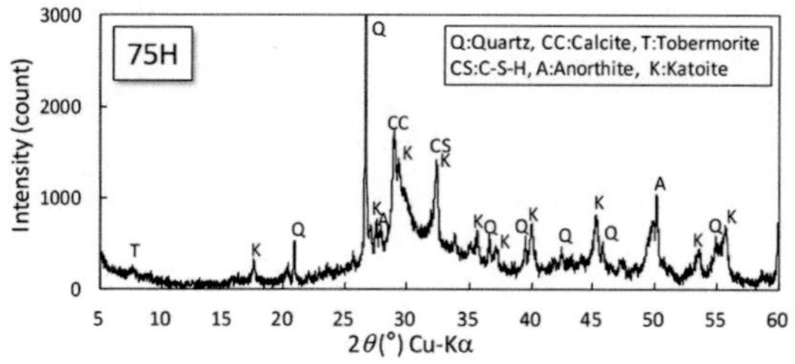

e) ^{75}H

Figure 4: *XRD analysis of aged concrete.*

It is considered that the C-S-H gel changes to crystalline C-S-H similar to tobermorite due to thermal influence. Longer exposure times of the older concrete specimens to thermal effects was associated with greater intensity of the diffraction line of the mineral. Crystalline C-S-H was not identified in sample ^{59}H despite the exposure to thermal effects. It is considered that this difference was caused by the temperature in the position of the samples in the foundations and the concrete materials in the samples (e.c., sort of aggregate). For sample ^{33}N, portlandite ($Ca(OH)_2$) was identified. On the other hand, it was not identified for samples ^{33}H, ^{60}H, ^{59}H and ^{75}H.

C/S distribution in aged concrete by EPMA shown in Figure 5. C/S value was obtained from the concentrations of Ca and Si by EPMA, and the C/S value of concrete cross section was mapped by color scale. White area indicates unhydrated cement minerals. Black area indicates aggregate or pore. Other areas under C/S = 1.8 is almost C-S-H. Table 3 shows the average C/S values of C-S-H part (pixels) extracted from EPMA analysis area. The average C/S value of C-S-H of ^{33}H was 0.84 which is close to 0.83, the composition of tobermorite. On the other hand, the C/S values of samples ^{33}N, ^{60}H and ^{59}H, were between 1.31 and 1.36, which is lower than the value at initial state of general OPC concrete (e.c., age 28 days). These results indicate that decreases in C/S are related to elapsed time since start of exposure to thermal effects.

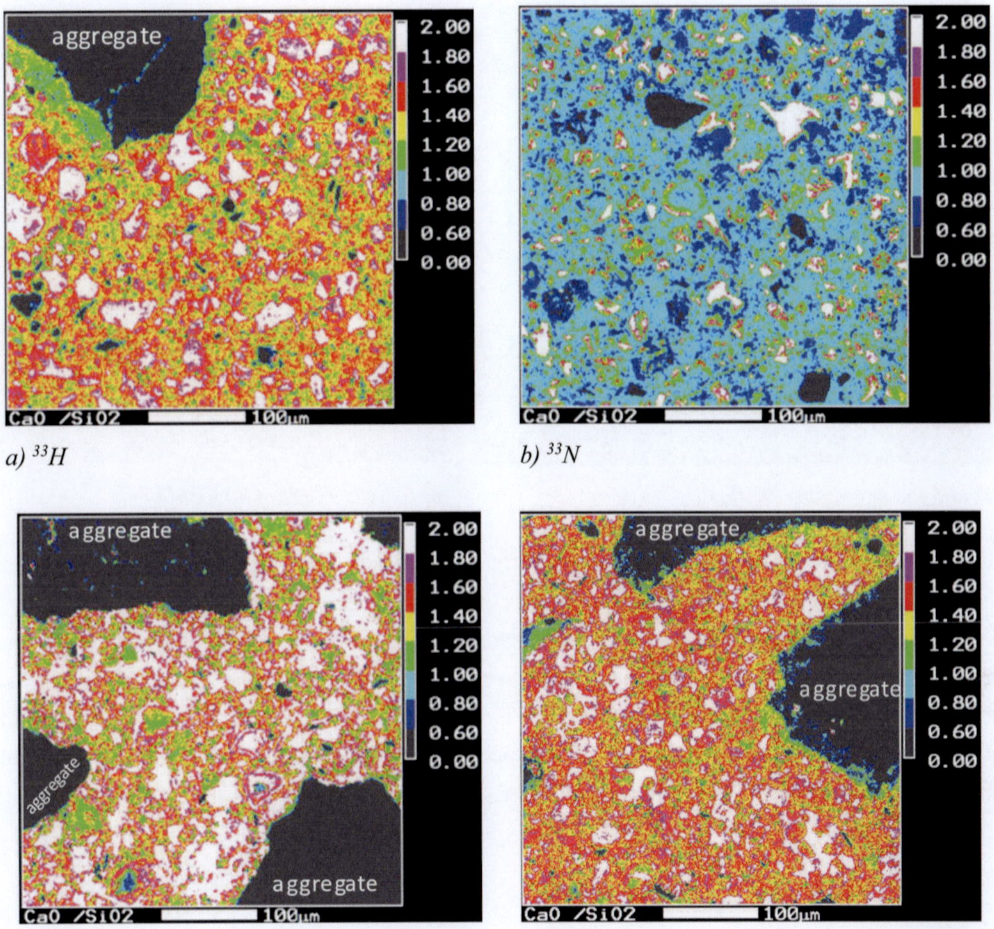

a) ^{33}H *b) ^{33}N* *c) ^{60}H* *d) ^{59}H*

Figure 5: *C/S distribution of aged concrete by EPMA.*

Table 3: *C/S of C-S-H (aged concrete).*

Specimen	C/S (average of 5 areas)
^{33}H	0.84
^{33}N	1.31
^{60}H	1.36
^{59}H	1.32

Figure 6a show observation around aggregate by BEI image and Figure 6b shows C/S distributions of same area by EPMA. Cement paste in contact with the aggregate has lower C/S than the paste that is farther away from the aggregate. It is considered that the dissolved Si from the quartz within the aggregate was supplied to formation of low C/S value C-S-H gel. In addition, it is considered that the reaction was accelerated by thermal effects. In sample ^{33}H ,^{60}H, ^{59}H and ^{75}H, portlandite was not identified. The phenomenon is considered that portlandite consumed to reacted with Si from aggregate to become C-S-H gel just like pozzolanic reaction.

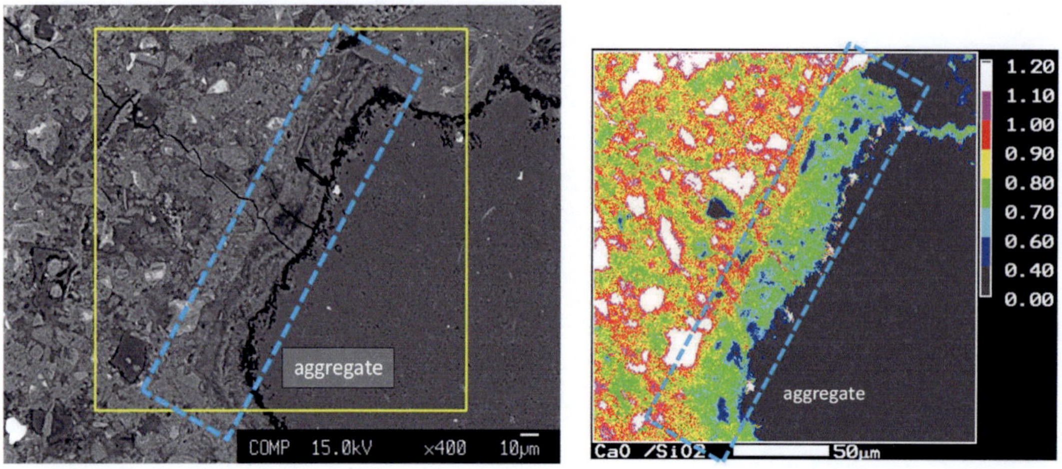

a) Around the aggregate (BEI) *b) C/S distribution by EPMA*

Figure 6: *C/S distribution of cement paste around aggregate by EPMA.*

Ca concentration and pH value of liquid phase of samples ^{33}H and ^{33}N in the dissolution test are shown in Figure 7. Dissolution of Ca from solid phase was less in sample ^{33}H than sample ^{33}N. The pH values of sample ^{33}H were 11 or slightly less, and pH values of sample ^{33}N were 12 or slightly less. It may be that concrete that is older than 50 years exposed to thermal effects may be expected to have a smaller impact on the bentonite alteration in the EBS than concrete exposed to non-thermal effects.

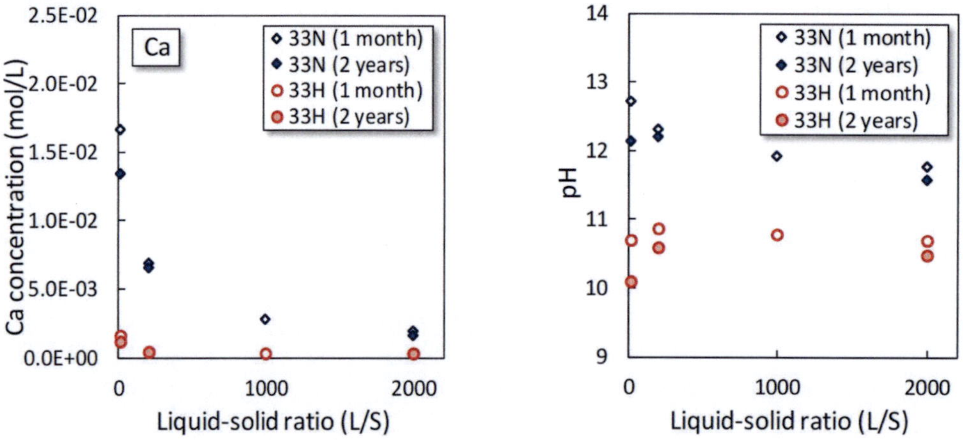

a) Ca concentration *b) pH value*

Figure 7: *pH value and Ca concentration of liquid phase (aged concrete dissolution test).*

Conclusions and Future work

In aged concrete which has aged about from 40 to 70 years, it was observed that C-S-H gel decreased C/S and changed to crystalline C-S-H. This observation is attributed to thermal effect. The crystalline C-S-H is similar to tobermorite. Due to these phenomena, Ca dissolution and pH value were decreased. Therefore, concrete that is older than 50 years and has been exposed to thermal alterations is expected to have a smaller impact on the bentonite alteration in EBS. Therefore, it is considered that these phenomena are necessary to properly consider in the long-term behaviour evaluation of EBS. In the future, it is necessary to clarify the condition of the phenomena (e.c., temperature) based on the results of the laboratory test.

Acknowledgement

This research is a part of the results of "Evaluation Experiments of Long-Term Performance of Engineered Barriers (FY2006, FY2007, FY2008 and FY2011)" and "Advancement of Processing and Disposal Technique for the Geological disposal of TRU Waste (FY2013 and FY2016)" under a grant from the Agency for Natural Resources and Energy (ANRE) in the Ministry of Economy, Trade and Industry (METI) of Japan.

The research leading to these results has been done in cooperation with the European Union's Horizon 2020 Research and Training Programme of the European Atomic Energy Community (EURATOM) (H2020-NFRP-2014/2015) under grant agreement n° 662147 (CEBAMA).

References

Atkins, M., Glasser, F.P., Mironi, L.P. (1991). The long-term properties of cement and concretes. MRS Proceedings, 212, 373-386.

Cong, X. and Kirkpatrick, J. (1996). ^{29}Si and ^{17}O NMR investigation of the structure of some crystalline calcium silicate hydrates. Adv., Cem., Bas., Mat., 3, 133-143.

Glasser, F.P. and Hong, S.Y. (2003). Thermal treatment of C-S-H gel at 1 bar H_2O pressure up to 200°C. Cem. Concr. Res., 33, 271-279.

Japan Atomic Energy Agency (JAEA) and The Federation of Electric Power Companies of Japan (FEPC) (2005). Second Progress Report on Research and Development for TRU Waste Disposal in Japan. JNC TY1400 2005-013 / FEPC TRU-TR2-2005-02.

Japan Cement Association (JCA) (1967). Report of join test on estimation of mixing ratio of hardened concrete (F-18) (*in Japanese*).

Kalousek, G.L. and Roy, R. (1957). Crystal chemistry of hydrous calcium silicates: II, Characterization of interlayer water. J. Am. Ceram. Soc., 40, 236-239.

Preliminary experimental results on the changes in microstructure, mineralogy and transport properties of Boom Clay - concrete interface

Quoc Tri Phung[1*], Stephane Gaboreau[2], Norbert Maes[1], Francis Claret[2]

[1] SCK•CEN, Belgian Nuclear Research Centre (BE)
[2] BRGM, Bureau de Recherches Géologiques et Minières (FR)

* Corresponding author: qphung@sckcen.be

Abstract

In this paper, preliminary experimental results on the changes occurring at the interface between Boom clay and cementitious materials due to chemical/physical interaction are presented. 14 years old in-situ interfaces as well as laboratory manufactured interfaces are currently under investigation. The *in-situ* interfaces were sampled using resin anchor concept both in mid-2016 and early 2017. A μ-tomography acquisition of the full size samples (~ 10 x 20 cm) were then performed to visualize the quality of the contact at the interfaces. Both *in-situ* interface samples have been impregnated with ^{14}C PolyMethyl MethAcrylate (PMMA) in order to perform autoradiography and examine the porosity change. Mineralogical maps done with an Electron Probe Micro-Analyser (EPMA) were also performed on concrete and Boom clay in the undisturbed area (distant from the interface). The same will be done in the near future at the interface to follow the mineralogical evolutions. In addition, microstructural characterization from nano to micrometer scale will be achieved using Transmission Electron Microscopy (TEM) and Focused Ion Beam / Scanning Electron Microscope (FIB / SEM). The lab manufactured interfaces were created by putting backfill concrete (high porosity) and Boom clay in contact in either accelerated percolation or batch-type experiments. The transport properties of concrete and clay (both newly made and *in-situ* samples) were measured by water permeability, diffusion of dissolved gases and water sorptivity using newly developed techniques.

Introduction

Despite the development of novel / advanced materials, cement-based materials remain by far the most widely used materials in construction over the world. Besides their classical use in construction, these materials can also be used for encapsulation of radioactive waste and as engineered barriers for disposal of radioactive waste, both in near-surface repositories for low / intermediate active waste and geological / deep repositories for high level long-lived radioactive waste (see Figure 1). The role of cement-based materials in these applications is not only focused on the retention of radionuclides but also in helping creating beneficial conditions for the waste package integrity (steel corrosion) and limiting the solubility of many cationic species because of its high-pH buffering capacity for a very long period.

Under its service environment, these concrete structures undergo complex chemical interactions with the geological matrices (e.g., clay, granite). The interaction processes are very slow but important for the long-term durability assessment (> 1,000 years). It is well documented that due to the large geochemical contrasts between cement-based materials and clay, chemical reactions are expected to induce modifications in both the chemical and physical properties of both the clayey and cement materials. The interaction disturbs the equilibrium between the pore solution of the cementitious materials and the solid phases of the cement matrix which results in dissolution and/or precipitation of minerals. The alteration is even more pronounced at the host rock side. The interaction typically results in alteration of mineralogy and microstructure which are mostly followed by alteration of its transport properties and influences the long-term stability of an engineered structure.

The Boom clay formation, which is a dark organic-rich and poorly indurated argillaceous clay with a high pyrite and glauconite content in the more silty layers, has been considered as a potential host rock for intermediate and high level radioactive waste in Belgium. Therefore the concrete-Boom clay interface has been chosen as a candidate to study. The concrete and Boom clay have been in contact since 2003 resulting in 14 year-interfaces stemming in the underground research laboratory (HADES) located at 225 m in depth at Mol, which are representative for disposal conditions. A procedure to sample these concrete-clay interfaces has been proposed in previous work (Phung et al., 2017). This paper aims at describing an integrated downscaling approach, with the goal to quantify and spatialize how the geochemical perturbation surrounding those interfaces will modify the microstructure of both engineered and natural barriers and thereby the effect on the transport properties. Preliminary experimental results are also reported. Furthermore, the laboratory experiments to mimic the real interaction conditions are also described in this paper.

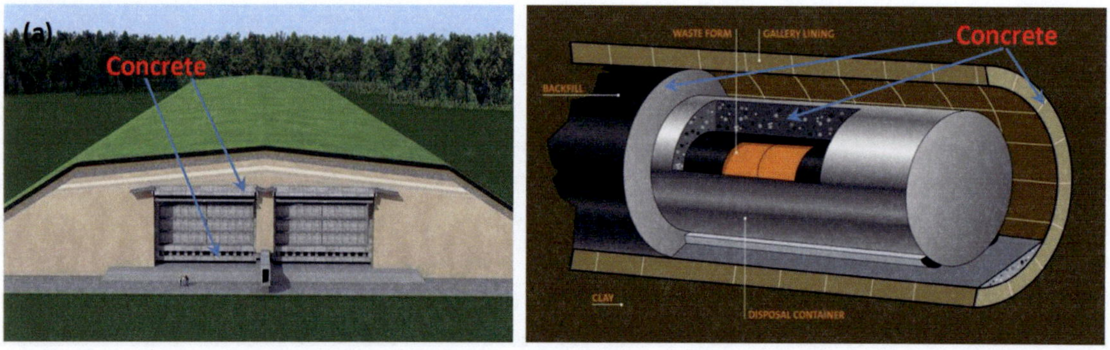

Figure 1: *Concrete used for encapsulation, backfill and construction in low level waste facility (a) and disposal gallery lining, buffer within Supercontainer, end-plug of gallery and backfill in high level waste facility - Supercontainer concept (b) [adapted from NIRAS' reports NIRAS (2012) and NIRAS (2011)]*

Materials and interface sampling

Examined materials

In-situ concrete

The tests are performed on high strength concretes (liner) used as lining material in the Connecting Gallery of HADES URL. The composition of concrete is shown in Table 1.

Table 1: *Mix compositions of in-situ concrete: SP - Superplasticizer.*

For 1 m^3 concrete	Cement (kg)	Fly ash (kg)	Coarse agg. (kg)	Fine agg. (kg)	Admixture		Water (L)
					SP (L)	µ-silica (kg)	
Connecting gallery	335 (CEM I)	115	1252	540	4.5	90	135

Cement pastes for laboratory studies

With the purpose of accelerating the interaction processes, cement pastes were made with a target high porosity for the laboratory experiments. The mixture compositions are mimicking backfill materials, but without superplasticizer (Table 2). Note that the backfill compositions for Belgian radioactive waste facilities have not yet fixed. The typical composition is composed of water, Portland cement, limestone filler, hydrated lime and to improve the workability some superplasticizer. Type I ordinary Portland cement (CEM I 52.5 N) was used.

Table 2: *Mix compositions of cement pastes for laboratory experiments.*

Cement type	Limestone filler/cement	Lime/cement	Water/cement
CEM I 52.5 N	3/1	0.377	1.5

Boom clay properties

The main minerals in Boom clay are quartz and clays, in different proportions. Table 3 presents the Boom clay mineralogical compositions (min - max ranges) at the Mol site (Honty and De Craen, 2012). Note that the same minerals are present over the whole Boom Clay formation although the proportions of the various minerals vary in the vertical profiles. The reference composition of Boom clay pore water at the Mol site in Belgium is also described in Table 3 (De Craen et al., 2004). The carbonate concentration in pore solution is quite high (~ 0.014 mol/L $NaHCO_3$) which may accelerate the carbonation of the concrete side.

Table 3: *Mineralogical and chemical properties of Boom Clay and its pore water.*

Mineralogical composition Boom clay %		Boom clay pore water	
Quartz	22 - 66	Na	359 mg/L
Na-plagioclase	0 - 6.3	K	7.2 mg/L
K-Feldspar	0.4 - 8	Ca	2.0 mg/L
Siderite	0 - 1.5	Mg	1.6 mg/L
Calcite	0 - 4.6	Fe	0.2 mg/L
Dolomite	0 - 1	Cl	26 mg/L
Apatite	0 - 0.9	SO_4	2.2 mg/L
Pyrite	0.3 - 5	HCO_3	879 mg/L
Illite / muscovite	5 - 37	DOC	115 mg/L
Smectite and illite-smectite	6.8 - 37	Eh (mV)	-274 mV *vs.* SHE
Kaolinite	2 - 14	pCO_2 (atm)	$10^{-2.62}$ atm
Chlorite	0.5 - 4	pH	8.5

Interface sampling

In-situ sampling

Boom clay-concrete interfaces were sampled in the HADES underground research facility (Mol, Belgium). Samples were taken from the Connecting Gallery, in which low-permeability high-pH concrete wedge-blocks have been in contact with Boom clay since 2003. Two drilling campaigns have been done using the anchor concept. In the first campaign, only two resin anchors were installed which resulted in some gaps at the interface (see Figure 2 and later in Figure 6). In the second campaign, four resin anchors were installed, which kept the interface contact much better preserved. Details of drilling procedure can be found in (Phung et al., 2017).

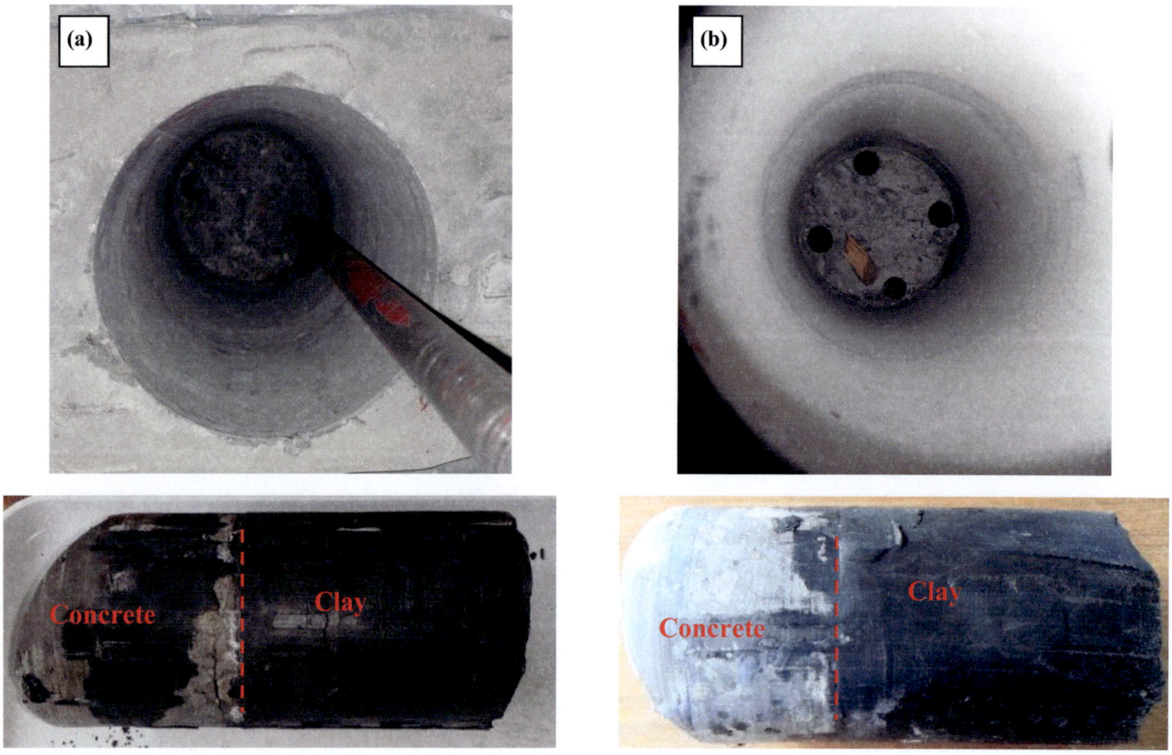

Figure 2: *Concrete-clay interface samples obtained by using 2 resin anchors (a) and 4 resin anchors (b).*

Lab made interface

Cement pastes were poured and cured in a cylindrical PVC tube with an inner diameter of 97.5 mm. The samples were then rotated during 24 hours to prevent segregation and subsequently cured under sealed conditions in a temperature controlled room (21 ± 1°C). The cured cement pastes were sawn into 45 mm thick slices.

Percolation test

The interface is created by putting plugs of Boom clay and hardened backfill paste in contact in a percolation cell which is a modified design of a permeability cell (Phung et al., 2013). In these experiments, the chemical conditions can be rigorously controlled. The conditions in the HADES underground research facility were mimicked by chemical compositions and advective flow of Boom clay pore solution. Partial CO_2 pressure at the underground was also simulated by bubbling 0.4% CO_2 gas (in argon background) into clay pore water solution which was injected in the percolation cell (from clay side) by a Syringe pump (Figure 3). The Syringe pump also enables recording the flow rate and cumulative pore water injected into the system. A pressure gradient of 3 bar was applied on the cell which accelerates the interaction. A series of 3 experiments were performed, which can be stopped as function of interaction time (3, 6 and 12 months). For each interaction time, samples will be used for measuring the changes in transport properties (i.e., permeability, diffusivity, sorptivity) and changes in chemical and microstructural properties. The outflow was also sampled regularly to follow up the changes in chemical compositions of the pore solution.

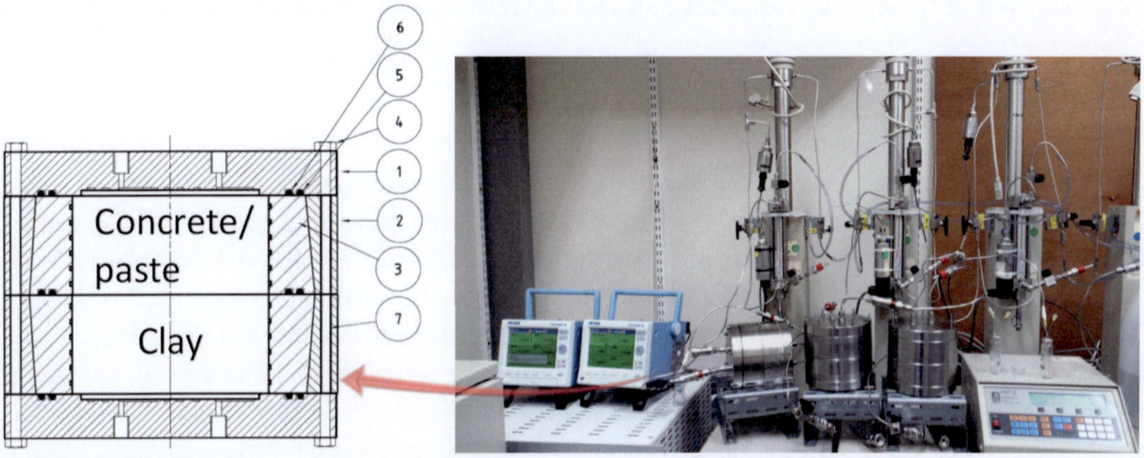

Figure 3: *Test setup for percolation experiment - the percolation cell (left) composed of: (1) steel lid, (2) steel body, (3) inner polycarbonate body, (4) bolt, (5) inner O-ring, (6) outer O-ring, and (7) threaded rod.*

Batch test

Batch type experiments were also performed in which a backfill concrete disc was immersed into a Boom clay slurry. The experiments were conducted in closed chambers which was modified from leaching tests (Phung et al., 2014b; Phung et al., 2015a). CO_2 0.4% was also bubbled to the chambers to mimic *in-situ* CO_2 concentration. The chemical composition of the slurry is followed regularly. The transport properties and microstructural evolution will be determined at the end of experiments (3, 6 and 12 months). These experiments are expected to induce more rapid cement degradation due to excess water compared to *in-situ* experiments. Note that all the samples need to be saturated following the saturation procedure proposed in (Phung et al., 2013). These saturated conditions simulate the real conditions in which concrete is getting saturated after a few decades in contact with the clay host formation.

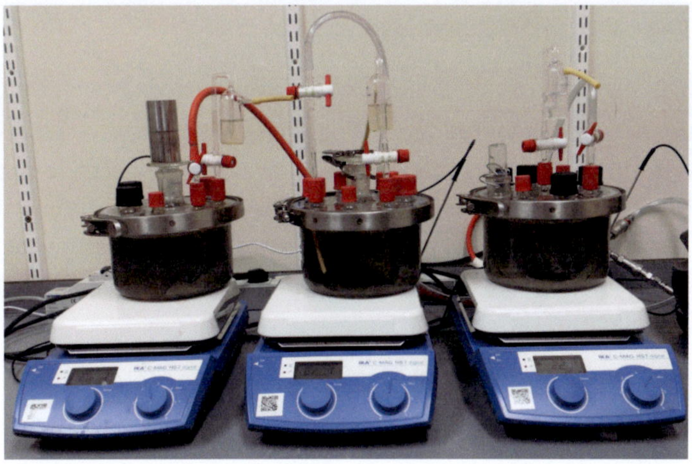

Figure 4: *Batch experiment setup.*

Quantification of transport properties and characterisation of microstructure and mineralogy

Transport properties

Permeability

The percolation cell was designed in a special way which allows determination of the composite permeability of clay and concrete and also permeability of separate materials by removal of concrete or clay part from the assembly. Prior to the percolation experiments, the permeability of cement paste and clay samples were determined separately using a controlled constant flow method as described in (Phung et al., 2012; Phung et al., 2013). The method is based on the application of a constant flow instead of a constant pressure, as used in traditional methods, which overcomes the problem of measuring extremely low flow rate. A constant pressure gradient was applied until the flow almost reached steady state, than the constant pressure mode was changed to the constant flow mode. The pressure was then measured until reaching the steady state in pressure, which is much more accurate than measuring flow rate. The pressure and water flow were controlled by precise Syringe pumps.

The composite permeability of clay and cement paste samples can be calculated at any specific time during the percolation experiments. At a certain interaction time, the percolation cell will be decoupled (separate clay and cement paste samples) for determination of the permeability of each material.

Diffusivity

Similar to permeability, the diffusion coefficients of initial materials was determined before and after a certain time of interaction. A through-diffusion methodology was used, allowing simultaneous determination of diffusivities of two dissolved gases (He and CH_4) in a single experiment as described in (Jacops et al., 2013; Phung, 2015; Phung et al., 2015). Note that thanks to its compatibility, the permeability / percolation cell can be easily disconnected and then reconnected to the diffusion setup. In order to interpret the experimental data, a 1D diffusive transport model has been used (Jacops et al., 2013), taking into account the drop of gas pressure during sampling.

Water sorptivity

Considering extremely slow concrete-clay interaction, the degraded depths, especially at the (high strength) concrete side, are expected to be limited. Initial observations on a few concrete samples taken from HADES underground research facility showed that the degraded depths were only 1 - 2 mm for 13 years of interaction, which are indeed not sufficient to perform permeability or diffusion measurements mentioned in previous sections. Therefore, a water sorptivity test was performed which is relevant to characterize the transport properties of an exposed surface of concrete or clay.

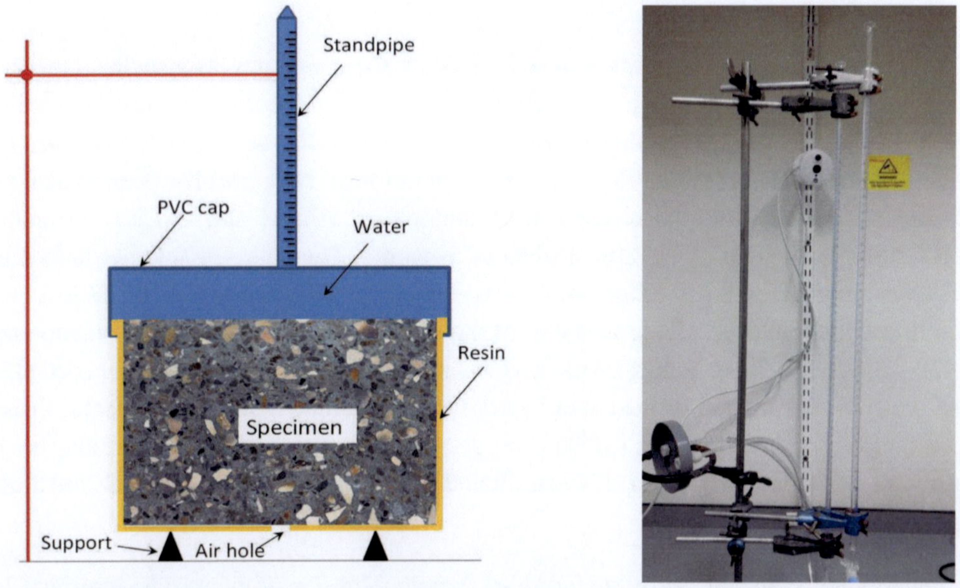

Figure 5: *Experimental setup to measure water sorptivity.*

The samples are dried in an oven at 50°C. This gentle temperature is chosen to, on one hand, reduce the conditioning duration and, on the other hand, prevent cracking to occur due to heating and decomposition of water in C-S-H phase. The water sorptivity setup is shown in Figure 5. In this setup, water is on top of the concrete surface, gravity also will play a role in the water penetration. However, water ingress into the sample is mainly driven by capillary forces as these are much larger than the gravity. The volume of water absorbed by the sample was determined by following the level of the standpipe, captured by an IP camera. The sorption coefficient is defined as Eq. 1.

$$I = \frac{W}{A} = S\sqrt{t} + I_0$$ ***Eq. 1***

where I = cumulative water absorption on tested surface [mm³/mm²]; W = volume of water absorption [mm³]; A = tested surface [mm²]; t = time variable [min]; S = sorption coefficient [mm/min$^{1/2}$]; and I_0 = initial water absorption [mm³/mm²]. Water sorptivity is one of the independent transport indicators. However, the water sorptivity is strongly related to (unsaturated) permeability (and thereby diffusivity) and capillary pressure which depends on microstructure and water content of the sample (Kelham, 1988).

Microstructure and mineralogy

A multi-scale investigation methodology, based on clay materials, has been developed during the European FP7 CATCLAY project (Gaboreau et al., 2016). The methodology integers several bulk macroscopic characterization techniques and imaging methods to display quantitative data from macroscopic to nanoscopic scale. The association of several 2D/3D techniques (mineral cartography, autoradiography, X-ray µ-tomography, SEM, FIB-nT and TEM) allows reaching a quantitative and spatial distribution of the mineralogy and the pore network, from nanometer to micrometer represented by several geometric key parameters such as the pore size distribution and the associated mineral distributions.

Microtomography acquisition

The sample, before any sub-sampling, was scanned by X-ray μ-tomography to visualize the interface and localize the region of interest (ROI). The μ-tomography acquisition of the data has been performed with a laboratory device, an EasyTom XL Duo developed by the company RX-solutions (Chavanod, France). A Hamamatsu sealed 150 kV microfocus X-ray source (up to 5 μm of focal spot size) has been used, coupled with a Varian PaxScan 2520DX detector (flat panel with amorphous silicon and a CsI conversion screen; 1,920 x 536 pixel matrix; pixel pitch of 127 μm; 16 bits of dynamic). The entire sample has been scanned in a continuously helicoidal mode (5,460 projections in 4 turns) with a spatial resolution of 61.38 μm. Achieving a higher resolution is not possible due to large diameter of the sample. Parameters of the acquisition are 140 kV (tube voltage), 430 μA (tube current), large spot mode (focal spot up to 50 μm), 8 frames per second, averaging of 20 frames per projections, filtration of the beam by a 1-mm Aluminum, a source-to-detector distance and a source-to-object distance of 376.708 mm and 182.066 mm, respectively. Reconstruction of the data has been done with the XAct software (RX-solutions) with a classical filtered back projection algorithm (Kak and Slaney, 2001) and correction of beam hardening.

This acquisition allows visualizing the resin embedded sample. The size of the sample is around 160 mm in height and 96 mm in diameter (Figure 6). Many cracks are observed in the clay and at the interface with an opening up to 5 mm.

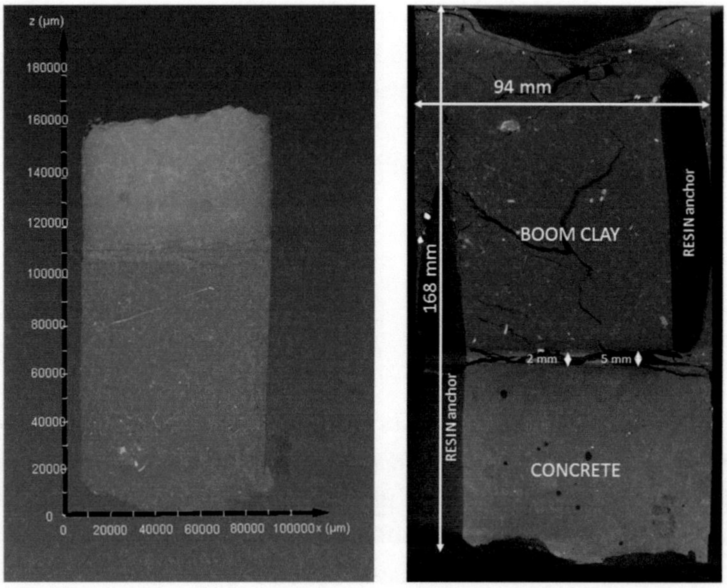

Figure 6: *3D visualisation of the concrete - Boom clay interface (left) and 2D slice in the yz plane showing the interface and the two resin anchors (right).*

Some 2D slices have been extracted from the 3D volume to visualize the interface between the concrete and the clay (Figure 7). The 2D slices are given in 3 different orientations (XY, YZ and XZ). These different views allow visualizing the cracks and the deformations of the sample and the areas of well-preserved contact (Figure 8).

The zones of well-preserved contact were localized close to the anchor, while in the middle of the sample, the opening reached 5 mm. It seems obvious that the anchor allows preserving the contact between clay and concrete.

The clay also presented many cracks all over the height of the drilled sample. A compression at the bottom of the clay (opposite to the interface) was also observed.

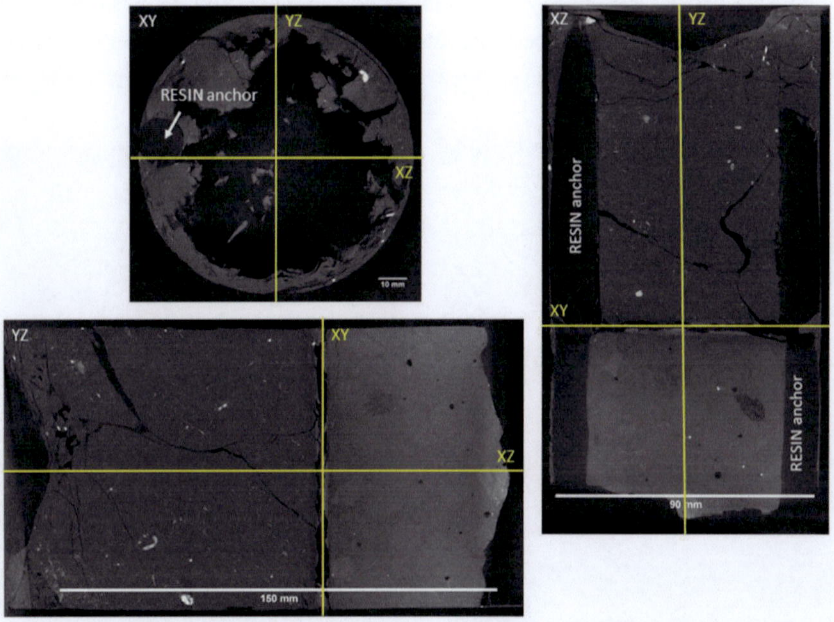

Figure 7: *2D slices extracted from the 3D μ-tomography volume in different orientations showing the interface between the concrete and the Boom clay.*

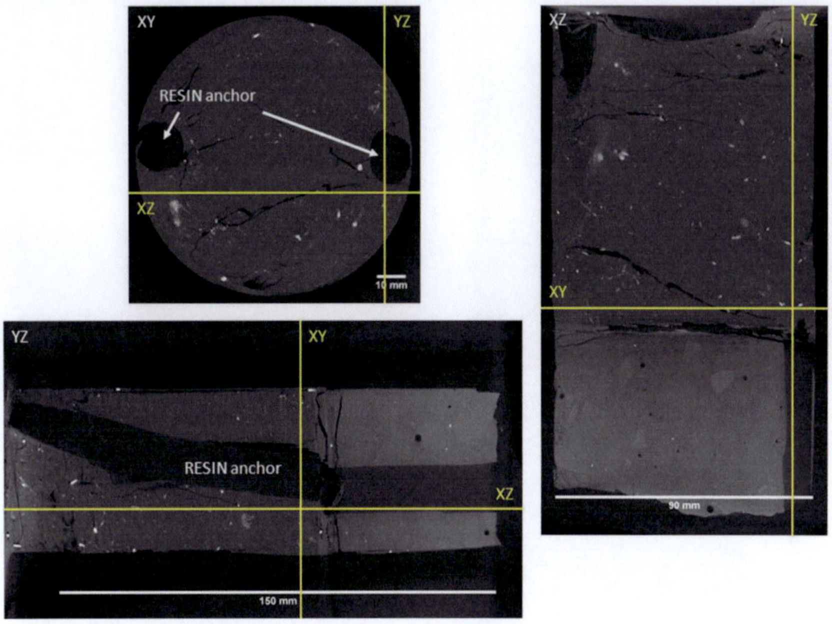

Figure 8: *2D slices extracted from the μ-tomographic volume in the different orientations showing the contact at the interface close to the resin anchor.*

A max intensity projection has also been performed (Figure 9) to visualize the phases with the highest density (pyrite, metallic hydroxides, etc.). This algorithm created an output image each of whose pixels contains the maximum value over all images in the stack at the particular pixel location. Over the clay plug, millimetric structures with tubular shape were observed (pyritized wormholes).

µ-tomography data allow the possibility to visualize the sample and to extract a localized sub-volume. Figure 10 displays a region of interest, close to an anchor that will be cut later on to prepare some samples for the autoradiography impregnation. This area represents a zone where contact seems to be preserved.

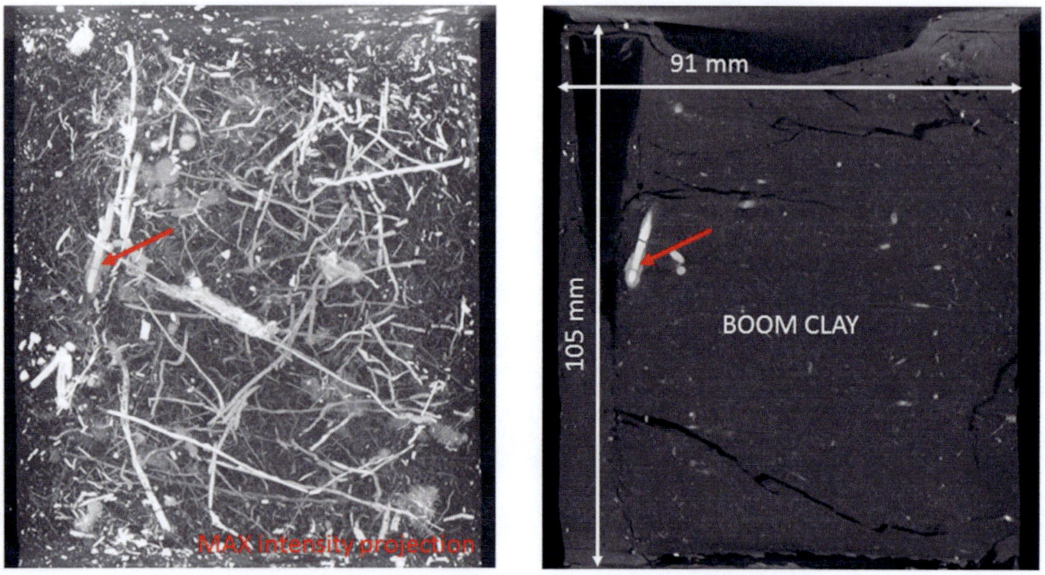

Figure 9: *2D slices illustrating the correspondance of the high density minerals with the max intensity projection.*

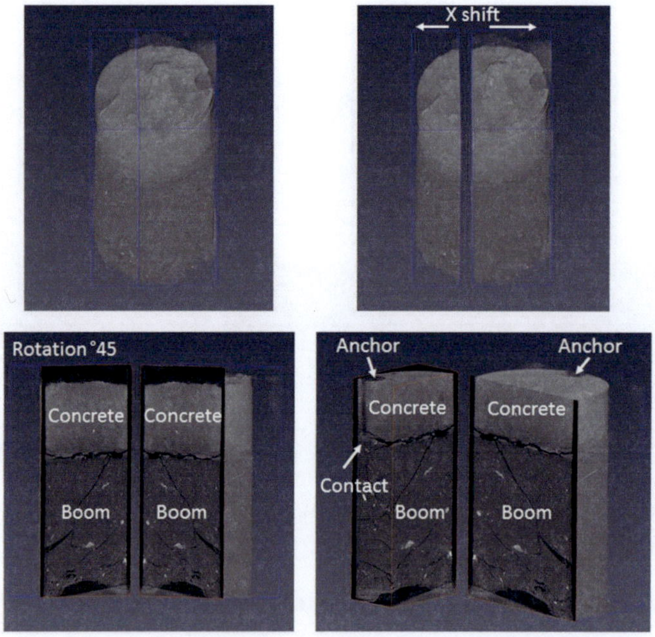

Figure 10: *3D visualization showing the subsampling procedure in order to localize and sample the contact at the interface.*

<u>*Autoradiography*</u>

Autoradiography acquisition was performed according to the method developed by (Prêt, 2003) for clay materials. The selected polished sections, from the 3D μ-tomography volume (Figure 10) were oven dried during 20 days at 75°C, under vacuum. The conditions used for the cement materials were chosen according to the work of (Gallé, 2001) which showed less impact of drying. The samples were then impregnated during 90 days with ^{14}C PMMA with tracer activity of 10 μCi/mL. The objective was to reach a total saturation of the pore space by the resin to obtain quantitative porosity measurements by autoradiography. The polymerization of the impregnated samples was done by ^{60}Co irradiation with a total dose of 120 kGy to polymerize the PMMA resin into the clay interlayer spaces (Prêt, 2003). The textural state of the impregnated samples was then expected to be close to the textural state encountered for samples in equilibrium with a water saturation of 98% (Prêt et al., 2010). This was mainly allowed by the properties of the resin (viscosity and dipole moment), that are close to those of water. For the sample, two polished sections were prepared with a low speed diamond saw by cutting in half the sample thickness. They were polished using diamond powders with grain sizes down to 0.05 μm.

The thick sections were then exposed on Kodak Biomax MR film during 7 days (Figure 11).

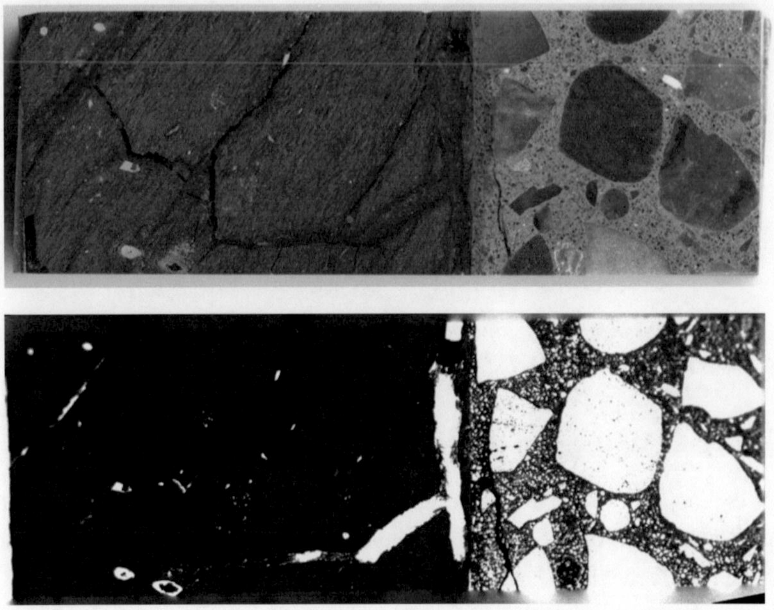

Figure 11: *Thick section of fully ^{14}C PMMA impregnated samples (top) and associated autoradiography (bottom).*

Exposure times were optimized according to different porosity of claystone and cement materials to acquire the best porosity contrast (Prêt, 2003). The autoradiographs were then digitized in 8-bit images with a resolution of 2,400 dpi (10.4 x 10.4 μm^2 pixel size).

The local blackening of the film (expressed by a local optical density) is proportional to the local concentration of radioactive resin. Sets of calibration sources of known activity were simultaneously exposed on the autoradiograph films to obtain the relationship between the local activities and optical densities by fitting the experimental data using the calibration function proposed (Hellmuth et al., 1993). With the calibration function and the tracer activity, a local mean porosity was then calculated for each pixel of the digitized autoradiographs using the relationships provided by (Sammartino et al., 2002) and using autoradio software (Prêt, 2003). Such local pixel porosity includes the connected micro-, meso-, and macro-porosity. For each autoradiograph, a combination of analysis methods was applied to reveal porosity distribution. Thus, porosity frequency histograms could be computed on different sub-area of the autoradiograph by counting the number of pixels which have similar porosity values. To localize the spatial heterogeneities, porosity maps (Figure 12) were then computed by displaying the local porosity of each pixel using a linear grey level scale ranging between two porosity values identified on the histogram.

Porosity profiles were also computed to reveal the porosity evolution as a function of two distal points. Profiles were calculated for 1 pixel along a straight line (1-pixel width) or dilating the straight line by a given number of pixels in width to reach average porosities with an improved statistical error (see later in Figure 15).

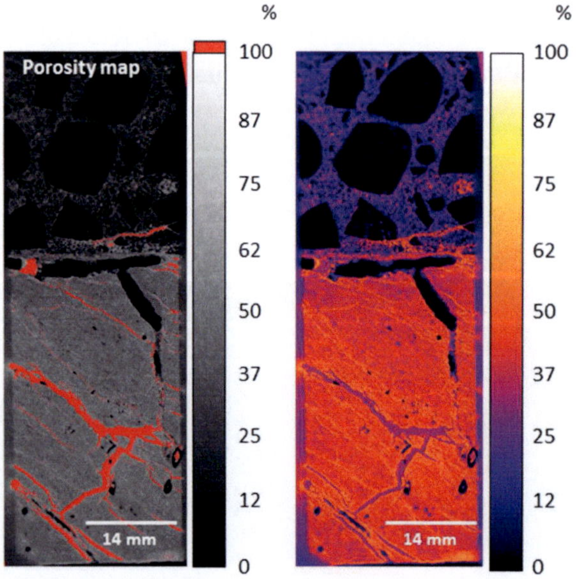

Figure 12: *Porosity map calculated from the autoradiography of the concrete - Boom clay interface - porosity maps are encoded in gray level and also in false color (LUT).*

<u>Mineral characterisation by EPMA mapping</u>

Mineralogical maps can be built based on the combination of a set of chemical element maps according to the method developed by (Prêt et al., 2010). Two maps have been acquired. The first map was measured on the concrete side far from the interface with a Field Of View (FOV) of 1 x 0.8 mm². The second one was acquired on the clay side; also far from the interface with a FOV of 1 x 1.5 mm² (Figure 13).

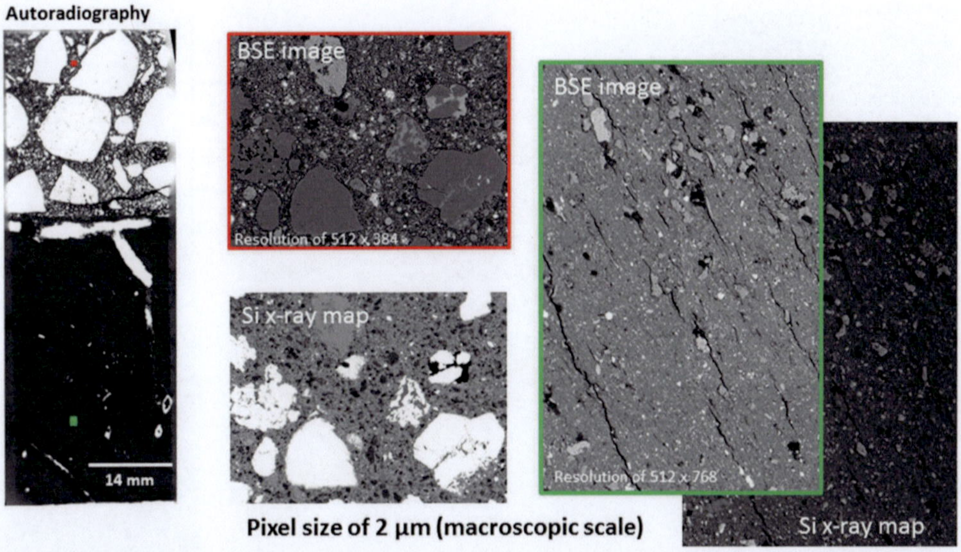

Figure 13: *BSE image and associated X-ray intensity map of the mapped area in the concrete and Boom clay - two maps are localized on the autoradiography (in red and green).*

The map obtained on the concrete was not representative for the complete mineralogy of the concrete since it was smaller than the aggregate size. X-ray intensity maps and a BSE image (Figure 13) were acquired with a Cameca SX Five EPMA equipped with five WDS and operating at 15 keV and 30 nA. To investigate the repartition of the 10 elements identified in the concrete, the analyzed area was scanned two times as the number of the simultaneously detected elements is constrained by the number of available spectrometers. Kα (Si, Al, Fe, K, Na, Ca, Mg, Ti, Mn, S) peak intensities were collected using large TAP, PET, and LiF monochromatic crystals, which allowed a high counting rate and a short dwell time (~ 4 s) for a quantitative point analysis. Such a dwell time is too time-consuming for mapping mode, so a shorter counting time of 100 ms per pixel was used. To reduce the total acquisition time (2 days), no background measurement and subtraction from X-ray emission peaks was performed as its contribution at short dwell times and for major concentrations was low (Goldstein et al., 1992) and its recording would have doubled the acquisition time. A PHIRHOZ matrix correction (Merlet, 1994) was then applied to provide a weight percentage for each element per pixel. The elemental maps were recorded by stage raster using a stationary beam, with a spatial resolution of 2 µm per pixel. X-ray map processing was then performed using the µMAPphase software (Prêt et al., 2010).

FIB / SEM and TEM

All the previous methods (µ-tomography, autoradiography, EPMA mineralogical mapping) allow quantifying and spatializing the total porosity and mineralogy at a macroscopic scale. The quantitative spatial distribution of the porosity and mineralogy will permit localizing the region of interest in order to subsample this area and decline techniques, such as SEM, FIB / SEM and TEM which enable to characterize those ROI at the scale of interest (meso and nanoscopic scale) in 2D and 3D.

A scheme is given in Figure 14 to describe the multi-scale approach used to characterize the concrete - Boom clay interface (results are not presented in this paper).

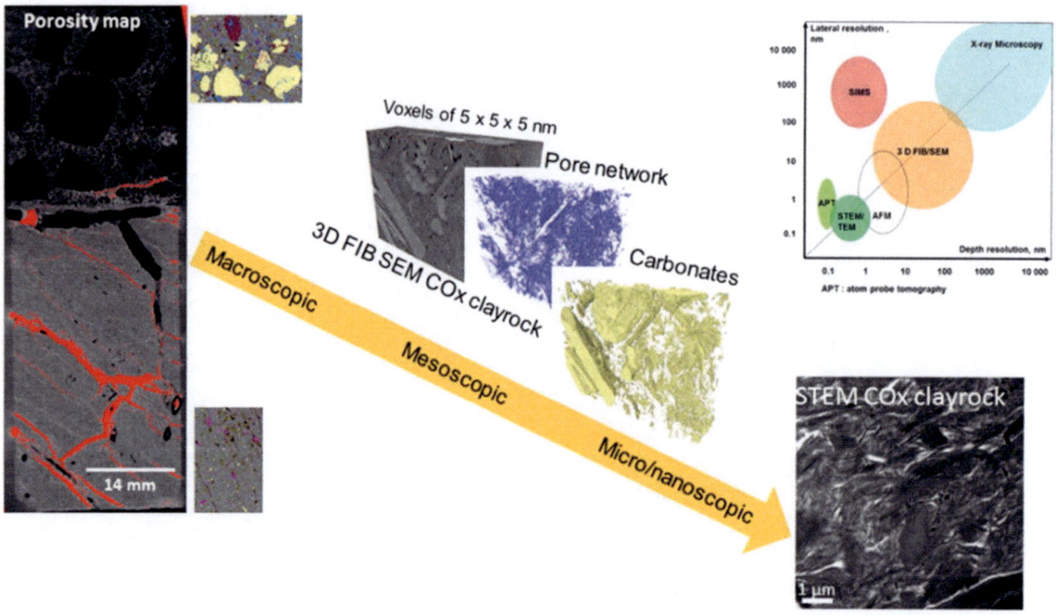

Figure 14: *Image illustrating the methodology which will be used to characterize the concrete-Boom clay interface from the macroscopic to nanoscopic scale.*

Preliminary experimental results

Permeability of concrete and Boom clay

Permeability (reported as hydraulic conductivity) of different testing materials are shown in Table 4. Three backfill concrete samples were measured in which two samples were cured in 4 months and one was cured in 8.5 months. As expected, the permeability of backfill materials was very large, in the order of 10^{-8} m/s, due to high w/c ratio. The permeability was reduced by a factor 2 when the curing time increased from 4 to 8.5 months indicating a continuous hydration. Lining concrete exhibited an extremely low permeability material. The permeability of two lining samples was quite similar and in the order of 10^{-14} m/s.

Table 4: *Mineralogical and chemical properties of Boom Clay and its pore water.*

Samples	Permeability, m/s	Curing time
Backfill 1	$1.25 \cdot 10^{-8}$	4 months
Backfill 2	$1.18 \cdot 10^{-8}$	4 months
Backfill 3	$0.59 \cdot 10^{-8}$	8.5 months
Clay 1	$2.43 \cdot 10^{-12}$	-
Clay 2	$2.46 \cdot 10^{-12}$	-
Clay 3	$2.32 \cdot 10^{-12}$	-
Liner 1	$1.58 \cdot 10^{-14}$	13 years
Liner 2	$1.27 \cdot 10^{-14}$	13 years

The permeability of Boom clay was in the range of $2.32 \cdot 10^{-12}$ - $2.46 \cdot 10^{-12}$ m/s for three measured samples which is consistent with the values reported in literature (Yu and Weetjens, 2009; Jacops et al., 2013).

Changes in microstructure at interface

The first results obtained by autoradiography are presented in Figure 15. The total porosities were measured by integrating the surfaces of each material (white squares). The total porosities were around 10% and 42% for concrete and Boom clay, respectively. Some profiles have also been plotted (Figure 15) to measure the evolution of the porosity in both material and mainly at the interface. The main problem with this first interface is the opening cracks at the interface on the clay side (epoxy resin) which complicates the interpretation of the measured data. However, the profile plotted from the concrete to the clay side (blue line) displays some porosity evolution. On the concrete side, the porosity evolves from 0 to 25% when the line intersects the non-porous aggregates and the hydraulic binder, respectively. Moreover the porosity increased at the concrete interface, which indicates dissolution of hydrated phases occurred.

Furthermore, two lines have been plotted parallel to the interface (profiles A and B), avoiding the intersection with aggregates, to confirm the evolution of the total porosity. The total porosity increased from 20 to 30% at the concrete interface. On the clay side, even if the porosity seemed to decrease at the interface, it is difficult to interpret this value, due to the presence of the large cracks. A second well preserved interface should be studied to confirm this evolution.

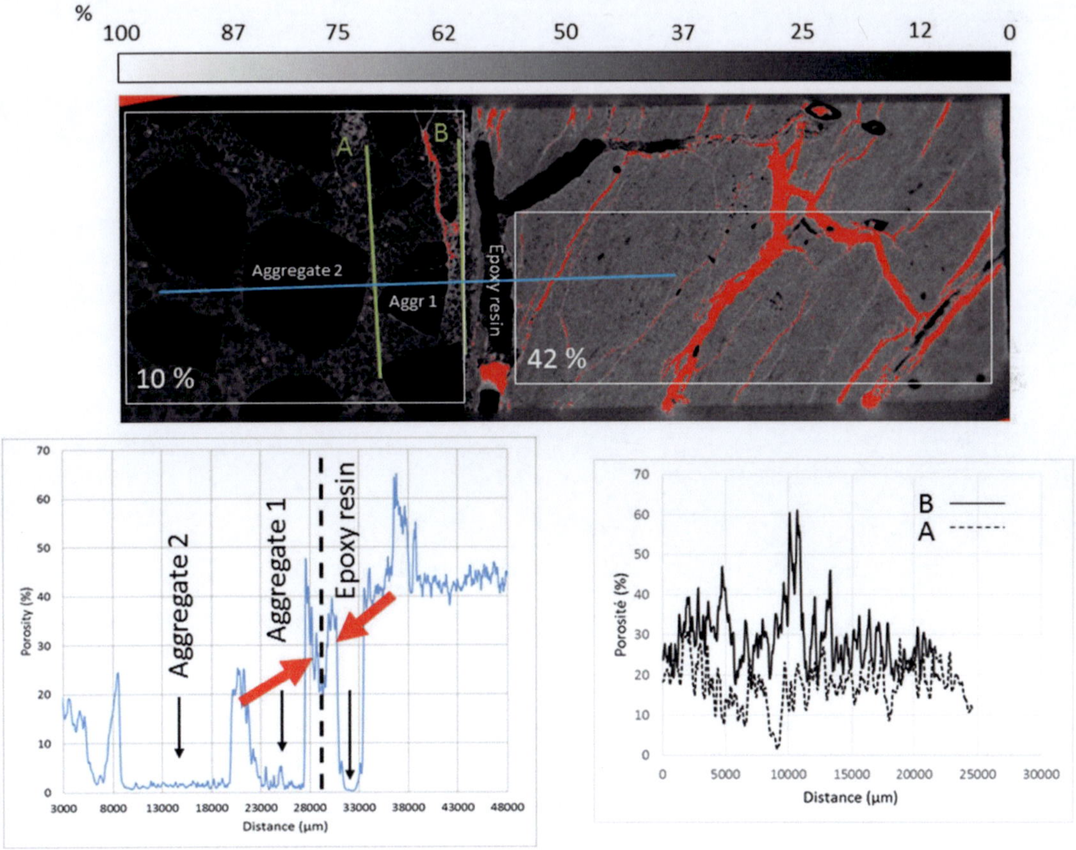

Figure 15: *Porosity maps of the concrete-Boom clay interface: Total porosity values are given for concrete and clay; Profiles are also computed through the interface (blue line) and in the concrete at different distances from the interface (lines A and B).*

Changes in mineralogy at the interface

Two mineralogical maps were acquired in both materials far from the interface. These mineralogical maps calculated from the quantitative X-ray intensity maps of the major elements constituting the materials are shown in Figure 16. These methods allow calculating the mass fraction of each minerals and phases constituting the materials. Not only the mass fraction of each phase, the chemical composition of each mineral and phase is also calculated by integrating all the pixels constituting a phase.

The same work will be done at the interface in the area where the porosity has changed to quantify the mineralogical evolution and characterize the chemical evolution.

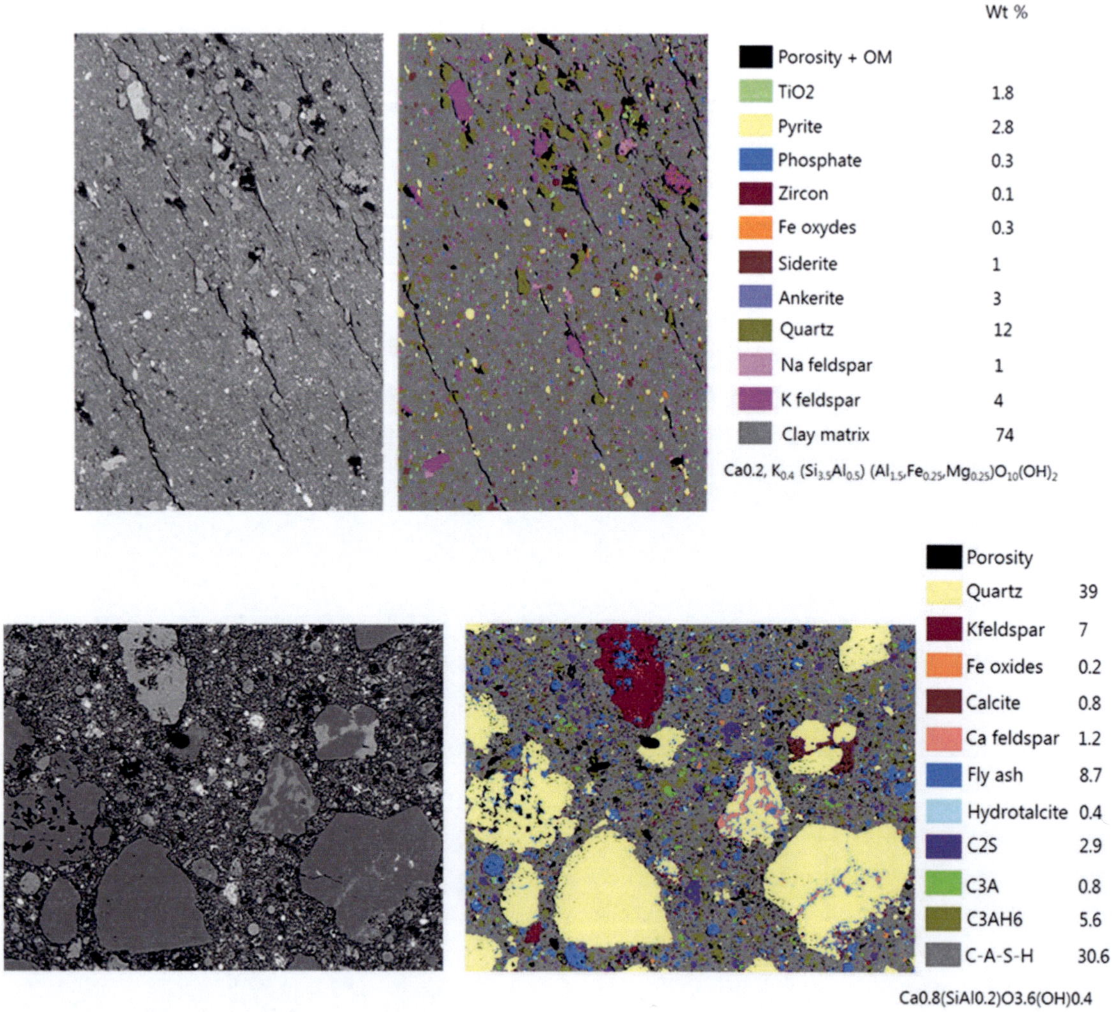

Figure 16: *Quantitative mineralogical map obtained from EPMA analyses for the Boom clay (top) and the concrete far from the interface (bottom).*

5. Conclusions and perspectives

This paper presents methodology and preliminary results on the study of interface interactions between concrete and Boom clay under real disposal conditions as well as under accelerated conditions. The global goal of this study focuses on changes occurring both on the concrete and clay sides with the aim to answer some key questions: (i) whether clogging occurs in the clay and/or concrete side or not; (ii) to which extent the permeability, transport properties and water sorptivity of the concrete are changed as a consequence of microstructural and mineralogical alterations; and (iii) whether the combined effect of carbonation and leaching could alter the microstructure and transport properties of the concrete in a different way as compared to the separate effects. The preliminary results on the alteration of porosity at the interface show an increase in total porosity of the concrete interface. This indicates that clogging at the concrete side might not be the case as was anticipated and also predicted in modelling approaches (Liu et al., 2014). The increase in porosity of the concrete interface can be interpreted as Ca-leaching (resulting in porosity increase (Phung et al. 2014a; Phung et al., 2015a) might dominate compared to carbonation (resulting in porosity decrease (Phung et al. 2015c; Phung et al., 2016a; Phung et al., 2016b) at the interface. Furthermore, the porosity in the clay seems to decrease indicating some precipitations occurring which may clog the clay pores in the long-term. More results in microstructural, mineralogical and transport properties changes will help clarifying these issues and answer the remaining questions to fulfil the goal of this project.

Acknowledgements

The research leading to these results has received funding from the European Union's Horizon 2020 Reasearch and Training Programme of the European Atomic Energy Community (EURATOM) (H2020-NFRP-2014/2015) under grant agreement n° 662147 (CEBAMA).

This work is performed in close cooperation with ONDRAF/NIRAS, co-funding this research, as part of the programme on geological disposal of high-level / long-lived radioactive waste.

References

De Craen, M., Wang, L., Van Geet, M., Moors, H. (2004). Geochemistry of Boom Clay pore water at the Mol site. SCK•CEN Scientific Report, SCK•CEN-BLG-990.

Gaboreau, S., Robinet, J.-C., Prêt, D. (2016). Optimization of pore-network characterization of a compacted clay material by TEM and FIB/SEM imaging. Microporous and Mesoporous Materials, 224, 116-128.

Gallé, C. (2001). Effect of drying on cement-based materials pore structure as identified by mercury intrusion porosimetry: A comparative study between oven-, vacuum-, and freeze-drying. Cement and Concrete Research, 31(10), 1467-1477.

Goldstein, J., Romig, A., Newbury, D., Lyman, C., Echlin, P., Fiori, C., Joy, D., Lifshin, E. (1992). Scanning electron microscopy and X-ray microanalysis. A text for biologists, materials scientists and geologists. Plenum Press.

Hellmuth, K., Siitari-Kauppi, M., Lindberg, A. (1993). Study of porosity and migration pathways in crystalline rock by impregnation with ^{14}C-polymethylmethacrylate. Journal of Contaminant Hydrology, 13(1), 403-418.

Honty, M. and De Craen, M. (2012). Boom Clay mineralogy-qualitative and quantitative aspects. Status 2011. SCK•CEN External Report, SCK•CEN-ER-194.

Jacops, E., Volckaert, G., Maes, N., Weetjens, E., Govaerts, J. (2013). Determination of gas diffusion coefficients in saturated porous media: He and CH_4 diffusion in Boom Clay. Applied Clay Science, 83-84, 217-223.

Kak, A.C. and Slaney, M. (2001). Principles of computerized tomographic imaging. SIAM.

Kelham, S. (1988). A water absorption test for concrete. Magazine of Concrete Research, 40(143), 106-110.

Liu, S., Jacques, D., Govaerts, J., Wang, L. (2014). Conceptual model analysis of interaction at a concrete-Boom Clay interface. Physics and Chemistry of the Earth, Parts A/B/C, 70-71, 150-159.

Merlet, C. (1994). An accurate computer correction program for quantitative electron probe microanalysis. Microchimica Acta, 114(1), 363-376.

NIRAS (2011). Afvalplan voor het langetermijnbeheer van geconditioneerd hoogradioactief en/of langlevend afval en overzicht van verwante vragen. NIROND 2011-02 N: 269.

NIRAS (2012). Synthese van het veiligheidsrapport voor de oppervlaktebergingsinrichting van categorie A-afval in Dessel. NIROND-TR 2012-17 N 174.

Phung, Q.T. (2015). Effects of Carbonation and Calcium Leaching on Microstructure and Transport Properties of Cement Pastes. PhD thesis, Ghent University.

Phung, Q.T., Maes, N., Claret, F., Gaboreau, S., Leysen, J. (2017). Methodology to study the changes in microstructure and tranport properties of the Boom Clay - concrete interface. Proceedings of the 1st CEBAMA Annual Workshop. KIT Scientific Publishing, KIT-SR 7734.

Phung, Q.T., Maes, N., De Schutter, G., Jacques, D., Ye, G. (2013). Determination of water permeability of cementitious materials using a controlled constant flow method. Construction and Building Materials, 47, 1488-1496.

Phung, Q.T., Maes, N., Jacques, D., De Schutter, G., Ye, G. (2015a). Investigation of the changes in microstructure and transport properties of leached cement pastes accounting for mix composition. Cement and Concrete Research, 79, 217-234.

Phung, Q.T., Maes, N., Jacques, D., De Schutter, G., Ye, G., Perko, J. (2016a). Modelling the carbonation of cement pastes under a CO_2 pressure gradient considering both diffusive and convective transport. Construction and Building Materials, 114, 333-351.

Phung, Q.T., Maes, N., Jacques, D., Jacop, E., Grade, A., De Schutter, G., Ye, G. (2015b). Determination of diffusivities of dissolved gases in saturated cement-based materials. Concrete Repair, Rehabilitation and Retrofitting IV. Sep 2015, 1019 -1027.

Phung, Q.T., Maes, N., Jacques, D., De Schutter, G., Ye, G. (2014a). Decalcification of cement paste in NH_4NO_3 solution: Microstructural alterations and its influence on the transport properties. 10th fib International PhD Symposium in Civil Engineering, Québec, Canada.

Phung, Q.T., Maes, N., Jacques, D., De Schutter, G., Ye, G. (2014b). Microstructural and permeability changes due to accelerated Ca leaching in ammonium nitrate solution. Concrete Solutions. Proceedings of Concrete Solutions, 5th International Conference on Concrete Repairm, 431-438.

Phung, Q.T., Maes, N., Jacques, D., De Schutter, G., Ye, G. (2015c). Evolution of Microstructure and Transport Properties of Cement Pastes Due to Carbonation under a CO_2 Pressure Gradient - A Modeling Approach. CONCREEP 10: Mechanics and Physics of Creep, Shrinkage, and Durability of Concrete and Concrete Structures.

Phung, Q.T., Maes, N., Jacques, D., De Schutter, G., Ye, G. (2016b). Effect of Limestone Fillers on Ca-Leaching and Carbonation of Cement Pastes. Key Engineering Materials, 711, 269-276.

Phung, Q.T., De Schutter, G., Maes, N., Jacques, D., Ye, G. (2012). Measuring permeability of cementitious materials. Concrete Repair, Rehabilitation and Retrofitting III, 287-295.

Prêt, D. (2003). Nouvelles méthodes quantitatives de cartographie de la minéralogie et de la porosité dans les matériaux argileux: application aux bentonites compactées des barrières ouvragées. PhD thesis, Poitiers.

Prêt, D., Sammartino, S., Beaufort, D., Meunier, A., Fialin, M., Michot, L.J. (2010). A new method for quantitative petrography based on image processing of chemical element maps: Part I. Mineral mapping applied to compacted bentonites. American Mineralogist, 95(10), 1379-1388.

Sammartino, S., Siitari-Kauppi, M., Meunier, A., Sardini, P., Bouchet, A., Tevissen, E. (2002). An Imaging Method for the Porosity of Sedimentary Rocks: Adjustment of the PMMA Method-Example of a Characterization of a Calcareous Shale. Journal of Sedimentary Research, 72(6), 937-943.

Yu, L. and Weetjens, E. (2009). Summary of gas generation and migration-Current State-of-the-Art. SCK•CEN External Report, SCK•CEN-ER-108.

Interaction between cement and Czech bentonite under temperature load and in in-situ conditions: results after first testing period

Radek Vašíček[1], Petr Večerník[2], Jaroslav Hloušek[1], Radek Červinka[2], Lucie Hausmannová[1], Václava Havlová[2]

[1] Centre of Experimental Geotechnics, Faculty of Civil Engineering, Czech Technical University in Prague (CZ)
[2] ÚJV Řež, a.s. (CZ)

* Corresponding author: radek.vasicek@fsv.cvut.cz

Abstract

All experimental works of both Czech partners (ÚJV Řež, a.s. and CEG CTU) within the CEBAMA project focuses on alteration and interaction studies on cementitious materials and bentonite. Laboratory works are based on ageing procedures among Czech bentonite, cement, low pH binder and groundwater from Underground laboratory Josef under high temperature. *In-situ* experiment deals with long-term behaviour of compacted bentonite, samples from cement paste and underground water. Duration of interaction is planned to be up to 27 months in case of laboratory experiments and 72 months in case of in-situ test. Geotechnical and geochemical analyses provide information for material characterisation and potential changes description (swelling pressure, hydraulic conductivity, liquid limit, mineralogy, bulk chemical analysis, cation exchange capacity and exchangeable cations, specific surface area for bentonite; uniaxial strength, mineralogy, bulk chemical analysis, pH of leachates, diffusion parameters of ^{3}H and ^{36}Cl for the cement samples). Ageing procedures and results from CEBAMA activities are complementary to and extend those gained already in 2009 - 2013. Laboratory ageing procedures for both types of cementitious materials are ongoing and further sampling is expected in August and September 2017.

This paper focuses on preparation of low pH reference cement paste samples (LPC, CEBAMA reference mix design, Vehmas et al., 2016) and evaluation of *in-situ* experiment in more detail (data set from 2010 - 2016). LPH paste samples, successively prepared by CEG CTU, provide comparable compressive strength results to those reported by VTT. The results gained on bentonite extracted from testing cartridges complete the 6-year long *in-situ* ageing test in the Underground laboratory Josef. It confirmed slow rate of mineralogical changes. The geotechnical parameters (hydraulic conductivity and swelling pressure) were also influenced in measurable way. Data is available for modelling purposes in WP3.

Introduction

The objective of both Czech partners (ÚJV Řež, a.s. and CEG CTU) within the CEBAMA project is to contribute to experimental study on interaction between cement based materials and bentonite. All experimental works are planned for alteration and interaction studies on cementitious materials and bentonite as representatives of engineered barrier materials according to the Czech concept. Low pH "reference cement paste" (according to Vehmas et. al., 2016) is also examined. Long-term laboratory and *in-situ* experiments are exploited.

Laboratory works are based on ageing procedures among Czech bentonite, cement CEM II, low pH binder and groundwater from Underground laboratory Josef under high temperature. *In-situ* experiment deals with long-term behaviour of compacted bentonite, samples from cement paste and underground water. In this case, bentonite cylinders and cement samples were inserted into perforated cartridges, which were emplaced in the boreholes in Underground laboratory Josef already in 2010. The summary of the experimental program is shown in Table 1. Detailed description of CEBAMA ageing procedures, examined parameters, test plan and composition of Bentonite 75 and groundwater Josef (GW Josef) can be found in proceedings of the 1st CEBAMA workshop (Večerník et al., 2016).

Results from CEBAMA activities are complementary to and extend those gained already in 2009 - 2013 (Vašíček et al., 2013). Therefore, selected results from both projects are presented here.

This paper focuses on explaining the preparation of LPC samples (i.e., CEBAMA reference mix design, Vehmas et al., 2016) and evaluation of *in-situ* experiment in more detail (data set from 2010 - 2016).

Table 1: *Summary of experimental program - long-term laboratory experiments and dismantling of in-situ experiment.*

	Ageing procedures	Bentonite B75	GW Josef	CEM II 42.5R	LPC	Temperature 95°C
Laboratory experiments	Bentonite suspension + CEM II (LPC) + 95°C	x	x	x	x	x
	Bentonite suspension + CEM II (LPC)	x	x	x	x	
Sampling intervals 9/18/27 months	GW Josef + CEM II (LPC) + 95°C		x	x	x	x
	GW Josef + CEM II (LPC)		x	x	x	
	Bentonite suspension	x	x			
	CEM II (LPC) + humid air (10°C)			x	x	
In-situ test, 72 months	Cartridge (bentonite)	x	x			
	Cartridge (bentonite + CEM II)	x	x	x		

Note: "Bentonite suspension" is a mixture of Bentonite B75 and GW Josef (ratio 1:5)

Materials and methods

LPC and its interaction

As part of inter-laboratory cooperation, low pH reference cement paste (VTT Finland; Vehmas et al., 2016) was examined. The experimental programme corresponds to main programme on OPC with reduced number of tests and ageing periods. Stability of behaviour of cement samples is determined by uniaxial compressive strength of thin samples (with 50 mm diameter and height of 8.2 mm) before and after 2 ageing periods. The leachate chemistry inside the ageing vessels (concentration of Ca^{2+}, Mg^{2+}, TDS, and pH) is regularly monitored. Quality of LPC material is sensitive to compliance of mixing procedure, ratio and type of components (plasticizers, silica fume, etc.). Verification of influence of plasticizer and mixing procedure had been performed before manufacturing of samples for ageing procedures.

Cement-bentonite from in-situ cartridges

The dismantling of *in-situ* experiments in the Underground laboratory Josef completed the long-term interaction experiment among Czech Ca-Mg bentonite "Bentonit 75" (denoted as B75 from 2010, Keramost, a.s.) cement material CEM II A-S 42.5R (denoted as CEM II, Lafarge Cement, a.s.) and groundwater from

Underground laboratory Josef (denoted as GW Josef) lasted 6 years (Vašíček et al., 2013). In general, two perforated stainless steel pipes ("cartridge sets") were used. Both cartridge sets were initially assembled from four sections ("cartridges"), each with 60 cm long space for samples (diameter 30 mm). One cartridge set (P1) was filled with bentonite only, another one (P3) was filled alternately by cement / bentonite cylinders. Bentonite B75 was compacted to 1,600 - 1,700 kg/m^3 in a shape of cylinders 7.8 cm long and with cement powder compacted to 1,800 kg/m^3 also in a shape of cylinders but 3.5 - 4.0 cm long. Cement samples were cured in GW Josef for 48 days before emplacement into the cartridges. The cartridge sets were installed into boreholes in the Josef Underground laboratory (Čelina section) in September 2010. They were partially pulled out at four time intervals (after 9, 20, 26 and 72 months). One cartridge was detached from the set in each step. Then materials were extracted and interaction among cement, bentonite and groundwater was examined by following laboratory analyses: mineralogical changes, cation exchange capacity and cations at exchangeable complex, specific surface area, groundwater composition, hydraulic conductivity and swelling pressure.

Detailed description of CEBAMA ageing procedures, methods, examined parameters and test plan can be found in Večerník et al. (2016).

Figure 1: *Extraction of the cartridge from the borehole (left) and cement / bentonite sample squeezed out from the cartridge (right).*

Experimental results and discussion

Preparation of low pH cementitious samples (LPC)

LPC paste was prepared according to VTT, as described by the composition and mixing procedure (Vehmas et al., 2016). For paste mixing, two tools were used at CEG CTU - a hand drill with extension for colour mixing and a "High speed mixer" (AKM-70D, according to VTT recommendation). Mixing took place according to the procedure in standard EN 196-1. The paste consisted of CEM I 42.5, silica, blast furnace slag, plasticizer, water and quartz filler at a ratio determined according to the recipe of the CEBAMA reference mix produced also at VTT. PVC cylindrical forms (d = 50.0 mm) were used and LPC paste was poured into them after mixing. After casting, the cylinders were stored for 3 days in humid conditions (100% RH, 10°C) then extracted by manual press and stored again in the same humid underground storage room for total 28 days. Then the cylinders were sliced into 8.2 mm thin discs that matured in underground storage until 45 days after casting. Batches of thin discs were inserted into the ageing vessels; water (and bentonite in some cases) was added and ageing procedures begun.

Verification of influence of plasticizer and mixing procedure

Before final mixing of LPC paste, eight manufacturing alternatives were tested. The target was to gain optimal procedure providing sufficient amount of samples with strength comparable to the values referred by VTT

(69 MPa after 7 days; Vehmas et al., 2017). The trials varied in plasticizer and various details in mixing procedure (two types of plasticizer from three sources were used - Ha-Be supplied by VTT, Ha-Be purchased on market and alternative one by DenBraven; procedure varied in total amount of mixed material, duration of pouring into the forms and time intervals between adding of particular components). All trials provided thin samples for uniaxial compressive strength test. Tests on thin samples (discs) does not provide results directly comparable to results obtained on samples with standard shape and size (cylinder or cube). Therefore, before comparison of results, the first mentioned ones must be converted. Already converted results are plotted in Figure 2. Detailed description of methodology is in ČGÚ (1987) and its application to CEBAMA needs in Večerník et al. (2016). Test no. 8 provided input values for ageing procedures on LPC.

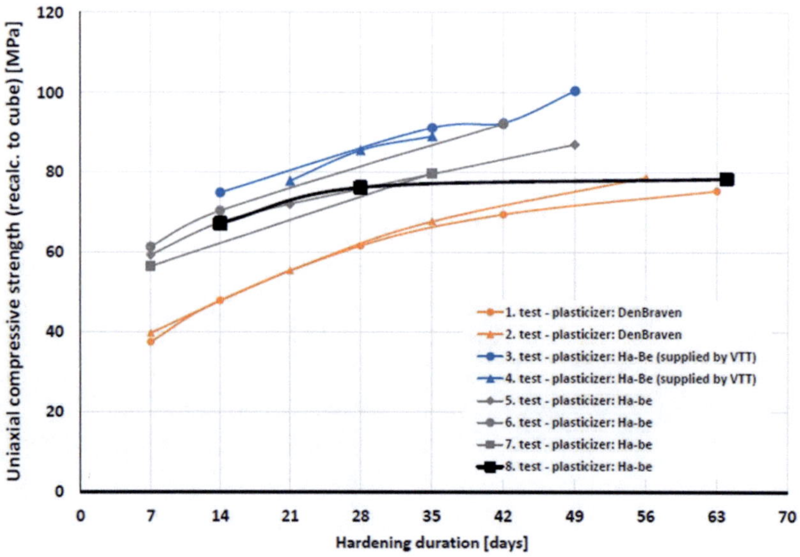

Figure 2: *Uniaxial compressive strength on thin LPC (paste) samples (recalculation to cubic shape). Comparison of 8 sets of mixing tests with variations in plasticizers and mixing procedure details.*

Measuring of pH of LPC leachates

Preliminary pH leachate tests (after 170 days since LPC mixing) were performed by the procedure described in Alonso et al. (2012) and are shown in Table 1. The first sampling period is planned after 300 days since LPC mixing in September 2017. Certain samples are being placed in the underground storage of samples in the Josef facility and thus exposed to humid air only (no immersion into water; temperature 10°C; relative humidity 95 - 100%; marked as "Curing 10°C" below in the Table 2). Two columns of pH values represent two steps in pH determination (suspension / leachate). Each of the samples was powdered and mixed with degassed distilled water for defined time period. The pH values were evaluated in this cementitious sample suspension and also in the leachate after centrifugation and filtration of sample suspension.

Table 2: *pH values of LPC samples at different conditions (170 days after mixing).*

LPC sample	pH in suspension	pH in leachate
Curing 10°C	12.4	12.4
Josef GW 10°C	12.3	12.3
Josef GW 95°C	11.1	11.0

Evaluation of in-situ experiments in the Underground laboratory Josef

Bentonite and cement samples were extracted from the cartridges and subsequently analysed. Hydraulic conductivity and swelling pressure tests were performed on samples cut from the extracted "cores" (see Figure 1, right). Dry densities of the cores were usually in range of 1,350 - 1,500 kg/m^3, i.e., roughly about 250 kg/m^3 lower than initial ones (1,600 - 1,700 kg/m^3), because cartridges did not protect bentonite against erosion through cartridge perforation.

Geochemistry

The main goal was to interconnect the geochemical and geotechnical changes of bentonite and cement material during the long-term interaction. The macroscopic mineralogical changes were not visible by X-ray diffraction, only one new phase was detected at cement / bentonite boundary - aragonite (after 72 months of interaction). Different situation was found in exchangeable cation occupancy (Figure 3). Even though the cation exchange capacity varied (68.8 ± 13.9 meq/100g) among the bentonite samples from the time series, the clear trend in exchangeable cation occupancy was visible. The dominant driving force for cation exchange was the composition of groundwater, which is Ca-Mg-SO_4-HCO_3 type (Večerník et al., 2016), and only minor effect was observed from alkaline front originated from cement paste. Originally, Na-Mg bentonite exchanged to Mg-Ca bentonite in the whole bentonite volume in all cartridges (P1 and P3). The simple equilibrium model of cation exchange in PHREEQC shows the same trend. Simulation of time evolution was done by four steps re-equilibrium with cation exchange with a new groundwater and stored exchange occupancy from previous step. The changes in clay exchangeable complex had probably an impact on geotechnical parameters.

Hydraulic conductivity

Figure 4 shows all results of hydraulic conductivity values in dependence on bentonite dry density for cartridges with B75 and B75 + CEM; with the same sampling time periods as used for geochemistry tests. A black line is an exponential regression of not-loaded bentonite material. It is clear that hydraulic conductivity decreased for all samples extracted from the cartridges.

Better understanding provides Figure 5, which shows time evolution of hydraulic conductivity as percentage of initial values. Each point on the graph represents average value of three particular results in each sampling campaign. Results from both types of cartridges (B75 and B75 + CEM) are plotted.

For the first sampling (after 9 months) the values of hydraulic conductivity decreases significantly, down to 50%. Then it increases back to 80 - 90% of initial value and remains stable already after 26 months. Based on change of exchangeable complex from Na-Mg to Mg-Ca inverse general trend was expected, where hydraulic conductivity should increase (e.g., Pusch, 2006) over the values for not loaded material. It should be noted that clearly increasing trend is visible after 9 months of interaction where minimal value can be found. Therefore, it is assumed that initial decrease of hydraulic conductivity can be caused by other effects than bentonite / water interaction. Unfortunately, there are no geochemical results available from 9 months sampling campaign to support or reject such idea.

It can be assumed that there are no differences in time evolution between results from two types of cartridges (bentonite / bentonite + cement). But it must be taken in account that length of the sample (20 mm) is much larger than thickness of bentonite visibly affected by cement - bentonite interaction (less than 1 mm).

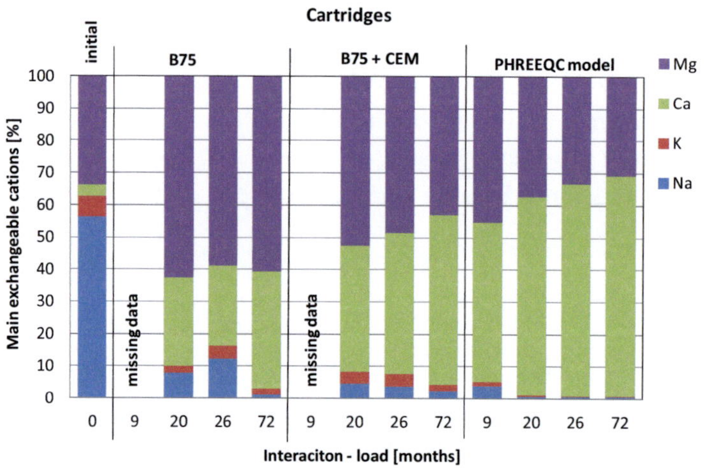

Figure 3: *The time change of main exchangeable cations occupancy for cartridges. B75 - only bentonite, B75 + CEM - sandwich of cement paste and bentonite, PHREEQC model - equilibrium model of cation exchange.*

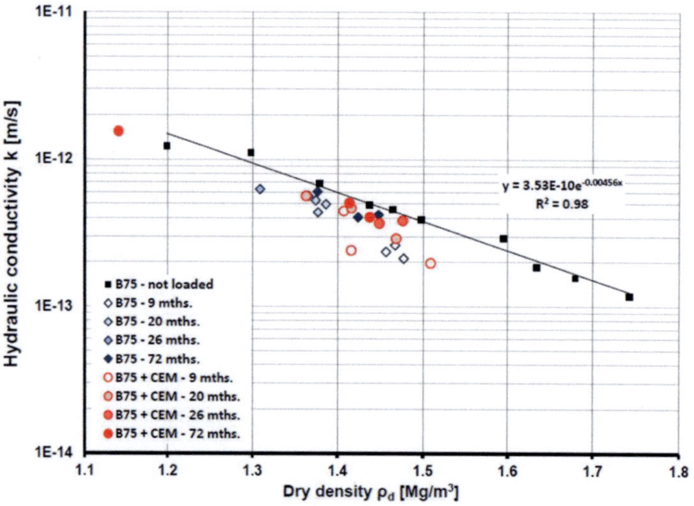

Figure 4: *Comparison of hydraulic conductivity of B75: Original, not loaded material (in black); in contact with cement (B75 + CEM, red) and pure bentonite (B75, blue).*

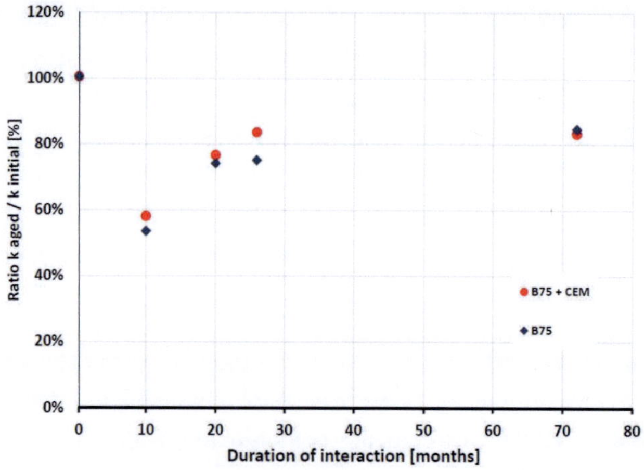

Figure 5: *Time evolution of ratio of hydraulic conductivity measured on aged samples and initial ones (each point represents average from 3 results; aged B75 in blue and B75 + CEM in red).*

Swelling pressure

The results presented in Figure 6 show swelling pressure values in dependence on bentonite dry density for cartridges with B75 and B75 + CEM; with the same sampling time periods as used for geochemistry tests. The black line (Figure 6) is an exponential regression of not-loaded pure B75 bentonite material.

Better understanding provides Figure 7, which shows time evolution of swelling pressure as percentage of initial values. Each point on the graph represents average value of three particular results in each sampling campaign. Results from both types of cartridges (B75 and B75 + CEM) are plotted.

In case of swelling pressure, the drop (Figure 7) is in good accordance with the change of exchangeable complex (Figure 3). Dominantly Mg-Ca occupation in clay exchangeable complex reflects the decrease of swelling pressure. The average change is approximately 20%.

It can be assumed that there are no significant differences between results from two types of cartridges (bentonite / bentonite + cement). Very low ratio of the affected bentonite layer and needed size (volume) of sample can be the reason.

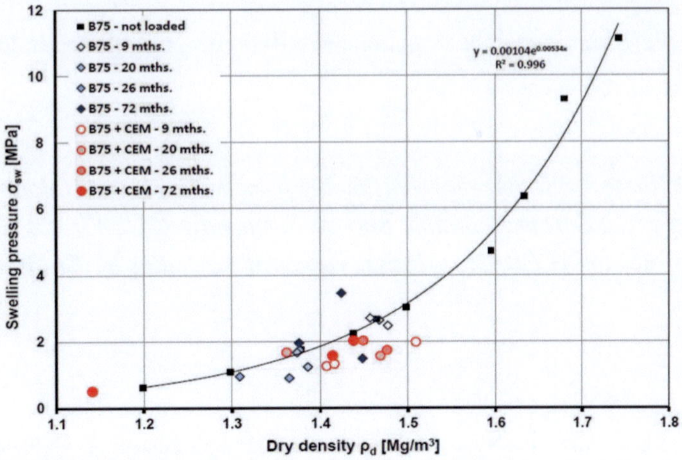

Figure 6: *Comparison of swelling pressure of B75: Initial (not loaded) material (in black); in contact with cement (B75 + CEM, red) and pure bentonite (B75, blue).*

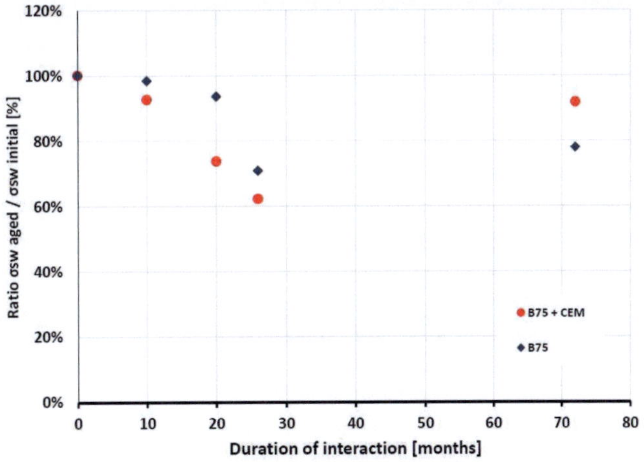

Figure 7: *Time evolution of ratio of swelling pressure measured on aged samples and initial ones (each point represents average from 3 results; aged B75 in blue and B75 + CEM in red).*

Conclusions and Future work

The long-term in-situ interaction experiments (cartridges) confirmed slow rate of mineralogical changes. After 72 months of interaction only minor changes were visible on the cement/bentonite boundary (new formation of aragonite). One remarkable parameter was replacement of Na cation by Ca and Mg cation in exchangeable complex of bentonite clay minerals. In this process, groundwater was dominantly involved, with minor effects of alkaline front from cement material. However, this effect may be associated with a small scale of experiment where a small amount of bentonite and cement material is in contact with relatively large amount of groundwater. Changes in exchangeable complex influenced the geotechnical parameters in measurable way (mainly the swelling pressure and partly the hydraulic conductivity).

Laboratory ageing procedures for both types of cementitious materials are ongoing. Results from first sampling of (9 months aged) OPC are known although not reported here; second sampling of OPC should take place in late August 2017. First sampling of LPC is expected in September 2017. The leachate chemistry inside the ageing vessels (concentration of Ca^{2+}, Mg^{2+}, TDS, and pH) is regularly monitored. It will support interpretation of results of laboratory analyses. At present, collected data are discussed with partner organization modellers (CTU FNSPE) as some of the results are inputs to CEBAMA modelling activities to better understand the long-term phenomena.

The laboratory experiments continue to be ongoing and will be further reported in the future.

Acknowledgement

The research leading to these results has received funding from the European Union's Horizon 2020 Reasearch and Training Programme of the European Atomic Energy Community (EURATOM) (H2020-NFRP-2014/2015) under grant agreement n° 662147 (CEBAMA). It uses research capacities of the large infrastructure RINGEN (MŠMT LM2015084).

References

Alonso, M.C., García Calvo, J.L., Walker, C., Naito, M., Pettersson, S., Puigdomenech, I., Cuñado, M.A., Vuorio, M., Weber, H., Ueda, H., Fujisaki, K. (2012). Development of an accurate pH measurement methodology for the pore fluids of low pH cementitious materials. SKB Report, R-12-02.

ČGÚ (1987). Metodiky laboratorních zkoušek v mechanice zemin a hornin (Methods of laboratory testing in soil and rock mechanics). Český geologický ústav (Czech Geological Institute).

ČSN (2005). ČSN EN 196-1. Metody zkoušení cementu: Část 1: Stanovení pevnosti (Methods of testing cement. Part 1 - Determination of strength) Prague: Czech Office for Standards.

Pusch, R. (2006). Mechanical properties of clays and clay minerals. *In:* Bergaya, F., Theng, B.K.G., Lagaly, G. (editors) Handbook of Clay Science. Elsevier Ltd.

Vašíček, R., Levorová, M., Červinka, R., Hausmannová, L., Kaisr, Z., Venkrbec, Z. (2013). Výzkum vlastností materiálů pro bezpečné ukládání radioaktivních odpadů a vývoj postupů jejich hodnocení. PODETAPA 4.2 Laboratorní výzkum tlumících, výplňových a konstrukčních materiálů. (Research on properties of materials for safe disposal of radioactive waste and development of methods for their assessment. Subtask 4.2 Laboratory research on buffer, backfill and other building materials). Czech project MPO TIP FR-TI1/362 (2009-2013).

Večerník, P., Hausmannová, L., Červinka, R., Vašíček, R., Roll, M., Hloušek, J., Havlová, V. (2016). Interaction between cement and Czech bentonite under temperature load and in *in-situ* conditions: an overview of experimental program. Proceedings of the 1st CEBAMA Annual Workshop. KIT Scientific Publishing, KIT-SR 7734.

Vehmas, T., Schnidler, A., Löija, M., Leivo, M., Holt, E. (2016). Reference mix design and castings for low-pH concrete for nuclear waste repositories. Proceedings of the 1st CEBAMA Annual Workshop. KIT Scientific Publishing, KIT-SR 7734.

Preliminary assessment of interaction between UK backfill cement material and groundwater

Rita G.W. Vasconcelos[1], Andres Idiart[2], Neil C. Hyatt[1], John L. Provis[1], Claire L. Corkhill[1*]

[1] NucleUS Immobilisation Science Laboratory, Department of Materials Science and Engineering, University of Sheffield (UK)
[2] Amphos 21 Consulting S.L. (ES)

* Corresponding author: c.corkhill@sheffield.ac.uk

Abstract

Nirex Reference Vault Backfill (NRVB) is a high pH cement being considered for use as a backfill material for the disposal of low heat generating wastes in a UK geological disposal facility in high strength crystalline rock. It is well understood that, as a consequence of the interaction with the geological environment (e.g., groundwater), the backfill material will evolve and age with time. In this paper, we present the results of preliminary leaching experiments, where NRVB is immersed in three different groundwater compositions representative of granitic, clay and saline groundwater. Some differences were observed after 35 days of leaching. The formation of more ettringite, mainly in the NRVB leached in clay groundwater, and the reduction of the presence of monocarboaluminate hydrate in the NRVB leached in granitic and clay groundwater, was visible. Taking into account the present results, several methodologies are being employed for longer-duration groundwater leaching experiments, to study the long-term mineralogy and geochemistry of this backfill material and to make predictive models.

Introduction

International consensus for the final disposal of radioactive waste is through geological disposal. In the United Kingdom (UK), the projected volume of intermediate level waste for disposal in a geological disposal facility (GDF) is approximately 450,000 m^3 (RWM, 2015). The GDF utilises a multi-barrier concept, where waste will be conditioned (e.g., by encapsulation with a cementitious grout), packaged in steel or concrete containers and placed in engineered vaults excavated hundreds of metres below the ground (NDA, 2010a and 2010b). In a hard rock, crystalline geology conceptual scenario, these vaults will be backfilled with a cementitious material, such as the Nirex Reference Vault Backfill (NRVB) (NDA, 2010b; Francis et al., 1997).

NRVB is a high pH cement composed of Portland cement, limestone flour (calcium carbonate) and hydrated lime (calcium hydroxide). This backfill material was especially designed to fulfil specific requirements, including maintaining a high pH that will minimise the solubility and transport of many radionuclides, providing high sorption capacity and minimising pressure build-up from gas through a high porosity microstructure (Francis et al., 1997).

Due to the required lifetime performance of a cementitious backfill material in a GDF (> 10,000 years), it is well understood that this material will evolve and age with time, as a consequence of interactions with the environment, especially groundwater (NDA, 2010b). When a cementitious backfill material is in contact with groundwater, it is expected that the dissolution of some phases will buffer the pH of the porewater. Figure 1 shows the predicted evolution of the pH of the cement pore solution after contact with pure water, with the consequent four stages of cement dissolution.

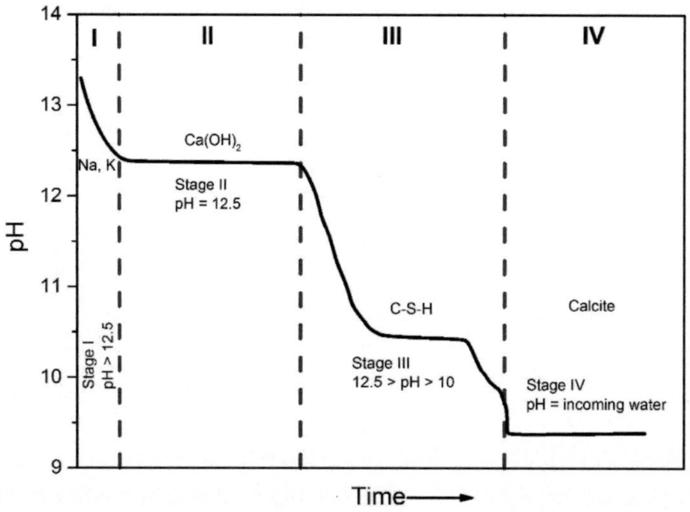

Figure 1: *Schematic representation of the evolution of pH at 25°C in cement pore fluid as a result of pure water leaching (from NDA, 2010b).*

This project aims to develop an understanding of how these interactions affect the mineralogical, geochemical and microstructural properties of NRVB, to support the development of a robust safety case for the disposal of radioactive waste in a GDF. We present here the results from preliminary leaching experiments, where three different groundwater compositions representative of granitic, clay and saline groundwater were used to study the interactions with NRVB.

Methods

Materials

A batch of NRVB paste was prepared using the formulation previously outlined in the literature (Francis et al., 1997): 450 kg/m^3 CEM I 52.5 N (sourced from Hanson Cement Ltd, Ribblesdale works); 170 kg/m^3 hydrated lime (sourced from Tarmac Cement & Lime) and 495 kg/m^3 limestone flour (sourced from National Nuclear Laboratory, NNL), with a water / solid ratio of 0.55. The cement paste was mixed for 5 minutes using Kenwood benchtop mixer and placed in centrifuge tubes at 40°C and 95% relative humidity, for 28 days.

For the leaching experiments, three different groundwater compositions were used. Different chemicals were mixed with distilled water to obtain the compositions shown in Table 1, which were based on literature examples (Gascoyne et al., 2002; Vinsot et al., 2008).

Table 1: *Groundwater composition utilised in this study.*

	Granitic Groundwater (mmol/L) (Gascoyne et al., 2002)	**Saline Groundwater (mmol/L)** (Gascoyne et al., 2002)	**Clay Groundwater (mmol/L)** (Vinsot et al., 2008)
Na	2.8	140	55
K	0.1	2.1	1.1
Ca	0.5	19.9	7.5
Mg	0.2	0.4	5.7
Cl	2.1	172.7	52.5
HCO_3	2.0	2.0	--
SO_4	0.1	4.0	15
pH	8.2	7.7	7

Leaching Experiment Methodology

After 28 days of curing, cylindrical monoliths were prepared, each with 15 mm height x 15 mm diameter. The ends of the cylinders were sealed with epoxy resin to allow only radial diffusion of the groundwater (necessary for future modelling of diffusion). Samples were placed on a Teflon basket, as shown in Figure 2. Duplicates of the samples were used and duplicate blanks (i.e., with no cement sample) were also prepared.

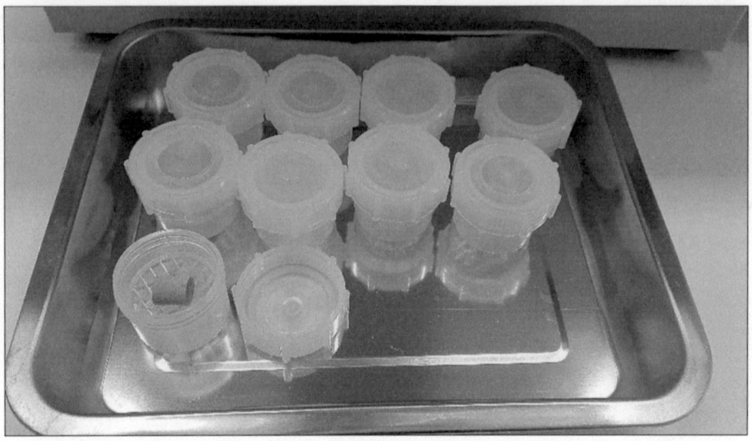

Figure 2: *Leaching experimental set-up showing 60 mL Teflon vessels, containing 50 mL groundwater with a cylindrical cement sample placed on a Teflon basket.*

Static dissolution experiments were conducted in 60 mL vessels in contact with 50 mL of each of the three types of groundwater, in an oven at 40°C (Figure 2). The presence of CO_2 was not controlled during leaching. Sampling was performed at specific intervals until 35 days (1, 3, 7, 14, 21, 28 and 35 days). During those sampling points the pH was measured (at room temperature), leachate was collected for chemical analysis (Inductively Coupled Plasma - Optical Emission Spectroscopy, ICP-OES) and the cement was placed in acetone to stop the hydration for post-leaching characterisation. The sampling was made in a controlled oxygen and nitrogen environment (presence of CO_2 was not controlled). Post-leaching characterisation of the cement paste included analysis of cement mineralogy by X-ray Diffraction (XRD) and cement porosity by Mercury Intrusion Porosimetry (MIP).

Analytical Methods

Analysis of major elements leached into solution was performed using a Thermofisher Scientific iCAP Duo6300 Inductively Coupled Plasma-Optical Emission Spectrometer, operating using a gold internal standard and ceramic torch. Samples were filtered (0.22 μm), diluted and acidified with 20 μL of conc. HNO_3 (99.999% purity) prior to analysis. X-ray Diffraction (XRD) analysis was performed on loose powder (< 63 μm) using a PANalytical X'Pert[3] Powder diffractometer utilising a Cu Kα source. Measurements were taken from 5° to 80° 2θ with a step size of 0.02° and 2 s counting time per step. Small pieces of hardened cement paste were prepared and placed into the sample holder of an Autopore V 9600 Mercury Intrusion Porosimeter (MIP) (Micromeritics Instruments) for total porosity analysis. The maximum pressure applied was 208 MPa, the surface tension was 485 mN/m and the contact angle was 130°.

Results

As expected from reactive transport modelling (Figure 3, courtesy of Amphos 21), an increase in the pH was observed when comparing the blank solutions for each groundwater and the groundwater solutions after contact with NRVB (Figure 4). After 35 days of leaching the pH of the groundwater solutions in contact with cement buffered to between pH 12 and 13, compared with pH values of pH 8.3, pH 7.6 and pH 6.8 for the granitic, saline and clay groundwater blanks, respectively.

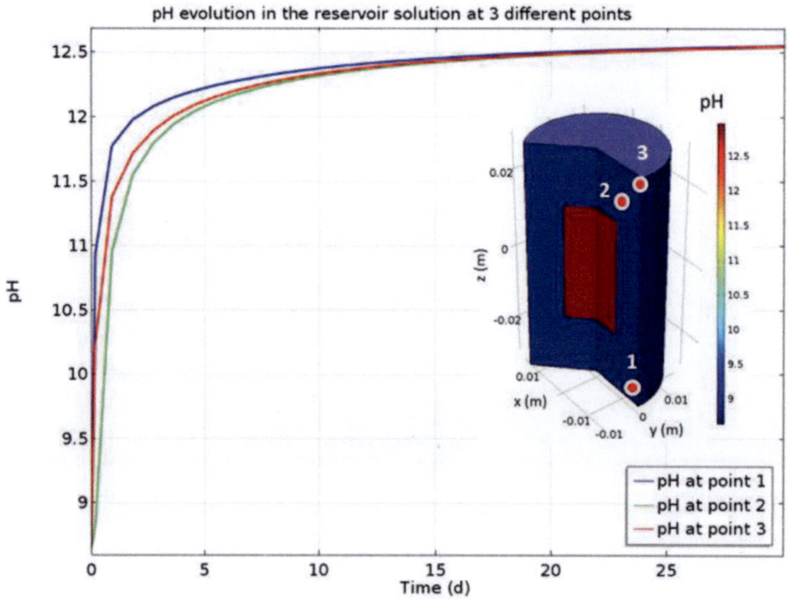

Figure 3: *Results of 2D axisymmetric reactive transport modelling with iCP of pH evolution of a granitic groundwater solution (50 mL) in contact with high-pH hardened cement paste (hcp) at different points, over 28 days. The initial pH distribution in the system is also depicted. The composition of the hcp sample corresponds to an OPC (CEM I), with a porosity of 30% and an effective diffusion coefficient of* $1.0 \cdot 10^{-12}$ *m²/s.*

An increase of the concentration of silicon in solution, relative to the blank samples, was observed (Figure 5). The variation observed in the results may be due to different rates of absorption and release of this element by NRVB. Despite the large scatter in the results, it can be observed that the release of silicon was similar for all groundwater compositions investigated.

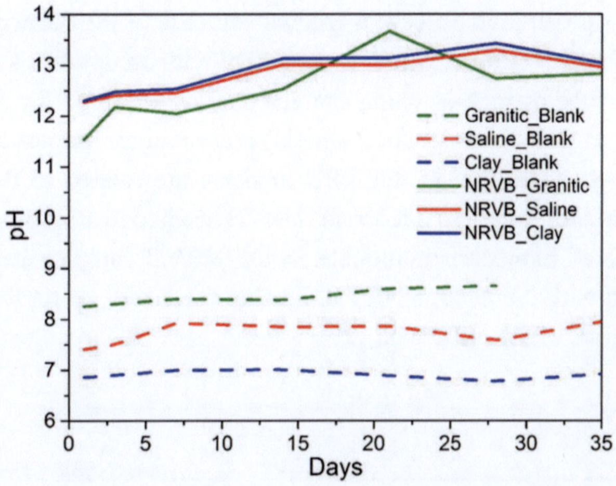

Figure 4: *pH measurements of the granitic, clay and saline groundwater blank solutions, and of the solutions from the NRVB leached in granitic, clay and saline groundwater. The measurements were taken during 35 days.*

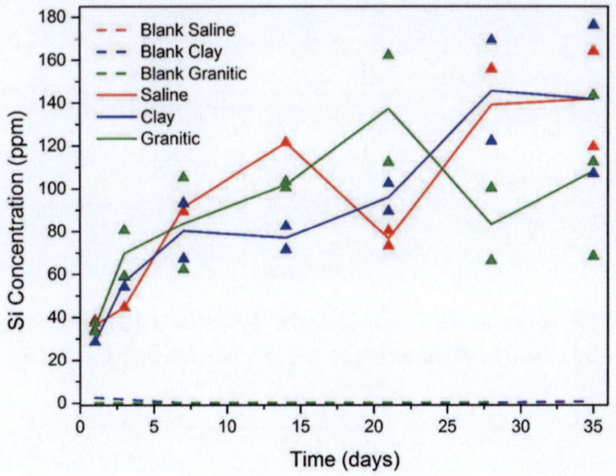

Figure 5: *Si concentration in solution (ppm), using ICP-OES, of the granitic, clay and saline groundwater blank solutions, and of the solution from NRVB leached in granitic, clay and saline groundwater. The measurements were taken during 35 days.*

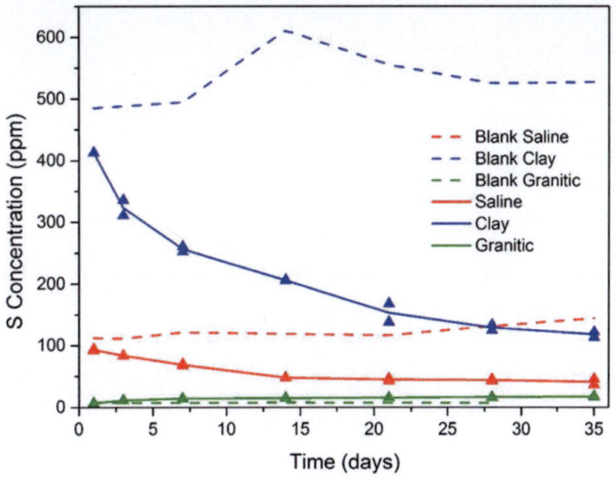

Figure 6: *S concentration in solution (ppm), using ICP-OES, of the granitic, clay and saline groundwater blank solutions, and of the solution from NRVB leached in granitic, clay and saline groundwater. The measurements were taken during 35 days.*

In the case of sulfur, it was possible to observe a gradual decrease of the concentration in solution (mainly in the saline and clay groundwaters) over time, which is consistent with an uptake of sulfur by NRVB, as shown in Figure 6. This uptake is confirmed by post-leaching characterisation of NRVB by XRD (Figure 7), which shows that excess ettringite (compared with a non-leached sample) is formed in samples exposed to the saline and clay groundwaters. Other differences identified in the XRD analysis are related to the intensity of the peaks, for example peaks indexed as portlandite are less intense for NRVB leached in all groundwater, whereas it is possible to observe the disappearance of monocarboaluminate in the NRVB samples leached with granitic and clay groundwater, when compared with the control NRVB. Further work will be performed in the quantification of these phases by Rietveld analysis of the XRD data.

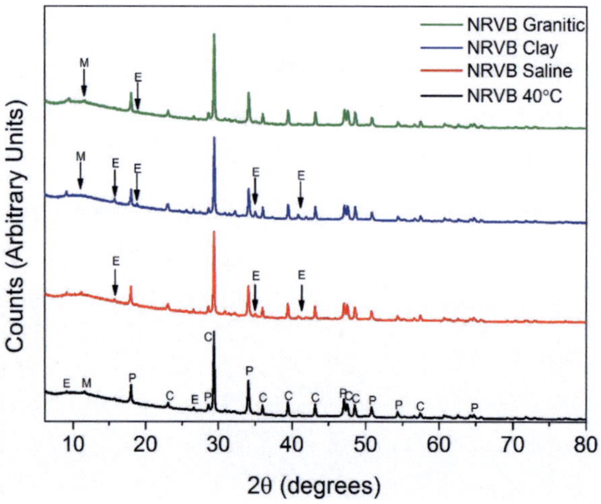

Figure 7: *XRD patterns of NRVB cured at 40°C for 28 days, NRVB leached for 35 days in granitic, clay and saline groundwater. E – ettringite; M – monocarboaluminate hydrate; P – portlandite; C – calcite.*

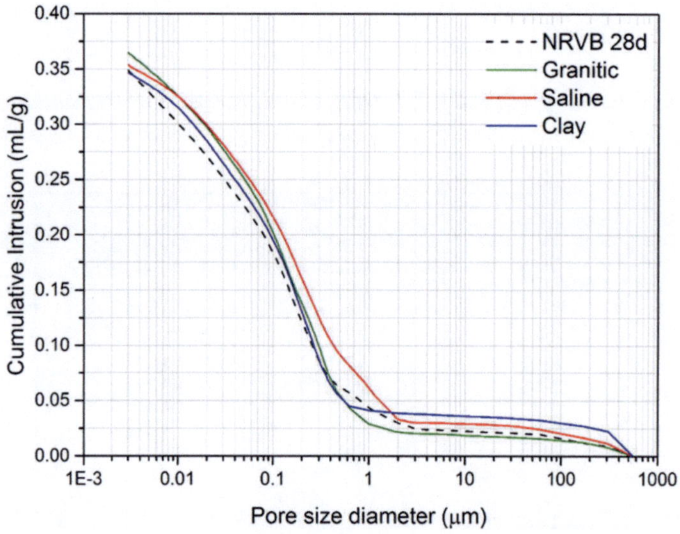

Figure 8: *Pore entry size distribution of NRVB cured after 28 days and NRVB leached for 35 days in granitic, clay and saline groundwater, using MIP.*

MIP was performed to study the difference in the porosity between the different samples. It is important to note that MIP measures the pore entry size of the cement and not the pores themselves, thus the technique is here being used as a comparison between pre- and post-leaching. The porosity obtained for NRVB cured for 28 days, before leaching, was 45 ± 1%. Similar results were obtained for the post-leaching samples, where for the NRVB leached with granitic, clay and saline groundwater the porosity obtained was 46, 45 and 42 ± 1%, respectively. Figure 8 shows the pore entry size distribution for all NRVB samples. The curves show a very similar pore entry size distribution between the samples leached with clay and granitic groundwater in comparison with control NRVB. Due to the low compressive strength of NRVB (ca., 8 MPa), this technique is not the most adequate to use to measure the pore size distribution of this material. The high pressures applied by MIP, lead to the collapse of the sample at an early stage. For this reason, future work will include the use of multiple porosity techniques, including X-ray computed tomography, to overcome this problem.

Conclusions and Future work

With these preliminary leaching experiments, it was possible to observe that some minor changes had occurred after 35 days of leaching. This included the dissolution and absorption of different elements, with consequent reduction of the presence of some phases (for example, monocarboaluminate hydrate) and increase of presence of other phases (for example, ettringite), as observed in ICP-OES and XRD results.

As future work, three different leaching experiments are proposed. Two are long-duration experiments (1 to 1.5 years), where controlled nitrogen or CO_2 environments will be applied. For one of the experiments, the cement will be placed in a 60 mL vessel and replacement of the groundwater will take place at specific time intervals. For the second experiment, the groundwater will not be replaced, and the cement samples will be placed in a 2 L vessel and only analysed at the end of the experiment. The third experiment will be a short-duration experiment, where the groundwater will flow through the cement. This experiment will be performed in collaboration with Amphos 21 and FZ Juelich. All the experiments will be performed at 40°C, as advised by the UK waste management organisation, Radioactive Waste Management (RWM) to be representative of the peak of a 50 year thermal transient after backfilling of the disposal area (NDA, 2010b). In all scenarios, cement samples will be analysed by XRD and porosimetry techniques. Selected samples will be chosen for XCT analysis and NMR. Solutions will be analysed by ICP-OES (total element concentrations), and ion chromatography (anion concentrations).

Acknowledgements

The authors wish to acknowledge funding for this research from Radioactive Waste Management, and funding from the European Union's Horizon 2020 Reasearch and Training Programme of the European Atomic Energy Community (EURATOM) (H2020-NFRP-2014/2015) under grant agreement n° 662147 (CEBAMA).

The authors would also like to acknowledge Dr. Andrea Hamilton and Dr. Nicolas Beaudoin, from University of Strathclyde, for the ongoing assistance with X-ray computed tomography (XCT) measurements.

References

Francis, A.J., Cather, R., Crossland, I.G. (1997). Nirex Safety Assessment Research Programme: Development of the Nirex Reference Vault Backfill. Report on Current Status in 1994. Nirex Report, S/97/014.

Vinsot, A., Mettler, S., Wechner, S. (2008). In situ characterization of the Callovo-Oxfordian pore water composition. Phys. Chem. Earth, 33, SUPPL. 1.

Gascoyne, M. (2002). Influence of grout and cement on groundwater composition. POSIVA Working Report, 2002-07.

Nuclear Decommissioning Authority (2010a). Geological Disposal: Generic Disposal Facility Designs. NDA Report, NDA/RWMD/048.

Nuclear Decommissioning Authority (2010b). Geological Disposal: Near-Field Evolution Status Report. NDA Report, NDA/RWMD/033.

Radioactive Waste Management (2015). Geological Disposal: The 2013 Derived Inventory. NDA Report, NDA/RWM/120.

Comparison of experimental and modelled pore solutions of low-pH ordinary portland cement based mix designs

Tapio Vehmas[1*], Markku Leivo[1], Erika Holt[1]

[1] VTT, Technical Research Centre of Finland Ltd. (FI)

* Corresponding author: tapio.vehmas@vtt.fi

Abstract

Posiva Oy is constructing one of the world's first long-term nuclear waste repositories in Finland. The safety of the nuclear waste repository is ensured with a combination of natural and engineered barriers. High-pH leachates originating from cementitious materials dissolution can potentially alter the engineered barrier systems performance. Prediction of the leachates pH and composition is important as the performance of engineered barrier system should not alter in the presence of cementitious materials. Geochemical modelling is a potential method to evaluate the pH and composition of the leachates during the lifetime of the repository. In the current study, pH, sodium, magnesium, aluminium, silicon, sulphur, calcium, iron and potassium concentrations were analysed from leachates originating cement pastes having various calcium/silica-ratios. The studied leachate mediums were ion exchanged water, saline groundwater and saline bentonite porewater. Saline groundwater and saline bentonite porewater were selected to present the potential environment in nuclear waste repository. Experimental results were compared to geochemical modelling results and the sources of deviations were investigated. According to results, geochemical modelling is a usable method to evaluate the solution compositions if the results are treated cautiously. The differences between measured and modelled results originated from kinetic factors, inaccurate ionic activities, precipitation of secondary reaction products and missing absorption mechanism. A critical parameter related to the cementitious materials pH was related to the model that is used for calcium-silicate-hydrate phases.

Introduction

Posiva Oy is constructing one of the world's first long-term nuclear waste repositories in Finland. The safety of the nuclear waste repository is ensured with a combination of natural and engineered barriers. Final nuclear waste disposal will take place in deep underground repository in Olkiluoto bedrock at a depth of 400 - 450 meters underground. The natural barrier consists of the surrounding Olkiluoto bedrock (migmatite mica gneiss, tonalite, granodiorite and coarse granite veins) and its inherent isolating properties (Lindberg, 2001). The engineered barrier system consists of water- and gas-tight sealed copper canisters with a cast iron insert and bentonite-based buffer and backfill. Concrete plugs are used for closure and sealing of the repository and the deposition tunnels (Palomäki et al., 2013; Keto et al., 2010). Concrete plugs are utilized to assure mechanical and hydrological isolation of the compartments of the repository. Isolation is needed because compartments have different states in terms of water saturation and pressures. Such difference could potentially induce undesired mass redistributions. Besides concrete plugs, cementitious materials are used in shotcrete for tunnel wall rock supports, rock-bolting grouts and injection grouts for fissure sealing. Only some of the cementitious materials will remain in the repository after final closure. Shotcrete is removed before the final closure with an assumed efficiency of 95% (Koskinen, 2013).

The amount of cementitious materials in the repository will be significant although some of the materials are removed before final closure. Approximately half of the cementitious materials that will remain in the closed repository are located in the access tunnels and consists of grouts, shotcrete remnants and concrete constructions. In deposition tunnels, cementitious materials are present in grouted fractures, rock bolt mortars and deposition tunnel end plugs. Cementitious materials will be in direct contact with backfill and closure materials. It is assumed that cementitious leachates will interact with backfill and closure materials but not to the extent that compromises the long-term safety of the barrier system. The bentonite buffer is not in direct contact with cementitious materials. However, the buffer may be affected due the migration of cementitious leachates through the bedrock fracture network. To meet the designed performance requirements during the lifecycle of the repository, the properties of the engineered barrier system should not be adversely altered due the presence of cementitious materials (Koskinen, 2013).

Cementitious materials high pH leachate is recognised as the main mechanisms that can potentially alter the engineered barrier systems performance. The pH of the cementitious materials leachate depends on the material composition and the surrounding environment. Various cementitious mix designs have been developed to minimize the pH of the leachates (Coumes et al., 2006; Vogt et al., 2009; Martino et al., 2011; Holt et al., 2014). The low-pH mix designs consist of low-alkaline Portland cements mixed with large amounts of pozzolanic materials. Pozzolanic materials are known to react with the hydration products that control the pH of the cementitious materials pore solution. Low-pH materials are still highly basic but the pH value is significantly lower than in traditional Portland cements.

The characteristics of the surrounding environment also affect the cementitious materials leachate pH. In a deep underground repository, the surrounding environment of the cementitious materials consists of groundwater and the engineered barrier system. In the engineered barrier system, clay based backfill- and buffer- materials are soluble components that will affect the surrounding environment. The pH of the cementitious materials leachates can be measured in various environments experimentally, providing exact information on a certain environment. As the environment alters throughout the groundwater composition and saturation degree of the clay-based materials, the experimental determination of the leachate pH becomes highly laborious and time consuming.

Another approach is to use geochemical modelling to provide information of the leachate composition and pH in the deep underground repository. The benefit of the modelling is that a large number of potential scenarios can be attained relatively easily. Geochemical modelling can be also integrated to the reactive transport modelling which enables the determination of the cementitious materials leachates in the various parts of the deep underground repository. A disadvantage of the modelling is the uncertainty of the results. Modelling results can be only as accurate as the used parameters. A modelling approach can lead to false assumptions on the leachate pH and composition if the used parameters do not describe the modelled conditions sufficiently.

The goal of the current paper is to compare experimental and modelled compositions of the cementitious leachates in various deep underground repository environments. On the basis of the comparison, accuracy of the modelling method is evaluated, potential sources of errors are identified and their effect on modelling applicability is discussed.

Surrounding environment of the cementitious materials in deep underground repository

Cementitious materials are in direct contact with the groundwater and potentially influenced with clay based backfill- and buffer materials. Cementitious materials in the repository are influenced with clay-based materials with varying degrees. Limiting scenarios are pristine groundwater and bentonite saturated groundwater. Cementitious materials behaviour was studied in both of these two limiting scenarios. Groundwater compositions in the Olkiluoto region are well identified and reported. In the current study, groundwater composition at the depth of final disposal site was selected to present the groundwater composition. Saline reference groundwater in anoxic conditions was selected from the literature (Hellä et al., 2014). Chemical composition of the selected groundwater is presented at the Table 1.

Defining the groundwater that is influenced by the clay-based materials of the engineered barrier system is more difficult task. Experimental data of groundwater composition that is influenced by the clay-based materials was not available. At least three models of saline groundwater compositions, saturated with bentonite exist. The latest of these models was selected to present the clay-saturated saline groundwater (Hellä et al., 2014). Chemical composition of the bentonite saturated saline groundwater is presented in Table 1.

Cementitious materials pore solutions were also studied in de-carbonated ion-exchanged water. Ion exchanged water presents the pure pore solution of cementitious materials. On the basis of three selected waters, various groundwater scenarios in deep underground repository can be deduced.

Table 1: *Elemental compositions of the solutions recipes. Bicarbonate is neglected due the de-carbonation of the used solutions.*

Element	Ion exchanged water (mmol/L)	Saline groundwater (mmol/L)	Saline bentonite porewater (mmol/L)
Na^+	0.0	208.80	499.00
Cl^-	0.0	412.91	340.10
K^+	0.0	0.54	2.55
Ca^{2+}	0.0	99.78	11.05
Mg^{2+}	0.0	2.30	10.49
Sr^{2+}	0.0	0.40	0.40
SO_4^{2-}	0.0	0.04	102.01
Br^-	0.0	1.31	1.31

Materials and Methods

Ordinary Portland Cement paste samples were manufactured from Anläggningscement (CEM I, Cementa Ab) and nanosized silica (Levasil 100/45 AzkoNobel) according to Table 2. Water/binder-ratio of the samples was 1. Paste samples were manually mixed for three minutes and the samples were stored in closed plastic vials for over 100 days. After a storage period, samples were finely ground (Herzog grinder) and the pH of the samples were measured according to method described in the reference (Alonso et al., 2012). pH was determined in three solutions, ion exchanged water, saline groundwater and saline bentonite pore water. Elemental composition of the solutions is presented in the Table 1. Solutions were prepared from ion exchanged water that was de-carbonated using ELE de-airing equipment. Chemicals used for solutions were purchased from Sigma Aldrich having purity per analysis.

Table 2: *Mix design of the studied pastes.*

CaO/SiO_2 (ratio)	m(OPC) (g)	$m(SiO_2)$ (g)
3.14	20.00	0.00
2.60	19.11	0.88
2.00	17.76	2.23
1.80	17.17	2.82
1.60	16.49	3.50
1.40	15.70	4.30
1.20	14.74	5.25
1.00	13.59	6.40
0.80	12.16	7.83
0.60	10.35	9.64
0.40	7.97	12.02
0.20	4.72	15.27

The liquid phase of the samples was further centrifuged (Universal centrifuge, 2,000 rpm, 10 minutes) and decanted. Ion concentrations of the liquid phase were determined with high resolution ICP-MS (Element 2™ from ThermoScientific)

Identical compositions of the paste samples were modelled by thermodynamic modelling. Gibbs energy minimization simulation was performed with GEMS-Selektor and CEMDATA14 database (Kulik et al., 2013; Wagner et al., 2012; *http://gems.web.psi.ch*). In the modelling, the ion activity coefficient was calculated according to Davies equation assuming thermodynamic equilibrium.

Results and discussion

Hydrated cement paste pH was measured in three solution described in Table 1. Results are presented in Figure 1. Comparison of the measured and modelled pH is presented in Figure 2. In ion exchanged water modelled and measured results are very close to each other at the pH values higher than 11.5. In pH values below 11.5 modelling predicted higher values than the measured values. In saline groundwater, measured and modelled results were comparable near pH 12.5. Below pH 12.5, modelled values were constantly lower than the measured pH values. In saline bentonite porewater, modelled pH values were higher than the measured values at pH values above 11.5. Below pH 11.5, modelled values were lower than the measured.

The differences between measured and modelled values can originate from various sources. Incomplete reaction degree of used colloidal silica would induce higher measured pH values in low pH region. Another potential source of error is the used model for calcium-silicate-hydrate. The used model assumes that the minimum CaO/SiO_2-ratio for calcium-silicate-hydrate is above 0.6, inducing constant pH values below CaO/SiO_2-ratio 0.6. In saline groundwater, the modelled pH values were constantly lower than the measured values, except in the very high or low CaO/SiO_2-ratios. Such difference could originate from inaccuracy ion activity coefficients or difference in kinetics due the salinity of the groundwater.

In saline bentonite porewater, pH values were higher than in ion exchanged water in both modelled and measured values. According to the chemical composition of saline bentonite porewater, the solution is saturated in respect to calcium sulphate. A minor increase of calcium would cause a precipitation of calcium and sulphate containing phases, increasing the solubility of calcium phases in cementitious matrix. As a consequence, the pH was higher than in ion exchanged water.

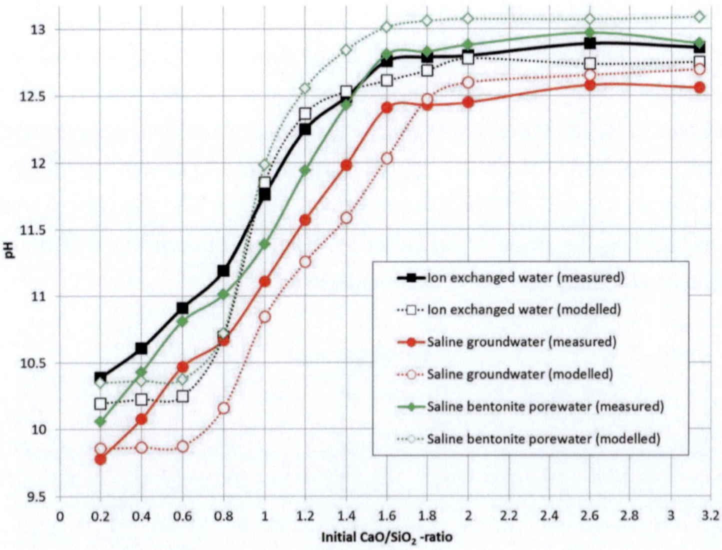

Figure 1: *Measured and modelled pH values of the studied samples.*

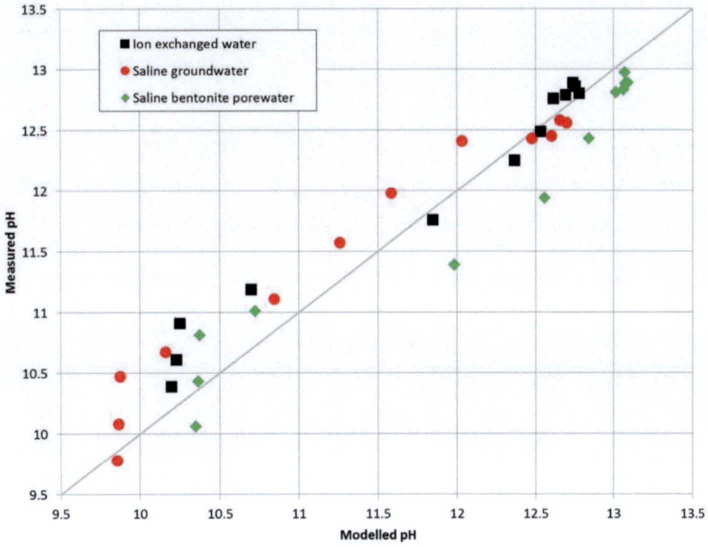

Figure 2: *Comparison of the measured and modelled pH values.*

Higher pH was observed in the measured and modelled samples. Modelled pH values were higher than the measured values. This difference can originate from two factors. First, the measured samples were not in thermodynamic equilibrium respect to precipitating phases. Secondly, the ion activities might not be accurate in the studied environment, causing elevated precipitation of calcium containing phases in modelling respect to the measured samples.

In low CaO/SiO_2-ratio samples, the measured pH values were lower than in ion exchanged water, indicating that the common ion effect dominated over the precipitation of secondary phases. In modelled samples the effect was opposite. This again can be explained by the above described two factors.

Concentrations of sodium, magnesium, aluminium, silicon, sulphur, calcium, iron and potassium were measured in the leachates from the samples. These values were compared to the modelled values. The results are presented in Figures 3 to Figure 5. In ion exchanged water, measured aluminium-, sulphur-, calcium- and sodium concentrations were relatively close to the modelled. Measured values of iron, magnesium and silicon were systematically higher than the modelled values. Largest difference was observed with iron. Precipitation of iron containing phases is probably a very slow process and the kinetic effect was not included in the modelling approach. Measured potassium values were lower than the modelled values. Lower measured potassium values probably originated from potassium absorption to calcium-silicate-hydrates which is a known phenomenon (Stade, 1989; Hong and Glasser, 1999; Lothenbach and Nonat, 2015). The modelling method did not include the absorption mechanism, explaining the observed difference. Sodium absorption was not observed in the samples and the modelled and measured values had good correspondence.

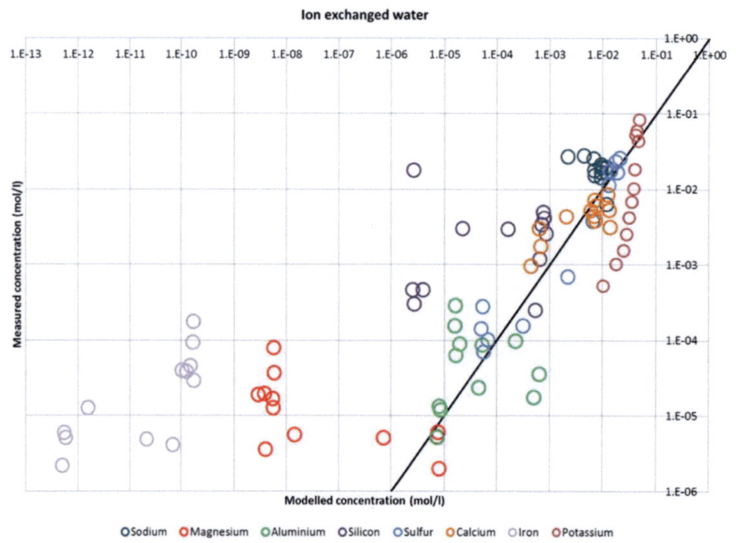

Figure 3: *Comparison of the measured and modelled concentrations in ion exchanged water.*

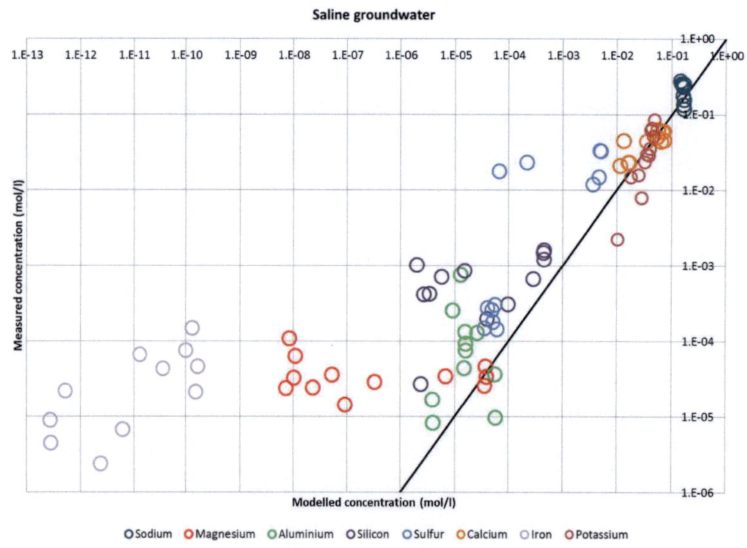

Figure 4: *Comparison of the measured and modelled concentrations in saline groundwater.*

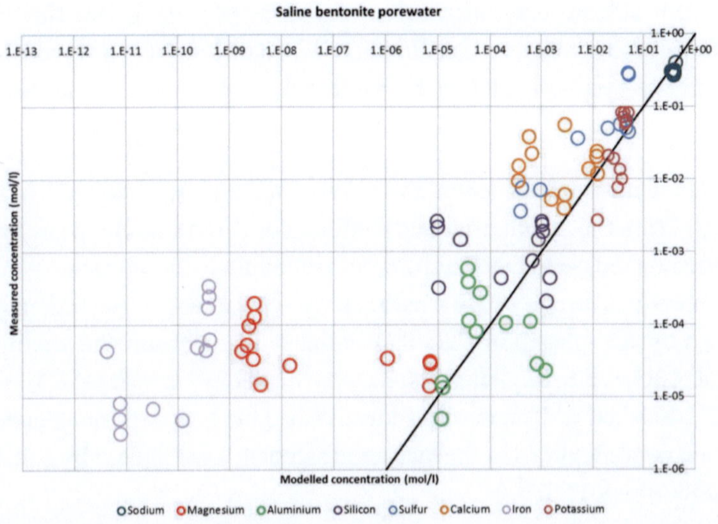

Figure 5: *Comparison of the measured and modelled concentrations in saline bentonite porewater.*

In saline groundwater, the ion concentrations had similar trends as in ion exchanged water. Measured iron concentrations were significantly higher than the modelled values. Measured potassium values were also systematically lower than the measured values. In general, the measured ion concentrations were slightly higher than the modelled values. In saline groundwater, the ionic strength was higher than in ion exchanged water. The possible explanation of the higher measured concentrations is that the ion activities were not as suitable in saline groundwater as they were in ion exchanged water.

In saline bentonite porewater, measured calcium and sulphur values were higher than the modelled values. The measured solutions were either supersaturated respect to calcium and sulphate or the selected activity model was not suitable to describe the ionic activities of calcium and sulphate. Measured iron concentrations were higher than the modelled values. Potassium absorption was observed in part of the samples.

Error sources

Error sources are classically separated into random errors and systematic errors. Random errors happen from time to time, causing deviation of the results. Systematic error is a constant error that affects each of the studied samples. In the current study, random errors cause the scattering of the data mainly due to the errors in experimental samples. Carbonation of the experimental samples is probably the most likely systematic error that could affect the experimental samples. The carbonation of the samples was minimized by using de-carbonated solutions and storing the samples in closed sample vials. Also the time between the grinding and the measurements was minimized. On these bases, it is assumed that the systematic errors originated solely from the modelling and the scattering of the results originates from the random errors in experimental samples.

In ion concentration measurements, iron concentrations of the measured samples were decades higher than the modelled concentrations. This difference was systematically observed in each of the three studied water compositions. In the modelled samples, most of the iron was precipitated as goethite and tri-calcium aluminium ferrite. The precipitation of iron containing phases is probably a very slow process that caused a systematic error between the measured and modelled samples.

Measured silicon concentrations were also higher than the modelled concentrations in high CaO/SiO_2-ratios (Figure 6). The measured silicon concentrations were approximately 0.003 mol/L, throughout the studied

CaO/SiO_2-range. Modelled silicon concentrations were $2.5 \cdot 10^{-6}$ mol/L in CaO/SiO_2 -ratios > 1.6. In CaO/SiO_2-ratios < 1.6, the modelled silicon concentrations were 0.001 mol/L. Observed difference could originate from the used calcium-silicate-hydrate model or the measured samples were not in thermodynamic equilibrium respect to silicon.

Measured and modelled aluminium concentrations situated quite close to theoretical line but the scattering of the results were high. The scattering could originate from random errors in the measured samples or systematic errors in modelling. Modelling indicated that the maximum aluminium concentration is observed near CaO/SiO_2-ratio 1 (Figure 6). In the measured samples such maximum was not observed. According to modelling, aluminium concentration is restricted by Aft -phases in CaO/SiO_2-ratios > 1.2, gibbsite and stratlingite in 1.2 > CaO/SiO_2 > 0.6 and kaolinite in CaO/SiO_2 < 0.6. Modelled results were higher in 1.2 > CaO/SiO_2 > 0.6 -region, and smaller in CaO/SiO_2-ratios >1.2. Modelled and the measured results had the best correspondence at the CaO/SiO_2 < 0.6. It might be that aluminium concentrations in the measured samples were limited by a phase that is not accounted in the modelling in CaO/SiO_2-ratios > 0.6.

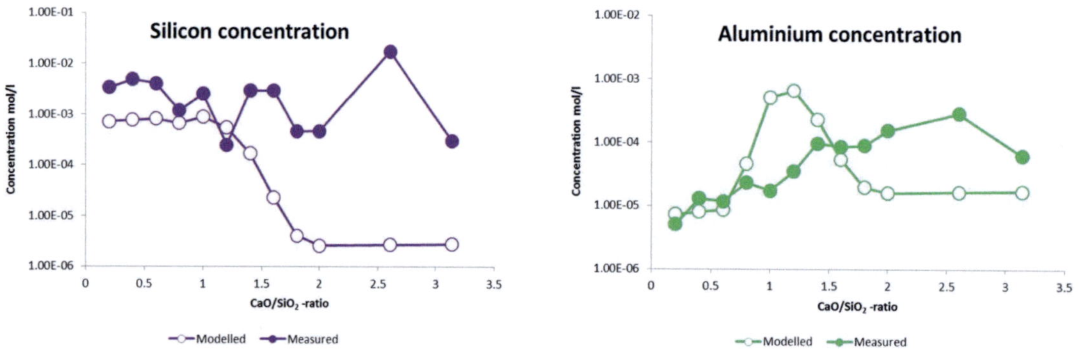

***Figure 6:** Measured and modelled silicon and aluminium concentrations.*

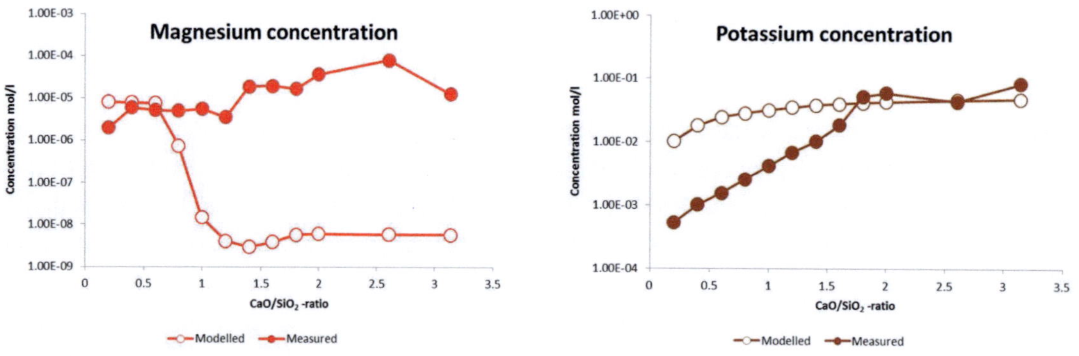

***Figure 7:** Measured and modelled magnesium and potassium concentrations.*

Magnesium concentrations (Figure 7) were related to the aluminium and silicon concentrations present in the system. In the modelling, magnesium was mainly bound to hydrotalcite ($Mg_4Al_2O_7$) in high CaO/SiO_2-ratios. In low CaO/SiO_2 ratios, aluminium became bound to kaolinite ($Al_2Si_2(OH)_4$), causing magnesium concentrations increase in solution. According to measured magnesium concentrations, hydrotalcite is not likely present in the experimental samples, as the magnesium concentrations were relatively constant in throughout the measured samples. Magnesium-silicate-hydrates (M-S-H) were not included in the modelling. Including M-S-H phases in the modelling could potentially improve the modelling accuracy related to the magnesium containing phases.

Measured potassium concentrations were lower than the modelled values (Figure 7). Potassium is known to absorb to calcium-silicate-hydrate that was not accounted in the modelling. Absorption process is able to explain the differences between the measurements and modelling.

In saline groundwater, measured concentrations were slightly higher than the modelled values. The difference could originate from the inaccuracy of the used ion activity model. Otherwise the trends were similar as in ion exchanged water.

In saline bentonite porewater, measured and modelled calcium concentrations diverged more than in ion exchanged water and saline groundwater. Measured calcium concentrations were higher than the modelled values. The difference was related to the precipitation sulphate containing phases. In the modelling, precipitation of sulphate containing phases increased the dissolution of calcium containing cementitious phases. In modelling, the increased dissolution of calcium containing phases caused significantly elevated pH values respect to ion exchanged water. Although the increase in pH was observed also in measured samples, the effect was smaller than in modelled samples. The difference could originate from kinetics of sulphate containing phase precipitation, or modelling incapability to estimate the activities of calcium and sulphate ions in solution.

Conclusions and Future work

Thermodynamic modelling predicted fairly well the experimentally measured pH values and ion concentrations in various solution compositions. The best agreement between the modelled and measured pH values were in ion exchanged water. The largest difference in pH-values were observed in low CaO/SiO_2-ratios. Modelling predicted constant pH values below CaO/SiO_2-ratios below 0.6. Measured values were not constant at the CaO/SiO_2-ratios below 0.6. The difference could originate from incomplete reaction degree of the colloidal silica. Another explanation is the formation of calcium-silicate-hydrates, having CaO/SiO_2-ratio lower than 0.6. In the future, reaction degree of the colloidal silica will be measured to determine the source of the deviation. In saline groundwater, measured pH values were higher than the modelled. As the pH is controlled by the common ion effect in saline groundwater, the difference could originate from the selected ion activity model. This subject will be evaluated in the future by comparing manually calculated common ion effect to modelled common ion effect. In saline bentonite porewater, modelled pH values were higher than the measured. The difference was related to the precipitation of sulphate containing phases.

Modelled and measured concentrations of sodium, magnesium, aluminium, silicon, sulphur, calcium, iron and potassium were compared. Best correspondence between measured and modelled concentrations was observed with ions that are highly soluble in the studied conditions. The correspondence impaired as the solubility of the ions became depended on the precipitation of secondary phases. The difference could originate from slow kinetics of the experimental samples and/or inadequate model of ion activities.

Absorption process was missing in the used geochemical modelling which caused systematic error to the modelled and measured potassium concentrations. However, the missing potassium absorption did not cause a drastic difference to the modelled and measured pH values. The critical parameter related to the cementitious materials pH seems to be the model that is used for calcium-silicate-hydrate.

According to results, geochemical modelling is a usable method to evaluate the solution compositions if the results are treated cautiously. Especially caution is needed when the solution composition is determined by secondary precipitates that do not naturally exist in hydrated Portland cement and the results are evaluated in the short time span. In infinitely long time span, thermodynamic modelling presents a potential scenario of various precipitates and solution compositions. However, infinitely long time span might be too long even respect to the lifetime of nuclear waste repository.

Acknowledgement

The research leading to these results has received funding from the European Union's Horizon 2020 Reasearch and Training Programme of the European Atomic Energy Community (EURATOM) (H2020-NFRP-2014/2015) under grant agreement n° 662147 (CEBAMA).

References

Alonso, M., Garcia Calvo, J., Walker, C., Naito, M., Pettersson, S., Puigdomenech, I., Cunado, M., Vuorio, M., Weber, H., Ueda, H., Fujisaki, K. (2012). Development of an accurate pH measurement methodology for the pore fluids of low pH cementitious materials. SKB Report, R-12-02.

Cau Dit Coumes, C., Courtois, S., Nectoux, D., Leclercq, S., Bourbon, X. (2006). Formulating a low-alkalinity, high-resistance and low-heat concrete for radioactive waste repositories. Cem. Concr. Res., 36(12), 2152-2163.

Hellä, P., Pitkänen, P., Löfman, J., Partamies, S., Vuorinen, U., Wersin, P. (2014). Safety Case for the Disposal of Spent Nuclear Fuel at Olkiluoto - Definition of Reference and Bounding Groundwaters, Buffer and Backfill Porewaters. Posiva Report, 2014-04.

Holt, E., Leivo, M., Vehmas, T. (2014). Low-pH concrete developed for tunnel end plugs used in nuclear waste containment. Concrete Innovation Conference (CIC2014).

Hong, S.-Y. and Glasser, F. (1999). Alkali binding in cement pastes: Part I. The C-S-H phase. Cem. Con. Res., 29(12), 1893-1903.

Kulik, D.A., Wagner, T., Dmytrieva, S.V., Kosakowski, G., Hingerl, F.F., Chudnenko, K.V., Berner, U. (2013). GEM-Selektor geochemical modeling package: revised algorithm and GEMS3K numerical kernel for coupled simulation codes. Comput. Geosci., 17, 1-24.

Keto, P., Dixon, D., Jonsson, E., Börgesson, L, Hansen, J., Gunnarsson, D. (2010). Assessment of backfill design for KBS-3V Repository. Posiva Working Report, 09-115.

Koskinen, K. (2014). Effects of Cementitious Leachates on the EBS. Posiva Working Report, 2013-4.

Lindberg, A. (2001). Biotiitin pysyvyys Olkiluodon loppusijoitusolosuhteissa -kirjallisuusselvitys. Posiva Työraportti, 2001-12.

Lothenbach, B. and Nonat, A. (2015). Calcium silicate hydrates: Solid and liquid phase compositions. Cem. Con. Res., 78, 57-70.

Martino, J., Dixon, D., Holowick, B., Kim, C.-S. (2011). Enhanced sealing project (ESP) Seal Construction and Instrumentation report. Geological Society, Special Publications, 400, 63-70.

Palomäki, J. and Ristimäki, L. (2013). Facility Description 2012: Summary Report of the Encapsulation Plant and Disposal Facility Designs. Posiva Working Report, 2012-66.

Stade, H. (1989). On the raction of C-S-H(di,poly) with alkali hydroxides. Cem. Con. Res., 19, 802-810.

Vogt, C., Lagerblad, B., Wallin, K., Baldy, F., Johansson, J.-E. (2009). Low pH self compacting concrete for deposition tunnel plugs. SKB Report, R-09-07.

Wagner, T., Kulik, D.A., Hingerl, F.F., Dmytrieva, S.V. (2012). GEM-Selektor geochemical modeling package: TSolMod library and data interface for multicomponent phase models. Can. Mineral., 50, 1173-1195.

A method for quantitative visualization of heterogeneous transport processes at the host rock - cement interface

Johannes Kulenkampff [1*], Marion Gründig[1], Stefan Gruhne[1], Johanna Lippmann-Pipke[1,2], Kyra Jantschik[3], Helge C. Moog[3]

[1] Helmholtz-Zentrum Dresden Rossendorf (DE)
[2] BGR, Bundesanstalt für Geowissenschaften und Rohstoffe (DE)
[3] GRS, Gesellschaft für Anlagen- und Reaktorsicherheit gGmbH (DE)

* Corresponding author: j.kulenkampff@hzdr.de

Abstract

We present a method for quantitative visualization and parametrization of heterogeneous transport processes at the host-rock - cement interface. As measuring method for spatiotemporal observation of the concentration of a radiotracer (^{22}Na), we apply high-resolution positron emission tomography (PET). For illustration, the study here was conducted on a generic laboratory sample with drill core dimensions of a halite - salt cement contact. By sequential PET-imaging over a period of 70 days, we could delineate the initial patchy tracer distribution and the diffusional propagation of the tracer concentration into the cement. After 10 days, we observed significant propagation of the tracer into the cement. From the propagation pattern, we derived a first approximation of the penetration depth.

Introduction

The zone at the host-rock - cement interface is principally a laterally extended highly structured thin volume. Its structure is caused firstly by the roughness of the contact itself, but also, because the host-rock in this zone is disturbed by excavation damage, and because the cement is initially a heterogeneous mixture of different components. Therefore, transport and exchange processes are to be considered as heterogeneous processes with strong spatial variations. Such processes should be characterized with tomographic methods, rather than by conventional determination of physical properties and transport experiments, because these miss local variations. This is why we developed our GeoPET-method (Kulenkampff et al., 2016a and 2016b), which directly yields spatial concentration distributions; and concentration is considered the most important variable for geochemical modelling.

Geochemical modelling relies on due parameterization of the model. We therefore developed methods for transport parameter extraction from measured spatiotemporal concentration patterns (Lippmann-Pipke et al., 2017).

The present study was conducted on an artificial salt rock - salt cement contact, where the joint between the cement and rock was designed to be sealed by brine injections. Our method was intended to confirm the tightness of the joint and to indicate the weakest part. This was accomplished by transport experiments parallel to the interface. Our particular example on an imperfectly sealed sample is demonstrating the suitability of the method for characterizing the properties of the contact zone and deriving diffusion parameters of the materials in this zone. The procedure is also applicable to clay systems that are in the focus of CEBAMA.

Sample and Procedure

An aged salt cement plug was inserted into an axial drill hole through a cylindrical halite drill core (Figure 1 and Figure 2). This compound sample was exposed to a confining pressure of 10 MPa while the contact was flushed with concentrated salt brine in an isostatic cell until it became impermeable by salt deposition. Afterwards this compound sample was cast with resin (Araldite) into an acryl glass cylinder. The characteristics of the sample and the development of the preparation procedure are described in Jantschik et al. (*in prep.*).

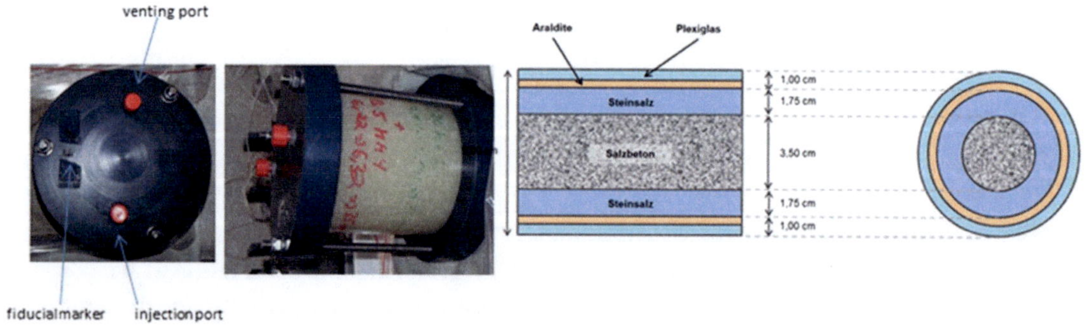

Figure 1: *Photo and design of the sample. After preparation of the sample, the PET-tracer is injected into the head space through the injection and venting port.*

Figure 2: *µCT-image of the sample with cut at the injection port. The joint between salt and cement, as well as the head space, are detectable, but below image resolution of 30 µm.*

The complete cylindrical sample was covered with end caps, leaving a small head space of 0.6 mL between sample and end caps that serves as tracer reservoir. Formation fluid (concentrated brine) was labelled with 3.16 MBq of ^{22}Na, which is a PET-tracer with a decay time of 2.6 years, and injected into this head space.

A total of 23 PET-frames were recorded with a frame length of 30 min and increasing time lag over a period of 70 days.

The PET-data were corrected and reconstructed according to the procedure in Kulenkampff et al. (2016a). More images and the details of the processing will be published in a future paper (Kulenkampff et al., *in prep.*). The total injected activity is known and remains constant within the field of view during the whole experiment. Activity calibration of the images is simply computed by normalizing the raw count rates to this injected activity value. This calibration was cross-checked with a point source of known activity (fiducial marker, see Figure 1) that was permanently fixed to the outside of the sample.

First Results and Outlook

An initial PET-image after injection shows penetration of the tracer into the joint (Figure 3). The tracer distribution in the joint is patchy, but appears continuous. Obviously, the sealing of the joint was not perfect, and the joint opened again during the period of some days, between sealing of the joint, when the sample became impermeable, and the time of tracer injection. Thus, tracer was sucked into the void space of the joint immediately after tracer injection on to the head space. This process - although undesirable in the real repository - is a proper example for demonstrating the suitability of the method.

The heterogeneous pattern is the source concentration distribution for diffusional spreading of $^{22}Na^{+}$.

Figure 4 shows three frames out of the suite of 23 PET-frames that suggest a shift with time of the activity towards the zone of the cement. A slight indication of increase of activity near the axis may be a noise effect and is not significant at this stage of data processing.

Figure 3: *Initial PET-image as volumetric presentation. Tracer distribution shortly after injection (red: high activity, blue: low activity, transparent: activity below noise threshold, uncalibrated). The sample dimension is outlined by the grey cylinder. The injection port is at the left-hand side.*

As an approximation, the complicated tracer distribution was simplified by computing circular mean values of the activity *vs.* radius for all times and all slices (Figure 5). This approximation corresponds to a homogeneous cylindrical source that diffuses in both directions, with convergent inward pathways into the cement and divergent outward pathways into the salt.

We further simplified the situation by defining the interquartile range of the distributions in Figure 5 as a measure of the depth of penetration. This interquartile range is the distance between the median and the 25%-, resp. 75% -value of the distribution. It is not physically motivated, but it is a statistical measure that is robust with respect to outliers and noise. Figure 6 shows the evolution of this depth with time. The initial measurements, from 4 to 20 h, should be disregarded, because they were taken with smaller frame lengths, and therefore are prone to noise. After 200 h, the red line (penetration depth in cement) significantly exceeds both, the noise level and the spatial resolution of the measurement, which means significant diffusion into the cement. The penetration depth increases to 5.5 mm after 70 days. In the same time, the blue line hardly exceeds the spatial resolution, although it increases steadily.

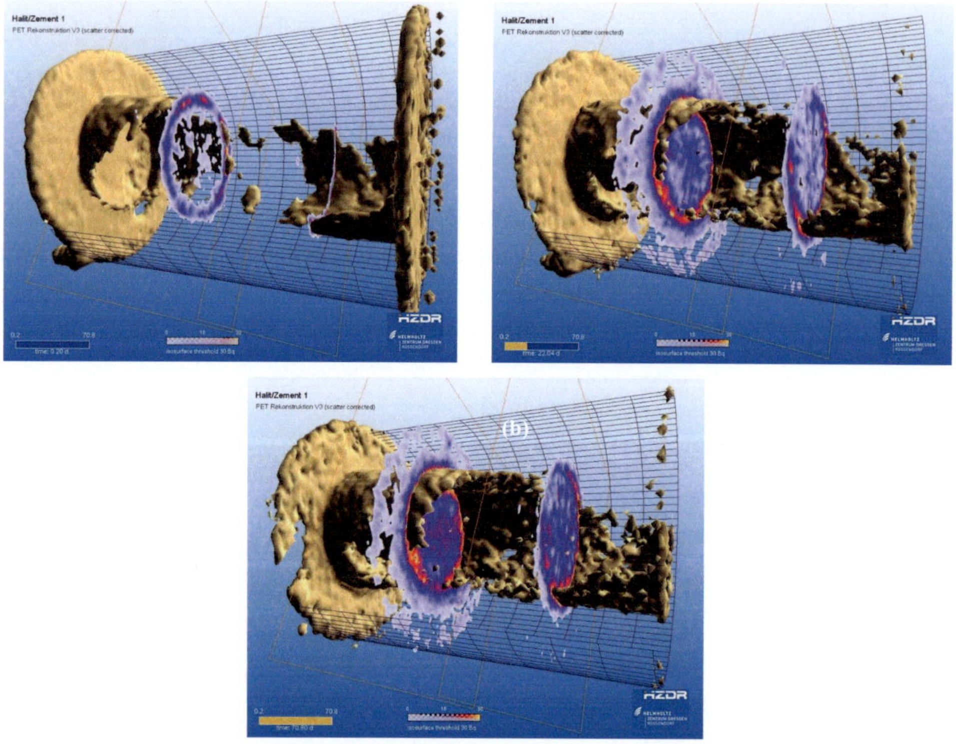

Figure 4: *Three PET time frames (a: directly after injection, b: after 22 and c: after 70 days): Tracer distribution iso-surface (30 Bq/voxel) with two cross sections.*

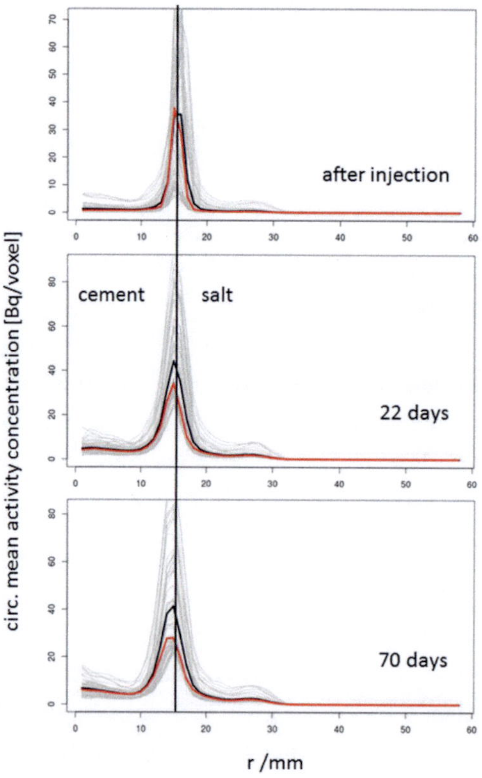

Figure 5: *Circular mean values vs. radius (grey: individual transaxial slices, black: mean value, red: median). The position of the maximum value is stationary and coincides with the position of the joint. The distribution broadens with time to the left-hand side (cement), whereas the right-hand side (salt) remains fixed.*

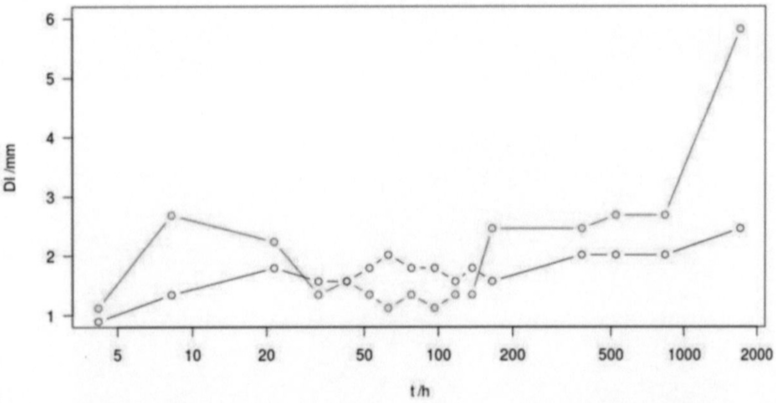

Figure 6: *Estimate of the tracer penetration depths vs. time from the spreading of the radial tracer distribution, red line: cement, blue line: halite.*

This so-defined robust penetration depth can be used in a simplistic approach for directly estimating a preliminary diffusion coefficient. Certainly, fitting the corresponding solution of the cylinder symmetric diffusion equation (according to Crank, 1975) to the original data set would be more correct, although it neglects the heterogeneous pattern of the source. Another option for parameter determination is optimization of a finite-element model, similar to Lippmann-Pipke et al. (2017). Heterogeneities will be considered as residuals from the best fit model. Please notice this is not only a very simple mathematical approach but also neglects any other potential driving forces, like suction due to undersaturation of the salt cement.

The distribution of the initial tracer distribution after seepage into the contact zone can be quantitatively evaluated with respect to the infiltrated volume of brine: Considering the activity concentration per voxel of the initial PET-image along the joint, $a(r_j, \varphi, z)$ (in cylinder coordinates, with $r_j = 17.5$ mm: radius of the joint), the thickness of the layer of labelled fluid in the joint can be estimated from the partial volume effect. The injected activity concentration is $a_0 = 3.16$ MBq/0.6 mL = 8 kBq/voxel. Then, with the base area of the voxel $q_v = (1.15 \text{ mm})^2$ the thickness δ of the layer can be calculated as Eq. 1.

$$\delta(\varphi, z) = \frac{a(r_j, \varphi, z)}{a_0 \cdot q_v} \qquad \textbf{\textit{Eq. 1}}$$

This thickness estimate is shown as cylinder map in Figure 7. Its maximum is in the order of 15 µm, and thus far below the spatial resolution of PET, even below the resolution that is achievable with µCT when the complete sample is in the focus. This achievement, the derivation of the thickness of a very thin film, is not in contradiction to the weak spatial resolution of PET in the order of 1 mm, because it is owing to the extreme sensitive measure of concentration, whereas the exact position is not considered. Of course, this procedure only applies to one single film in one voxel, and microporosity is not considered.

In summary, PET yields quantitative spatiotemporal tracer distributions from which it is possible to derive mean quantitative transport parameters and to estimate their spatial variations. In our example we could estimate the diffusional penetration of the tracer into the material in the complex and heterogeneous situations of the cement-rock contact zone. This is accomplished non-destructively and without the need to wait for tracer break-through, which would be required in diffusion cells. It was also possible to estimate the thickness of very thin films or thin fractures anywhere in the material, even when their thickness is far below the resolution of PET, or even µCT. We also outlined tools for computing mean diffusion coefficients from process tomography with PET. Further 3D-image analysis will then enable to qualify heterogeneous effects and to provide worst case estimates.

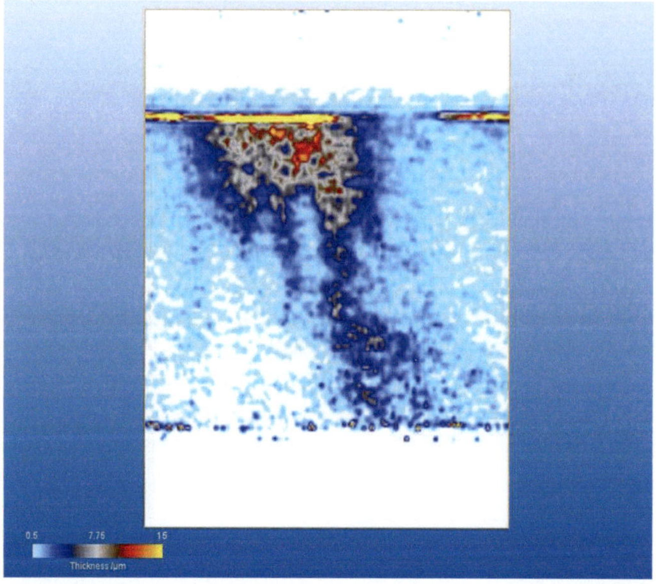

Figure 7: *Cylindrical map of the thickness $\delta(\varphi,z)$ of the fluid film, derived from measured activity concentration referring to the initial concentration.*

Future work will include both, develop of computational methods for parameter derivation, in order to provide a quantitative laboratory test method for the transport along and across the critical contact between the natural and engineered barrier in repositories, and application on the actual materials considered in CEBAMA. This laboratory testing method will be applicable for basic scientific understanding of transport processes at the contact zone, estimation of enhanced engineered barrier design, and for quality assurance on samples from the repository.

Acknowledgement

The research leading to these results has received funding from the European Union's Horizon 2020 Research and Training Programme of the European Atomic Energy Community (EURATOM) (H2020-NFRP-2014/2015) under grant agreement n° 662147 (CEBAMA).

References

Crank, J. (1975). The mathematics of diffusion. Clarendon Press, Oxford.

Lippmann-Pipke, J., Gerasch, R., Schikora, J., Kulenkampff, J. (2017). Benchmarking PET for geoscientific applications: 3D quantitative diffusion coefficient determination in clay rock. Computers & Geosciences, 101, 21-27.

Jantschik, K., Kulenkampff, J., Lippmann-Pipke, J., Moog, H.C. *(in prep.)* Investigation the sealing capacity of the contact seam of a cement-based sealing element in rock salt. To be published in Cement and Concrete Research.

Kulenkampff, J., Gründig, M., Zakhnini, A., Lippmann-Pipke, J. (2016a). Geoscientific process monitoring with positron emission tomography (GeoPET). Solid Earth, 7, 1207-2015.

Kulenkampff, J., Mäder, U., Gründig, M., Eichelbaum, S., Lippmann-Pipke, J. (2016b). PET/CT during degradation processes at the cement-clay interface and derivation of process parameters. Proceedings of the 1st CEBAMA Annual Workshop. KIT Scientific Publishing, KIT-SR 7734.

Kulenkampff, J., Jantschik, K., Gründig, M., Moog, H.C., Lippmann-Pipke, J. *(in prep.)* Laboratory determination of transport properties at the interface of rock salt and salt cement with PET processs tomography and µCT. To be published in Geotechniques.

High pH concrete - FEBEX bentonite interface reactions: from months to decades and from cm^3 experiments to m^3 in-situ scenario

Jaime Cuevas[1], María Angulo[1], Daniel González-Santamaría [1], Javier González-Yelamos[1], Raúl Fernández[1], Almudena Ortega[1], Ana Isabel Ruiz[1]

[1] UAM, Universidad Autónoma de Madrid (ES)

* Corresponding author: Jaime.cuevas@uam.es

Abstract

During the second year of the Cebama project, the activities performed by the research team working at Universidad Autónoma de Madrid (UAM) have been focused in characterization tasks at different scales of *in-situ* and laboratory experiments with regard to FEBEX bentonite-concrete interaction. The gained experience of handling and testing procedures for *in-situ* taken samples at the FEBEX tunnel at Grimsel test site (GTS) have been applied to characterize high quality over-cored samples with preserved 13 years interfaces. The 10 years HB6 CIEMAT cell containing CEM-I concrete-bentonite interface under a thermal gradient has been dismantled and it is in progress the study for mineralogical, geochemical am microstructural features. Complementary, the first six small-scale bentonite-concrete surface reactivity SERIE experiments, running during 4 - 6 months, have been also dismantled. Their interface chemistry and microstructural aspects are being observed for different cement paste-mortars including CEM-I (high-pH), CEM-I + silica fume (low pH) and CEM-II-a-L. The spatial and time scales for the studied scenarios ranged from m^3/decade (FEBEX-GTS), hundreds cm^3/decade (HB6) to cm^3/0.5 - 1 year (SERIE). This is a first effort to compare the ongoing results.

Introduction

The UAM contribution to CEBAMA project is to be developed by the achievement of two main objectives: (1) to perform the geochemical study of high pH concrete (OPC)-FEBEX-bentonite at *in-situ* and long-term experiments, and (2) to determine the characteristic surface interface reactivity of compacted FEBEX bentonite induced by OPC based concrete pore water (high and low pH).

The aim of the study is to relate the interface reactivity of the concrete-bentonite system from micro to macro experiments. Three testing scenarios display the framework of the research. *In-situ* concrete-bentonite interfaces, taken during the dismantling of the FEBEX tunnel at the GTS, represents a 13 years aged interface within a 50 m^3 experiment (10 m^3 concrete plug and 40 m^3 heated bentonite) hosted in a granitic rock. HB6 is part of a series of HB tests (1 - 5 already dismantled) implemented by CIEMAT and consists of a bentonite cylinder column (7 cm length) which hydrates through a high pH OPC concrete disc (3 cm). At the opposite side the bentonite face a hot steel plate maintained at 100ºC. HB6 represent a 385 cm^3 10 years experiment (115 cm^3 concrete and 270 cm^3 bentonite). In both cases the concrete-bentonite interface was < 40ºC. Finally, groundwater infiltration experiments through cylinder casted cement mortars (1 cm length, high pH and low pH) were put in contact with compacted bentonite (0.8 length column) at ambient temperature (20 ± 3ºC), during a period of time sufficient to collect water at the bentonite end (a few cm^3 aliquots). The experiment represent a 5.5 cm^3 0.5 - 1 year experiment

(3 cm^3 and 2.5 cm^3 bentonite); bentonite dry density was near 1.6 g/cm^3 in all of the interfaces. Details on experiments setups have been described in Cuevas et al. (2016) and Fernández et al. (2017). It comes evident that cement material / bentonite volume ratio increased from macro to micro, thus representing more specifically the interface location.

The purpose of this report is to follow, at different scales, most of the processes identified at the concrete-bentonite interface as defined in the state of the art document (Vehmas and Holt, 2016): Ca-carbonation of the concrete and concrete-bentonite interface (XRD mineralogy and stable isotopes); Formation of C-A-S-H and secondary ettringite at the bentonite and concrete interface, respectively (SEM-EDX phase analysis); and Mg-perturbation (EDX-chemical profiles, XRD, TG).

Reactions characterization at different scales

Ca-carbonation of the concrete and concrete-bentonite interface

The different scales, sampling and real aspects of the studied experiments can be taken form Figure 1. One of the first visible alteration of the 10 - 13 years concrete bentonite contact consisted in a whitish rim of millimetre size. This rim consisted of calcite and CSH rich material attached to the concrete side both in overcoring (three interfaces, 50 mm concrete and 50 mm bentonite lengths for a ¼ of 125 mm diameter core under study) and HB6 (one cell) cell material, which is not visible at macroscopic scale in the small SERIE experiments (six experiments now dismantled and six more running up to 12 - 18 months). Concrete in HB6 experiment did not present calcite initially. This allowed detection of this carbonate as a new precipitated phase at near 0.5 mm in the concrete side of the interface (Figure 2). In the overcoring interface it is not possible to precisely detect new calcite precipitates because the cement paste had initially calcite. Nevertheless, this carbonation affects mainly concrete interface because bentonite at a distance longer than 1 mm from the concrete interface does not presented an apparent increase of calcite nor any impact in the general aspect of the XRD profiles. Table 1 compare isotopic signatures of carbonates in one overcoring and HB6. The presence of lighter carbon and also important oxygen fractionation near both interfaces indicates the precipitation of dissolved (lighter) C both in concrete and bentonite. The particular oxygen dynamics should be followed in additional interfaces (Fernández et al., 2017). In the case of the small SERIE experiments, carbonation can be traced by GIXRD and by the micrometric enrichment of Ca at the mortar interfaces detected mainly by chemical analysis and mapping.

Formation of CASH and secondary ettringite

Both *in-situ* and long term HB-6 experiments are characterized by the detection of Al-S ettringite phases and carbonates at the concrete interface and, at the same time, the presence of Mg-silicate phases (without any or very poor in Ca) coexisting with C-A-S-H or C-S-H-montmorillonite mixtures at the bentonite side. The original 14.5 Å basal spacing of montmorillonite is visibly decreased (to 12.5 - 13 Å) at the interface and up to 1 - 2 cm deep in the bentonite (Figure 2). This effect has to be further resolved as it can be a consequence both of sodium exchange or C-S-H-montmorillonite interaction (Fernández et al., 2016). Anyway, montmorillonite remained expansive in EG solvation so it had minor alteration at 10 - 13 years scale. Figure 3 illustrate the very interface obtained by a precise fresh fracture of a pristine recent cut sample. A series of photographs prove the net separation of the materials and the existence of discrete phases with very different chemistry (Mg-silicate separated from calcium phase). The HB6 studied interface show different Mg-silicate phases of columnar (serpentine-like) or lamellar (2:2 or 2:1:1 sheet silicate), and the presence of C-A-S-H phases, although not associated to characteristic morphologies became represented in XRD data.

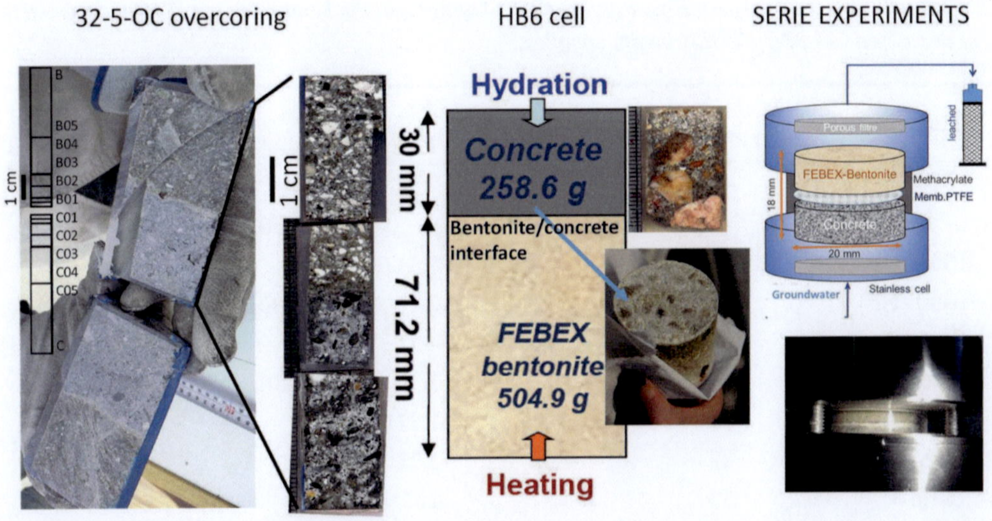

Figure 1: *Experiments at different scales and details of sampling and images of the studied materials.*

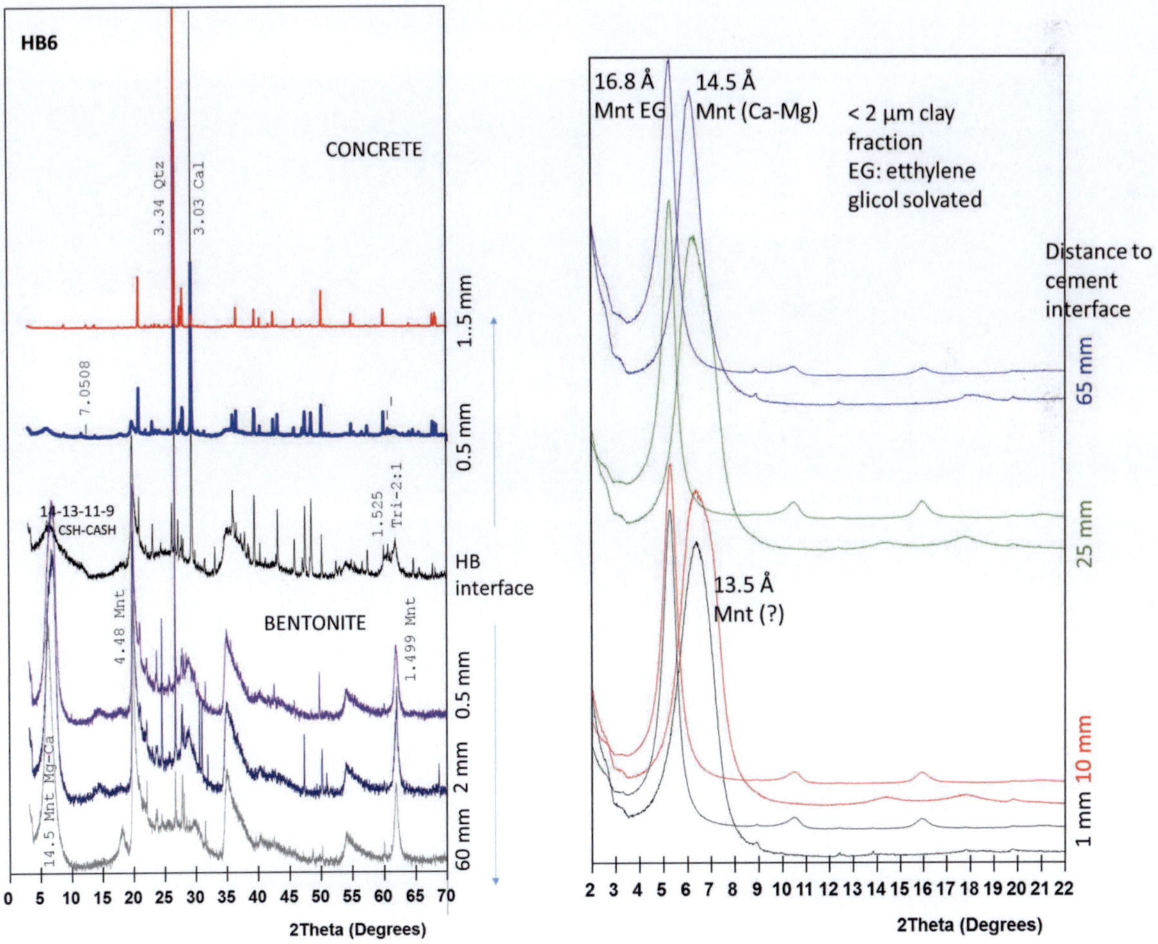

Figure 2: *XRD patterns illustrating bentonite and concrete interface characterization in HB6 10 years experiment. Qtz: quartz, Cal: calcite, Tri- 2:1: characteristic peak for tri octahedral sheet silicate.*

Table 1: *Stable Isotopic signatures δ permil variations) for ^{18}O and ^{13}C in the bentonite side of HB6 experiment and at near the interface in overcored CC-32-5 FEBEX in-situ samples.*

HB samples	Distance to interface (mm)	δ¹⁸O VPDB	δ¹³C VPDB	OC samples	δ¹⁸O VPDB	δ¹³C VPDB
	-45			OC-C	-6.81	-1.45
	-1			OC-C01	-8.29	-9.47
HB6 1-1-1	1	-2.17	-9.34			
HB6 1-1-2	3	-6.24	-2.21	OC-B01	-7.35	-2.93
HB6 1-2	7	-5.83	-4.31			
HB6 1-3	11	-6.85	-3.71	OC-B02	-7.13	-3.58
HB6 1-4	15	-6.04	-1.87			
HB6 2-2	22	-6.53	-2.21			
HB6 3	37	-5.51	0.03			
HB6 4	62	-5.72	-3.28	OC-B	-6.21	-1.98
HB6 5	65	-8.35	-2.55			
FEBEX		-7.32	-1.23			

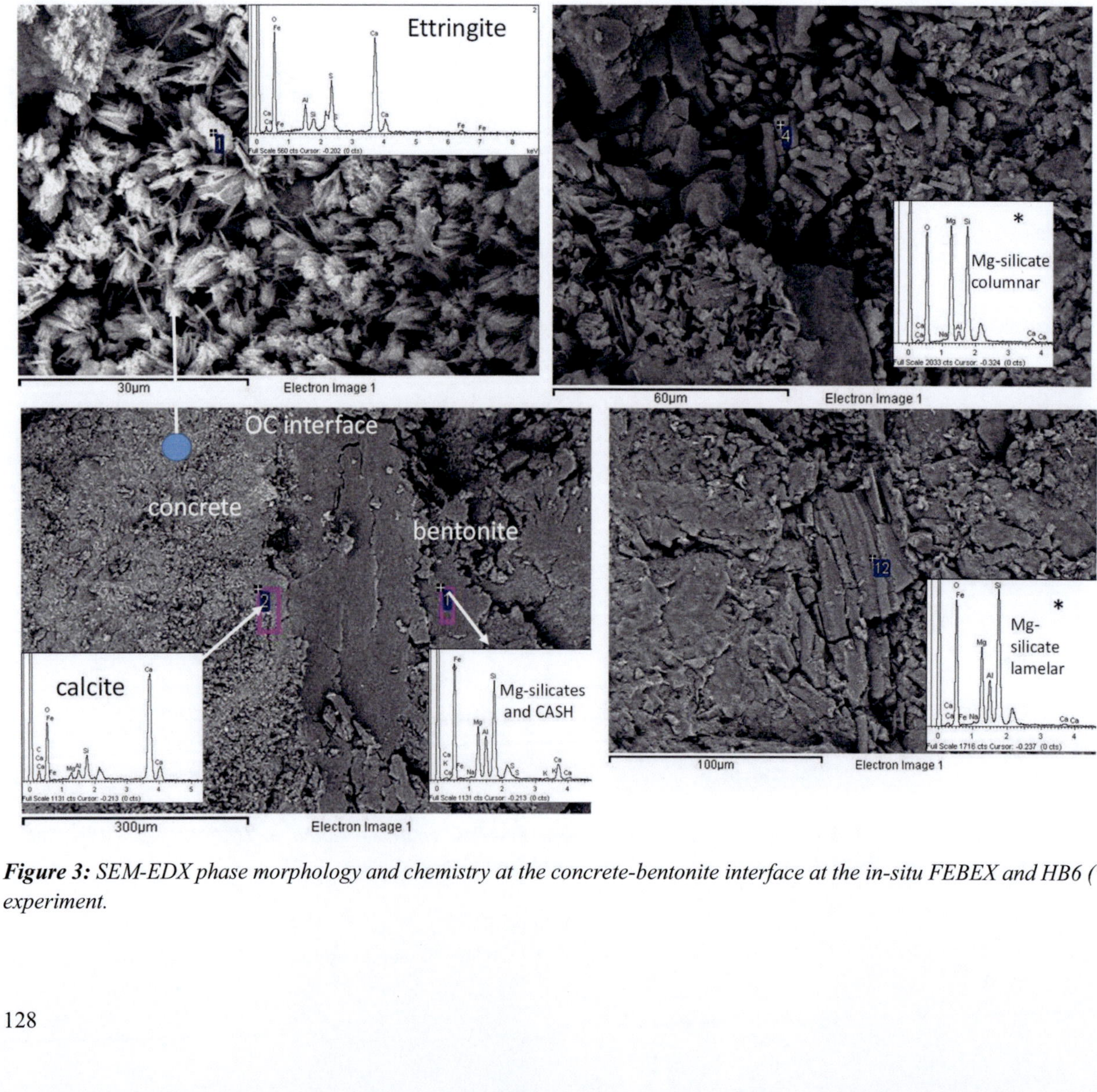

Figure 3: *SEM-EDX phase morphology and chemistry at the concrete-bentonite interface at the in-situ FEBEX and HB6 (*) experiment.*

Mg-perturbation and chemical profiles

The use of different cement pastes in the small SERIE reactivity experiments has opened the first chance to compare several cement-bentonite interface dynamics and to detect chemical fronts not clearly identified from the SEM-EDX analysis of mineral phases and bulk XRD determinations. In the SERIE experiments, Mg-phases concentration is dependent on the type of mortar in contact with bentonite (Figure 4, left). < 0.5, 0.5 and 1.0 mm thickness Mg fronts are developed for CEM-I, CEM-II and lpH experiments, respectively (orange dots). A < 0.1 mm Al front penetrates into the mortar at the expense of an Al depletion near the bentonite in all the cases. Compared to CEM-II 13 years *in-situ* FEBEX experiment, taking into account the difficulties in the isolation of cement matrix spots or simply homogeneous zones for analysis in the *in-situ* concrete, the fronts at *in-situ* have display some larger displacements. The Mg front affected 2.5 mm bentonite thickness and Ca seems to be unevenly depleted in the concrete side (green line). The Mg front displaces the calcium in the bentonite contact, which confirms the absence of Ca in Mg-silicate concentration zones shown in the EDX analyses of mineral phases shown in Figure 3.

The ongoing studies regarding the analyses of long time SERIE (> 1 year) additional experiments and also of the distinct *in-situ* overcored interfaces, will all provide new information on how the scale of observation can affect the spatial magnitude and the minerals formed in the observed reaction fronts. As an example, Mg concentration at the lpH SERIE experiment is related to brucite precipitation near the concrete side (Figure 5); however, brucite is not observed in long term *in-situ* experiments, where Mg-silicates are formed instead.

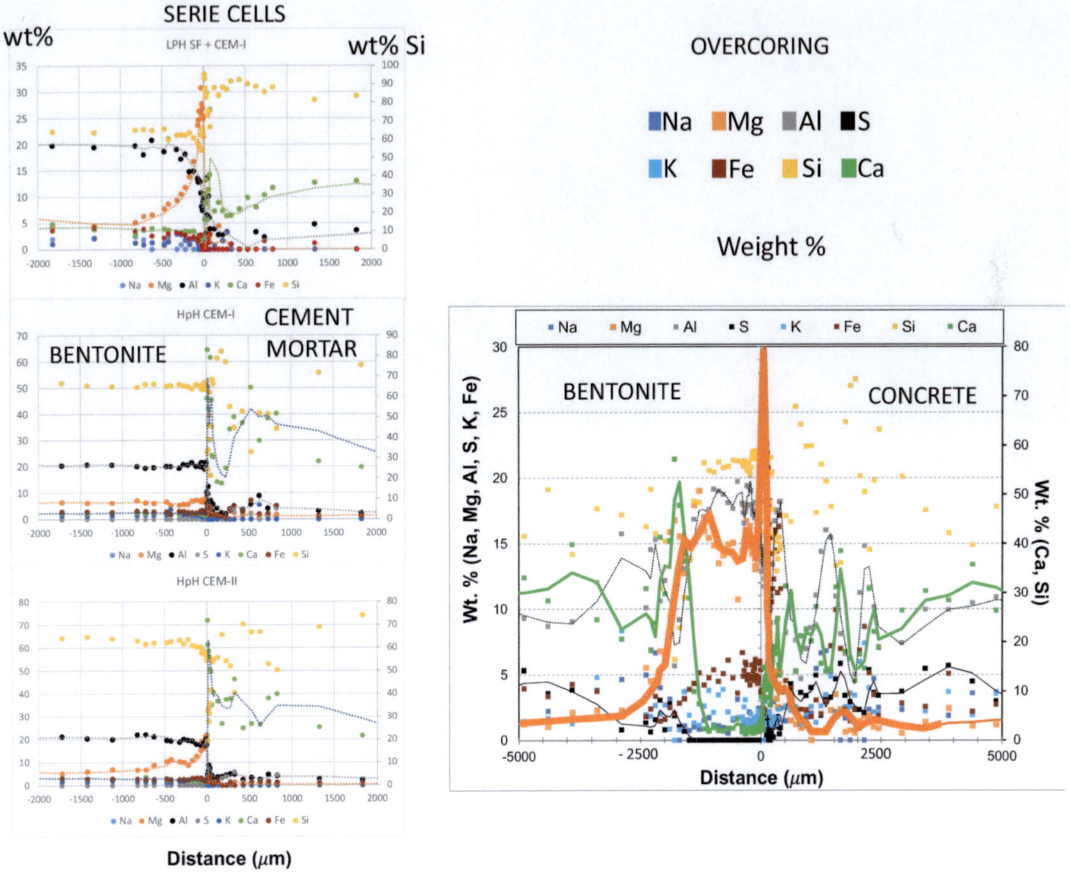

Figure 4: *SEM-EDX chemical profiles of a detailed section within the cement-bentonite interface comparing small SERIE reactive transport experiments to an overcored sample. Mg chemical front is outlined in orange colour.*

The mineral phases identified in the SERIE experiments in the cement (photographs in Figure 5) or bentonite interfaces are being characterized by grazing incidence XRD. Brucite is identified in the concrete, when the penetration of then X-rays are deep (0/2Θ). GIXRD can better characterize C-S-H phases in bentonite. C-S-H compositions taken from undifferentiated matrix of the cement side were also determined and are shown in Figure 5. Ca/Si of 1.1 to 0.7 (lpH) are consistent with the presence of CSH tobermorite types.

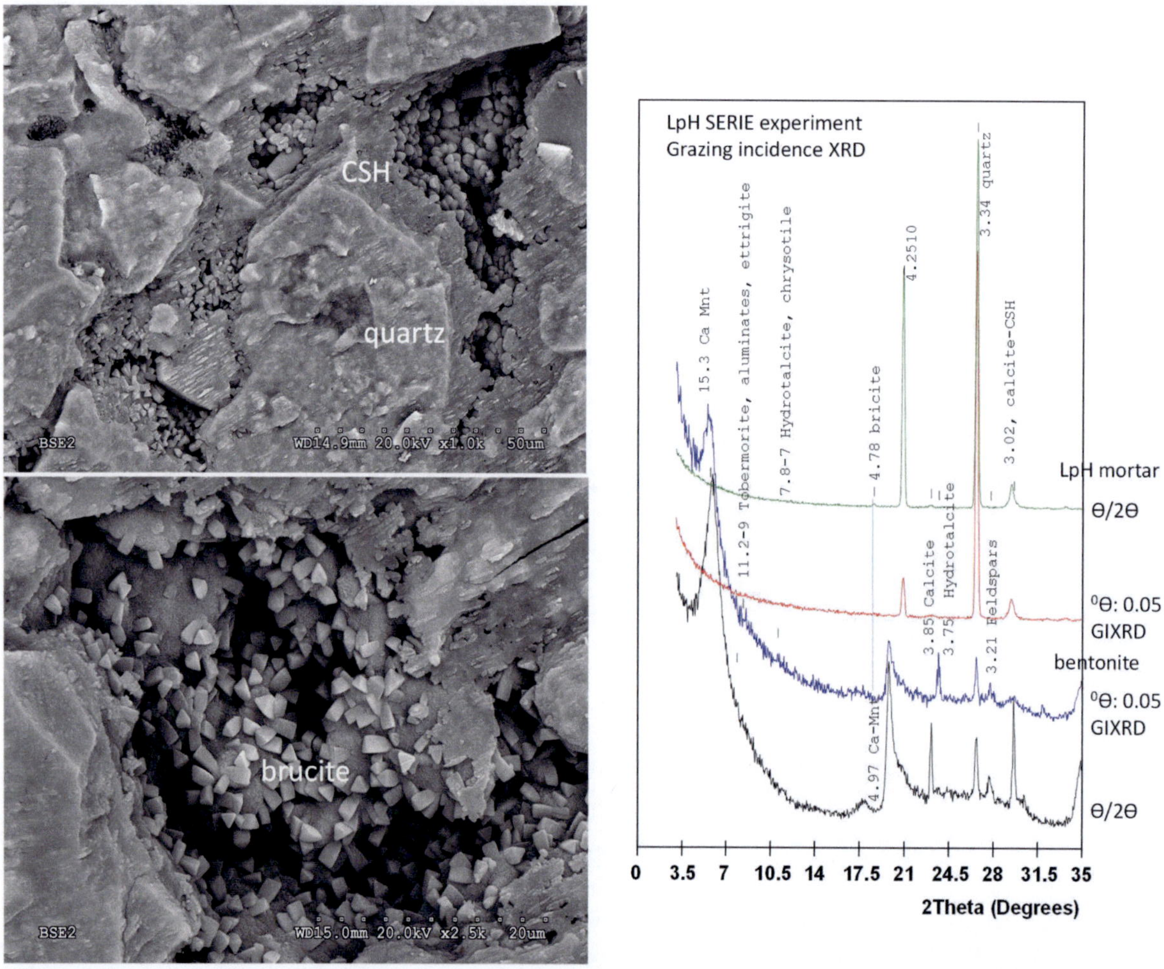

Figure 5: *Mineral phases at the concrete interface preserved in SERIE experiments. Identification by GIXRD.*

Table 2: *CSH compositions in SERIE experiments. Atomoc percentajes.*

	C (%)	O (%)	Mg (%)	Al (%)	Si (%)	Ca (%)	S (%)	Fe (%)	Ca/Si
LpH		70.7	6.9	1.2	12.1	8.5		0.5	0.7
Sdeviation		0.9	4.0	0.4	2.3	3.0		0.1	0.1
CEM II	23.2	54.2		0.3	10.6	11.8			1.1
Sdeviation	1.6	1.6		0.1	0.6	0.8			0.0
CEM I		76.9		0.4	10.5	11.4	0.4	0.4	1.1
Sdeviation		1.5		0.1	0.9	0.6	0.0	0.0	0.0

Conclusions and Future work

The characterization of different scale experiments considering basically high-pH concrete bentonite interface is under development and many sets of data have been acquired. The expected reaction paths advanced in the state of the art study are consistent with the results summarized in this report. The reaction fronts at different scales of time and space are very promising to develop reaction concepts and support modelling in the future. All the experiments under consideration are now available as we have already dismantled the HB6 10 years experiment at CIEMAT. Thus, the on-going scheduled work is: (1) the study of two more overcored *in-situ* FEBEX interfaces (6 - 42 months); (2), the sequential sampling and characterization of SERIE experiments (6 - 30 months); (3) the HB6 cell (10 years characterization (12 - 38 month); and later on: (4), the conceptual modeling and further data interpretation in order to achieve the integration for local to global (SERIE-HB6-*in-SITU*) geochemistry scale. Inputs for future model validation are going to be fulfilled in the last 38 - 46 months of the CEBAMA project. However, intermediate results are now available.

Acknowledgement

The research leading to these results has received funding from the European Union's Horizon 2020 Reasearch and Training Programme of the European Atomic Energy Community (EURATOM) (H2020-NFRP-2014/2015) under grant agreement n° 662147 (CEBAMA).

We want to thank specially to Enrique Rodríguez and all the people of the central research services of the Faculty of Sciences of UAM (SIdI) for their fruitful aid and work.

References

Cuevas, J., Fernández, R., Ruiz, A.I., Ortega, A., González Yélamos, J., González Santamaría, D. (2016). Definition of sampling and characterization of in-situ FEBEX-OPC concrete contact and design of new experiments studying surface interface reactivity. Proceedings of the 1st CEBAMA Annual Workshop. KIT Scientific Publishing, KIT-SR 7734.

Fernández, R., Ruiz, A.I., Cuevas, J. (2016). Insight into the formation of C-A-S-H phases as a consequence of the interaction between concrete or cement and bentonite. Clay Minerals, 51, 223-235.

Fernández, R., Torres, E., Ruiz, A.I., Cuevas, J., Alonso, M.C., García-Calvo, J.L., Rodríguez, E., Turrero, M.J. (2017). Interaction processes at the concrete-bentonite interface after 13 years of FEBEX-Plug operation. Part II: Bentonite contact. Physics and Chemistry of the Earth, Parts A/B/C, 99, 49-63.

Vehmas, T. and Holt, E. (2016). WP1 Experimental studies - State of the art literature review. CEBAMA Project Deliverable, D 1.03 (M09 - Feb 2016).

Ageing of high and low pH concretes after short and long periods in contact with groundwaters

María Cruz Alonso[1*], José Luis García Calvo[1], Virtudes Flor Laguna[1], Álvaro Fernández[1]

[1] CSIC, Institute of Construction Science Eduardo Torroja (ES)

* Corresponding author: mcalonso@ietcc.csic.es

Abstract

During the second year of the Cebama Project, the work performed by CSIC was focused in two directions: 1) Study of the long-term interaction of concrete with groundwaters: i) samples of the FEBEX shotcreted concrete plug, after 13 years in the Grimsel Test Site conditions under the interaction of the granite and bentonite porewaters, at the concrete / bentonite contact and ii) concrete of the HB6 cell, after 10 years operation with the interaction of clayey and bentonite porewater, and 2) short-term interaction, under percolation tests, of aged or new high-pH (hpH) and low-pH (lpH) concretes. The general objective is to analyse the performance of different concrete types in real and simulated deep geological repository conditions.

In the long term studies, the interaction of the concrete of the Febex concrete with the bentonite porewater has caused additional ettringite formation due to the sulphate penetration from the saturated bentonite. This ettringite formation has been evidenced from the concrete / bentonite Interfacial Transition Zone (ITZ) towards 1 cm inside the concrete. Alterations of the C-S-H enriched in Al and portlandite dissolution was also observed. Additionally, Cl diffusion from bentonite up to 5 cm inside the concrete was evident with Friedel salt formation. The pore structure of the concrete was affected with increase of micropores. In HB6 concrete the clayey and bentonite waters lead also to new ettringite formation, pH decrease and portlandite dissolution.

In the short-term percolation tests a decrease in the pH of the percolated liquid was observed from hpH concrete with a sharp decay to that of Grimsel granite water pH. While the pH of percolated liquid remains unchanged with the lpH concrete. The first results of the short-term interaction between concrete and granite groundwater showed the formation of more ettringite in FEBEX hpH as consequence of the particular concrete composition that favours the movement of sulphates and Al to form new ettringite. Although the pore pH decreases below portlandite stability the bulk concrete still remains alkaline. On the contrary in Äspö lpH no new solid phases have been detected and alkalies are not leached. Longer and deeper studies are in way.

Introduction

The effect of groundwater interaction with concretes has been classified at two different time scales:

Long-term interaction: two systems (age and test conditions described in Table 1) are being evaluated:

1. Concrete from Febex Test: Grimsel granitic groundwater / shotcreted high pH concrete / bentonite interfaces are under study, after 13 years of exposure. The Febex system has been extensively described in Alonso et al. (2017). Figure 1 shows the shape of the samples (left) and their location

(right) in the Grimsel concrete plug. Two pieces of overcoring and six short cores drilled around the overcorings were used for characterisation, from bentonite ITZ up to 7 cm inside the concrete plug. Also, a sample taken far from the bentonite and Grimsel groundwater interface was used to represent the non-affected concrete as a reference in the present work (in the part of the section 1 furthest from the bentonite (≈ 1 m) and in the centre of the concrete plug).

2. Concrete from HB6 cell: simulated clayey groundwater / high pH concrete / bentonite interfaces are under study, after 10 years of exposure. HB6 cell is one of the 6 cells of the HB laboratory experiment described in Turrero et al. (2016). Figure 2 (left) shows the piece of concrete used for characterisation of groundwater interactions.

Short-term interaction: Two types of groundwaters: 1) Grimsel real site water, and 2) simulated clayey water, are being tested for the interaction with different high and low pH concretes identified in Alonso et al. (2016). The percolation leaching test protocol is the same as described in García Calvo et al. (2010). Table 2 includes data from initial concretes and percolation test conditions. The initial pH in concrete porewater was measured according to Alonso et al. (2012). Three samples for each groundwater / concrete system are under study to evaluate the interaction processes. Different pressures are being used accordingly with the type of concrete. The variables characterised are: periodic analyses of the leachates (volume of eluted water, hydraulic conductivity, ions released and pH). At the end of each percolation test (up to now, only the FEBEX hpH-1 and Äspö lpH-1 samples have been removed), the chemical, mineralogical and microstructural changes in the concretes are analysed (porosity, pore water pH, soluble ion content, XRD, TG/TDTA, SEM-Back-scattering and EDX) at several distances from water inlet. Each concrete cylinder was sectioned as shown in Figure 2 (right). Additionally, concretes not exposed to water were characterized as a reference of the initial state.

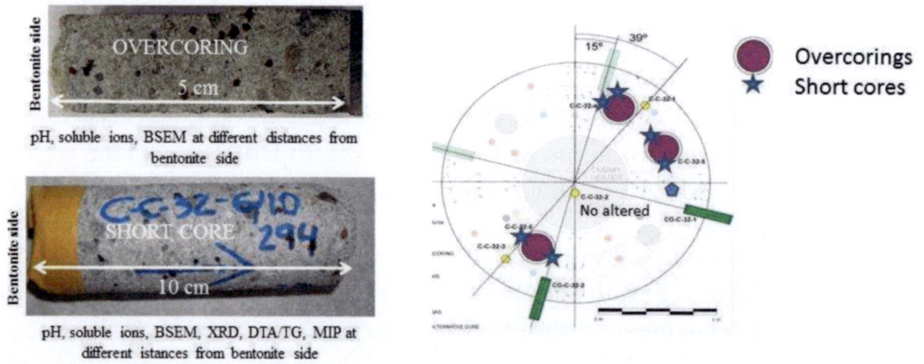

Figure 1: *Concrete samples from FEBEX system (left) and location in the concrete plug (right).*

Figure 2: *Concrete sample from HB6 cell (left) and concrete sample from percolation test (right).*

Table 1: *Concrete Samples studied at long term.*

Concrete identification	Binder	Concrete Age	Groundwater
Febex *Shotcrete (hpH)*	CEM II A-L + 6.5% nanosilica	13 years	Granite
HB6 cell *(hpH)*	CEM I 45 R-SR	10 years	Simulated Clayey

Table 2: *Concrete types used in short term test, percolation test conditions.*

Concrete identification	Binder	Concrete Age (before starting the test)	Initial pore concrete pH	Initial porosity (%)	Groundwater	Pressure (bar)
Aged: Febex *Shotcrete (hpH)*	CEM II A-L + 6.5% nanosilica	13 years	12.8	11.2	Granite	0.5
New: HB6 *(hpH)*	CEM I 45 SR	4 months	12.6	11.3	Clayey	2
Aged: Äspö *Shotcrete (hpH)*	60% OPC + 40% SF	9 years	9.2	19.4	Granite	0.5
Aged: Äspö *(lpH)*	60% OPC + 40% SF	9 years	10.2	10.3	Clayey	2
New: VTT *(lpH)*	CEM I + SF+ BFS	6 months	11.5	2.3	Granite	8
New: VTT *(lpH)*	CEM I+ SF+ BFS	9 months	11.5	2.3	Clayey	8

hpH = high pH, lpH = low pH, OPC = Ordinary Portland Cement, SF = silica fume, BFS = blast furnace slag

Long-term interaction of concrete with groundwater in real exposure conditions

Ageing of shotcreted concrete plug of Febex test: Grimsel granitic groundwater / high pH concrete / bentonite interfaces

The pH measurements of the pores of the concrete indicate a decrease in the first cm in contact with the bentonite but still remain alkaline (11.5 - 12). At distances further than 1.5 cm the pH stabilises above 12.5. Within this cm, the portlandite content is lower than in the reference. An increase of the ettringite content is evident close to the ITZ due to the interaction with the bentonite porewater. In combination with this process, higher soluble SO_3 content are also accumulated within this first cm which has contributed to the new ettringite formation at the interface with the bentonite. The chloride ion seems to compensate the SO_3 entrance, and it penetrates faster inside the concrete, up to even 5 cm, and Friedel salt formation is observed. EDX analyses of the solid phases also shows a similar trend for both elements, S and Cl, as shown in Figure 3 (left), confirming of the penetration of both ions from the bentonite. The C-S-H at the bentonite ITZ is significantly altered. Incorporation of Al and Mg in the C-S-H gels has been observed. The progressive alteration of the initially formed C-S-H gels is observed at the bentonite-concrete interface, as shown in Figure 3 (right), with progressive formation of massive ettringite but decomposition of the C-S-H that enriches in Al and decreases of Si content.

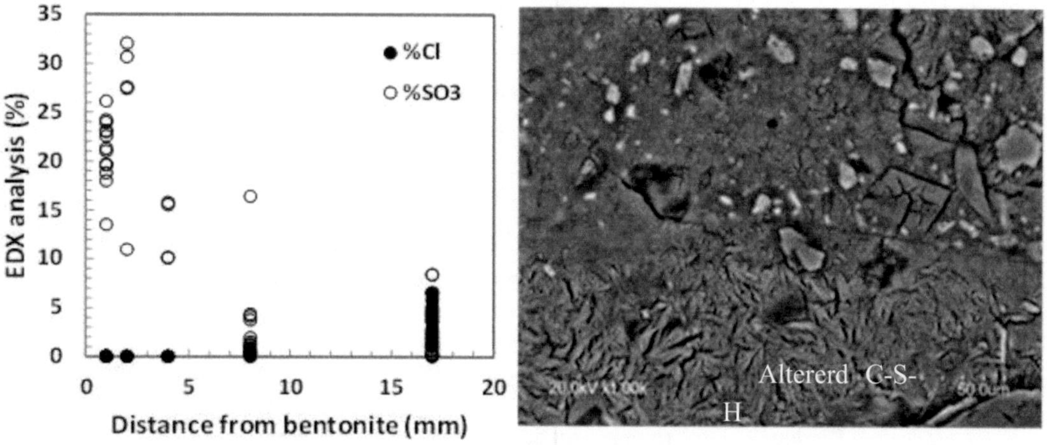

Figure 3: *SO_3 and Cl content in cement paste (EDX analysis) (left). BSEM image of the altered CSH at the ITZ concrete / bentonite (right).*

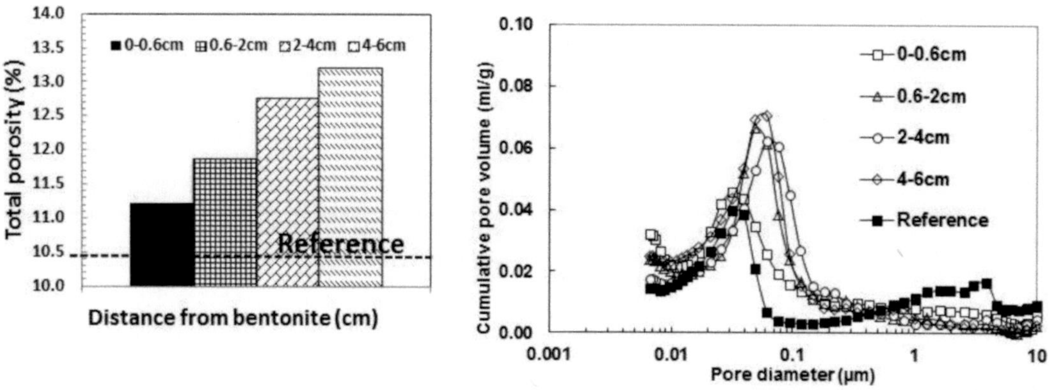

Figure 4: *Total porosity (left) and pore size distribution (right) of the concrete of FEBEX system from bentonite.*

In addition and as a consequence of the changes in cement paste composition described, increase in total porosity is observed (Figure 4, left) with respect the non-altered concrete (reference), that even increase at higher distance from bentonite ITZ to inside. The capillary pores in the region 0.04 - 0.4 µm are those more altered.

Ageing of HB6 cell: simulated clayey groundwater / high pH concrete / bentonite interfaces

In the HB6 concrete after 10 years interaction changes in the microstructure as function of the distance with the bentonite and clayey waters were detected and three regions of influence have been identified from bentonite contact to clayey water: 1) 0 - 1 cm, 2) 1 - 2 cm and 3) 2 - 3 cm. The pH of the concrete is below that of a $Ca(OH)_2$ saturated solution, 12.5 (pH = 11.4 at bentonite interface and 11.6 at clayey water interface were measured). In the middle of the sample there is a lower effect of the bentonite and the clayey water inlet, and therefore, a slightly higher pH (11.7) is measured. In the three zones, the XRD and dTG confirm the absence of portlandite (see Figure 5, left). However, the presence of ettringite is confirmed. Thus, although the cement used had low C_3A content, both the bentonite porewater and the clayey water have contributed to the formation of ettringite. Probably the C_4AF content could be participating in the new ettringite formation that is a stable phase up to pH 10.7 (Gabrisova et al., 1991). This new ettringite formation is quite homogenous throughout the sample, as can be seen in Figure 5 (right).

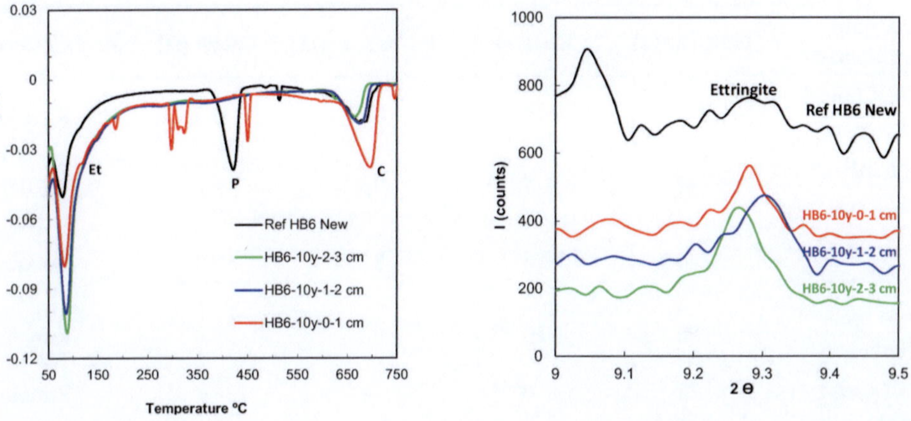

Figure 5: *dTG (left) and XRD (right) of the HB6 concrete cell. Distances from bentonite vs. Reference New HB6.*

Soluble ions confirm the penetration of SO_3 and Al ions in the concrete pores due to the interaction with the bentonite and clayey waters. The interaction enriches the concrete pore fluid in Ca, Si and Mg. A movement and decrease of the alkalis, especially in the zone near the clayey water inlet was detected. Further data of HB6 cell characterization (BSEM and MIP results) are under way to further understand the interactions of both the bentonite and clayey groundwater.

Short term interaction of aged and new concretes with granite-Grimsel and simulated clayey waters

Table 3 compiles the hydraulic permeability coefficient (K), the pH of the first percolated waters and after long time of percolation, together with the pH of the concrete porewater of two samples after percolation test. From the K coefficient different aspects are highlighted as a function of the type of concrete: a) the shotcreted concretes are more permeable to water entrance than the same based concrete (in case of the shotcreted Äspö concrete two orders of magnitude more permeable than non-shotcreted Äspö concrete), b) a relationship exists between K and porosity, the less porous concrete (VTT) shows the lowest K. The technique employed for the placement of the concrete significantly influences the groundwater interaction performance.

The pH and the ionic composition of the percolated waters are being periodically measured. The initial pH leachates are also affected by the type of concrete, the alkaline plume is detected in the hpH concretes. In FEBEX hpH, there is a significant release of the alkalis from the concrete matrix which causes a progressive decrease in the pH of the percolated water. By contrast, in Äspö lpH concretes, alkalis are not removed and even retention of Na takes place, as shown in Figure 6 (left). The pH of the percolated water in both concretes evolves and approaches to the Grimsel groundwater pH after ≈ 3 months of test, also in Table 3.

Table 3: *Main data from percolation tests: hpH, lpH and granite and clayey water interaction.*

Sample identification	Time (days)	K (m/s)	Init leach pH	Leach pH	End Concrete pH
FEBEX hpH-1* *(shotcrete)*	185	$8 \cdot 10^{-11}$	11.8	8.5	12.02/12.07/11.92
FEBEX hpH-2 *(shotcrete)*	170	$3 \cdot 10^{-11}$	12.2	9.4	Running

Sample identification	Time (days)	K (m/s)	Init leach pH	Leach pH	End Concrete pH
FEBEX hpH-3 *(shotcrete)*	273	$2{\cdot}10^{-11}$	12.2	9.1	Running
Äspö lpH-1* *(shotcrete)*	155	$1.3{\cdot}10^{-9}$	7.8	8.6	9.34/9.04/9.02
Äspö lpH-2 *(shotcrete)*	280	$3{\cdot}10^{-12}$	8.3	8.1	Running
Äspö lpH-3 *(shotcrete)*	190	$6{\cdot}10^{-10}$	8.6	8.4	Running
VTT lpH-1	121	$5.4{\cdot}10^{-15}$	--	--	Running
VTT lpH-2	149	--	--	--	Running
VTT lpH-3	149	$1.5{\cdot}10^{-13}$	--	--	Running
HB6 hpH-4	72	$4{\cdot}10^{-12}$	11.5	11.5	Running
HB6 hpH-5	72	$5{\cdot}10^{-12}$	10.8	10.8	Running
HB6 hpH-6	72	$1{\cdot}10^{-13}$	--	--	Running
Äspö lpH-4	65	$4{\cdot}10^{-12}$	--	--	Running
Äspö lpH-5	65	$1{\cdot}10^{-13}$	--	--	Running
Äspö lpH-6	65	$3{\cdot}10^{-11}$	8.6	8.7	Running
VTT lpH-4	65	$3{\cdot}10^{-13}$	--	--	Running
VTT lpH-5	77	$4{\cdot}10^{-13}$	--	--	Running
VTT lpH-6	73	$2{\cdot}10^{-13}$	--	--	Running

*removed

The Ca leaching is also dependent with the type of concrete; while Ca is not released from the concrete in FEBEX hpH, in Äspö lpH some Ca is removed. Another interesting aspect to observe from the leachates of both concretes is the Sulphate and Al, Figure 6 (right). Extra SO_3 is removed in both types of concretes, being more relevant in Febex hpH, while Al is leached only from the FEBEX hpH concrete but not in the Äspö lpH.

The tests using clayey waters, with more dense concretes than those from shotcrete application, have not allowed up to collect enough water, so the decay of the initial alkaline plume is not detected in the HB6 concrete. On the contrary, low pH values of eluted water are measured with Äspö lpH concrete from the beginning. The particular case of lpH VTT concretes, the densest of those under study ($K < 10^{-14}$) have only allowed to percolate few mL of water after 3 and 6 months, both with granite and clayey waters. A new more porous VTT concrete is under study to allow to characterise the interaction of this concrete type with groundwaters.

Two samples (Febex hpH-1 and Äspö lpH-1 concretes) were removed from the percolation system after more than 1.5 L percolation of water through the pores. The DTA/TG and XRD results of FEBEX hpH-1 shows that portlandite is not present after the test along the concrete sample (5 cm) due to the leaching with ganitewater that causes the pH decrease. Besides the peak related to ettringite, from XRD, increase along the thickness of the concrete. Although the reason is not yet clearly understood, it can be suggested that leaching of sulphates from the upper to the lower part of the concrete sample can take place. Later on the formation of ettringite is possible thanks to the unusual excess of Al in the cement paste, probably due to the accelerator used during shotcreting. The higher presence of sulphates and Al in FEBEX pore water is confirmed in the leachates. Clearly, the content of alkalies and Ca in the concrete pores is significantly removed after the leaching test (Figure 7, right). However, more detailed studies are needed to confirm this information and the consequences related to the amount and type of additives used during the production of concretes.

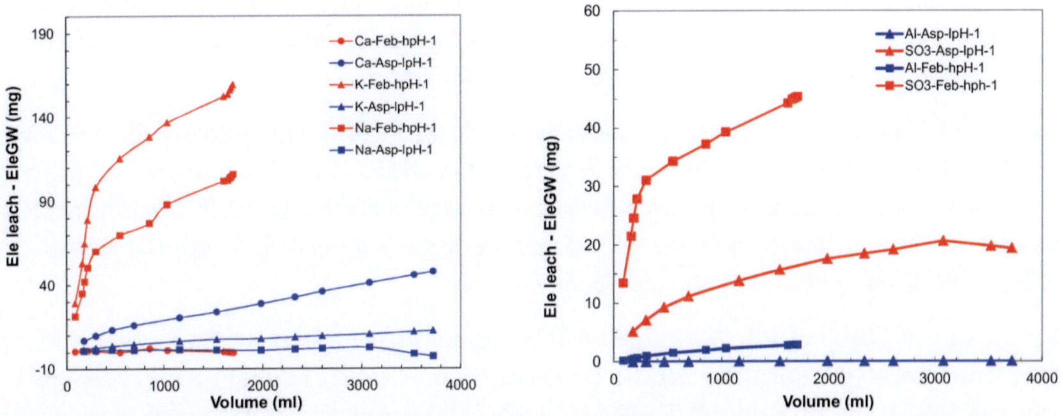

Figure 6: *Na, K, Ca (left) and Al, SO_3 (right) evolution of the percolated waters with Grimsel groundwater.*

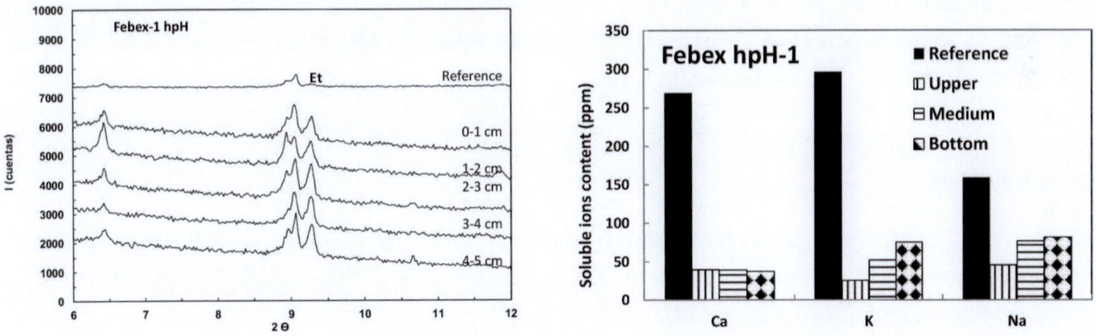

Figure 7: *XRD (left) and soluble ions content (right) of FEBEX hpH-1 in different zones.*

In the Äspö lpH-1 case the soluble ion analysed confirm the retention of Na and Ca in the concrete during the test, as can be seen in Figure 8. DTA/TG results show a significant additional formation of calcite throughout the entire sample. By contrast, neither ettringite or portlandite were detected along the sample. Further data of the characterization of Febex hpH-1 and Äspö lpH-1 (BSEM-EDX analysis and MIP results) are underway.

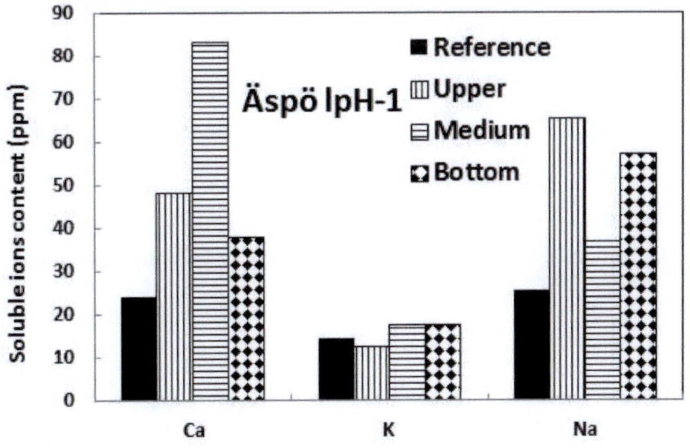

Figure 8: *Soluble ions content in different zones of Äspö lpH-1.*

Conclusions and Future work

- After long periods in contact with groundwater, the concrete of FEBEX system shows additional ettringite formation and portlandite dissolution due to its interaction with the bentonite porewater near to the interface. The concrete of HB6 cell also shows portlandite dissolution and additional formation of ettringite due to the interaction with bentonite and clayey waters.
- After short periods of accelerated groundwater attack, in FEBEX hpH samples the low ionic of the Grimsel granitic water causes the initial release of the alkalis from the concrete and the portlandite dissolution, while in Äspö lpH samples there is no release of alkalis. In FEBEX hpH-1sample after the test, portlandite is not present after the test and more ettringite is formed. In Äspö lpH-1 there is a retention of Na and K in the concrete cement paste during the test.
- Future work of CSIC in CEBAMA project will be focused on completion of the analysis of the ageing at long term through the complete characterization of the overcorings in the FEBEX system (pH, soluble ions and BSEM) and the HB6 cell (BSEM-EDX and MIP) and at short term through the evolution of the percolation tests and the characterization of the concretes after the test with Grimsel granitic water or clayey water in different concretes. New VTT samples with higher porosity than the ones from Ref lpH VTT will be included. Also, no-stationary natural ion diffusion tests with simulated bentonite water are in way in samples extracted from FEBEX concrete core, the aim is to understand the first stages of concrete interaction with the bentonite water.

Acknowledgement

The research leading to these results has received funding from the European Union's Horizon 2020 Reasearch and Training Programme of the European Atomic Energy Community (EURATOM) (H2020-NFRP-2014/2015) under grant agreement n° 662147 (CEBAMA).

References

Alonso, M.C., García Calvo, J.L., Petterson, S., Puigdomenech, I., Cuñado, M.Á., Vuorio, M., Weber, H., Ueda, H., Naito, M., Walker, C., Takeshi, Y., Cau-dit-Coumes, C. (2012). Round robin test for defining an accurate protocol to measure the pore fluid pH of low-pH cementitious materials. *In:* Bart, F., Cau-dit-Coumes, C., Frizon, F., Lorente, S. (Eds.) Cement-Based Materials for Nuclear Waste Storage. Springer, 251-259.

Alonso, M.C., García Calvo, J.L., Flor Laguna, V. (2016). Characterisation of concrete aging due to the interaction with groundwaters in contact with different engineering barrier system (EBS). Proceedings of the 1st CEBAMA Annual Workshop. KIT Scientific Publishing, KIT-SR 7734.

Alonso, M.C., García Calvo, J.L., Cuevas, J., Turrero, M.J., Fernández, R., Torres, E., Ruiz, A.I. (2017). Interaction processes at the concrete-bentonite interface after 13 years of FEBEX-Plus operation. Part I: Concrete alteration. Physics and Chemistry of the Earth.

García Calvo, J.L., Hidalgo, A., Alonso, C., Fernández Luco, L. (2010). Development of low-pH cementitious materials for HLRW repositories. Resistance against ground waters aggression. Cement and Concrete Research, 40, 1290-1297.

Gabrisova, A., Havlica, J., Sahu, S. (1991). Stability of calcium sulphoaluminate hidrates in wáter solutions with various pH values. Cement and Concrete Research, 21, 1023-1027.

Turrero, M.J., Torres, E., Garralón, A., Gómez, P., Sánchez, L., Peña, J., Buil, B., Escribano, A., Durán, J.M., Domínguez, R. (2016) Interaction concrete / FEBEX bentonite: outlining experimental conditions and characterization approaches at laboratory and in situ scale. Proceedings of the 1st CEBAMA Annual Workshop. KIT Scientific Publishing, KIT-SR 7734.

Geochemical evolution of cementitious materials in contact with a clayey rock at 70°C in the Tournemire underground research laboratory

Alexandre Dauzeres[1*], Philippines Lalan[1,2], Laurent De Windt[2], Valéry Detilleux[3]

[1] IRSN, Institute for Radiological Protection and Nuclear Safety (FR)
[2] MINES ParisTech (FR)
[3] BELV (BE)

* Corresponding author: alexandre.dauzeres@irsn.fr

Abstract

Radioactive wastes in future clayey underground disposal sites will potentially induce, in function of the architecture chosen, a temperature increase at the interface between the cementitious materials and the host rock. To understand the evolution of cementitious materials (Portland and low-pH cements) in this environment, an *in-situ* specific device named CEMTEX (CEMent Temperature EXperiment) was developed in the IRSN Underground Research Laboratory in Tournemire (France). CEM I and two types of low-pH cement pastes were put in contact with the clayey rock under water-saturated conditions at 70°C for 9 experiments.

The first six experiments, three for CEM I and three for the first low-pH cement paste (Mix of CEM I, silica fume and fly ashes) were started in 2012 for a duration of 1, 2 and 5 years. In April 2016, three other experiments were started with a second type of low-pH cement paste (Mix of CEM III, silica fume and filler), based on a self-consolidating concrete (SCC) (Poyet et al., 2014) for three different durations: 1, 2 and 3 years.

This publication is dedicated to the characterization of the high-pH CEM I (OPC) / clayey rock interfaces by XRD, and SEM-EDS for the three durations and by TGA, X-ray microtomography and autoradiography on a sample impregnated with ^{14}C-PMMA, for the interface after one year of interaction. After one, two and five years of interaction, partial decalcification and diffuse carbonation (calcite precipitation) was observed in the cement, globally resulting in a total porosity decrease. Close to the interface, in the heterogeneity linked to the initial bedding of the rock, a layer rich in potassium, silicon and aluminium is systematically formed, constituted of zeolites as phillipsite, well-crystallized C-S-H as tobermorite (only after one year), and C-(A)-S-H.

Introduction

In France, the decision has been taken to design a radioactive waste disposal facility in a natural clayey environment with favourable radionuclide containment properties. However, for stability reasons during the excavation work and for facilitating the installation of radwastes, galleries excavated in clayey rocks must generally be mechanically supported by concrete components. The contact between the clayey rock and the concrete inevitably leads to an alkaline plume spreading from the concrete toward the host rock, while a multi ionic attack occurs from the clayey pore water against the cementitious material side.

Several types of cementitious binders are selected as references today for the design of deep disposal facilities: an ordinary Portland cement and two low-pH binders. The low-pH binders were formulated to induce a less aggressive alkaline plume mainly thanks to a lower pore water pH. Over the last decades, interest has grown about the geochemical behaviour of cementitious materials in a clayey environment. Studies have focused mainly on the physico-chemical evolution of clay under alkaline conditions and not on the material interface. Only a handful of studies have discussed the interactions at the interface between cement binder and claystone. Such interfaces have been created at ambient temperature by pouring cement mixes into boreholes (Gaboreau et al., 2011; Jenni et al., 2014; Bartier et al., 2013; Dauzères et al., 2016) or by putting disks of material into contact in transport cells (Dauzères, 2010). In the current design of the French radioactive waste deep disposal facility (Andra, 2005), the thermal transient, due to the presence of heat-emitting waste, is expected to entail a temperature increase of up to 70°C in the concrete plugs located in the high-level waste cells and potentially also in the intermediate long-lived waste cells. Based on the existing literature, only a few laboratory studies have focused on the impact of such a temperature level on the evolution of the interface between natural rock and anthropic material, and even fewer in-situ studies have been carried out.

The CEMTEX experiment (CEMent Temperature EXperiment) (Lalan et al., 2016) aims to fill the lack of knowledge about the geochemical and micro-structural evolutions of these interfaces under representative thermo-hydraulic conditions (i.e., 70°C and full water saturation), which should provide input for future studies on the durability of cementitious materials. Within this framework, six *in-situ* experiments (three with CEM I and three with a low-pH cement) have been set up in IRSN's underground research laboratory in Tournemire (France) in 2012, before the starting of the CEBAMA project and three other with a second low-pH cement paste were started inside the CEBAMA project in 2016. This paper presents quickly the experimental device already describe in previous papers (Lalan et al., 2016; Dauzères et al., 2016b) and focus mainly on recent results obtained on the first experiments putting in contact high pH CEM I (OPC) and clayey rock after 1, 2 and 5 years of interaction.

Experimental device

Nine downward vertical boreholes of 1 m depth and 25 cm of diameter were drilled into the argillite in the Tournemire tunnel (6 in 2012 and 3 in 2016). This depth was chosen to ensure that the devices were located out of the excavation disturbed zone (EDZ) created while the gallery was dug. The bottom of each borehole was polished with a specific tool and all the dust was carefully vacuum cleaned to obtain an as perfect as possible contact surface between the rock and the cementitious material. Three cups filled with water and humidity probes were placed into the borehole, which was then tightly closed to resaturate the rock by vapour phase. The rock was considered saturated when the measured relative humidity within the borehole reached a value of 99.5%. In each borehole, the relative humidity reached 97.5% and 99.5% after 7 and 45 days, respectively. To force a one-dimensional mass transport (vertical) across the interface and to protect the borehole sides from hyperalkaline solution, a PVC tube (25 cm outer diameter, 0.5 cm thick, 125 cm height) was placed within the borehole (Figure 1). A rubber seal was placed at the bottom of the PVC tubing to seal the bottom of the tube from the borehole wall.

At the same time, heater devices were built outside the boreholes (Dauzères et al., 2016b). Four temperature sensors (PT 100) were attached to a single support in order to monitor the temperature during the experiment. Two sensors were close to the centre of the coil, one at 1 cm height from the future interface (M1) and the other at 20 cm (M0). The two other sensors were placed just inside the coil, again at 1 cm (M2) and 20 cm (M3). The coil top and the temperature sensor support were fixed to a PVC shell of 20 cm diameter. A rubber seal provided sealing between the argillite borehole bottom and this latter PVC shell. After the argillite saturation, the entire device was placed into the borehole and then connected to a boiler via 12 mm diameter PFA pipes (perfluoroalkoxy) and to the temperature acquisition system (ALMEMO). Once all these steps had been achieved, the cement paste was poured onto the heater device and the temperature sensors to fill the full inner PVC volume constituting a 30 cm height OPC and low-pH paste plug. A small amount of cement paste was also poured between the two PVC shells over a few centimetres.

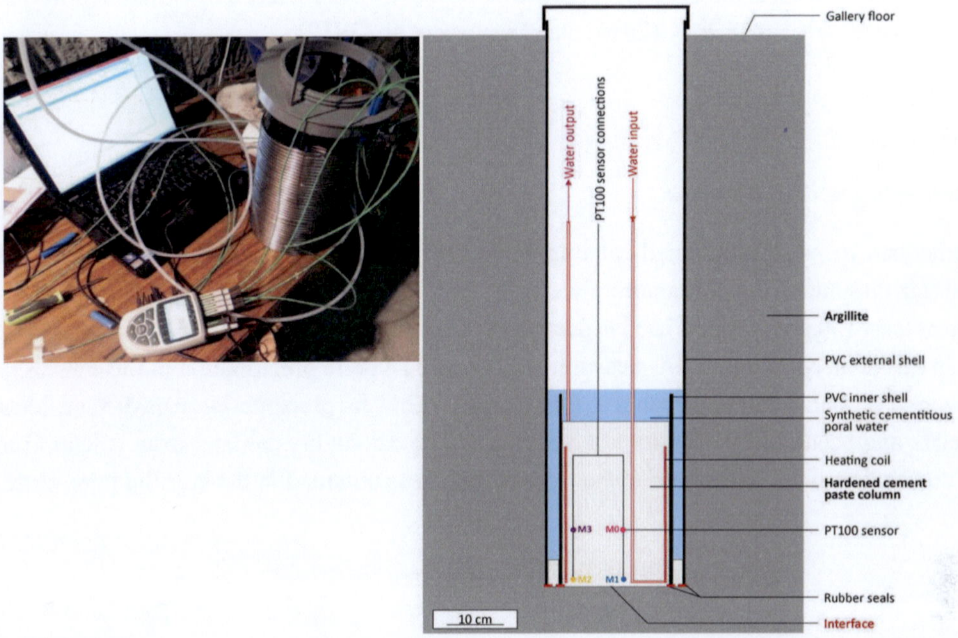

Figure 1: *View of the heating coil device connected to the temperature monitoring system (a) and outline of the CEMTEX in-situ device (b).*

One month after the beginning of the cement hydration, heating was started while the device was being maintained under the equilibrium solution at the good pH value with the considered cementitious material in order to ensure water saturation throughout the test. The water reservoir is trapped between the cement paste and a layer of Resin to limit the evaporation of water. The system is closed and protected.

The materials

Three types of cementitious binders were placed in the experiments:

- An Ordinary Portland Cement, Sulphate Resisting Portland Cement (OPC-SRPC) from Val d'Azergues (France, Lafarge®) was used. The cement paste had a water/cement ratio of 0.42. (Lalan et al., 2016; Dauzères et al., 2016b). It was put in place in 2012 in 3 tests. The results of these tests are presented in this paper.
- A low-pH binder, chosen currently as a reference material by Andra, composed by a mix of CEM I Le Teil (Lafarge) at 37.5 wt.%, silica fume at 32.5 wt.% (Condensil) and fly ashes (Calcia) at 30 wt.%. A super plasticizer type CHRYSO® Fluid Optima 175 was added (Codina et al., 2008). These formulation was used in the three other tests started in 2012 with a water on binder ratio (w/b) equal to 0.4.
- The low-pH binder formulation chosen for the 3 new tests launched in 2016 is detailed in Poyet et al. (2014) (DOPAS European project - FSS experiment). It was chosen as formulation with the lower pH in the pore solution after 28 days of hydration (based on this study). It is a mix of CEM III Rombas (Calcia) at 20 wt.%, Silica Fume (Condensil) at 20 wt.% and 60% of calcareous filler (Carmeuse).

The Super Plasticizer Glenium 537 is added (2.3 wt.%). The equivalent w/b ratio (reactivity index is equal to 1 for the cement and the silica fume and 0.2 for the fillers) used is equal to 0.60.

- The clayey rock is the Toarcian argillite in the Tournemire URL. The different properties of this rock are detailed in Lalan et al. (2016) and Tremosa et al. (2012).

The results

The main mineralogical evolutions

EDS elementary maps showed small calcium depletion in the cement paste after every test. TGA and XRD analyses exhibit dissolution of portlandite always inferior to 1 mm (Figure 2). After one year, Portlandite is dissolved on at least 100 µm (method used is described in Le Saoût et al., 2008) and C-S-H was probably partially decalcified in this area. XRD and TGA measurements showed calcite precipitation at the cement paste side even though no calcite crystals were observable at the interface. Thus its precipitation was diffuse, located within the material matrix and occurred on 700 µm after one year. The calcite crystallized from calcium ions released by portlandite dissolution and a part would come from carbonates contained in the argillite pore water.

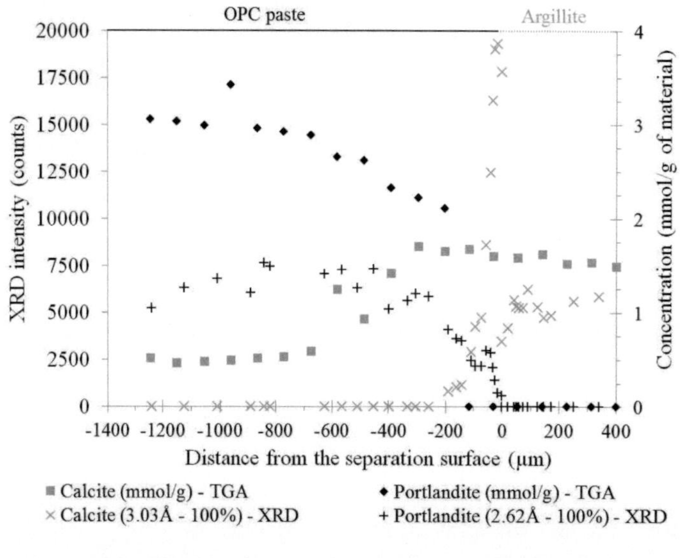

Figure 2: *TGA and XRD analyses of the CEM I / clayey rock interface sample after 1 year of interaction.*

Our first observations by SEM after one year of interaction showed the presence of honeycomb structure pierced by prismatic minerals while other crystals have crystallized among the honeycomb structure (Lalan et al., 2016). The honeycomb structure is often described as typical of C-S-H. Moreover prismatic minerals presented twin planes. This feature can be displayed by phillipsite crystals. These formations are observed at 1, 2 and 5 years of interaction. After 2 years, an homogeneous layer of zeolite, measuring between 30 and 50 µm-thick is precipitated in the clayey rock bedding, close to the interface with the cement paste (Figure 3).

The presence of zeolites is confirmed by the XRD analyses (Figure 4) on the interface after 5 years of interaction. The XRD outlined patterns of phillipsite over a thickness of about 60 microns around the separation plan after one year and more than 100 µm after 2 years and 5 years. This layer was also observed on EDS elementary maps where calcium and silicon intensities were relatively high. After one year of interaction, tobermorite was identified clearly by XRD but never at 2 and 5 years of interaction. TEM images and diffraction confirmed the presence of tobermorite and C-S-H after one year (Lalan et al., 2016) in this area. Possibly, the precipitation of zeolite already in progress at 2 and 5 years profits from the tobermorite dissolution and from the aluminium and the potassium coming respectively from the clay dissolution and the cement pore solution.

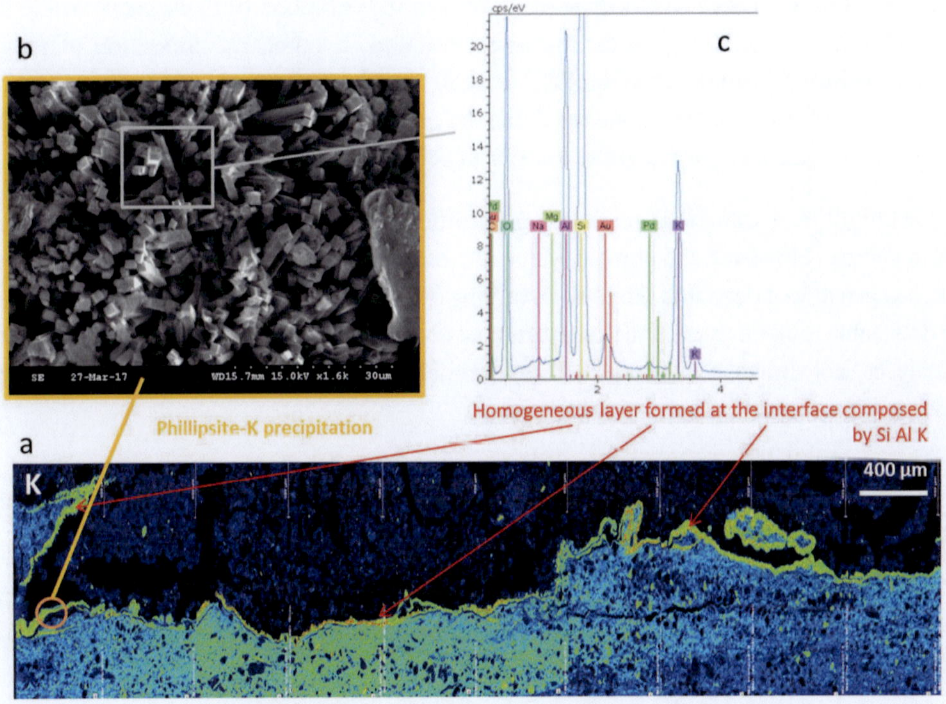

Figure 3: *SEM-EDS analyses of the CEM I/clayey rock interface after 2 years of interaction: a) set of EDS mapping of potassium on polished section; b) SE picture of zeolite crystals on fresh surface; c) chemical analyses of the zeolite crystals.*

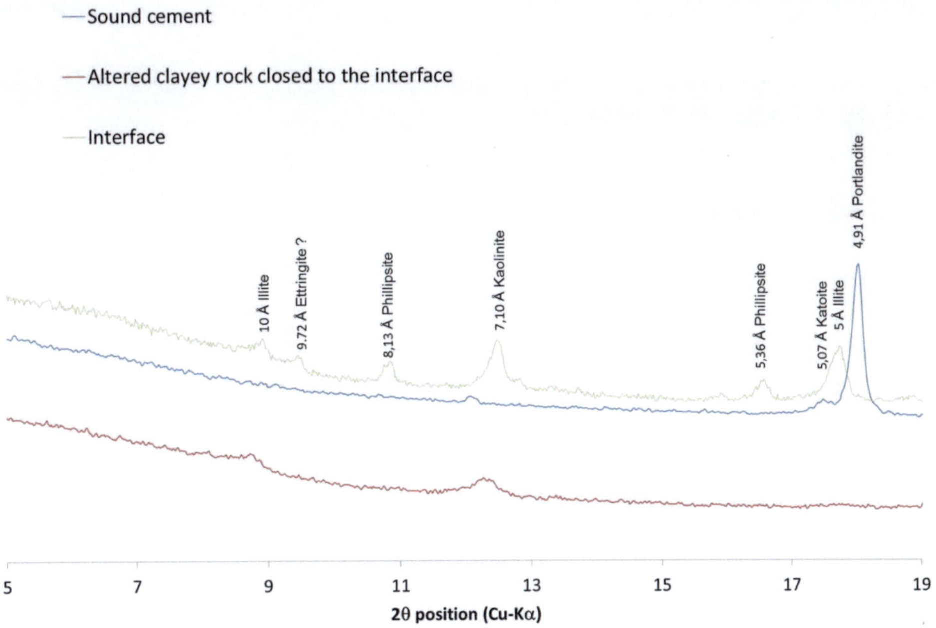

Figure 4: *XRD analyses (focus on the phillipsite main peaks) of a pure CEM I cement paste, the interface area composed by zeolite and the altered clayey layer in contact with the cement paste, after 5 years of interaction.*

The autoradiography analyses carried out on the one-year test show the decrease of the total connected porosity of the cementitious material to one centimeter, from 34% (initial porosity) to 27% in contact with the interface. Overall, the alteration of the cement paste led to a decrease in the total connected porosity (Figure 5). Precipitation processes such as the formation of C-A-S-H, zeolites or carbonation lead to a reduction of the total connected porosity of the cement paste. These processes seem to govern the evolution of the system, unlike the process of decalcification. The increase of temperature at these interfaces favoured the formation of clogging phases, contrary to what has already been observed at 20°C or at 50°C. (Gaboreau et al., 2011; Dauzères, 2010). As this porosity decrease remains low, it can be concluded that the precipitation of the C-S-H and zeolite ribbon close to the interface does not constitute a barrier to the transfer of solutes.

Concerning the argillite, it seems that, regarding the profile (Figure 5), a reduction of the porosity at the surface is observed as in cement. However, the porosity value at 1 cm depth is between 14 and 15%. This means that the alkaline plume has penetrated deep into the rock disturbing the clayey sheets causing a porosity increase without, however, any detectable mineralogical transformations, as observed in the CI experiment in the Mont-Terri URL. The current analyses (not shown, see Lalan, 2016) of strontium isotopes seem to show that the depth reached by the alkaline plume is much greater than the centimeter.

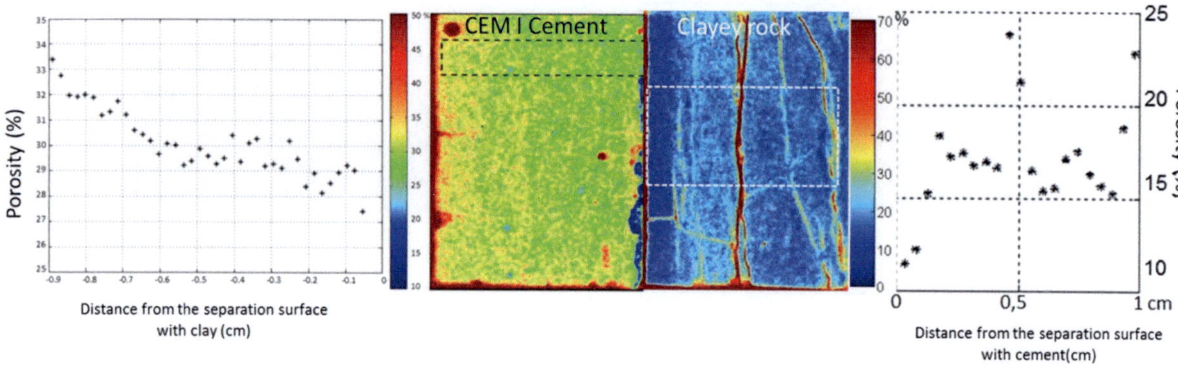

Figure 5: *Total porosity profile around the CEM I / clayey rock interface after 1 year of interaction, measured by autoradiography on doped sample with* ^{14}C*-PMMA.*

Conclusions and Future work

The CEMTEX project is still in progress and the first step of the work concerning the characterization of CEM I cement paste / clayey rock interface is almost finished. The important results in terms of exchange of soluble species, mineralogy changes causing microstructure modifications are summarized in the Figure 6. The new important result in term of reactive mechanism is the systematic precipitation of zeolite at 70°C after 1, 2 and 5 years of interaction. This kind of precipitation was never observed in the previous study putting in contact CEM I cementitious material and clayey rock in representative conditions of geological disposal. The second important result is the very low degradation of the CEM I material. The temperature seems favoured the carbonation causing a small decreasing of the total porosity contrary to the results in previous study at 20°C (Dauzères, 2010; Gaboreau et al., 2011). The future work as part of the CEBAMA project in the next year will focus on the low-pH / clayey rock interface.

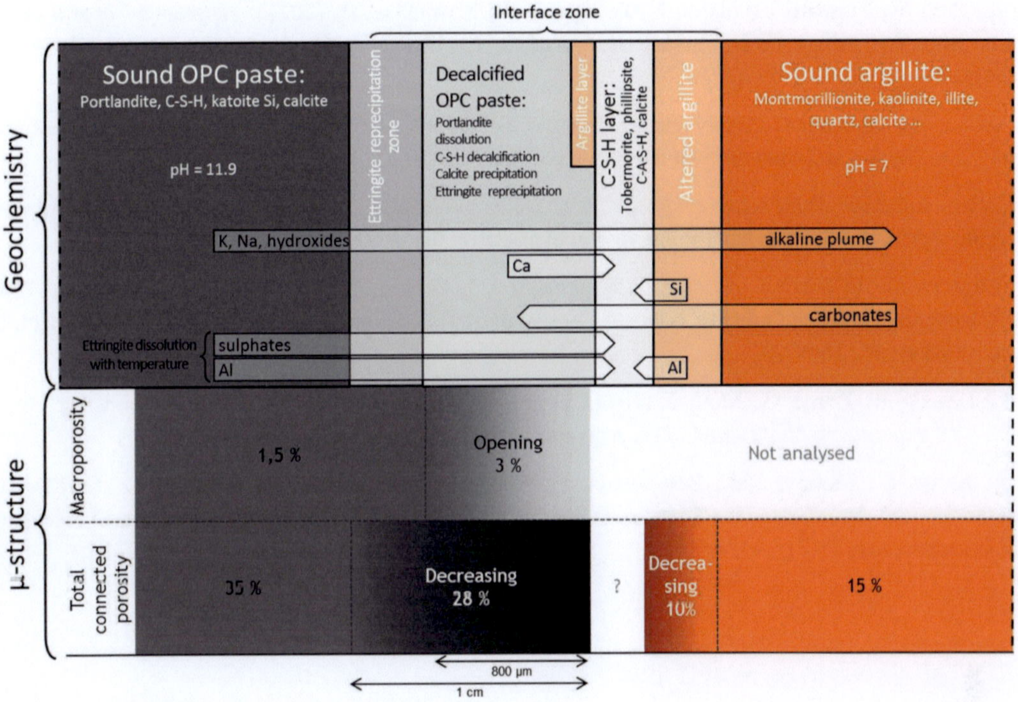

Figure 6: *Summary diagram of the main mechanisms observed at the CEM I cement paste / clayey rock interface at 70°C.*

Acknowledgement

The research leading to these results has received funding from the European Union's Horizon 2020 Reasearch and Training Programme of the European Atomic Energy Community (EURATOM) (H2020-NFRP-2014/2015) under grant agreement n° 662147 (CEBAMA).

References

Andra (2005) Dossier 2005 Argile - Evaluation de la faisabilité du stockage radioactif en formation argileuse. Site de Meuse/Haute-Marne, Andra.

Bartier, D., Techer, I., Dauzères, A., Boulvais, P., Blanc-Valleron, M.-M., Cabrera, J. (2013). *In-situ* investigations and reactive transport modelling of cement paste / argillite interactions in a saturated context and outside an excavated disturbed zone. Applied Geochemistry, 31, 94-108.

Codina, M., Cau-Dit-Coumes, C., Le Bescop, P., Verdier, J., Ollivier, J.P. (2008). Design and characterization of low-heat and low-alkalinity cements. Cement and Concrete Research, 38(4), 437-448.

Dauzères, A., Le Bescop, P., Sardini, P., Cau Dit Coumes, C. (2010). Physico-chemical investigation of clayey/cement-based materials interaction in the context of geological waste disposal: Experimental approach and results. Cement and Concrete Research, 40, 1327-1340.

Dauzères, A., Achiedo, G., Nied, D., Bernard, E., Alahrache, S., Lothenbach, B. (2016). Magnesium perturbation in low-pH concretes placed in clayey environment-solid characterizations and modeling. Cement and Concrete Research, 97, 61-72.

Dauzères, A., Lalan, P., De Windt, L., Detilleux, V., Bartier, D., Techer, I. (2016b). Geochemical evolution of cementitious materials in contact with a clayey rock at 70°C. Proceedings of the 1st CEBAMA Annual Workshop. KIT Scientific Publishing, KIT-SR 7734.

Gaboreau, S., Prêt, D., Tinseau, E., Claret, F., Pellegrini, D., Stammose, D. (2011). 15 years of *in-situ* cement-argillite interaction from Tournemire URL: Characterisation of the multi-scale spatial heterogeneities of pore space evolution. Applied Geochemistry, 26, 2159-2171.

Jenni, A., Mäder, U., Lerouge, C., Gaboreau, S., Schwyn, B. (2014). *In-situ* interaction between different concretes and Opalinus Clay. Physics and Chemistry of the Earth, Parts A/B/C, 70-71, 71-83.

Lalan, P. (2016). Influence d'une température de 70°C sur la géochimie, la microstructure et a diffusion aux interfaces béton/argile: experimentation en laboratoire, *in-situ* et modélisation. PhD Thesis.

Lalan, P., Dauzères, A., De Windt, L., Bartier, D., Sammaljärvi, J., Barnichon, J.-D., Techer, I., Detilleux, V. (2016). Impact of a 70°C temperature on an Ordinary Portland Cement paste / claystone interface: an *in-situ* experiment. Cement and Concrete Research, 83, 164-178.

Poyet, S., Le Bescop, P., Touze, G., Moth, J. (2014). Formulating a low-alkalinity and self-consolidating concrete for the DOPASS-FSS experiment. NUWCEM 2014, Avignon.

Tremosa, J., Arcos, D., Matray, J.M., Bensenouci, F., Gaucher, E.C., Tournassat, C., Hadi, J. (2012). Geochemical characterization and modelling of the Toarcian/Domerian porewater at the Tournemire underground research laboratory. Applied Geochemistry, 27(7), 1417-1431.

Cebama reference mix design for low-pH concrete and paste, preliminary characterisation

Tapio Vehmas*[1], Markku Leivo[1], Erika Holt[1], María Cruz Alonso[2], José Luis García[2], Álvaro Fernández[2], Matthew Isaacs[3], Emily Rastrick[3], David Read[3,4], Radek Vašíček[5], Jaroslav Hloušek[5], Lucie Hausmannová[5], Petr Večerník[6], Radek Červinka[6], Václava Havlová[6], Steve Lange[7], Martina Klinkenberg [7],Dirk Bosbach [7], Guido Deissmann[7], Vanessa Montoya[8], Naila Ait Mouheb[8], Christian Adam[8], Dieter Schild[8], Thorsten Schäfer[8]

[1] VTT, Technical Research Centre of Finland Ltd. (FI)
[2] CSIC, High Research Council of Spain. Inst. of Construction Science Eduardo Torroja (ES)
[3] University of Surrey (UK)
[4] National Physical Laboratory (UK)
[5] Centre of Experimental Geotechnics, Faculty of Civil Engineering, Czech Technical University in Prague (CZ)
[6] ÚJV Řež, a.s. (CZ)
[7] Forschungszentrum Jülich GmbH, Institute of Energy and Climate Research (IEK-6): Nuclear Waste Management and Reactor Safety (DE)
[8] KIT-INE, Karlsruhe Institute of Technology-Institute for Nuclear Waste Disposal (DE)

* Corresponding author: tapio.vehmas@vtt.fi

Abstract

A reference low-pH concrete and paste mix were manufactured within the Cebama-project. Reference mixtures were casted at VTT in March 2016 and were distributed among some of the Cebama partners. Reference mix designs will be used by different partners as a common material to study their behaviour in contact with waters of different composition and interaction with radionuclides. Additionally, these materials will be used for model calibration. This article provides a summary of the characterization methods used by different partners and gives an overview of the experiments which will be made in the future with the reference materials.

Introduction

A reference low-pH concrete and paste mixture for the Cebama-project was manufactured in March 2016 by VTT and distributed to some of the partners. Reference mix designs enable comparison of various research methods applied in the Cebama-project. Mix designs of the reference concrete (RCM) and reference paste (RPM) are presented at Table 1. The basis of the Cebama reference concrete was the low-pH concretes developed in the FP7 Dopas-project (Holt and Koho, 2016). The composition of the Cebama RCM and low-pH concretes have identical mix design of fillers and coarser aggregates. Also the CaO/SiO_2 -ratio of the mix designs is identical. The only factor that differ is the binder composition. The Cebama RCM has a ternary binder composition (cement 105 kg/m^3, silica fume 110 kg/m^3 and blast furnace slag 65 kg/m^3), whereas the ternary low-pH concrete from the

Dopas-project has a cement 105 kg/m^3, silica fume 91 kg/m^3 and fly ash 84 kg/m^3. The binder composition of binary low-pH concrete from the Dopas -project consisted of cement 120 kg/m^3 and silica fume 80 kg/m^3. More detailed information of the sample castings and mix design is described in the 1st Annual Workshop Proceedings of Cebama-project (Vehmas et al., 2016).

This contribution provides an overview of the preliminary results performed to characterize the RCM and RPM. The article presents preliminary results from the partners and summarizes the upcoming studies.

The data from the reference mixture assessments will be supplied for WP3 - modelling and various parameters can be extracted from the data. The studies also provide common references in methodologies which allow better comparison to locally characterized materials by different partners.

Table 1: *Mix designs of Cebama reference concrete- and paste-mix designs.*

Materials	Concrete (RCM)	Paste (RPM)
CEM I 42.5	105 kg/m^3	468 kg/m^3
Silica fume	110 kg/m^3	491 kg/m^3
Blast furnace slag	65 kg/m^3	290 kg/m^3
Quartz filler	116 kg/m^3	517 kg/m^3
0 - 1 mm	168 kg/m^3	-
0 - 8 mm	770 kg/m^3	-
8 - 16 mm	532 kg/m^3	-
16 - 32 mm	396 kg/m^3	-
Water (effective)	120 kg/m^3	312 kg/m^3
Superplasticizer*	16.80 kg/m^3	75 kg/m^3
Water / binder -ratio	0.43	0.25

*Superplasticizer was naphthalene sulfonate -based Pantarhit LK FM from HaBe.

Research methods

Cebama reference concrete (RCM) and paste (RPM) were analysed with multiple methods (Table 2). Authors' contributions have been labelled with the superscript corresponding to the numbers described in the address lines. Besides the authors, University of Sheffield (USFD), Autonomic University of Madrid (UAM) and The French Geological Survey (BRGM) are performing additional studies related to the reference mix designs. The French Geological Survey results of spectral induced polarization (SIP) for reference -concrete (RCM) and -paste (RPM) are presented in a separate article (Huisman et al., 2017).

Table 2: *Cebama reference concrete and paste characterization methods.*

Quality	Analyzer	Paste (RPM)	Concrete (RCM)
	Fresh-stage properties		
Workability	VTT[1]	yes	yes
Air Content	VTT[1]	no	yes
Heat of hydration	USFD	yes	no
Setting	USFD	yes	no

Quality	Analyzer	Paste (RPM)	Concrete (RCM)
	Mechanical properties		
Compression strength	VTT[1], USFD, CTU[5]	yes	yes
	Chemical composition		
X-ray diffraction (XRD)	KIT[8], USFD, JUELICH[7], SURREY[3], CSIC[2], UAM	yes	yes
X-ray fluorescence (XRF)	SURREY[3]	initial materials	
Thermogravimetry (TG/DTA)	KIT[8], USFD, CSIC[2], UAM	yes	yes
^{29}Si and ^{27}Al MAS NMR	KIT[8]	yes	no
Energy dispersive microscopy (SEM, Back Scattering+ EDS)	KIT[8], CSIC[2], USFD, JUELICH[7]	yes	yes
Pore solution pH	KIT[8], VTT[1], CSIC[2]	yes	yes
	Microstructure		
Scanning electron microscopy (SEM)	KIT[8], CSIC[2], USFD, JUELICH[7]	yes	yes
Porosity	CSIC[2], USFD, UJV[6]	yes	yes
X-ray computed tomography	USFD	yes	no
	Transport properties		
Leaching	VTT[1]	yes	yes
Percolation	USFD, SURREY[3], CSIC[2], CTU[5], UJV[6]	yes	yes
Diffusion	JUELICH[7], CSIC[2], UAM, KIT[8]	yes	yes
	Other		
Density	VTT[1], USFD	yes	yes
Spectral induced polarization (SIP)	JUELICH[7] and BRGM	yes	no

Results

Results from the reference concrete and paste mix designs from Cebama partners are presented in the next chapters. The results are divided into concrete and paste studies as their properties are not identical. During the castings of the reference mix designs, it was decided to obtain the excellent mechanical properties of reference concrete to the reference paste. As a consequence, the chemical properties of the concrete and paste can differ due the lower total water content used in the Cebama reference paste mix design.

Reference concrete mix design (RCM) studies

Studies related to the characterization of the RCM have been mainly performed by CSIC and VTT and are briefly described below.

CSIC

All tests carrying out by CSIC relate to the concrete reference mix (RCM). Tests refer to:

1. Identification of chemical / microstructure evolution during hydration by XRD, DTA/TG, BSEM + EDS, MIP (mercury intrusion porosimetry) and pH measurement.

2. Characterisation of alteration transport properties through ground water percolation column test (with Granite Grimsel water and simulated clayey water) and natural Cl diffusion test.

In addition to RCM studies, CSIC is studying a similar concrete mix dosage (in cement + supplementary cementitious materials + aggregates content) to the RCM but higher w/b ratio = 0.6 (air content = 2.7% and low density = 2,320 kg/m^3). This new concrete will be identified as RCM-LD (Low Density). The aim of having this new concrete is to favour the transport properties that allow to identify changes in processes during the Cebama project. Characterisation of the initial materials, percolation and Cl diffusion tests will also be carried out in this RCM-LD concrete.

The tests conditions and progress are detailed in Table 3.

Table 3: *Test conditions and progress of the initial (Init) characterization of the RCM and RCM-LD.*

Concrete type	**Init. pH** (Alonso et al., 2012)	**Init. water soluble pore Ion**	**Init XRD**	**Init DTA/TG**	**Init BSEM+EDS**	**Init MIP**
RCM*	yes	yes	yes	yes	yes	yes
RCM-LD**	yes	In progress	yes	In progress	In progress	In progress

*age from casting: 5 months, **age from casting: 2 months

The pH in both concretes is below that from a $Ca(OH)_2$ saturated solution (12.5), values measured are: RCM (11.47 ± 0.03) and RCM-LD (11.57 ± 0.06). XRD tests were carried out of enriched powder samples of cement paste removing the coarse aggregates. XRD results do not reveal the presence of portlandite. Sulfoaluminate phases, as ettringite, are also not evident at this age of hydration. The calcium-silicate-hydrate (C-S-H) crystalline phases are not clearly identified due to the interference with aggregates. The absence of portlandite is also confirmed by TG/DTA. The BSEM images in RCM show a dense microstructure and good aggregate-cement paste of interfacial transition zone (ITZ). The EDS reveal a mean CaO/SiO_2 ratio = 0.55 ± 0.17 and a CaO/Al_2O_3 ratio = 8 ± 4.6. MIP confirms the low total porosity in RCM (2.3%) with pores distributed below 0.02 μm, (mean pore size 0.01 μm), all this agree with the high density of this concrete (2.42 g/mL).

Table 4 summarizes the tests carried out to characterise the evolution of transport properties in RCM and RCM-LD. The first results with RCM, during 100 days of operation in percolation tests at 8 bars show high resistance to percolation of water due to its low porosity. The hydraulic conductivity from the few ml of water percolated with time gives values between 10^{-13} to 10^{-14} m/s. The pH of the first eluted water shows values of 8.1 with clayey water.

Table 4: *Transport properties characterisation tests.*

Concrete type	**P (bar)**	**Percolation** 3 alteration ages	**Cl natural Difussion** At 12, 24 months
RCM	8	In progress (100 d)	Future
RCM-LD	1	In progress (7 d)	Future

VTT

VTT has measured the fresh-stage properties of the reference concrete (RCM). Slump of the fresh reference concrete was 180 mm. Density of the fresh concrete was 2,450 kg/m^3 with air content 0.9% (SFS-EN 12350-7,

2009). Compression strengths and densities of the reference concrete at various ages are presented in Table 5. Compression strength results and densities of Dopas-project binary and ternary low-pH concretes are also presented in Table 5 for comparison.

Table 5: *Compression strengths and densities of RCM, Binary- and Ternary mixtures developed in the Dopas-project.*

Age	Compression strength	Density
	Cebama Reference Concrete (RCM)	
7 d	47.0 MPa	2,420 kg/m³
28 d	80.0 MPa	2,430 kg/m³
424 d	88.5 MPa	2,420 kg/m³
	Binary low-pH Concrete	
91 d	95.5 MPa	2,420 kg/m³
483 d	112.5 MPa	2,420 kg/m³
	Ternary low-pH Concrete	
91 d	80.0 MPa	2,400 kg/m³
475 d	100.0 MPa	2,420 kg/m³

pH-development of RCM was determined according to the method described in the literature (Alonso et al., 2012). Results are presented in Table 6. The results are accompanied with pH-values of Binary- and Ternary low-pH concretes.

Table 6: *Measured pH-values of RCM, along with Binary- and Ternary- mixtures developed in the Dopas-project.*

Age	pH
Cebama Reference Concrete (RCM)	
8 d	12.16
19 d	11.94
28 d	11.92
428 d	11.47
Binary low-pH Concrete	
59 d	11.65
91 d	11.55
486 d	11.29
Ternary low-pH Concrete	
59 d	11.59
91 d	11.59
486 d	11.29

Reference paste mix design (RPM) studies

Studies related to the characterization of the RCM concrete have been mainly performed by KIT, SURREY, CTU, UJV, JULICH, and VTT and are described briefly below.

KIT

KIT-INE has mainly focused on the chemical and microstructure characterization of the cement paste using X-ray diffraction (XRD), thermogravimetric - differential thermal analysis (TG-DTA), ^{29}Si and ^{27}Al magic angle spinning nuclear magnetic resonance (^{29}Si and ^{27}Al MAS NMR) and scanning electron microscopy - energy dispersive X-ray spectroscopy (SEM-EDX). The analysis of the pH of the solid was also measured using the *ex-situ* leaching method described by Alonso et al. (2012).

The analysis of the pH of pore water results in a value around 11.7. The X-Ray diffractogram of the paste identified quartz as the main crystalline phase present in the solid. C-S-H phases could not be detected in any of the samples probably due to its amorphous nature. Furthermore, ettringite ($Ca_6Al_2(SO_4)_3(OH)_{12}\cdot 26H_2O$) has been identified as a minor phase. TG- DTA shows a loss of mass around 100 - 300°C which is attributed to C-S-H phases and ettringite. However, in this temperature range the mass loss can also be attributed to water bound to mineral surfaces (10.9%) making a quantification of the of C-S-H phases difficult. The weight loss at 450°C which is indicative for portlandite is not observed confirming the absence of this solid phase.

The pastes were also analyzed by SEM-EDX to observe the morphology and chemical composition of the various hydrated phases. C-S-H phases, non-reacted silica fume (with a characteristic spherical shape), quartz and feldspar are detected. Additionally, Ca/Si-ratio of the C-S-H phases are determined which is presently not possible by the other techniques used in this study. SEM-EDX revealed low Ca/Si-ratios between 0.5 and 0.7. To complete the interpretation of the data and to show the homogeneity of the sample at the microscale range, elemental maps of silicon, calcium, iron, magnesium, aluminum and sulphate have been produced (Figure 1). The various mineral phases present are characterized by the local concentrations of calcium, silicon and iron.

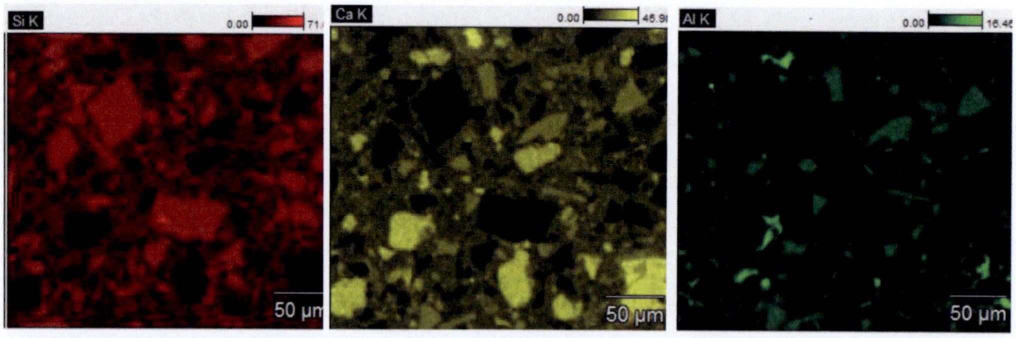

***Figure 1:** SEM-EDX elemental maps of silicon (red), calcium (yellow), and aluminum (green).*

Solid state NMR-spectroscopy was applied to the samples. The main advantage of solid state NMR spectroscopy is its ability to detect nuclei in amorphous phases and its possibility to provide structural information. The ^{29}Si MAS NMR spectra is depicted in Figure 2. A broad signal in the chemical shift range between -75 and -100 ppm is the main signal of the spectrum and can be assigned to the Si present in the C-S-H phases overlapped with the signal of feldspar (-95 to -100 ppm). The signal at -69 ppm results from unreacted clinker (alite). The overlap of the signals does not permit an unambiguous deconvolution.

The broad feature at -110 ppm is characteristic of silica fume not reacted during the hydration of the samples and quartz (Lothenbach et al., 2012).

The ^{27}Al-NMR spectra (also depicted in Figure 2) present two different signals at 63.0 and 16.7 ppm. Al in a tetrahedral environment- is generally observed between 50 and 80 ppm, while octahedral Al resonates between -20 and 20 ppm. Figure 2 shows that Al is in an octahedral and tetrahedral coordination. The observed tetrahedrally coordinated ^{27}Al resonances are associated with the aluminum in the bridging position of the C-S-H phases and feldspar and the octahedral ^{27}Al to the presence of ettringite with probably some substitution of the Al by Fe.

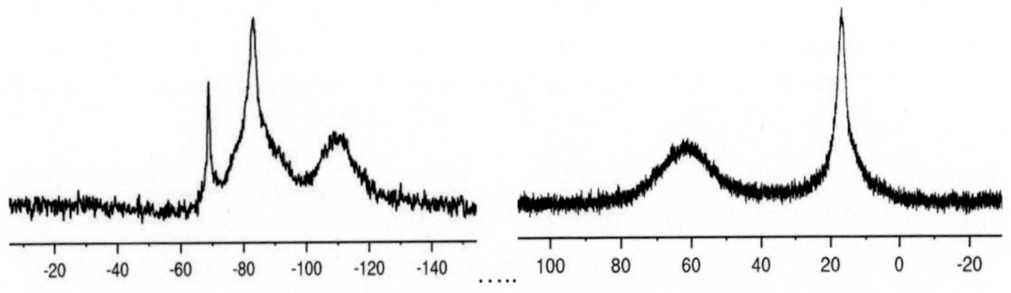

Figure 2: *^{29}Si MAS NMR spectra (left) and ^{27}Al MAS NMR spectra of the cement paste.*

SURREY

The bulk composition of the reference mix design binder used to manufacture the RPM was analysed by X-ray fluorescence, the results of which are summarised in Table 7.

Table 7: *XRF data providing the bulk composition of the binders used to form the Cebama reference paste and concrete.*

Binder	CaO (%)	SiO_2 (%)	Al_2O_3 (%)	SO_3 (%)	MgO (%)	Fe_2O_3 (%)	K_2O (%)	TiO_2 (%)	MnO (%)
Silica Fume	1.46	93.10	1.44	0.47	0.88	0.91	1.73	-	-
CEM I 42.5	67.72	17.60	3.42	3.81	0.6	5.17	1.3	0.17	0.21
Blast furnace slag	43.13	32.3	9.85	3.68	7.40	0.74	1.20	1.36	0.34

Blocks have been cast to determine the interaction of the Cebama reference paste with three selected groundwaters: granitic, saline (Gascoyne, 2002) and clay (Vinsot et al., 2008). Changes in groundwater composition will be measured over an extended period and cement samples will be taken periodically to observe mineralogical changes to existing phases and the formation of new phases. Analyses will be performed utilising XRD, SEM, RAMAN, EXAFS and XANES.

CTU - UJV

Laboratory work is based on ageing procedures among Czech bentonite, low-pH binder and groundwater from the underground laboratory Josef under high temperature. The ageing procedures on low-pH reference paste samples are ongoing since December 2016. Laboratory analyses of cementitious materials mainly include uniaxial strength test on thin samples, mineralogy, pH of leachates and diffusion properties. More detail can be found in CTU and UJV contribution to the 2nd CEBAMA Annual Workshop (Vašíček et al., 2017).

Because the interaction occurs mainly on interfaces (i.e., cement sample surface), specific shape of the samples is used to magnify this effect when uniaxial strength is tested. Thin plates (cylinders with diameter of 50.0 mm and height of 8.2 mm) and corresponding non-standard punch test method are being used. Due to specific shape and large amount of samples needed CTU decided to prepare them at their own laboratory according to procedure provided by VTT. Therefore, the first step focused on testing of eight manufacturing alternatives. The tests varied in plasticizer and various details in mixing procedure (two types of plasticizer from three sources were used - Ha-Be supplied by VTT, Ha-Be purchased on market and alternative one by DenBraven; procedure varied in total amount of mixed material, duration of pouring into the forms and time intervals between adding of particular components).

All mixing tests provided thin samples for punch compressive strength test. At least nine samples were used as one sample set. Results (standard variation of datasets 3 - 12%) were recalculated to the cubic shape of the sample and plotted in Figure 3. Test no. 8 provided input values for future comparison of results from ageing procedures.

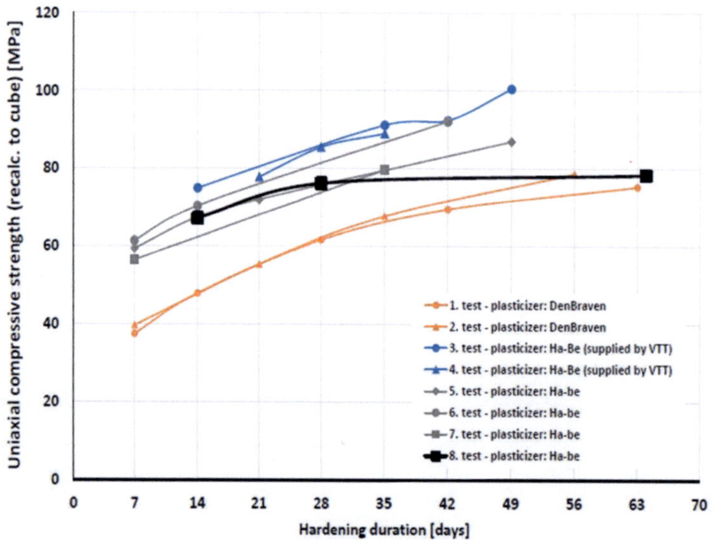

Figure 3: *Uniaxial compressive strength on thin samples (recalculation to cubic shape). Comparison of 8 sets of mixing tests with variations in plasticizers and mixing procedure details.*

Up to now, only the first measurement of cement paste pH in leachate is complete. Preliminary pH leachate tests (after 170 days since LPC mixing) were performed following the procedure described in (Alonso et al., 2012) and are shown in Table 8. First sampling of the reference paste samples is expected after 9 months of interaction - in September 2017. Then all of the analyses described in Table 2 will be performed.

Table 8: *pH values of Cebama reference paste samples at different conditions (170 days after mixing).*

paste sample	pH in suspension	pH in leachate
Curing 10°C	12.4	12.4
Josef GW 10°C	12.3	12.3
Josef GW 95°C	11.1	11.0

Note: Josef GW - groundwater from the Josef Underground Facility

VTT

VTT measured fresh-stage properties of the Cebama reference paste. Cement paste workability was characterized using a Haegermann flow table. Haegermann flow of the reference paste was 190 mm (DIN EN 1015, 2007). Compression strength, density and pore solution pH of the reference paste is presented in Table 9.

Table 9: *Compression strength, density and pore solution pH of RPM at various ages. Pore solution pH was measured according to method described in the literature (Alonso et al., 2012).*

Age	Compression strength	Density	pH
7 d	72.0 MPa	2,100 kg/m^3	-
28 d	111.0 MPa	2,150 kg/m^3	12.22
423 d	-	-	11.56

JUELICH

Complementary to experiments addressing the diffusion of safety relevant radionuclides in the Cebama reference paste, JUELICH characterized its microstructure and phases distribution. Here, we focused on the identification of phases and reductive species providing potential sorption sites for the radionuclides under investigation. Single components of the binder (quartz filler, silica fume, blast furnace slag and CEM I 42.5 MH/SR/LA) were analysed by SEM/EDX and XRD for their chemical and mineralogical composition. Scanning electron microscopic (SEM) investigations including SEM/EDX mapping and XRD were performed to analyse the microstructure of and phase distribution about 400 days after casting of the samples. XRD data on the grinded VTT reference paste show, beside intense reflexes of the quartz filler, only weak reflexes of crystalline hydration phases, mainly ettringite, and amorphous C-S-H. The microanalytical investigations reveal the presence of unhydrated clinker material, mainly C_2S, unreacted quartz filler, and blast furnace slag (Figure 4). The silica fume was found to be the main source of carbon present in the reference cement paste. Iron rich phases such as Fe-oxides and -sulphides were identified that can provide for the immobilisation of redox-sensitive radionuclides in the hardened cement paste. The generated data on the phase distribution in the reference paste will be further processed to provide pseudo-colour 2D maps of the phase distribution.

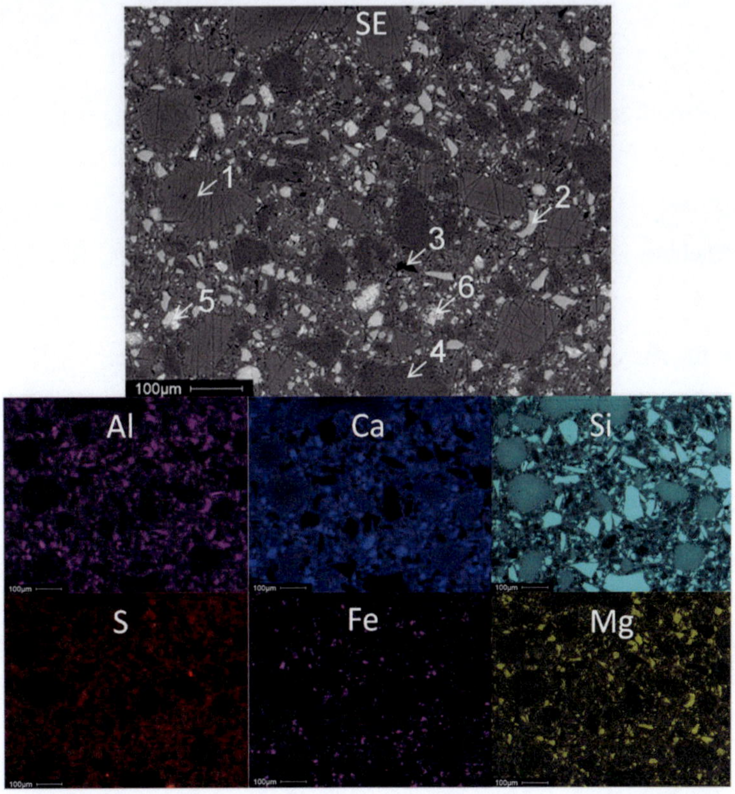

Figure 4: *SEM/EDX mapping of the Cebama reference paste. 1) fumed silica, 2) blast furnace slag, 3) carbon, 4) quartz, 5) iron oxide, 6) unhydrated clinker.*

Conclusions

Chemical characterization of the RCM determined by CSIC using XRD and EDX shown that portlandite and sulfoaluminate phases, such as ettringite, were not present. The EDX performed by CSIC reveal a low CaO/SiO_2 ratio = 0.55 ± 0.17 and enrichment in Al, CaO/Al_2O_3 ratio = 8 ± 4.6.

According to the pH measurements in CSIC, the pore solution pH of the reference concrete (pH = 11.47) was below the saturation pH of calcium hydroxide (pH = 12.40). In VTT's measurement the reference concrete pH was 11.47 at the age of 428 days which is in excellent agreement with the CSIC results.

Regarding transport parameters, MIP confirms the low total porosity in RCM (2.3%) with pores distributed below 0.02 μm, (mean pore size 0.01 μm), all this agree with the high density of this concrete (2,420 kg/m^3) measured by VTT and CSIC. The compression strength measured by VTT is 88.5 MPa after 424 days of aging. The concrete also present a high resistance to percolation of water, hydraulic conductivity 10^{-13} to 10^{-14} m/s and the pH of eluted water was close to neutral and far from the alkaline plume of OPC (Cau Dit Coumes et al., 2006).

Chemical characterization of the RPM determined by JUELICH and KIT identified ettringite in the paste samples. The observations further support the anticipated lower reaction degree of the paste samples compared to the concrete samples.

Reference paste pH measured in CTU-UJV presented pore solution pH 12.4. VTT measured pore solution pH 11.56 which is in agreement with the pH measured in KIT (pH = 11.7). CTU-UJV performed experiments in 10°C temperature whereas VTT and KIT experiments were performed in room temperature.

Future work

Concrete (RCM)

CSIC will study the evolution of hydration at further ages, evolution of interaction with granite and clayey groundwaters through percolation tests and transport Cl diffusion through natural tests, using RCM and RCM-LD-samples.

VTT will perform leaching experiments to the Cebama reference concrete, Binary- and Ternary- low-pH concretes.

KIT will characterize the chemical and microstructure of the concrete sample.

Paste (RPM)

KIT will study the diffusion of ^{36}Cl, ^{129}I, HTO and Be on RPM.

SURREY will use radioactive tracers (^{3}H, ^{14}C, ^{35}S and ^{36}Cl) to provide information on reaction mechanisms and kinetics governing incorporation of target elements into the cement matrix. Autoradiography will be used to measure migration rates of the radioactive tracers. Advection experiments will be carried out as described in (Felipe-Sotelo et al., 2016) where the radioactive tracers can allow an unambiguous determination of the uptake and movement of ions of interest.

Additionally, sorption / desorption experiments are in progress to study the uptake / retention of selected radionuclides (^{36}Cl, ^{79}Se, ^{129}I). These experiments should allow the determination of uptake and release kinetics as well as the partitioning of the radionuclide between solid and solution. Analysis of the solids by XRD, SEM and EXAFS will provide insight to the distribution of radionuclides on and within the phases present.

These experiments will provide an understanding of the mobility of the target radionuclides through the paste. Through-diffusion experiments are being carried out in the manner described by Felipe-Sotelo et al. (2017). These experiments involve cylindrical blocks being cast, a well drilled into the centre of each block and the radionuclide of interest spiked into the central well. The well is then sealed and the block placed in equilibrated water. Breakthrough of the radionuclide is measured by monitoring the surrounding equilibrated water and disribution of radionuclide within the cement can be determined by autoradiography once the experiment has concluded.

JUELICH will study small monolithic samples of the Cebama reference paste in in-diffusion experiments using ^{226}Ra, $^{99}Tc(VII)$, molybdate and iodide; the effective diffusion coefficient of the paste will be determined by diffusion experiments with HTO in cooperation with SURREY.

UAM will perform mineralogy analysis using XRD with Rietveld approach and thermal analysis.

They will also conduct a small transport experiment including FEBEX-bentonite.

Acknowledgement

The research leading to these results has received funding from the European Union's Horizon 2020 Reasearch and Training Programme of the European Atomic Energy Community (EURATOM) (H2020-NFRP-2014/2015) under grant agreement n° 662147 (CEBAMA).

References

Alonso, M.C., García Calvo, J.L., Petterson, S., Puigdomenech, I., Cuñado, M.Á., Vuorio, M., Weber, H., Ueda, H., Naito, M., Walker, C., Takeshi, Y., Cau-dit-Coumes, C. (2012). Round robin test for defining an accurate protocol to measure the pore fluid pH of low-pH cementitious materials. *In:* Bart, F., Cau-dit-Coumes, C., Frizon, F., Lorente, S. (Eds.) Cement-Based Materials for Nuclear Waste Storage, Springer, 251-259.

Alonso, M.C., García Calvo, J.L., Walker, C., Naito, M., Pettersson, S., Puigdomenech, I., Cuñado, M.A., Vuorio, M., Weber, H., Ueda, H., Fujisaki, K. (2012). Development of an accurate pH measurement methodology for the pore fluids of low pH cementitious materials. SKB Report, R-12-02.

Cau Dit Coumes, C., Courtois, S., Nectoux, D., Leclercq, S., Bourbon, X. (2006). Formulating a low-alkalinity, high-resistance and low-heat concrete for radioactive waste repositories. Cem. Con. Res., 36, 2152-5163.

DIN EN 1015-3:2007 (2007). Methods of test for mortar for masonry, -Part 3: Determination of consistence of fresh mortar (by flow table).

Felipe-Sotelo, M., Hinchliff, J., Field, L.P., Milodowski, A.E., Holt, J.D., Taylor, S.E., Read, D. (2016). The solubility of nickel and its migration through the cementitious backfill of a geological disposal facility for nuclear waste. J. Hazard Mater., 314, 211-219.

Felipe-Sotelo, M., Hinchliff, J., Field, L., Milodowski, A., Preedy, O., Read, D. (2017). Retardation of uranium and thorium by a cementitious backfill developed for nuclear waste disposal. Chemosphere, 179, 127-138.

Gascoyne, M. (2002). Influence of grout and cement on groundwater composition. Posiva Working Report, 2002-07.

Holt, E. and Koho, P. (2016). D4.5 POPLU Experimental Summary Report. European Commission Community Research, DOPAS (Contract Number FP7 -323273), Deliverable n°4.5.

Huisman, J., Zimmerman, E., Gaboreau, S., Leroy, P., Claret, F., Tournassat, C., Lutzenkirchen, J. (2017). Spectral Induced Polarization measurements of low-pH concrete. Proceedings of the 2nd CEBAMA Annual Workshop (*this document*).

SFS-EN 12350-7:en (2009). Testing of fresh concrete. Part 7: Air content, pressure methods.

Vašíček, R., Večerník, P., Hloušek, J., Červinka, R., Hausmannová, L., Havlová, V. (2017). Interaction between Cement and Czech Bentonite under Temperature Load and In *In-Situ* Conditions: Results after First Testing Period. Proceedings of the 2nd CEBAMA Annual Workshop (*this document*).

Vinsot, A., Mettler, S., Wechner, S. (2008). *In-situ* characterization of the Callovo-Oxfordian porewater composition. Phys. Chem. Earth, 33, S75-86.

Vehmas, T., Schindler, A., Löija, M., Leivo, M., Holt, E. (2016). Reference mix design and castings for low-pH concrete for nuclear waste repositories. Proceedings of the 1st CEBAMA Annual Workshop. KIT Scientific Publishing, KIT-SR 7734.

Building blocks on molybdenum retention processes in cement systems. First experimental results from solubility and sorption onto pure cement phases

Marta López-García[1], Javier Olmeda [1], Mireia Grivé [1*]

[1] Amphos 21 Consulting S.L. (ES)

* Corresponding author: mireia.grive@amphos21.com

Abstract

The present work reports the experimental work carried out during the last year within the WP2 in the frame of CEBAMA Project. Namely, this S&T contribution presents the synthesis of pure phases (C-S-H and AFt phases), synthesis and solubility determination of Powellite, and on-going experiments of Mo adsorption kinetics onto C-S-H gels and AFt phases. The characterization techniques used (XRD, SEM-SEI EDX, thermogravimetric measurements…) allowed confirming the good synthesis of both, C-S-H gels and Aft solid phases. Powellite ($CaMoO_4$) was also successfully synthesized and the determination of its solubility product was in very good agreement with the ones reported in the literature.

Introduction

The final disposal of radioactive wastes produced by nuclear power programmes, industry, medicine and research include cement based materials as the main near field barriers in the repository. The chemical processes occurring within the near-field due to metal corrosion and concrete degradation, as well as interactions with groundwater flowing through the facility will determine the long-term chemical evolution of the repository near-field. Molybdenum-93, an activation product from the steel with a half-life of 4,000 years has emerged as an important radionuclide in the safety assessments being one of the dominating contributors to the total radiation dose some thousands of years after closure (Lindgren et al., 2007).

Mo is a redox-sensitive element whose aqueous speciation is dominated by the thermodynamically highly stable molybdate (MoO_4^{2-}). This species is typically considered to dominate even under moderately reducing conditions. Molybdates of the alkali metals and ammonium are quite soluble in water; the same does not apply to calcium molybdate (Powellite), which is known as the pure solubility controlling phase of molybdates at neutral to high pH ranges.

The main aim of this work is to improve the understanding of the governing mechanisms of molybdenum retention in cementitious environments. Specifically, this work focuses on Mo retention processes onto relevant hydrated cement phases and alteration products.

The literature review has revealed that even though both, ettringite (AFt-SO_4) and monosulphate (AFm-SO_4) can be considered viable cement constituents for oxyanion immobilization, the extent of ettringite uptake capacity strongly depends on the size and charge similarities between sulphate and substituting oxyanions (Baur et al., 2003; Keller et al., 2002; Ochs et al., 2015).

While ettringite and monosulphate may serve as viable mechanisms for oxyanion immobilization, several factors must be considered though when the formation of these phases is proposed to be the main immobilization factor of oxyanions as their stability depends on several physico-chemical conditions, such as pH, temperature, presence of sulphate and other competing anions, among others.

Concerning C-S-H phases, there are studies showing that these phases can immobilize oxyanions (CrO_4^{2-} and AsO_4^{3-}) by replacing silicate in their structure, which suggests that these phases could also be an important sink for molybdate. In this context, zeta potential could be an important factor limiting the mechanism and thus the extent of molybdate retention. Therefore, three C-S-H phases with C/S ratio of 0.8, 1.2 and 1.5 have been selected to be studied in this work, as they show different ζ values and different Ca concentration in solution.

The present work reports the experimental work carried out during the last year. More specifically, the synthesis of new pure phases, first results concerning solubility of Powellite and on-going experiments concerning Mo adsorption kinetics onto C-S-H gels and AFt phases.

Experimental work

C-S-H phases

Synthesis and characterization of C-S-H phases

Two C/S ratios (0.8 and 1.2) were prepared and their synthesis and characterization was detailed in Grivé et al. (2016). One additional C-S-H gel with C/S ratio 1.5 was also foreseen to evaluate how Mo retention might be affected by a higher Ca concentration in solution.

The methodology followed to prepare the C-S-H with C/S ratio of 1.5 is as follows. Inside a glove-box with nitrogen-saturated atmosphere and constant temperature of 20ºC, different amounts of SiO_2 and CaO were appropriately mixed up to obtain the targeted C/S ratio for a solid/liquid ratio of 20 g/L. In this case, 0.125 g of SiO_2 and 0.175 g of CaO were put in different polyethylene tubes with 15 mL of degassed deionized water (DDW). The suspensions were left to react for 90 days in an end-over-end rotary shaker. During this time the pH of the suspension was continuously monitored. A constant value of 12.3 was obtained. The solid was then separated from liquid by filtration. Three representative samples were selected to carry out the analysis in both solid and liquid phases.

Quantitative composition and pH of the aqueous phase were determined for selected samples. Silicon and calcium concentration in C-S-H mother solutions were respectively determined by ICP-MS (model 7500cx, manufactured by Agilent Technologies Inc.) and Ionic Chromatography (ICS-2000 system, manufactured by Dionex). Results from these analyses are included in Table 1.

Table 1: *Comparison of the analysed aqueous composition of the new synthesized C-S-H equilibrated solutions from this work with the ones reported in the literature.*

C/S ratio	[Ca] mM	[Si] mM	pH	Reference
1.4 ± 0.1	16.4 ± 1.2	0.040 ± 0.005	12.3 ± 0.1	This work
1.51	15.141		12.19	Cong and Kirkpatrick (1996)
1.50		0.023		Greenberg and Chang (1965)
1.49		0.018		Chen et al. (2004)

The analysed Ca and Si concentration in solution was employed to calculate the C/S ratio present in the solid phase. Considering the initial concentration of each element and subtracting the analysed concentration after equilibrium, the C/S ratio obtained was 1.4 ± 0.1, which is close to the expected 1.5 ratio. Since the obtained C/S ratio was not 1.5 but 1.4, from now on we will consider that the new gels have a C/S ratio of 1.4.

There is a number of studies dealing with similar systems. Table 1 contains reported data on C-S-H aqueous composition for equal/similar C-S-H C/S ratios. As can be seen, there are few studies analyzing both the concentrations of Ca and Si in solution and pH for a given C/S ratio. The concentrations and pH values measured in the present study are in agreement with those reported in bibliography.

XRD pattern from the dried synthesised C-S-H samples was obtained in a X-ray diffractometer model PANalytical X'Pert PRO MPD Alpha1. For the sake of comparison, data reported in previous S&T contribution (Grivé et al., 2016), by Taylor (1997) and by Roosz (2016) have been included in Figure 1.

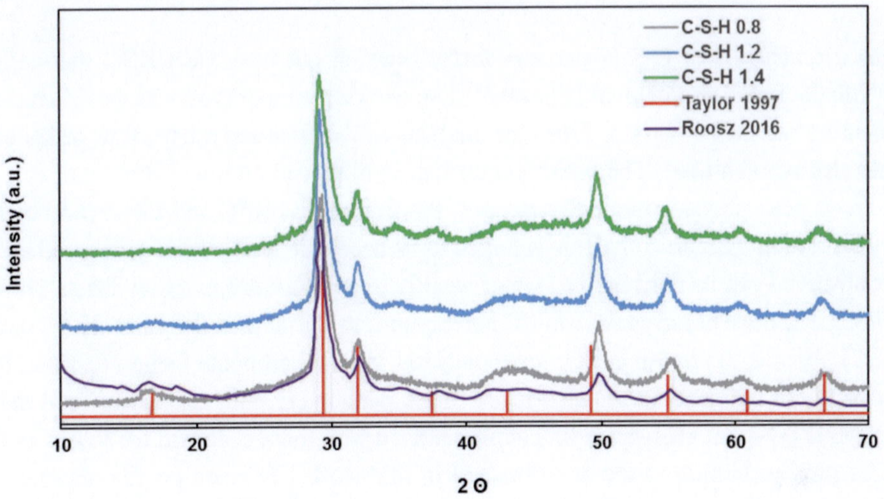

Figure 1: *X-ray powder diffraction spectra of dried synthesised C-S-H gel. XRD pattern reported by Taylor (1997) and for Roosz (2016) for C-S-H has been included (red and purple lines).*

The XRD pattern in this work for C-S-H gels of C/S = 1.4 is similar to the ones reported in bibliography by Taylor (1997), Roosz (2016) and Pointeau (2000) as well as in the previous S&T report for C-S-H 0.8 and 1.2 (Grivé et al., 2016). As can be seen in Figure 1, two typical intense peaks at 29 and 50 2Θ degrees were found. This pattern is very common on C-S-H XRD characterizations and it indicates the similarities between C-S-H gel and tobermorite structures (Grangeon, 2013a and 2013b).

Scanning electron microscopy (SEM) was additionally used to obtain topographic images (SEI) of the sample surface as well as the different chemical composition displayed on a specific surface spot (EDX). This technique offered rather information of the materials micro-texture and composition. The analysis of C-S-H samples was performed using a Field Emission Scanning Electron Microscope (FE-SEM - ZEISS Ultraplus) coupled with a EDX (Energy Dispersive X-ray Spectroscopy) analyser for semi-quantitative chemical composition analysis (X-Max EDX detector, from OXFORD Instruments).

Several SEM images were obtained at different resolution (up to 5,000x) around the C-S-H samples. Figure 2 shows some of the micrographs taken at 500x and 5,000x. Determination of Ca and Si through EDX analysis confirmed the presence of these elements as mains constituents of C-S-H surface.

a) 
b) 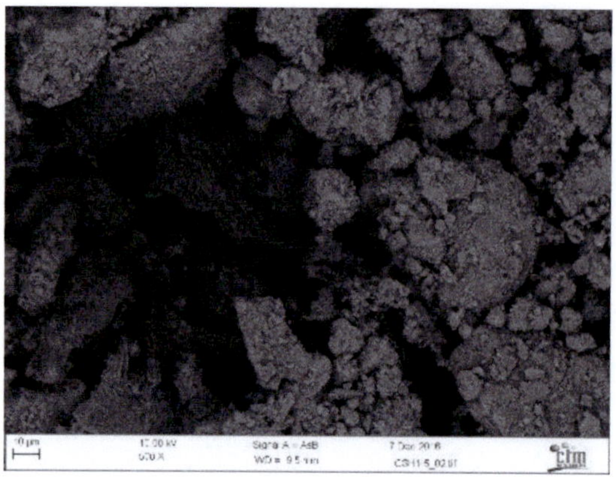

Figure 2: *Micrographs of synthesised C-S-H gels; a) 5 kx and b) 0.5 kx.*

Thermogravimetric analysis of C-S-H gel was further carried out in a TGA/DSC thermal analyser model StrareE (Mettler Toledo, S.A.E. Barcelona) (Figure 3). The working temperature was established between 30 and 1,000ºC with a heating rate of 10 ºC/min. From the analysis of the first derivative, four peaks corresponding to four weight losses can be identified. The main weight loss is observed around 100ºC, and can be attributed to water loss. This broad peak presents two deformations, the first one at 50ºC and the second one around 150ºC. Water loss through thermogravimetric analysis is linked to its energetic interactions with solid surfaces, therefore the two-shoulder observed can be attributed to water weakly bound to more external sheets (50ºC) and strongly bound water (150ºC). The third broad peak at 450ºC corresponds to Portlandite dehydration (El-Jazairi and Illston, 1977; Taylor, 1997). Finally, the fourth peak corresponds to a typical carbonate weight loss and fifth peak can be attributed to calcite, these two peaks account for low mass percentage with respect to total mass sample. The results obtained from DTG-TGA analysis in this work matched well those obtained for Roozs et al. (2016) where they reported TGA profiles similar to the one obtained in this work. Obtained profile also matched the results obtained for previously synthesized C-S-H gels with C/S ratio 0.8 and 1.2 (Grivé et al., 2016).

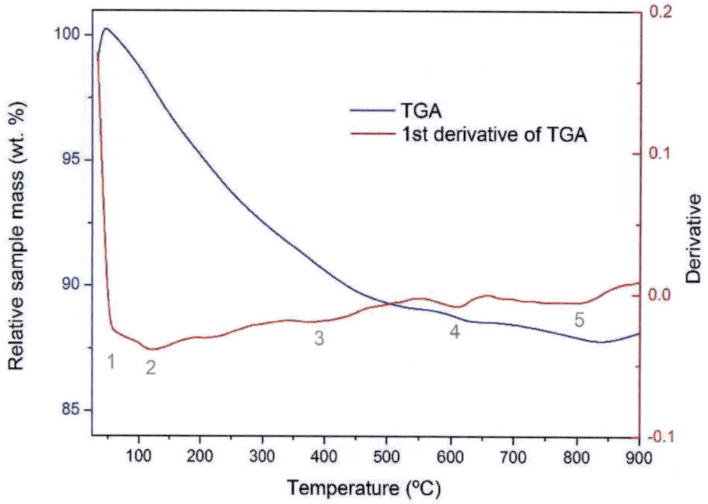

Figure 3: *Thermogravimetric analysis for C-S-H gel with C/S ratios 1.5 (solid line). Blue line stands for mass loss and red line thermogravimetric derivative.*

AFt phase (ettringite)

Synthesis and characterization of AFt phase

AFt-SO_4 phase (ettringite) was synthesized following the procedure described by Perkins (1999) with some modifications. Inside a glovebox with nitrogen-saturated atmosphere and constant temperature of 20ºC, 10.05 g of CaO (Alfa Aesar Gmbh, 99.95% purity) were mixed under magnetic stirring with 750 mL of DDW using a HDPE flask. In the same way, 19.86 g of $Al_2(SO_4)_3 \cdot 18\ H_2O$ (Scharlau, S.A., extra pure) were dissolved in 300 mL of DDW. Once dissolved, both solutions were mixed inside a 2 L HDPE bottle, 450 mL of DDW were added and the whole mixture was magnetically stirred. Then, 11.5 mL of a NaOH 0.1 M solution were added to obtain a final pH of 10. The mixture was heated at 60ºC under continuous stirring for 48 h and then vacuum-filtered through a 0.45 μm filter to separate liquid from solid. The former was collected in a HDPE bottle inside the glovebox until used. The latter was dried and stored at room temperature in a desiccator with N_2 stream and a saturated $CaCl_2$ solution (HR ~ 30%). Once the solid was completely dried, it was ground and sieved to obtain a particle size below 200 μm.

Both, solid and liquid phases were analysed to ensure that AFt phase was successfully synthesized. Characterization techniques included ICP-MS, IC, XRD, TGA, and SEM-SEI/EDX analysis. Calcium, aluminium and sulphate concentrations in ettringite equilibrated porewater solution were respectively determined by ICP-MS (model 7500cx, manufactured by Agilent Technologies Inc.) and Ionic Chromatography (IC) (ICS-2000 system, manufactured by Dionex). The obtained values for dissolved species present in AFt equilibrated porewater as well as the pH are summarised in Table 2.

Table 2: *Calcium, aluminium and sulphate concentration in AFt-SO_4 equilibrated porewater.*

Eq. porewater (mM)	[Ca]	[Al]	[SO_4^{2-}]	pH
AFt-SO_4	3.91	0.483	6.72	10.2

Solid phase of synthesised AFt-SO_4 was analysed in order to characterize its structure and composition. Note that careful handling was required to avoid solid interaction with atmospheric carbon dioxide. Figure 4 shows the XRD spectra of the synthesised (dried) ettringite (blue pattern) compared to a X-ray pattern reported by Goetz-Neunhoeffer et al. (2006) (red pattern) for the same mineral. Characteristic peaks of AFt pure phase were found in XRD spectra obtained for AFt synthesized in this work.

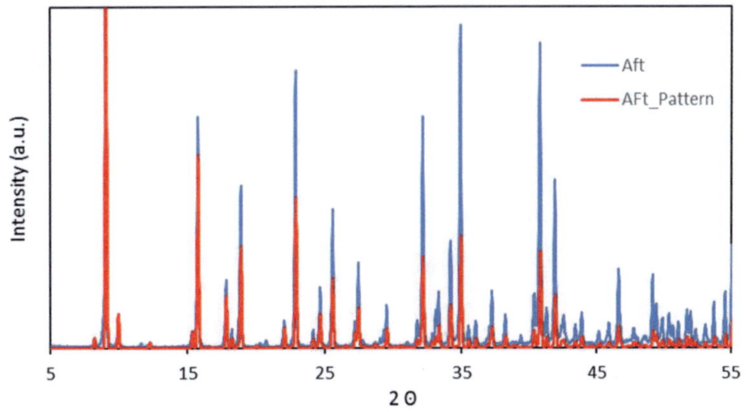

Figure 4: *XRD patterns of AFt phase obtained in the present work (blue line) and reported by Goetz-Neunhoeffer et al. (2006) (red line).*

Scanning electron microscopy (SEM-SEI) was used to obtain morphologycal information of AFt surface as well as different chemical composition displayed on a specific surface spot by EDX. Images obtained for AFt phases are shown in Figure 5. As can be seen, typical needle crystals of AFt were observed, which suggested the presence of this phase on the surface of the synhtesized solid.

a) b) c)

Figure 5: *Micrographs of synthesised AFt phase; a) at 0.5 kx; b) 5.0 kx; c) 10.0 kx.*

Results from the EDX analysis performed onto different spots of the powdered solid sample allowed determining that O, Ca, S and Al are the major constituents of AFt surface.

Thermogravimetric analysis was also carried out and weight losses are represented in Figure 6. Ettringite has a hexagonal prismatic shape, the columns of $[Al(OH)_6]^{3-}$ octahedra are linked together by calcium and hydroxide ions while sulphate and water molecules are located on the outer surface of the columns. The water between the columns is lost around 100ºC, as shown in Figure 6, while the water from the dehydroxylation of the aluminium hydroxide is removed between 200 and 400ºC. These results are in good agreement with other works reported in bibliography (e.g., Scrivener et al., 2016). As ettringite is thermally unstable at $T > 50$ºC in contact with atmosphere according to Jiménez et al. (2015), during the decomposition process not only water and hydroxyl losses occurs but also structure changes may take place. Thermal treatment eventually leads to the formation of an amorphous product known as metaettringite, a partially dehydroxilated product (Zhou et al., 2001).

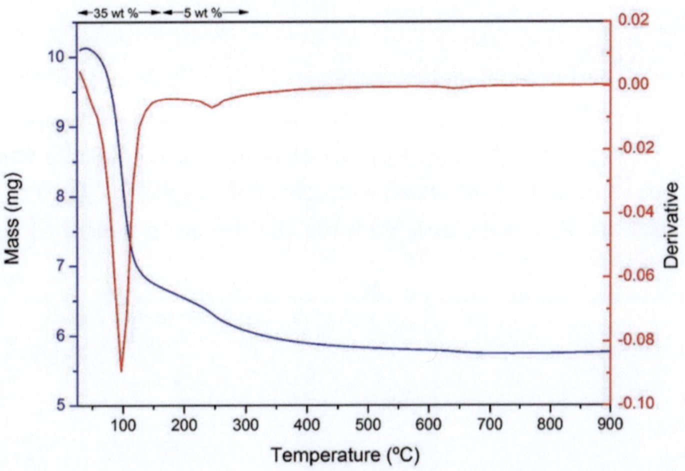

Figure 6: *AFt thermogravimetric analysis. Blue line represents mass loss and red line thermogravimetric first derivative.*

Calcium molybdate (Powellite)

Powellite ($CaMoO_4$) is considered as the main solubility controlling pure phase of molybdenum in repository environments. As the precipitation of this mineral could play an important role on molybdenum immobilization, it is interesting and important to study its solubility under different conditions to deeply understand its solubility controlling role prior to start the sorption studies foreseen. In this sense, and as a first step, the synthesis of this phase was carried out to be then used in solubility experiments. With the aim of acquiring a reference solubility product of Powellite under different media, these experiments entailed the study of solubility in degassed deionized waters (DDW) at fixed ionic strength and in AFt-SO_4 equilibrated porewaters.

Synthesis and characterization of Powellite

Powellite was synthesized following the procedure described by Zhang (2000). The synthesis was performed by titrating a 500 mL of molybdic acid solution 0.025 M with a $Ca(OH)_2$ saturated solution at room temperature (25 ± 2ºC) to reach a pH equal to 6.5. Then, the suspension was heated to 80ºC and we kept this temperature under continuous stirring for 30 min. Precipitation of Powellite was observed at this stage. The precipitated crystals were allowed to age in the mother liquid for two days and then vacuum-filtered through a 0.45 µm nylon filter. The white solid was then rinsed with water, followed by ethanol and left to dry up inside a desiccator to constant weight. After that, the solid was ground with an agate mortar and stored in a polyethylene flask until use.

Powellite was characterized by XRD; the obtained spectrum is shown in Figure 7. As can be seen, a clearly defined crystal structure was found in agreement with other XRD spectra found in literature (Ryu et al., 2007).

Study of Powellite solubility

Solubility experiments were performed with the synthesised Powellite sample from undersaturation direction. The experiments were performed by contacting 0.06 g of solid with 20 mL of solution at the desired pH value. The mixtures were stirred in high density polypropylene (DHPE) tubes (50 mL) under constant temperature (25 ± 2ºC), inside a glove box in order to minimize the effects of atmospheric CO_2. Two different solutions were used in the solubility experiments: degassed deionized water (DDW) and a typical AFt equilibrated porewater.

Study of Powellite solubility

Solubility experiments were performed with the synthesised Powellite sample from undersaturation direction. The experiments were performed by contacting 0.06 g of solid with 20 mL of solution at the desired pH value. The mixtures were stirred in high density polypropylene (DHPE) tubes (50 mL) under constant temperature (25 ± 2°C), inside a glove box in order to minimize the effects of atmospheric CO_2. Two different solutions were used in the solubility experiments: degassed deionized water (DDW) and a typical AFt equilibrated porewater.

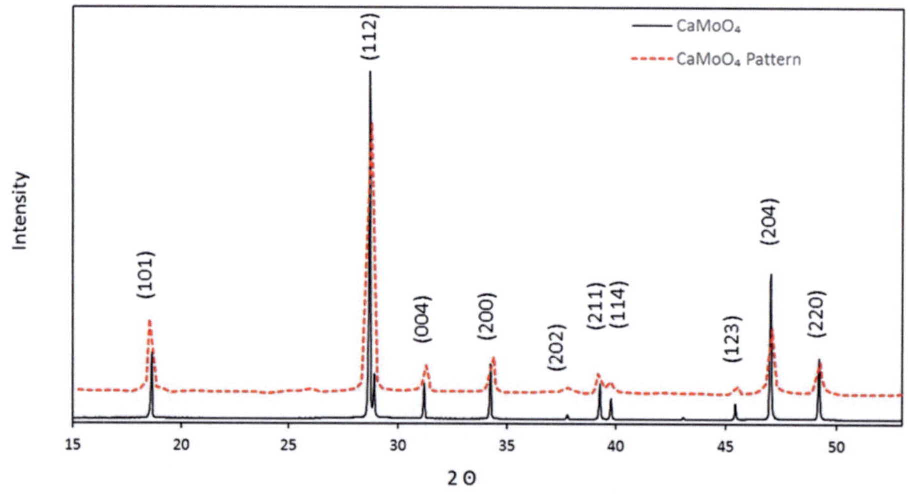

Figure 7: *XRD pattern for synthesized $CaMoO_4$. Black line corresponds to Powellite synthesized in this work and red line to the XRD data reported in Ryu et al. (2007).*

Solubility tests in DDW

DDW with controlled ionic strength ($NaClO_4$, I_m = 0.5 M) was used to study Powellite solubility in order to determine whether it is in the range of previous reported solubilities. Different solutions were prepared wherein the pH was adjusted by adding few amounts of either carbonate free NaOH or HCl 0.5 M solutions. Then, $CaMoO_4$ was added to each solution and placed in an end-over-end shaker. pH was monitored on alternate days until equilibrium was reached. Equilibrium criteria was set achieving constant pH measurements (with differences lower than 0.1 units) within 48 h. Once equilibrium was reached, aliquots of each tube were subtracted, filtered with 0.22 µm nylon filters and acidified with ultra-pure HNO_3 to determine Mo concentration by ICP-MS. Major cation and anion concentrations were also measured by IC.

Figure 8 shows the solubility curve obtained in this study. As can be seen, the calculated solubility (with PREEQC 3.3.8 program with the ThermoChimie database version 9.b.0 with the SIT approach (*https://www.thermochimie-tdb.com/*) can well explain the obtained experimental results.

Solubility tests in AFt porewater

Powellite solubility in AFt equilibrated porewater was also determined. As in the previous case, experimental concentrations of Mo and major anions and cations in solution were measured. Table 3 shows the quantitative composition of aqueous phase.

Table 3: *Quantitative composition of AFt porewaters in equilibrium with Powellite.*

Eq. porewater (mM)	[Ca]	[Al]	$[SO_4^{2-}]$	[Mo]	pH
AFt	4.13	0.487	8.19	$1.8 \cdot 10^{-2}$	10.6

In order to obtain a solubility curve of Powellite subject to pH variation, different amounts of HCl 0.5 M were added to the (initially) AFt porewater. It is worth to mention that even if AFt phase is not stable at pH lower than the initial pH (~ 10.2), the experimental conditions used in this part of the solubility experiments were performed for the sake of comparison with the solubility tests in DDW and to study the effect of Ca on the solubility of this element. Figure 9 compares the solubility of powellite in DDW (blue) with the one obtained in the Aft porewater (orange). Experimental measurements are plotted (dots) and their corresponding calculated solubilities (lines) are superposed with comparison purposes. As can be seen, the solubility of Powellite is lower in AFt modified porewater with respect to that determined in DDW. This decrease is attributed to the common-ion effect as AFt equilibrated porewater contains up to 3.91 mM of Ca in solution (see Table 2).

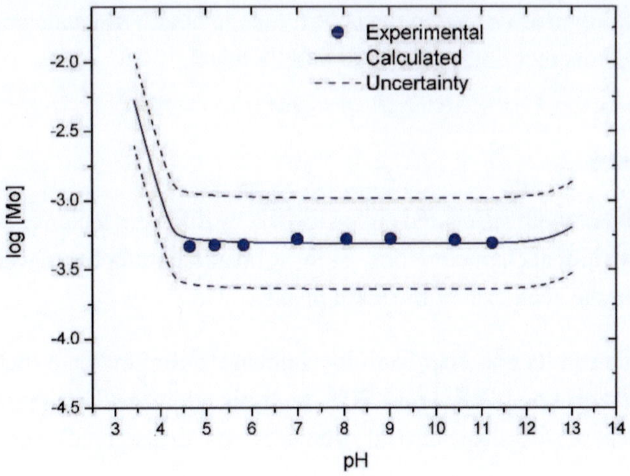

Figure 8: *Experimental and calculated solubility of $CaMoO_4$ as a function of pH solution obtained at 25°C, I_m=0.5 M.*

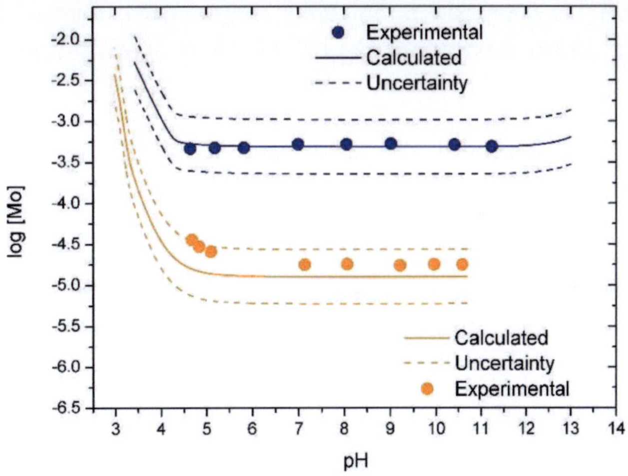

Figure 9: *Experimental solubility (dots) and calculated solubility (lines) of Mo as a function of pH in DDW ionic media (blue) and in modified AFt porewaters (orange).*

On-going experiments

Kinetic study of Mo adsorption onto cementitious pure phases

Adsorption onto C-S-H gels

Experiments on adsorption kinetics are currently being conducted using the synthesised C-S-H gels previously described, with selected C/S ratios (0.8, 1.2 and 1.4). Batch-sorption tests are being carried out in duplicate HDPE 15 mL tubes with screw closure wherein C-S-H gels were previously synthesized. According to the obtained solubility results, an initial Mo concentration of 10^{-3} M was used to avoid Powellite precipitation. Different tubes were prepared in order to study Mo adsorption at increasing times (5, 10, 20 and 28 days).

Adsorption onto AFt-SO_4 phase

Adsorption kinetic experiments are also performed onto AFt-SO_4 phase to get insight into its adsorptive properties towards oxyanions. The procedure followed is similar to that described above for C-S-H phases. In this case, 0.9 g of solid is poured in 15 mL of AFt mother solution previously obtained from the synthesis. After that, 150 µL of 100 ppm Mo stock solution are added to the HDPE tube to obtain Mo concentration of 10^{-3} M. Different tubes were prepared for targeted contact times (5, 10. 20 and 28 days).

Conclusions and future work

Different C-S-H gels have been synthesized and characterized by different techniques which allowed obtaining their main structural and compositional characteristics. Both, solid and liquid phases were analysed and the results allow to confirm the success on the synthesis of the solid phases.

AFt was synthetized and the results obtained from its characterization indicate that a single pure phase with high degree of crystallinity was obtained. XRD and TGA analysis additionally supported this confirmation and SEM-SEI images showed typical needle-shape crystal structures, which also confirmed the existence of pure AFt phase.

Powellite was synthetized and solubility experiments were conducted in both DDW with controlled ionic strength media and AFt porewater composition. Solubility measurements allowed selecting the most suitable experimental conditions for further adsorption experiments onto pure cement phases avoiding Powellite precipitation.

On-going experiments include adsorption kinetics of Mo onto synthesized C-S-H and AFt phases. Once selected equilibrium time will be attained, Mo concentration in solution will be determined as well as major species concentrations. Solid phases will also be characterized.

Future work will include adsorption isotherms which will be obtained by contacting pure phases with different Mo initial concentrations at a given time obtained from kinetic experiments. Desorption experiments will also be performed in order to study the reversibility of Mo adsorption process(es).

Acknowledgements

The research leading to these results has received funding from the European Union's Horizon 2020 Reasearch and Training Programme of the European Atomic Energy Community (EURATOM) (H2020-NFRP-2014/2015) under grant agreement n° 662147 (CEBAMA).

The authors would like to acknowledge Andra for the valuable input and support.

References

Baur, I. and Johnson, C.A. (2003). The solubility of selenate Aft (3CaO Al_2O_3 3$CaSeO_4$ 37.5H_2O) and selenate-AFm (3CaO Al_2O_3 $CaSeO_4$ xH_2O). Cement and Concrete Research, 33(11), 1741-1748.

Chen, J.J., Thomas, J.J., Taylor, H.F., Jennings, H.M. (2004). Solubility and structure of calcium silicate hydrate. Cement and Concrete Research, 34(9), 1499-1519.

Cong, X. and Kirkpatrick, R.J. (1996). ^{29}Si MAS NMR study of the structure of calcium silicate hydrate. Advanced Cement Based Materials, 3(3), 144-156.

El-Jazairi, B. and Illston, J.M. (1977). A simultaneous semi-isothermal method of thermogravimetry and derivative thermogravimetry, and its application to cement pastes. Cement and Concrete Research, 7(3), 247-257.

Goetz-Neunhoeffer, F. and Neubauer, J. (2006). Refined ettringite ($Ca_6Al_2(SO_4)_3(OH)_{12}\cdot 26H_2O$) structure for quantitative X-ray diffraction analysis. Powder Diffraction, 21, 4-11.

Grangeon, S., Claret, F., Lerouge, C., Warmont, F., Sato, T., Anraku, S., Numako, C., Linard, Y., Lanson, B. (2013a). On the nature of structural disorder in calcium silicate hydrates with a calcium/silicon ratio similar to tobermorite. Cement and Concrete Research, 52, 31-37.

Grangeon, S., Claret, F., Linard, Y., Chiaberge, C. (2013b). X-ray diffraction: a powerful tool to probe and understand the structure of nanocrystalline calcium silicate hydrates. Acta Crystallographica Section B Structural Science, Crystal Engineering and Materials, B69, 465-473.

Greenberg, S.A., Chang, T.N., Anderson, E. (1960). Investigation of colloidal hydrated calcium silicates. I. Solubility products. The Journal of Physical Chemistry, 64(9), 1151-1157.

Grivé, M., López, M., Olmeda, J. (2016). Synthesis and characterization of C-S-H gels for molybdenum sorption studies in cementitious media. Proceedings of the 1st CEBAMA Annual Workshop. KIT Scientific Publishing, KIT-SR 7734.

Jiménez, A. and Prieto, M. (2015). Thermal stability of ettringite exposed to atmosphere: Implications for the uptake for harmful ion by cement. Environmental Science and Technology, 49, 7957-7964.

Keller, I.R.B. (2002). The Immobilisation of Heavy Metals and Metalloids in Cement Stabilised Wastes: A Study Focussing on the Selenium Oxyanions SeO_3^{-2} and SeO_4^{-2}. Doctoral dissertation, Universität Zürich.

Lindgren, M., Pettersson, M., Wiborgh, M. (2007). Correlation factors for C-14, Cl-36, Ni-59, Ni-63, Mo-93, Tc-99, I-129 and Cs-135. In operational waste for SFR 1. SKB Report, R-07-05.

Ochs, M., Mallants, D., Wang, L. (2015). Radionuclide and Metal Sorption on Cement and Concrete. Vol. 9999, Springer.

Perkins, R.B. and Palmer, C.D. (1999). Solubility of ettringite ($Ca_6[Al(OH)_6]_2(SO_4)_3$ z $26H_2O$) at 5-75°C. Geochimica et Cosmochimica Acta, 63(13/14), 1969-1980.

Parkhurst, D.L. and Appelo, C.A.J. (2013). Description of input and examples for PHREEQC version 3.3.8 computer program for speciation, batch- reaction, one-dimensional transport, and inverse geochemical calculations: U.S. Geological Survey Techniques and Methods, book 6, chap. A43.

Pointeau, I. (2000). Etude mécanistique et modelisation de la rétention de radionucléides par les silicates de calcium hdratés (CSH) des ciments Andra. PhD thesis.

Taylor, H.F.W. (1997). Cement chemistry. Thomas Telford.

Ryu, J.H., Choi, B.G., Yoon, J., Shim, K.B., Machi, K., Hamada, H. (2007). Synthesis of $CaMoO_4$ nanoparticles by pulsed laser ablation in deionized water and optical properties. Journal of Luminescence, 124, 67-70.

Roosz, C., Gaboreau, S., Grangeon, S., Prêt, D., Montouillout, V., Maubec, N., Ory, S., Blanc, P., Vieillard, P., Henocq, P. (2016). Distribution of Water in Synthetic Calcium Silicate Hydrates. Langmuir, 32(27), 6794-6805.

Scrivener, K., Snellings, R., Lothenbach, B. (2016). A Practical Guide to Microstructural Analysis of Cementitious Materials. CRC Press.

Zhang, M. (2000). Incorporation of oxyanionic B, Cr, Mo and Se into Hydrocalumite and Ettringite: Application to cementitious materials. PhD Thesis, University of Waterloo, Ontario. Canada.

Zhou, Q. and Glasser, F.P. (2001). Thermal stability and decomposition mechanisms for ettringite at < 120°C. Cement and Concrete Research, 31, 133-1339.

The use of a flow-through reactor to study molybdenum oxianion exchange on hydrated calcium aluminate

Nicolas C.M. Marty[1*], Sylvain Grangeon[1], Francis Claret[1]

[1] BRGM (FR)

* Corresponding author: n.marty@brgm.fr

Abstract

This document presents preliminary results obtained during a Mo/Cl exchange experiment performed on an AFm sample, in alkaline conditions. Exchange processes were studied using flow-through experiments. A Mo-rich solution was percolated through a reactor initially filled with a Mo-depleted solution and where AFm-Cl had been introduced. Outlet-solution chemistries were monitored over time. Reacting material was carefully examined prior and after experiments in order to follow potential mineralogical transformations. Both modelling of chemical data and XRD analyses indicate that the exchanger composition is mainly dominated by Mo at the end of the flow-through experiment. Moreover, modelling exercise highlight that Cl/OH exchange process must be taken into account to explain the observed Cl behaviour. Preliminary results demonstrate that the apparatus combined with XRD analyses and numerical modelling are powerful tools to study such exchange reactions. The experiment allows the determination of Mo/Cl and OH/Cl exchange constants and the estimation of anionic-exchange capacity (AEC) of the studied material.

Introduction

In most of the designs of deep underground radioactive waste disposal, cementitious materials will be used to build access structures, galleries, vaults and packages for wastes. Cement materials have essentially mechanical function (e.g., low permeable barriers that retard radionuclide migration), but can also sorb radionuclides. Indeed, several studies report cationic and anionic sorption on concrete, cement as well as on pure cement phases (e.g., Aimoz et al., 2012; Atkinson and Nickerson, 1988; Baur and Johnson, 2003). In this framework, European project CEBAMA addresses key issues relevant for long-term safety and key scientific questions related to the use of cement-based materials in nuclear waste disposal applications (*https://www.cebama.eu/*). Hydrated calcium aluminate (AFm) are layered double hydroxides found in cementitious environments. They are foreseen to play a pivotal role on the fate of anion, through sorption/incorporation mechanisms. Retention and incorporation capacities are certainly driven by the crystallographic structure of AFm, including crystal size, and nature of the layer charge (e.g., isomorphic substitutions, vacancies).

BRGM contribution to the Work Package 2 (WP2) of the European project Cebama is focused on various aspects of AFm behaviour in cement-based materials, with the global aim of better understanding the role played by this phase in retaining anions from migrating. In this view, exchange constants for ^{99}Mo on AFm-Cl (i.e., Friedel's salt, $Ca_4Al_2Cl_2O_6 \cdot 10H_2O$) are determined, and the crystal structures of the resulting phases are studied. The ^{99}Mo is widely used as a precursor of the ^{99m}Tc and ^{93}Mo in activation product of the spent nuclear fuel (Ma et al., 2017). Oxyanions MoO_4^{-2} appears to be the main species and very mobile in near-neutral to basic conditions. Estimation of reliable exchange constants is a prerequisite to the thermodynamic modelling of anion migration in cements. Additionally, crystal structures of the resulting exchanged phases allow for a better understanding of the mechanisms of interaction between AFm and anions at the atomic scale, information which is crucial to our capacity to model retention phenomena. Ma et al. (2017) have recently proposed a model for the retention of MoO_4^{-2} in presence of AFm-Cl. Authors have demonstrated that batch experiments were lead to the formation of AFm-MoO_4 concomitant to the disappearance of AFm-Cl in medium Mo concentrations (i.e., ranging from 0.003 to 1 mM). To go further in these investigations, Cl^-/MoO_4^{-2} exchange experiments were studied in alkaline conditions using a different but complementary approach: flow-through experiments were performed with a synthetized AFm-Cl as starting material and outlet-solution chemistries were monitored over time. Experiments were carried out in medium Mo concentrations (i.e., 0.6 to 0.9 mM) taking care to stay in far-from-equilibrium conditions (i.e., where precipitation of AFm-MoO_4 is unexpected).

This report presents the first results obtained on a Mo/Cl exchange experiment in alkaline conditions. Flow-through experiments were performed with a synthetized AFm-Cl as starting material and outlet-solution chemistries were monitored over time. The reacting material was carefully examined prior and after the experiments in order to follow potential mineralogical transformations.

Materials

AFm-Cl sample

To avoid carbonation of the solid phases (e.g., Goñi and Guerrero, 2003), all syntheses were performed in a N_2 filled glove box, using ultra-pure water (resistivity = 18.2 MΩ·cm) which was degassed prior to introduction in the glove box. The synthesis protocol involved mixing of C_3A and $CaCl_2 \cdot 2H_2O$ (1:1 molar ratio) at room temperature, following Balonis and Glasser (2009), and allowing the reaction to run for 15 days, with periodical shaking. After 15 days of maturation, synthetized AFm-Cl were filtered (using a 0.45 µm Millipore HVLP membrane filters) and dried (using a lyophilizer Christ Beta 2 - 8).

Reacting solutions

Two solutions were made: (i) "an input solution" enriched in Mo which was injected inside the reactor and (ii) an "initial solution" which filled initially the reactor. Solutions were prepared using ultra-pure water and were bubbled with $N_2(g)$ during 24 hours before the starting of flow-through experiments. The objective here was to avoid the carbonation of the reacting material during the experiments. Chemical characteristics of solutions are given in Table 1.

Table 1: *Solution compositions.*

	pH	Cl (mg/L)	Mo (mg/L)	Ca (mg/L)	Al (µg/L)	Na (mg/L)	K (mg/L)
Initial solution	12.29	165.1	0.07	507.5	600	0	195.5
Input solution	12.26	167.9	70.68	496.8	263	115.4	0

Note that Na and K were respectively added as tracers to the input and the initial solutions. Moreover, high pH conditions (made using CaO) and Al supply limited the alterations of AFm during the flow-through experiments. Nonetheless, saturation indices of solutions with respect AFm-Cl, AFm-OH and AFm-MoO_4 indicated that all the considered AFm phases were under saturated and therefore, AFm precipitations were not expected from reacting solutions.

Flow-through experiments

Experimental apparatus

A first exchange experiment was carried out on AFm-Cl using flow-through reactors at room temperature (Figure 1). The total volume of the reactor was about 84 mL. The input solution was continuously bubbled under an $N_2(g)$ flux to avoid the carbonation of the reacting material. The inlet solution was injected through the reactors using a peristaltic pump (Watson Marlow, 205U). A constant flow rate of about 2.2 mL/min was imposed. A magnetic stirrer rotated on an axle maintained the AFm-Cl in suspension and has prevented any grinding of the material between the bar and the bottom of the reactor.

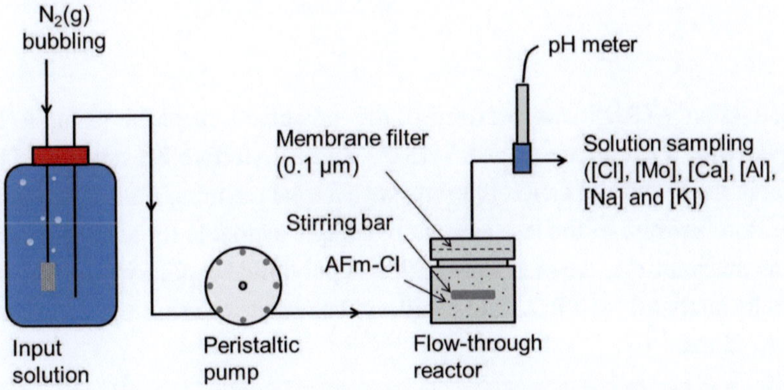

Figure 1: *The experimental apparatus. Flow rates, pH as well as Cl, Mo, Ca, Al, Na and K concentrations were monitored as function of time.*

Initial mass, flow rate and experiment duration are reported in Table 2.

Table 2: *Experimental condition.*

Exp. n°	Initial Mass (g)	Flow rates (mL/min)	Duration (h)
1	0.4987	2.23 - 2.16	6.9

Solution sampling

As shown on Figure 1, outlet solutions were filtered through a 0.1 µm membrane before being collected. The fluid sampling allows the monitoring of solution chemistries (Cl, Mo, Al, Ca, Al, Na, and K concentrations) and flow rates as function of time.

Solid sampling

The solid suspension was collected and filtered at 0.1 µm at the end of experiments. This provided enough solid so that characterization could be done with various methods (for instance XRD analyses). Note that retrieved materials were systematically stored in a N_2 glovebox to avoid any mineralogical transformations.

Analytical procedure

Solution analysis

As reported on Figure 1, the pH was measured online using a by-pass. The apparatus (a Metrohm electrode hooked up to a Mettler Toledo pH meter) was calibrated before the experiment. Ion chromatography was used for Cl analysis (Thermo-Dionex ICS3000), allowing detection limits of 0.5 mg/L. A part of the collected fluids was acidified using nitric acid (65% Suprapur®). Ca, Na, and K concentrations were then measured by ICP-AES (OPTIMA 5300 DV, Perkin Elmer); the detection limits were 0.5 mg/L for the three elements. Al and Mo concentrations were measured using an ICP-MS (NEXION 350 X, Perkin Elmer) with a detection limit of 0.5 and 0.05 μg/L, respectively.

Solid analysis

Powder X-ray diffraction (XRD) was performed the powdered material using a Bruker D8 Advance diffractometer equipped with a Cu anode (λ = 1.5418 Å) and a LynxEye SE detector. Data were acquired in continuous scan mode, in the 5 - 90° 2θ CuKα interval, with a total counting time of 4 hours, and were averaged every 0.02° 2θ CuKα. Both because of the low amount of sample available for analysis and to limit at maximum all possible carbonation mechanisms, samples were sealed in polyimide capillary (internal diameter ~ 1.5 mm) in the glove-box prior to be analysed by XRD. The capillary was rotated during the experiment to limit potential preferential orientation effects.

Modelling of flow-through experiment

Modelling strategy

The modelling of flow-through experiments was performed using PhreeqC3 (Parkhurst and Appelo, 2013). Flow-through experiments were simulated respecting the mass balance between the input flow rate, the homogeneous water volume in the reactor with the corresponding residence time and the output sample solution volume. The procedure described in Appelo website was followed (*http://hydrochemistry.eu/exmpls/cstr.html*); each pore volume was thereafter subdivided into 50 steps using mixing factors.

Thermodynamic database

The THERMOCHIMIE database, version 9, was used for geochemical calculations (Giffaut et al., 2014). The database is freely available in different formats at *https://www.thermochimie-tdb.com.*

Exchange reactions were implemented following the formalism of PhreeqC. The exchange between AFm^+ and Cl^- is assumed as reference (arbitrary choice). Therefore, the logarithm of the exchange constant is set to zero:

$$Afm^+ + Cl^- = AfmCl$$

$$log\_k = 0$$

where Afm^+ is the exchanger and Cl^- refers to the exchangeable anion.

The exchange reaction between OH^- and AFm^+ were taken into account. A rough estimation (i.e., fitted by hand) of the logarithm of the exchange constant has been obtained from the experimental data:

$$Afm^{+} + OH^{-} = AfmOH$$

Similarly, exchange reaction between AFm^{+} and MoO_4^{-2} were also implemented:

$$2Afm^{+} + MoO_4^{-2} = Afm_2MoO_4$$

As discussed below, the value of estimated exchange constant depends on modelling assumptions (i.e., consideration or not of exchange reaction between OH^{-} and AFm^{+}).

Results and discussion

Solution chemistry evolution over time

The evolution of the outlet concentration of inert tracers was first modelled to ensure that mass transport was correctly simulated. Evolutions of experimental and modelled Na and K concentration as function of time are reported on Figure 2. Experimental data and numerical results are in satisfying agreement, thus providing confidence in our modelled transport and reaction parameters.

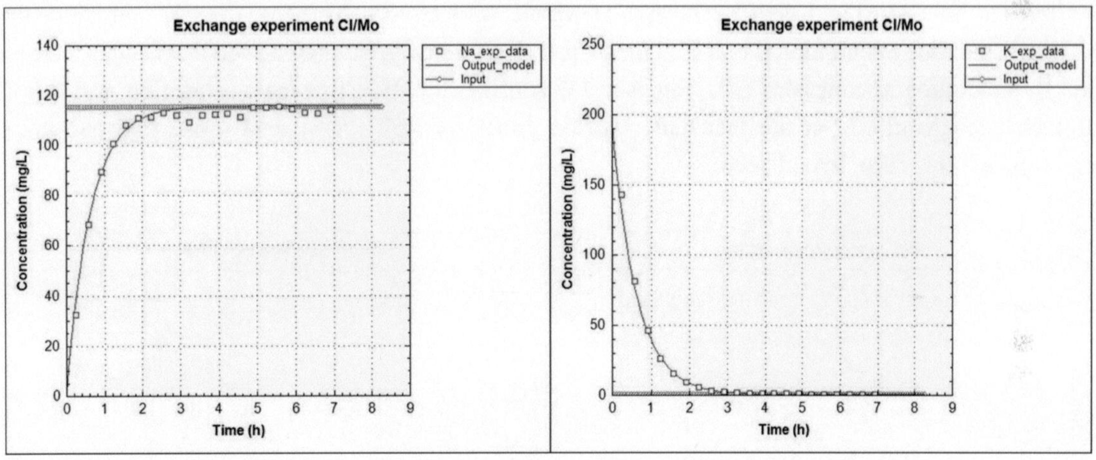

Figure 2: *Evolutions of experimental and modelled Na and K concentrations as function of time.*

A first simulation was performed without Cl/OH exchange. Evolution of Mo and Cl concentrations over time is reported on Figure 3. Experimental data indicated that the duration of the experiment was too short to observe a Mo plateau (i.e., for which output Mo concentration would equal input Mo concentration). The Mo behavior was correctly simulated using a value of 1.0 for the logarithm of the Mo/Cl exchange constant. The calculated AEC of 235.8 meq / 100 g of AFm was lower than the theoretical value which can be calculated from the mineral formula of the AFm-Cl (i.e., 337 meq / 100 g of AFm). Nonetheless, the fitting of exchange parameters was performed on a dataset were Mo plateau was not observed and, consequently, additional experiments and fitting models are required to increase confidence in exchange parameters. Moreover, the first model was not able to reproduce the high Cl release in the first hours of experiment (Figure 3).

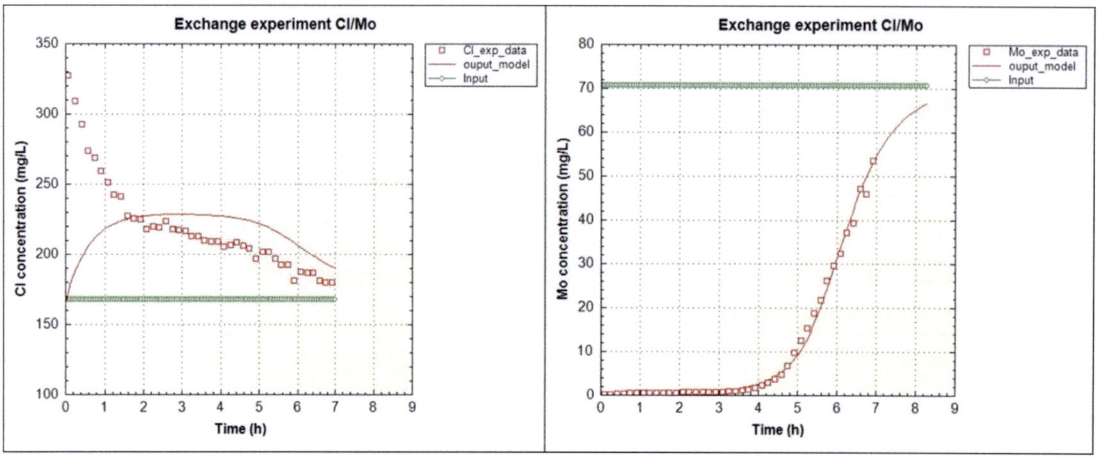

Figure 3: *Evolutions of experimental and modelled Cl and Mo concentrations as function of time. Numerical results have been obtained without the consideration of OH/Cl exchange.*

To tackle the issue of the Cl behavior, a second simulation was performed considering possible exchange between Cl and OH. Evolutions of experimental and modelled Cl and Mo concentrations as function of time are reported on Figure 4. Both Mo and Cl behaviors were correctly simulated. Best fit values were -0.6 and 1.6 for the logarithms of the OH/Cl and Mo/Cl exchange constants, respectively. Note that the AEC remained the same whatever the simulation assumption (i.e., 235.8 meq / 100 g). Obviously, estimated Mo/Cl constants were strongly impacted by modelling assumptions (i.e., log_k = 1.6 considering OH/Cl exchange reaction and log_k = 1.0 without such consideration) but the fact that chloride profile is well reproduced taken into account OH/Cl exchange supports this latter hypothesis.

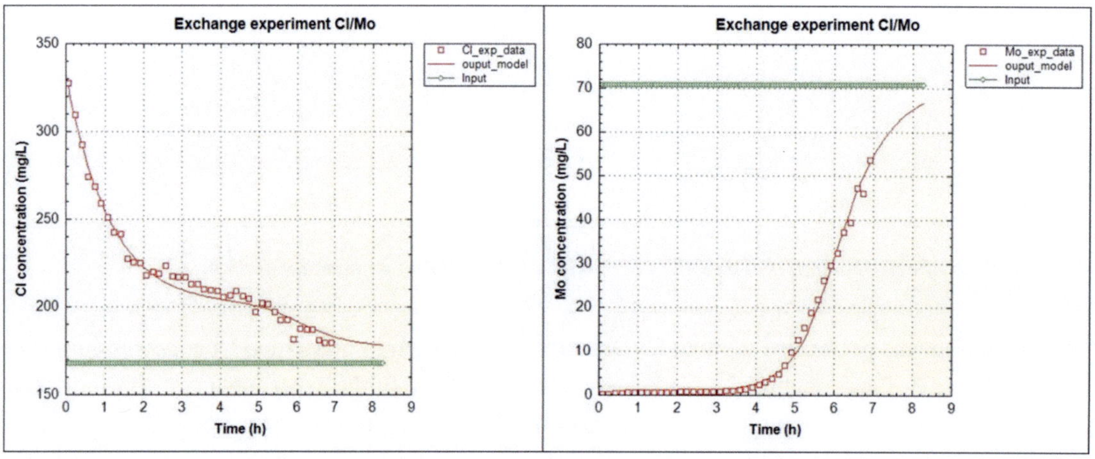

Figure 4: *Evolutions of experimental and modelled Cl and Mo concentrations as function of time.*

Calculated anionic composition at the end of the experiment is reported on the Table 3. The Cl in exchangeable position has been almost totally replaced by Mo after 6.9 h hours of experiments. The final composition is in agreement with XRD analyses (see next section *Mineralogical transformations*).

Table 3: *Calculated final exchanger composition (Phreeqc modelling).*

Species	Equivalent fraction
Afm_2MoO_4	$9.17 \cdot 10^{-1}$
AfmOH	$4.46 \cdot 10^{-2}$
AfmCl	$3.89 \cdot 10^{-2}$

Additionally, evolutions of experimental and modelled Al and Ca concentrations as function of time are reported on Figure 5. Results indicate that output concentrations are close to input concentrations. Therefore, the AFm dissolution can be neglected and the amount of exchanger was assumed to be constant during the whole duration of the experiment.

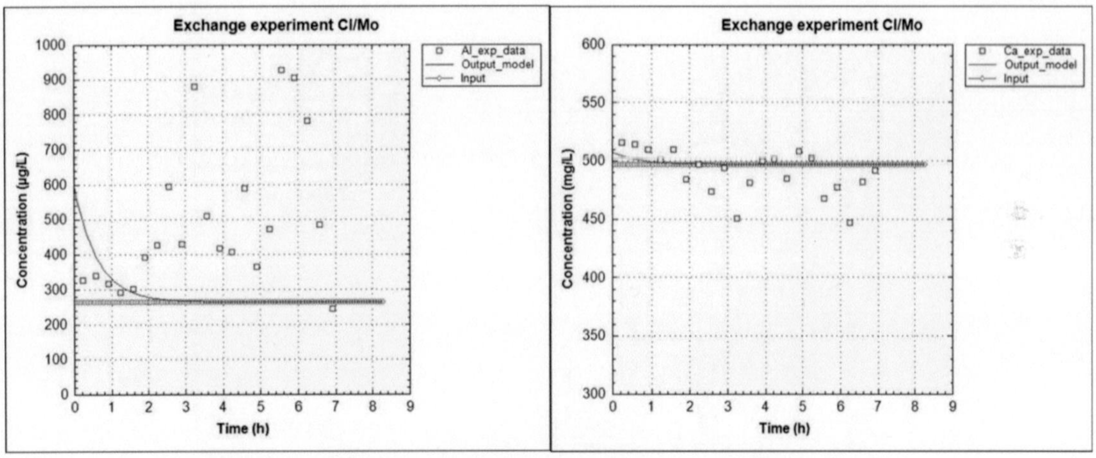

Figure 5: *Evolutions of experimental and modelled Al and Ca concentration as function of time.*

The pH evolution is reported on Figure 6. Recorded values are very constant over time. Weakest pHs that can be distinguished at the beginning of experiment are due to OH/Cl exchange processes.

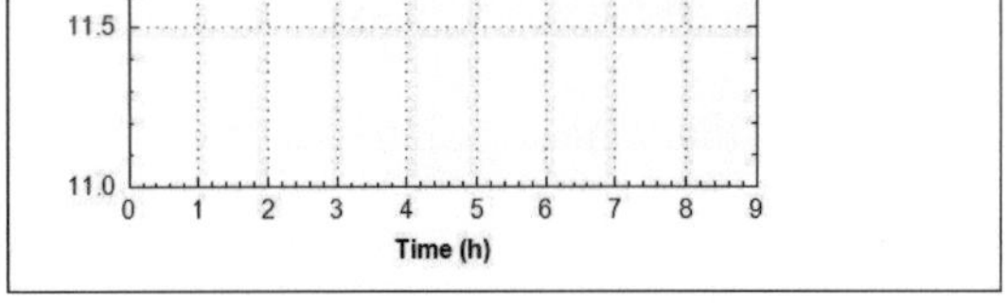

Figure 6: *Evolution of experimental and modelled pH as function of time.*

Mineralogical transformations

Upon exchange with the Mo, the diffraction pattern of the AFm sample underwent significant changes, in particular at low angle region (Figure 7). In the fresh samples, one maxima at 11.52° 2θ CuKα (7.7 Å) as well as its harmonics could be observed. These d-spacing value was in reasonable agreement with one expected for AFm-Cl (~ 7.8 Å - 11.3° 2θ CuKα). After exchange with Mo, the layer-to-layer distance increased as other maxima became apparent at lower angular values. Indeed, three maxima at 9.20, 10.55 and 11.52° 2θ CuKα (equivalent to 9.6, 8.4 and 7.7 Å) became apparent. The shortest distance can be attributed to hydrated AFm-Cl, while the two other ones are compatible with fully and partially exchanged AFm-MoO_4 (Ma et al., 2017).

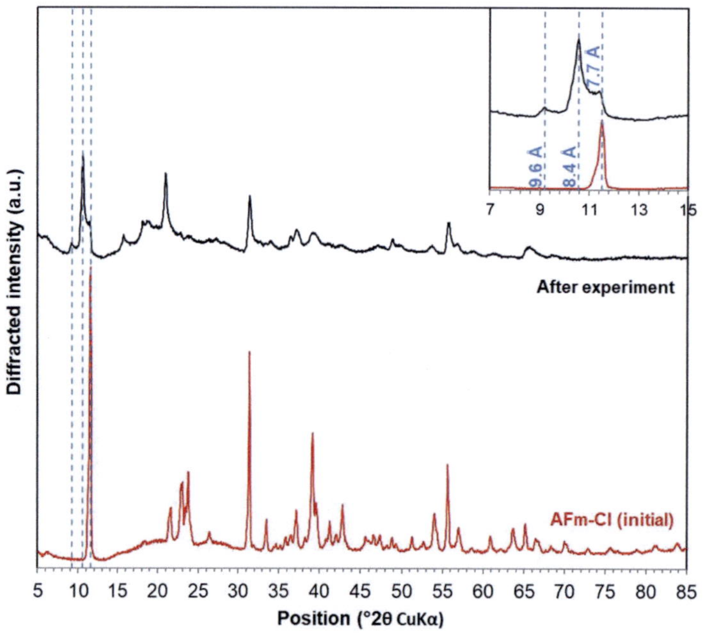

Figure 7: *X-ray diffraction patterns of initial freeze-dried AFm-Cl sample (bottom; red solid line) and of Mo-exchanged AFm-Cl (top, black solid line). Patterns were shifted along the y-axis to ease data comparison. The maximum at ~ 5.8° 2θ CuKα as well as the broad maximum at ~ 15 - 25° 2θ CuKα originate from the polyimide capillary.*

The presence of a minor peak that can be attributed to an AFm structure having interlayer Cl is compatible with the results from chemical measurements which showed that Cl was not fully exchanged by Mo (Table 2).

Conclusions and Future work

An integrated study using flow-through reactors, modelling and XRD analyses appears to be a powerful tool to investigate the exchange reactions of AFm. More specifically, the following conclusions can be drawn:

- the modelling exercise indicates that exchange kinetics can be neglected,
- the apparatus allow the fitting of Mo/Cl and OH/Cl exchange constants as well as AEC,
- OH/Cl exchange reaction should be taken into account to reproduce correctly the Cl behaviour,
- the AEC of AFm is a priori high. A first estimation tends to indicate that its value was lower than the theoretical value calculated from the mineral formula,

- the fitted value of Mo/Cl exchange constant was affected by the consideration (or not) of OH/Cl exchange process.

Future work will be focussed on the performing of additional flow-through experiments (anion of interest will be studied, in the Se/Mo/Cl competition system). More specifically, new experiments will be carried out until the Mo-plateau. Moreover, in regard to the modelling works, exchange parameters will be optimised using PEST (Watermark Numerical Computing, *http://www.sspa.com/pest*).

Acknowledgement

The research leading to these results has received funding from the European Union's Horizon 2020 Reasearch and Training Programme of the European Atomic Energy Community (EURATOM) (H2020-NFRP-2014/2015) under grant agreement n° 662147 (CEBAMA).

References

Aimoz, L., Wieland, E., Taviot-Gueho, C., Dahn, R., Vespa, M., Churakov, S.V. (2012). Structural Insight into Iodide Uptake by AFm Phases. Environmental Science & Technology, 46, 3874-3881.

Atkinson, A. and Nickerson, A.K. (1988). Diffusion and sorption of cesium, strontium, and iodine in water-saturated cement. Nuclear Technology, 81, 100-113.

Balonis, M. and Glasser, F.P. (2009). The density of cement phases. Cement and Concrete Research, 39, 733-739.

Baur, I. and Johnson, C.A. (2003). Sorption of selenite and selenate to cement minerals. Environmental Science & Technology, 37, 3442-3447.

Giffaut, E., Grivé, M., Blanc, P., Vieillard, P., Colàs, E., Gailhanou, H., Gaboreau, S., Marty, N., Madé, B., Duro, L. (2014). Andra thermodynamic database for performance assessment: ThermoChimie. Applied Geochemistry, 49, 225-236.

Goñi, S. and Guerrero, A. (2003). Accelerated carbonation of Friedel's salt in calcium aluminate cement paste. Cement and Concrete Research, 33, 21-26.

Ma, B., Fernandez-Martinez, A., Grangeon, S., Tournassat, C., Findling, N., Claret, F., Koishi, A., Marty, N., Tisserand, D., Bureau, S., Salas-Colera, E., Elkaïm, E., Marini, C., Charlet, L. (2017). Evidence for multiple sorption modes in layered double hydroxides using Mo as structural probe. Environmental Science & Technology, 21, 5531-5540.

Parkhurst, D.L. and Appelo, C.A.J. (2013). Description of input and examples for PHREEQC version 3 - A computer program for speciation, batch-reaction, one-dimensional transport, and inverse geochemical calculations. U.S. Geological Survey Techniques and Methods, book 6, chap. A43.

Stuctural investigation on the uptake of long-lived saftey relevant radionuclides by cementitious materials

Steve Lange[1*], Matthew Isaacs[1,2], Martina Klinkenberg[1], David Read[2,3], Dirk Bosbach[1], Guido Deissmann[1]

[1] Forschungszentrum Jülich GmbH, Institute of Energy and Climate Research (IEK-6): Nuclear Waste Management and Reactor Safety (DE)
[2] University of Surrey (UK)
[3] National Physical Laboratory (UK)

* Corresponding author: s.lange@fz-juelich.de

Abstract

The uptake of the long-lived radionuclides ^{226}Ra and ^{99}Tc, as well as iodide and molybdate ions by various model phases representative for phases present in hydrated cements was investigated. The model phases were synthesized under argon atmosphere, using well established procedures (e.g., Atkins et al., 1991 and 1992; Baur et al., 2004). Sorption and uptake kinetics by the model phases were studied in static batch experiments under anoxic conditions. Experiments performed for up to 60 days indicate a strong uptake of radium by calcite and CSH phases. Uptake studies performed on iodide revealed the incorporation into AFm-SO_4. In contrast, no significant uptake of technetium, present as $^{99}Tc(VII)$ in solution, by the various model phases was observed. Microanalytical investigations on the uptake of molybdate by AFm phases provided evidence of a molybdate substitution in AFm-SO_4.

Introduction

Cement-based materials are widely used in nuclear waste management, for example for the solidification of low and intermediate level wastes, as barrier materials in nuclear waste repositories, or in certain waste containers. The migration behavior of radionuclides in cementitious materials is controlled by radionuclide solubility phenomena at high pH conditions, diffusion, interface processes such as surface complexation, or incorporation of radionuclides into solid phases, including the formation of solid solutions. Within the framework of CEBAMA, we study the uptake of selected safety relevant long-lived fission and decay products such as ^{129}I, ^{99}Tc and ^{226}Ra in cementitious systems, using advanced micro analytical and spectroscopic tools. The objective of these investigations is to enhance the mechanistic understanding of the radionuclide uptake and retention in cementitious materials and to evaluate the relevance of cement alteration processes, such as carbonation, on the solid speciation of radionuclides in aged concrete. To unravel the contributions of individual hydrate phases on the radionuclide retention behaviour of complex multiphase cementitious materials, we adopt a bottom-up approach studying the radionuclide interaction with synthesized phases present in cementitious materials (model phases), such as calcium silicate hydrates (CSH) with different Ca/Si ratios, monosulphate (AFm) and ettringite (AFt) phases, in addition to hardened cement pastes of different composition (e.g., CEM I and CEM V). Here, we report on the uptake of ^{226}Ra, ^{99}Tc, iodide and molybdate by single model phases representative for phases present in hydrated cements.

Materials and Methods

Synthesis of model phases

The various model phases used in the present study were synthesized using well established procedures. Phase synthesis and sample preparation was carried out in an argon glove box (< 10 ppm CO_2); details on the chemicals used are compiled in Table 1. Calcium silicate hydrates (CSH) were prepared according to the procedure of Atkins et al. (1992). To achieve CSH with different CaO/SiO_2 ratios, different amounts of CaO, typically 1.8 g up to 3.4 g (weighted after 2 hours of calcination at 1,000°C), were mixed with SiO_2 in the desired ratio and suspended in water to achieve a water / solid ratio of 20. Ettringite ($Ca_6Al_2(SO_4)_3(OH)_{12}\cdot 26H_2O$) synthesis was carried out according to Atkins et al. (1991) and Baur et al. (2004), respectively. CaO was suspended in water and cooled to 4°C. Subsequently, a stoichiometric solution of $Al_2(SO_4)_3\cdot 18\ H_2O$ (Ca/Al = 3) was mixed with the CaO suspension. The suspension was stirred for 4 hours at approx. 4 °C and cured for one month at room temperature. C_3A ($3CaO\cdot Al_2O_3$), used as a reagent, was synthesized according to the method described by Atkins et al. (1991). Stoichiometric amounts (Ca/Al = 1.5) of CaO and Al_2O_3 were mixed and calcined at 1,000°C for 2 hours in a Pt/Au (5% Au) crucible and ignited at 1,450°C for one hour. The product was crushed and well ground using a mortar and ignited in a second cycle at 1,450°C for 48 hours. AFm ($Ca_4Al_2(X^{2-})(OH)_{12}\cdot 6H_2O$) synthesis was done according to Baur et al. (2004). Stoichiometric amounts (Ca/Al = 2) of freshly prepared C_3A and $CaSO_4\cdot 2H_2O$ / $CaCO_3$ / $CaMoO_4$ were mixed and suspended in water at 4°C and stirred for 4 hours. Afterwards the product was cured for 1 - 5 months at room temperature. The products were washed by centrifugation (30 min at 4,500 rpm) and resuspended in water three to five times, followed by vacuum filtering using a Büchner funnel and a paper filter (Whatman). Drying was carried out over $CaCl_2$ using a desiccator under continuous evacuation.

Table 1: *Details of chemical substances used in the synthesis of the model phases.*

Compound	Product description	Provider
CaO	puriss., 96 - 100.5% ex ignited substance	Sigma Aldrich
$Al_2(SO_4)_3\cdot 18\ H_2O$	AnalaR normapur	VWR chemicals
SiO_2	fumed silica 395 m²/g	Sigma Aldrich
Al_2O_3	99.5% purity based on trace metals analysis	Sigma Aldrich
$CaSO_4\cdot 2H_2O$	analytical grade, EMSURE	Merck Millipore

Artificial cement pore water (ACW)

ACW representative for young cementitious materials (i.e., pH ~ 13.5; Wieland et al., 1998) was prepared by filtration and dilution of highly concentrated alkali hydroxide solutions (NaOH and KOH), stored over $Ca(OH)_2$ to precipitate remaining carbonates, following the procedure of Sipos et al. (2000). The obtained solution containing 0.114 mol/L NaOH and 0.18 mol/L KOH was saturated with $Ca(OH)_2$ and filtered prior to use.

Phase characterization

Purity and composition of the final products was confirmed by XRD using either a D8 Advance with a θ - θ geometry or a D4 Endeavour with a θ - 2θ geometry (Bruker AXS GmbH) employing CuK_α-radiation (λ = 0.15406 nm), and by XRF-analyses, conducted by an external service company (Terrachem GmbH), respectively. Phase morphology and sample microstructure was studied by SEM analyses performed with a Quanta 200F from FEI equipped with a field emission cathode. Energy dispersive spectroscopy (EDS) was performed with an Apollo X Silicon Drift Detector (SDD) from EDAX. Point, multipoint measurements, line scans, element mapping and particle analysis were carried out with the Genesis software. Samples were prepared on sticky carbon pads which were glued to an aluminium sample holder. SEM analysis was carried out under low vacuum (60 Pa), where a coating of the samples was not required. Transmission electron microscopy (TEM), selected area electron diffraction (SAED) and energy dispersive X-ray spectroscopy (EDS) were performed on a FEI Technai G2 F20 transmission electron microscope at the Ernst-Ruska-Centre for Microscopy and Spectroscopy with Electrons (ER-C) at Forschungszentrum Jülich. The transmission electron microscope was operated at 200 kV accelerating voltage with a total beam current of about 2 nA. The samples present in form of few micrometre sized crystallites were supported by a hexagonal copper grid for analyses. TEM images and SAED patterns were recorded from parts of the crystallites exposed over the free grid space with the incident electron beam perpendicular to the platelet normal. EDS spectra were recorded with the samples tilted by 15° towards the spectrometer.

Sorption experiments

Batch sorption experiments were carried out in 20 mL LDPE bottles using a L/S-ratio of 200 L/kg. Weighted amounts of dried solids were equilibrated with deionized water prior to use. The suspension was stored at room temperature for up to 14 days under anoxic conditions to achieve equilibrium between aqueous and solid phases. After equilibrium was reached, the liquid was separated from the solids by filtration. Weighted amounts of fresh solids were added to this solution and stored for seven days before the tracer was added for sorption studies. The tracer concentrations used were $c(^{226}Ra^{2+}) = 10^{-8}$ mol/L, $c(^{99}TcO_4^-) = 10^{-7}$ mol/L, $c(I^-) = 10^{-7}$ mol/L and $c(MoO_4^{2-}) = 10^{-7}$ mol/L. The sorption experiments were conducted for up to 60 days, until a plateau was reached. Tracer sorption to the reaction vessels and filters was tested and found to be negligible. The radionuclide uptake by the respective solids is described in terms of the distribution ratio R_d between solid and liquid phase according to:

$$R_d = \frac{A_{ini} - A_t}{A_t} \cdot \frac{V}{m} \qquad \textbf{\textit{Eq. 1}}$$

with,

A_t: concentration at time t in liquid phase (Bq or M)
A_{ini}: initial concentration in liquid phase (Bq or M)
V: volume of liquid phase (L)
m: mass of solid phase (kg)

Solution analyses

Concentrations of ^{226}Ra were determined by γ-spectroscopy (186 keV γ-line) using 500 µL sample aliquots in a 2 mL borosilicate glass placed in a high-purity germanium coaxial N type detector system (type: EGC 35-195-R), from Eurisys Mesures, equipped with a spectrometer system from EG & G Ortec. Analysis of the spectra was performed with the GammaVision® Modell A66-B32 software (version 5.20). Liquid scintillation counting (LSC) measurements of ^{99}Tc were performed using a 1220 Ultra low level Quantulus™ LSC device (Perkin Elmer), equipped with the Winq software (version 1.2). Aliquots of 50 µL were diluted in a 20 mL polyethylene Vial with 15 mL Hionic-Fluor or Ultima Gold™ LSC-cocktail (Perkin Elmer). Concentrations of inactive elements were determined by ICP-MS. Aliquots of the individual phases were diluted to a concentration of the stable elements to a range of 1 to 100 ppb and measured by ICP-MS using an Elan 6100 DRC (Perkin Elmer) equipped with the Elan software (version 3.4).

Results

A variety of cement hydration phases were synthesized to cover the range of solids resulting from the hydration of ordinary Portland cement (CEM I) or composite cements (e.g., CEM III, CEM V). When hydrated, cementitious systems consist of manifold hydration phases (e.g., Taylor, 1997), however, predominated by calcium silicate hydrates (C-S-H phases), portlandite, and calcium-aluminate/ferrate compounds (AFm/AFt). Here we focus on systems containing calcium, aluminum, silicon, sulfate and in some cases carbonate. Thus at present our model phases do not account for the possible substitution of iron for aluminum in AFt and AFm phases, as well as for the substitution of magnesium for calcium resulting in MSH and the presence of hydrotalcite in hydrated cements.

The progressive alteration and leaching of cementitious materials (e.g., by groundwater) proceeds in various stages (cf., Atkins and Glasser, 1992; Glasser, 2011; Drace and Ojovan, 2013). First, the dissolution of alkali hydroxides (stage I) leads to a decrease of the initial pore water pH > 13 to 12.5, a calcium hydroxide buffered system that can persist over very long time periods (stage II). After portlandite depletion in stage III, the pH will be buffered in the range between 12.5 and 10 by the progressive incongruent dissolution of CSH-phases (i.e., preferential release of calcium), lowering the Ca/Si-ratio over time and reducing the pH value at which the pore water is buffered. To study the uptake of radionuclides by cementitious systems at different stages of alteration batch sorption experiments were performed at different pH conditions. Experiments performed with the artificial cement pore water at pH 13.5 are deemed to be representative to address the uptake of safety relevant radionuclides in stage I. Later stages of cement alteration achieved by calcium leaching are studied with aqueous solutions that are in equilibrium with the solids studied. Furthermore, we included synthetic calcite and a carbonate containing calcium-aluminum-LDH (AFm-CO_3) as model phases representative for carbonated cementitious materials.

Uptake experiments performed with technetium (^{99}Tc(VII)) indicated negligible Tc-uptake by the various model phases. Considering the analytical errors, in all experiments the measured activity of ^{99}Tc in solution remained at the level of the initial tracer concentration (10^{-7} mol/L). This suggests that Fe^{2+} and/or Fe^{2+}-containing phases, which were not studied here, may play a key role in technetium retention in cementitious system. According to Allen et al. (1997) these reducing agents are capable to form TcO_2 or TcS_2-like species, leading to a decrease of the technetium mobility in the cement pore solution, due to the significantly lower solubility of Tc(IV) compared to Tc(VII). As expected, the uptake of radium by CSH was found to be strong, with eq. R_d values (reached after 60 days) exceeding 1,000 L/kg. The R_d was found to depend on the Ca/Si ratio of the CSH phases as well as on pH. CSH with lower Ca/Si-ratios showed a stronger uptake of radium, which can be explained in terms of electrostatic sorption. The zeta potential of CSH is zero for Ca/Si = 1.2; at Ca/Si > 1.2 the zeta potential is positive, whereas the negative charge at Ca/Si < 1.2 allows for a more pronounced uptake of cations (cf., Atkins et al., 1992). The observation of higher radium uptake at lower Ca/Si ratios can also be explained in terms of cationic exchange, due to the significantly lower competition with calcium cations in solution at lower Ca/Si ratios that outweighs the lower number of Ca-sites at the surface (cf., Tits et al., 2006). At the high pH conditions in the ACW representative for young cementitious materials (pH ~ 13.5), the radium uptake by the CSH was generally lower than at the equilibrium pH of the CSH phases (CSH 0.9: pH 11.9; CSH 1.4: pH 12.2) in the systems. The uptake of ^{226}Ra in systems with AFt and AFm phases (R_d < 150 L/kg) was generally significantly lower compared to CSH. Experiments with $CaCO_3$ revealed a slightly more pronounced radium uptake (R_d 200 to 260 L/kg), indicating an additional Ra-retention mechanism in carbonated cementitious materials. According to Jones et al.

(2011), the most likely mechanism for radium uptake onto calcite at low concentrations is thought to be coprecipitation within a Ca/Ra-carbonate.

The uptake mechanism for iodine was found to depend essentially on the aqueous (redox) speciation. Structural incorporation of iodide (I^-) was observed for AFm-SO_4 (R_d = 500 L/kg). On the other hand, the uptake of iodate (IO_3^-) by AFt as well as by AFm-SO_4, led to the formation of an iodate-substituted ettringite, formed either by anion exchange or phase transformation. Moreover, we observed that the uptake of iodide by AFm-phases is dependent on the nature of the anion complex, with stronger iodide uptake by AFm-SO_4 compared to AFm-CO_3. Ettringite, known for its capability to incorporate various monovalent and bivalent oxo-anions by solid-solution formation, showed a stronger affinity to incorporate iodate than molybdate. In general, the uptake experiments performed with molybdate revealed no significant uptake by most of the model phases tested; only a slight uptake of molybdate by AFm-SO_4 was observed (R_d = 60 L/kg). According to Kindness et al. (1994) molybdate could substitute for sulfate in the AFm structure. XRD studies on the AFm-phases used in the batch sorption experiments showed an increase in the basal spacing compared to pure AFm-SO_4. This could indicate the structural incorporation of MoO_4-ions in the AFm-structure, since the size of the molybdate oxo-anions (Mo-O bond length ~ 1.77 Å) is larger than of SO_4-ions (S-O bond length ~ 1.47 Å). However, the increase in the basal spacing could also be attributed to an increasing number of water molecules in the interlayer.

A specific synthesis was performed by adding CaO and Na_2MoO_4 in stoichiometric amounts into water (w/s = 10) to study the structure of a (pure) molybdate AFm phase. Na_2MoO_4 and $Ca(OH)_2$ react to powellite and NaOH, resulting in the high pH environment required for AFm formation. The subsequent addition of C_3A led to a product with hexagonal plate like morphology (Figure 1). SEM/EDS measurements revealed a stoichiometric composition of Ca:Al:Mo of 4:2:1. The splitting of the basal reflex at 2Θ = 8.9° in the XRD patterns, indicate the presence of two different hydration states. However, some reflexes in the pXRD patterns cannot be related to the AFm structure.

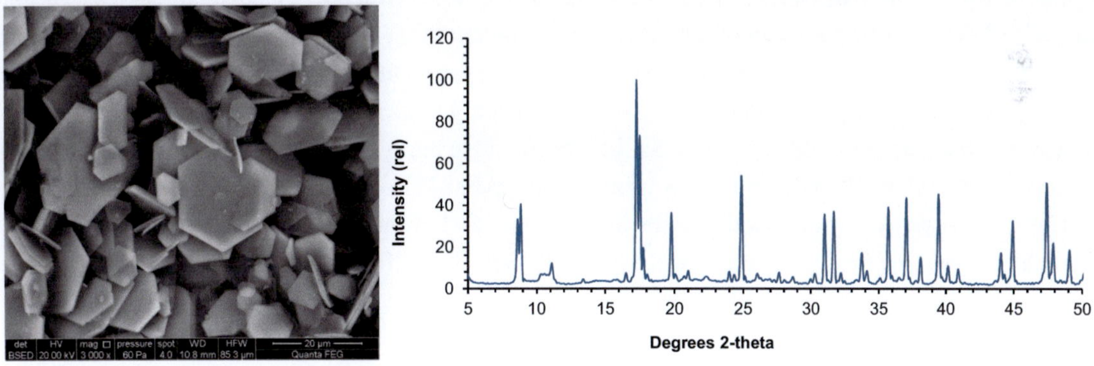

Figure 1: *SEM picture and pXRD pattern of the synthesised sheet like material.*

Thus a TEM study was performed to obtain more information on the structure of the synthesized layered material and to get an idea of the symmetry. Electron diffraction patterns were obtained along the 001 zone axis, due to the preferred orientation of the platelets. Despite the similarities in the SEM images of all platelets and their morphology, different electron diffraction patterns were obtained (Figure 2). A minority of the platelets are showing a hexagonal or trigonal symmetry (Figure 2, left), whereas the majority show a triclinic symmetry (Figure 2, center). Some platelets were found to overlap with a slight rotation (Figure 2, right). Moreover, small amounts of powellite, one of the educts used in synthesis, were found.

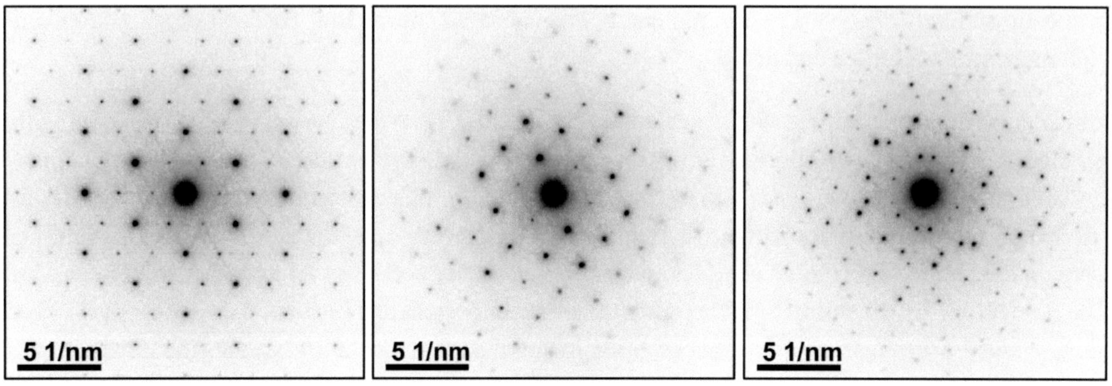

***Figure 2:** Example SAED patterns recorded from individual platelets.*

The TEM study showed that the synthesized material was not single-phased AFm-MoO_4 and the observed unidentified pXRD reflexes belong to a solid with triclinic symmetry. A potential candidate for this phase could be a molybdate-bearing phase similar in structure to the so-called U-phase ($4CaO \cdot 0.9Al_2O_3 \cdot 1.1SO_3 \cdot 0.5Na_2O \cdot 16H_2O$) synthesized by Li et al. (1996 and 1997). This U-Phase shows strong similarities in X-ray diffraction patterns compared to the additional product obtained in our AFm-MoO_4 synthesis, although it contains sulfate anions instead of molybdate.

Conclusions and future work

Sorption experiments addressing the uptake of radium, iodine, molybdenum and technetium by single model phases representative for phases in hydrated cements were performed under anoxic conditions. A significant uptake of radium by CSH due to cation exchange and a pronounced uptake by calcite, potentially due to coprecipitation effects, were observed. The radium uptake in systems with artificial cement waters representative for young alkali-rich cement pore waters was lower, due to competition effects with Na^+ and K^+. Microanalytical investigations revealed an uptake of molybdate by AFm-SO_4 by structural incorporation into the interlayer. The uptake experiments performed with pertechnetate indicated only a weak interaction with the hydration phases. The uptake mechanism for iodine was found to depend essentially on the aqueous speciation. Structural incorporation of iodide in the interlayer was observed for AFm-SO_4. Uptake of iodate by AFt as well as by AFm-SO_4, led to the formation of an iodate-substituted ettringite. Future work will focus on the uptake of ^{79}Se by the model phases and in-diffusion experiments (e.g., with ^{226}Ra, ^{79}Se, ^{99}Tc, iodide, and molybdate) using hardened cement pastes (CEM I and VTT CEBAMA reference mix).

Acknowledgement

The authors would like to thank Juri Barthel from Ernst Ruska-Centre (Forschungszentrum Jülich GmbH) for providing excellent TEM data and for scientific support.

The research leading to these results has received funding from the European Union's Horizon 2020 Research and Training Programme of the European Atomic Energy Community (EURATOM) (H2020-NFRP-2014/2015) under grant agreement n° 662147 (CEBAMA).

References

Allen, P.G., Siemering, G.S., Shuh, D.K., Bucher, J.J., Edelstein, N.M., Langton, C.A., Clark, S.B., Reich, T., Denecke, M.A. (1997). Technetium speciation in cement waste forms determined by X-ray absorption fine structure spectroscopy. Radiochimica Acta, 76, 77-86.

Atkins, M., Macphee, D., Kindness, A., Glasser, F.P. (1991). Solubility properties of ternary and quaternary compounds in the calcia-alumina-sulfur trioxide-water system. Cement and Concrete Research, 21, 991-998.

Atkins, M. and Glasser, F.P. (1992). Application of portland cement-based materials to radioactive waste immobilization. Waste Management, 12, 105-131.

Baur, I., Keller, P., Mavrocordatos, D., Wehrli, B., Johnson, C.A. (2004). Dissolution-precipitation behaviour of ettringite, monosulfate, and calcium silicate hydrate. Cement and Concrete Research, 34, 341-348.

Drace, Z. and Ojovan, M.I. (2013). A summary of IAEA coordinated research project on cementitious materials for radioactive waste management. *In:* Bart, F., Cau-dit-Coumes, C., Frizon, F., Lorente, S. (Eds.) Cement-based materials for nuclear waste storage, Springer, 3-11.

Glasser, F.P. (2011). Application of inorganic cements to the conditioning and immobilisation of radioactive wastes. *In:* Ojovan, M.I. (ed.) Handbook of advanced radioactive waste conditioning technologies. Woodhead, 67-135.

Jones, M.J., Butchins, L.J., Charnock, J.M., Pattrick, R.A.D., Small, J.S., Vaughan, D.J., Wincott, P.L., Livens, F.R. (2011). Reactions of radium and barium with the surfaces of carbonate minerals. Applied Geochemistry, 26, 1231-1238.

Kindness, A., Lachowski, E.E., Minocha, A.K., Glasser, F.P. (1994). Immobilization and fixation of molybdenum (VI) in portland cement. Waste Management, 14, 97-102.

Li, G., Le Bescop, P., Moranville, M. (1996). The U phase formation in cement-based systems containing high amounts of Na_2SO_4. Cement and Concrete Research, 26, 27-33.

Li, G., Le Bescop, P., Moranville-Regourd, M. (1997). Synthesis of the U phase ($4CaO\cdot0.9Al_2O_3\cdot1.1SO_3\cdot0.5Na_2O\cdot16H_2O$). Cement and Concrete Research, 27, 7-13.

Sipos, P., May, P.M., Hefter, G.T. (2000). Carbonate removal from concentrated hydroxide solutions. Analyst, 125, 955-958.

Taylor, H.F.W. (1997). Cement Chemistry. Thomas Telford Publishing.

Tits, J., Iijima, K., Wieland, E., Kamei, G. (2006). The uptake of radium by calcium silicate hydrates and hardened cement paste. Radiochimica Acta, 94, 637-643.

Wieland, E., Tits, J., Spieler, P., Dobler, J.P. (1998). Interaction of Eu(III) and Th(IV) with sulfate-resisting portland cement. MRS Proceedings, 506, 573-578.

C-14 sorption on CEM I: Effect of hardened cement pastes degradation on C-14 uptake

Crina Bucur[1*], Ionut Florea[1], Nicoleta Deneanu[1], Relu Dobrin[1], Alice Dinu[1]

[1] Institute for Nuclear Research (RO)

* Corresponding author: crina.bucur@nuclear.ro

Abstract

Cement based materials are used as engineered barriers in a geologic repository for spent fuel and long lived intermediate level waste. These materials exhibit an increased uptake for the ^{14}C contained in the disposed waste.

In this study the $^{14}CO_3^{2-}$ sorption on non-degraded and degraded hardened cement pastes (HCP) was evaluated. The non-degraded HCP used in the sorption tests was received in 2016 from Armines while the degraded states were synthetises by hydrolysis leaching process in degassed deionized water (for free of portlandite states) and by accelerate chemical degradation in ammonium nitrate solution (NH_4NO_3) for the degradation state free of ettringite.

Preliminary results indicate that after less than 40 days the sorption equilibrium was achieved both for non-degraded state and for degraded ones. For this contacting time the percentage uptake for the non-degraded HCP was higher than 80% and the distribution ratio higher than 1,000 L/kg. For the most degraded state the percentage uptake as well as the distribution ratio are much lower (around 30% and 116 L/kg, respectively).

The experiments are on-going and the sorption isotherms will be derived for all HCPs and the distribution ratio will be assessed. Also, the influence of the solid to liquid ratio in the batch experiments on the distribution will be evaluated at the equilibrium and desorption tests will be carried out.

Introduction

In Romania, the long-lived intermediate level waste (LL-ILW) and CANDU spent fuel are foreseen to be disposed of in the future geological disposal.

The main source of ^{14}C in the LL-ILW to be disposed of in the Romanian geologic repository is the spent ion exchange resins generated mainly in the Moderator System and in the Primary Heat Transport System of the two units of Cernavoda NPP. The dominant ^{14}C species in spent ion exchange resins is inorganic carbonate $^{14}CO_3^{2-}$. Other LL-ILW to be geological disposed of consists in the pressure tubes and calandria tubes that could contain Carbon-14 both in inorganic and organic forms.

In form of dissolved carbonate or bicarbonate, inorganic ^{14}C is strongly immobilized in cemented waste forms. Under strongly alkaline conditions, the dominant species of inorganic ^{14}C is $^{14}CO_3^{2-}$. The carbonate ions can either precipitate as $CaCO_3$ or sorb onto cement phases.

There are reported in the literature results of numerous studies on ^{14}C sorption on hardened cement paste and mortars (Allard et al., 1981; Bayliss et al., 1988; Bradbury and Sarott, 1994; Ochs et al., 2016; Wang et al., 2009). Experimental data regarding the effect of the cement degradation in the disposal conditions on radionuclide sorption are limited and the experiments that are on-going in RATEN ICN in the frame of CEBAMA - WP2 are specifically oriented to assess the influence of HCP degradation on ^{14}C sorption as well as on ^{226}Ra. This paper presents the preliminary results of the on-going experimental programme to assess the ^{14}C sorption on CEM I HCP both in non-degraded and degraded stages.

Materials and methods

Preparation of fresh and degraded HCP

In this study, a hardened cement paste (HCP) based on CEM I (Ordinary Portland Cement) was used. This HCP were prepared with a water/cement ratio of 0.38 and it was received in 2016 from ARMINES were it was kept for almost 7 years after setting in saturated portlandite water (pH = 12.4) at 20°C. Between receiving time and using it in the sorption experiments, the CEM I HCP were also kept in saturated portlandite water at room temperature (20 ± 3°C) to avoid its carbonation.

The hydrated paste was crushed and sieved under nitrogen atmosphere (using nitrogen of 99.99 grade purity) and the 125 - 250 μm fraction (noted as C0) was kept both for the ^{14}C sorption experiments and to prepare the degraded states of the cement paste.

To prepare the three main degradation states (noted as C1, C2 and C3, respectively), a procedure similar with that described by Pointeau et al. (2004) were applied. The first two degradation states (C1 and C2) were obtained by hydrolysis leaching process in degassed deionized water (DDW was prepared by bubbling nitrogen during 24 hours through deionized water) at two solid to water ratios: ~ 3.7 g/L for the first degradation state and 0.4 g/L for the second one. A more aggressive degradation protocol, also described by Pointeau et al. (2004), was applied to obtain the third degradation state. The cement paste was first mixed with ammonium nitrate solution (NH_4NO_3) to accelerate the chemical degradation. After the established contact time, the cement paste was washed four times with DDW (until the pH of solution was constant) and after that it was mixed with DDW for calcium leaching.

Details on the experimental conditions used to obtain the three degradation states are given in Table 1.

Table 1: *Experimental parameters for the synthesis of the degraded cement samples.*

Sample ID	Degradation state	Solid/liquid (DDW or 0.4 mol/L NH_4NO_3), g/L	Contacting time, days
C1	1st state	3.7 (0.74 g C0 + 200 mL DDW)	15
C2	2nd state	0.4 (0.8 g C0 + 2000 mL DDW)	15
C3	3rd state	7.8 (1.2 g C0 + 153 mL NH_4NO_3)	4
		0.01 (1.2 g Co + 100 mL DDW)	14

The water content of the non-degraded cement paste (C0) and synthesized degraded cement pastes (C1, C2, C3) was calculated based on loss on ignition analysis at 1,100°C for 1 hour.

Cement pastes and equilibration waters characterization

In order to characterize the HCP samples, XRD and TGA investigations were performed. The XRD analyses were carried out on dried samples from C0, C2 and C3 powders using an X'Pert PRO MPD diffractometer. Due to some technical problems with the diffractometer C1 sample was not analysed yet. For phase identification, data were collected over the 2θ range of 10 - 70°. The TGA analyses were performed using TG/DTA simultaneous measuring instruments model DTG - 60H.

The XRD spectra (Figure 1) are going to be construed using High Score Plus Program and ICDD PDF4 + 2015 data base in order to assess the main hydration compounds in the HCP samples that could influence the ^{14}C uptake.

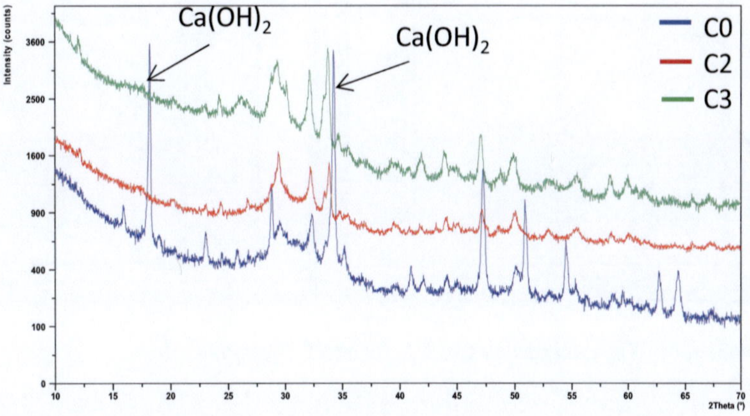

Figure 1: *XRD spectra of no degraded HCP (C0) and the two degraded phases (C2 and C3).*

Preliminary XRD results indicate that C2 and C3 samples are free of Portlandite. This assumption is also sustained by the TGA analyses that show that the mass loss between 400 - 600°C (the rage characteristic for Portlandite decomposition) is lower in C1, C2 and C3 compared to C0 (Figure 2).

Also, the TGA results show that the calcite content of the samples are quite low indicating that carbonation did not occur. The very low amount of calcite measured by this method could be formed during samples drying in an oven with uncontrolled atmosphere, as a sample preparation step for the TGA and XRD analysis.

Batch sorption experiments

The influence of cement degradation on ^{14}C uptake were studied by batch sorption experiments performed on non-degraded cement paste (C0) and three degraded cement pastes (C1, C2 and C3).

The batch tests were carried out with a solid to liquid ratio of 4 g/L (0.048 g of cement to 0.012 L of contacting ^{14}C-labelled solution) (expressed as dry mass determined by ignition of HCP at 1,100°C), at room temperature (23 ± 3°C) in polypropylene centrifuge. Before use, the PP tubes were prewashed in 0.1 M HNO_3 and thoroughly rinsed with deionized water. All experiments were carried out in duplicate samples and cement paste weighting, batch preparation, and supernatant sampling were performed under nitrogen atmosphere. Blank batches were prepared for measuring the ^{14}C content in the cement pastes and also the ^{14}C sorption on the tubes walls was checked.

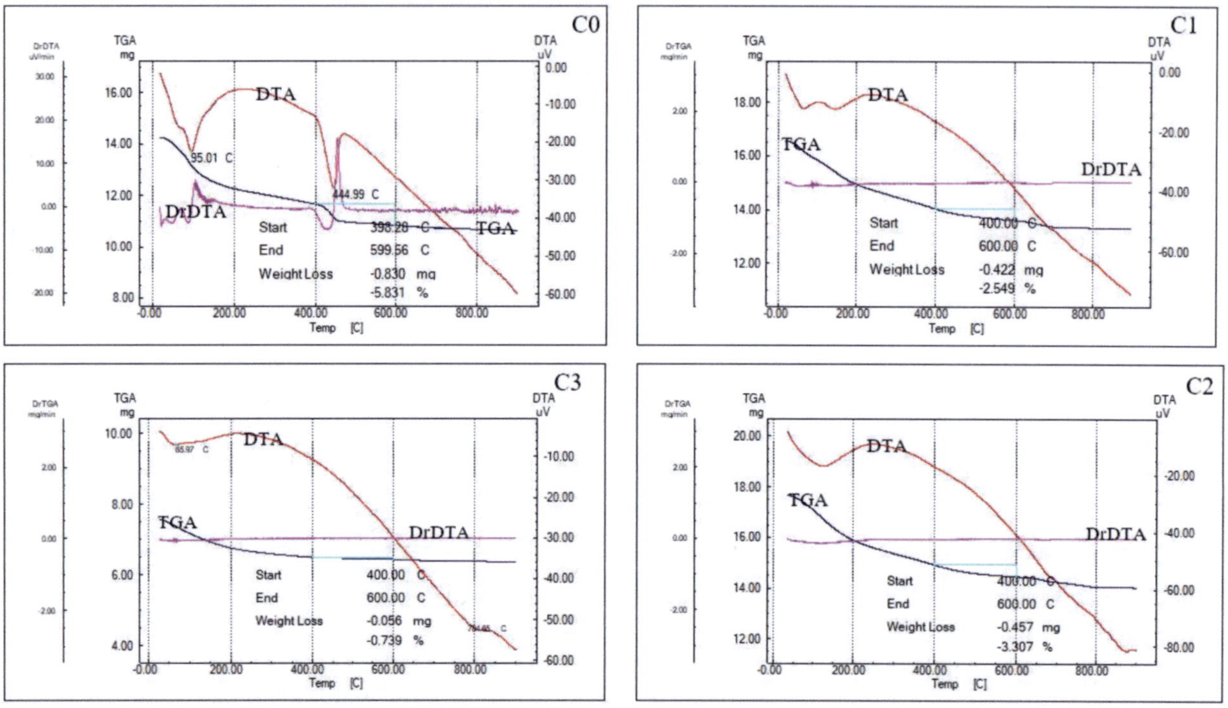

Figure 2: *The curves obtained by TGA analyses on C0, C1, C2 and C3 samples.*

A stock solution with ^{14}C in carbonate form ($^{14}CO_3^{2-}$) was used to prepare de contact solutions for batch tests. For C0, the stock ^{14}C solution was diluted in artificial cement pore water with the composition given in Tits et al. (2003), while for the sorption tests with C1, C2 and C3 cement pastes the contacting waters used to obtain the respective degraded states were considered to be in equilibrium with the degraded cement pastes and were used to prepare de contact solutions spiked with $^{14}CO_3^{2-}$.

The higher ^{14}C activity in the contact solutions was 227 Bq/mL that is equivalent to a $^{14}CO_3^{2-}$ concentration of $1.2 \cdot 10^{-7}$ M, and it was assumed it is one order of magnitude lower than the solubility of carbonate in portlandite saturated water that is around $6 \cdot 10^{-6}$ M (Pointeau et al., 2004). To confirm that the total carbon content was below this solubility limit the total carbon content of the contacting solution will be measured. Before use in the batch tests the ^{14}C spiked solutions were filtered through 0.45 μm Millipore filters and the initial activity of ^{14}C were measured in the filtered solutions and this value were considered for further computation of the percentage uptake and distribution ratio.

Since the distribution coefficient is defined at equilibrium, assuming that the system is reversible and it is independent of the contaminant concentration in the aqueous phase, four sets of sorption experiments are on-going for each cement paste:

- kinetic tests to assess the time needed to reach the phases equilibrium;
- tests to derive the sorption isotherms for three different initial concentration of $^{14}CO_3^{2-}$ in contacting solution (the initial ^{14}C activity ranging between $1 \cdot 10^2$ Bq/mL and $5 \cdot 10^2$ Bq/mL);
- tests to assess the influence of the solid to liquid ratio (S/L) on the $^{14}CO_3^{2-}$ uptake (for S/L between 1 and 10 g/L);
- desorption tests to confirm the sorption reversibility;

- tests to assess the experimental uncertainties (8 batch tests carried out for C0 cement paste with the same S/L ration and $^{14}CO_3^{2-}$ concentration).

During the equilibration time, the test tubes were placed on an orbital shaker at 100 rpm. After equilibration, the test tubes were centrifuged at 11,000 rpm for 40 min. The supernatant was separated and the residual activity of ^{14}C in the batch solutions was measured by liquid scintillation counting (LSC) using Hionic Fluor scintillation cocktail. For each batch solution, two parallel LSC counting vials were measured with a sample to scintillation cocktail ratio of 1 to 10.

Results and discussions

The experimental results obtained from the kinetic tests on the cement pastes are expressed as percentage uptake given by the Eq. 1.

$$A_{ti} = \frac{C_{aq}(ti) - C_0}{C_0} x100 \qquad \textbf{\textit{Eq. 1}}$$

where: A_{ti} is the amount of the tracer sorbed at time t_i (%), $C_{aq}(t_i)$ is the tracer activity in the aqueous phase at time t_i (Bq/mL), and C_0 is the initial tracer activity in the contacting solution (Bq/mL).

The preliminary results obtained for $^{14}CO_3^{2-}$ uptake on C0 indicate that after 40 days almost 84% of the tracer seems to be sorbed (Figure 3). Between 30 and 40 days the percentage uptake slightly increased from around 80% to 84% indicating that the batches are nearly to the equilibrium time.

A high percentage sorption was also obtained for the C1 sample (~ 88% after 30 days) and for C2 sample (~ 70% after 30 days) but for the most degraded state only 32% of the tracer was sorbed after 30 days.

The kinetic experiment will continue up to 200 days to establish the equilibrium time at which the batches prepared for the sorption isotherm will be sampled to measure the residual ^{14}C activity in aqueous phase.

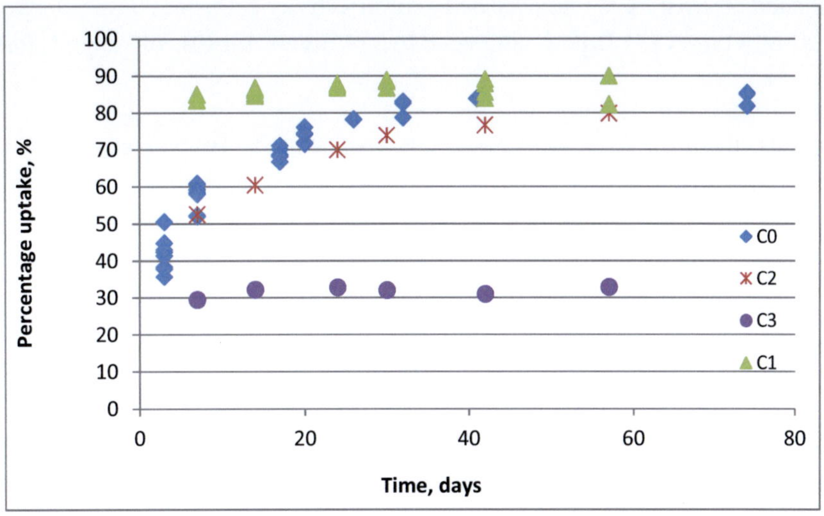

Figure 3: *The evolution of the $^{14}CO_3^{2-}$ percentage uptake on non-degraded and two degraded cement pastes.*

The distribution ratio was computed based on Eq. 2 for each sampling time.

$$R_d = \frac{A_{sorbed}}{C_{aq}} = \frac{C_0 - C_{aq}}{C_{aq}} \times \frac{V}{m} \qquad \textit{Eq. 2}$$

where: A_{sorbed} is the amount of the tracer sorbed (Bq/g), C_{aq} is the tracer activity in the aqueous phase (Bq/mL), C_0 is the initial tracer activity in the contacting solution (Bq/mL), V is the volume of the tracer solution in the batches (mL) and m is the mass of the cement paste (g).

As it was expected, the distribution ratio increased during the kinetic study (Figure 4).

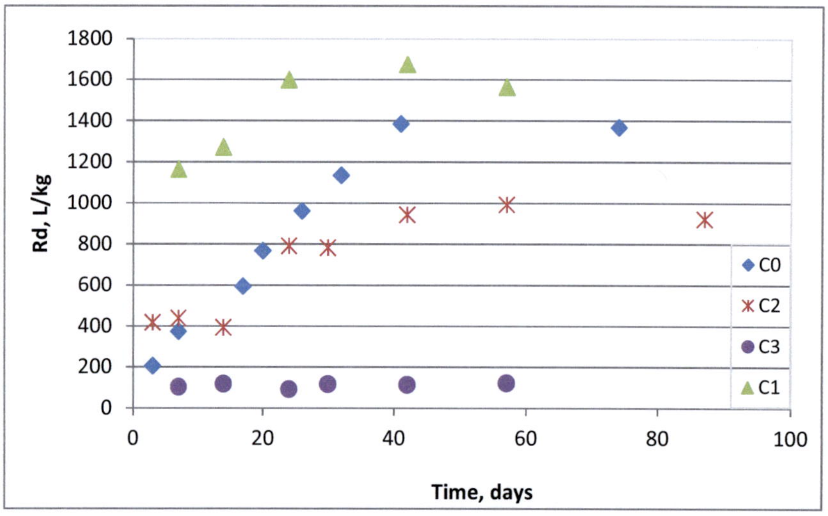

Figure 4: *The evolution distribution ratio for* $^{14}CO_3^{2-}$ *on non-degraded and two degraded cement pastes.*

The distribution ratios estimated based on the data obtained after 40 days for the non-degraded state and 30 days for degrades states indicate an important decrease of this parameter for the most degraded cement paste and a slightly decrease for the free of Portlandite one (C2).

The ^{14}C labelled solution used for C1 batches had a lower activity compared to the contact solution prepared for C0, C2 and C3 cement pastes (21 Bq/mL compared to 227 Bq/mL for C0, 144 Bq/mL for C3 and 46 Bq/mL for C2).

The kinetic tests carried out for C1 cement paste indicate that the percentage absorption is even higher than for C0. Supplementary investigations have to be carried out to explain these first experimentally results.

Conclusions and Future work

The data presented in this paper are preliminary experimental data of the on-going sorption tests carried out to assess the influence of the cement paste degradation on $^{14}CO_3^{2-}$ uptake.

The kinetic tests show that after 40 day the equilibrium time was reached both for non-degraded HCP and for degraded ones. For this contacting time the percentage uptake for the non-degraded HCP was higher than 80% while the distribution ratio is around 1,380 L/kg. For the C2 sample, the percentage uptake after 30 days is around 75% (slightly lower than for the non-degraded state after the same contacting time) while the distribution ratio is around 804 L/kg. For the most degraded state both the percentage uptake and the distribution ratio are much lower (around 30% and 116 L/kg, respectively).

The batches prepared to get the sorption isotherms are going to be sampled and the distribution coefficients will be derived from these isotherms.

Also, the phase separation for the batches prepared to assess the influence of the S/L ratio on the $^{14}CO_3^{2-}$ uptake will be sampled and desorption tests will be performed.

For all cement waters [Ca], [Si], [Al], [Na] and [K] concentration are going to be assessed by emission spectrometry (ICP-OES). Also, subsamples of the cement pastes will be mineralized by acid microwave digestion and the resulted solutions will be measured by ICP-OES.

The interpretation of the XRD spectra is on-going to assess the main hydration compounds in the HCP samples used for ^{14}C sorption experiments.

Acknowledgement

The research leading to these results has received funding from the European Union's Horizon 2020 Reasearch and Training Programme of the European Atomic Energy Community (EURATOM) (H2020-NFRP-2014/2015) under grant agreement n° 662147 (CEBAMA).

References

Allard, B., Torstenfelt, B., Andersson, K. (1981). Sorption studies of $H^{14}CO_3^-$ on some geological media and concrete. MRS Proceedings, 3, 465-473.

Bayliss, S., Ewart, F.T., Howse, R.M., Smith-Briggs, J.L., Thomason, H.P., Willmott, H.A. (1988). The solubility and sorption of lead-210 and carbon-14 in a near-field environment. MRS Proceedings, 112, 33-42.

Bradbury, M.H. and Sarott, F.A. (1994). Sorption databases for the cementitious near-field of a L/ILW repository for performance assessment. PSI Bericht, 95-06 and Nagra Technical Report, NTB 93-08.

Ochs, M., Dirk, M., Wang, L. (2016). Radionuclide and Metal Sorption on Cement and Concrete. Springer International Publishing.

Pointeau, I., Landesman, C., Giffaut, E., Reiller, P.E. (2004). Reproducibility of the uptake of U(VI) onto degraded cement pastes and calcium silicate hydrate phases. Radiochimica Acta, 92, 645-650.

Pointeau, I., Coreau, N., Reiller, P.E. (2008). Uptake of anionic radionuclides onto degraded cement pastes and competing effect of organic ligands. Radiochimica Acta, 96(6), 367-374.

Tits, J., Jakob, A., Wieland, E., Spieler, P. (2003). Diffusion of tritiated water and $^{22}Na^+$ through non-degraded hardened cement pastes. Journal of Contaminant Hydrology, 61(1), 45-62.

Wang, L., Martens, E., Jacques, D., De Canniere, P., Berry, J., Mallants, D. (2009). Review of sorption values for the cementitious near field of a near-surface radioactive waste disposal facility. NEA/RWM/R(2012)3.

Radionuclide through-diffusion experiments in partially saturated carbonated and non-carbonated hardened cement paste using the osmotic technique

Sabrina Rasamimanana[1*], Katy Perrigaud[1], Solange Ribet[1], Nicolas Bessaguet[1], Bernd Grambow[1], Catherine Landesman[1]

[1] SUBATECH/ARMINES, Institut Mines Télécom Atlantique, Université de Nantes, CNRS (FR)

* Corresponding author: rasamima@subatech.in2p3.fr

Abstract

A specific experimental set-up using the osmotic technique has been developed in order to determine diffusion parameters (porosity and diffusion coefficient) in partially saturated conditions. The diffusion of tritiated water (HTO) has been investigated to validate this technique before an investigation of the diffusion properties of inorganic carbon-14 in carbonated and non-carbonated hardened cement paste. Carbonated samples were prepared in an accelerated carbonation device developed at Subatech (Relative humidity 55%, P_{CO2} 3%). The samples were characterized by different methods (mass monitoring, inorganic carbon measurement, XRD analysis) to ensure a stabilize state of carbonation before the diffusion experiments.

Introduction

Cemented wastes may remain partially unsaturated for a long time in their disposal locations (many thousands to tens of thousands of years in clay rock) and carbon-14 species (aqueous and volatile, organic and inorganic) when released from waste may react with the unsaturated cement materials. The carbonation process, describing the reactivity of inorganic carbon (CO_2) from porewater or from the atmosphere with cement materials, occurs at the outer surface of cement under saturated conditions (clogging), but also under unsaturated conditions. This process may modify the mechanical properties of the cement material and may lead to a change of the transfer conditions of the solutes and the mobile radionuclides present in waste packages.

The objective of the present work is to investigate the diffusion properties of radionuclides in partially saturated carbonated and non-carbonated hardened cement paste (HCP). For Subatech, the experimental part of the CEBAMA project started in September 2016. This progress report presents the work performed during the last eight months. It aims at (i) describing the preparation of the materials used (non-carbonated or carbonated HCP), (ii) characterizing the carbonated samples, (iii) validating the osmotic technique on tritiated water (HTO) through-diffusion in HCP, which is a necessary step before an investigation can start on the diffusion properties of inorganic carbon-14.

Materials and methods

Solid samples preparation

The experiments are carried out on HCP samples prepared from a BFS cement, CEM V/A ROMBAS (Calcia, France), with a water-to-cement ratio (w/c) of 0.4 L/kg. The cement paste was prepared by Macé et al. (2015). Samples were cured under humid conditions (RH > 98%) for 200 days. They were stored, in their containers, covered by a thin layer of artificial cement porewater (ACW) solution (Table 1) to avoid dehydration and atmospheric carbonation.

Table 1: *Composition of extracted cement porewater, ACW (Macé et al., 2015).*

Cement porewater composition (mmol/L)					
Na^+	K^+	Ca^{2+}	Cl^-	SO_4^{2-}	pH
79	291	2.1	0.57	0.96	13.5

Samples for through-diffusion experiments were prepared by sawing (with a diamond wire saw) HCP cylinders into 2.5 mm thick disks. These disks are then polished to remove the streaks introduced during cutting and eliminate the calcite layer formed in contact with air. The disks were eventually stored in ACW solution and under Argon atmosphere until being used.

ACW was prepared, following the composition in Table 1, in an inert glove box under Argon in order to avoid atmospheric carbonation. Ultrapure deionized water (MilliQ, 18.2 MΩ/cm) decarbonated under vacuum (t = 2 hours) and high purity commercial salts were used. The composition and pH of ACW was checked by ion chromatography (Metrohm 850 pro IC chromatograph) and pH meter (Mettler Toledo, glass electrode, 3 mol/L KCl).

Water suction controlled by osmosis (Lagerwerff et al., 1961; Zur, 1966)

A suction potential for water is generated at the water/solid interface by an osmosis process between the water in the cement pore space and a highly concentrated solution of polymer (polyethylene glycol, PEG). The sample is separated from the PEG solution by a semi-permeable membrane which is permeable to water molecules and all dissolved species except PEG. The exclusion of PEG from the sample pore water produces a chemical potential imbalance between the pore water and the external solution. This osmotic suction has the effect of keeping the sample unsaturated. The imposed suction potential (Ψ in MPa, Eq. 1) is directly correlated to the mass concentration of PEG (kg of PEG/kg_{water}) by the following parabolic relation (Delage et al., 1998):

$$\Psi = 11 \times [PEG]^2 \qquad \textbf{\textit{Eq. 1}}$$

The main drawback of the technique is the limitation of the pressure which can be exerted, which is less than 10 MPa. A mass concentration of 0.95 kg of PEG/$kg_{solution}$ was chosen for reaching a maximal suction potential of 9 MPa (Delage et al., 1998; Pham et al., 2007; Savoye et al., 2010). Then, the correspondent degree of water saturation (S_w) of the cement sample pore space can be determined by performing petrophysical measurements (total porosity and volumetric water content).

The choice of the molar weight of PEG is driven by practical considerations linked to the ease of sampling in diffusion cells. PEG 6,000 (g/mol molecular weight) was then chosen because of its low viscosity (50 MPa.s at 20°C). PEG solutions were prepared with commercial PEG 6,000 (purity > 99%, Sigma-Aldrich) diluted in ACW solution. The cut-off of the semi-permeable membrane was chosen in order to prevent the intrusion of PEG in the sample. A regenerated cellulose dialysis membrane (Spectra-Por 3,500 Da, Spectrum laboratory) with a 3,500 g/mol molecular weight cutoff (MWCO), a pore diameter of 24 Å, and good chemical resistance to alkaline solution has then been selected.

Porosity and degree of saturation measurements

The porosities of the samples were determined by weight difference of initially water saturated samples after drying the samples in an oven for three different temperatures (60°C, 80°C and 105°C), until reaching a constant weight (successive mass difference < 0.05% over 2 weeks). The total porosity of a sample (Ø) is defined as the ratio of void volume to total volume and can be measured by water porosimetry as follows (Eq. 2):

$$\emptyset = \frac{m_s - m_d}{m_s - m_h} \qquad \textbf{\textit{Eq. 2}}$$

Where m_s is the water saturated mass (in g), m_d is the dried mass (in g) and m_h is the hydrostatic mass (mass of the sample measured in water in hydrostatic conditions).

The degree of saturation (S_w) corresponds to the ratio of the volumetric water content (w) over the total porosity (Ø). The volumetric water content (w) is defined by the ratio of the mass of water to the dry mass of the sample and can be calculated using the following equation (Eq. 3, Drouet et al., 2015):

$$w\,(T, RH) = \frac{d_s}{d_s - \emptyset}\left[\frac{\emptyset}{d_s} + \frac{\Delta m}{m}(T, RH)\right] \qquad \textbf{\textit{Eq. 3}}$$

Where d_s is the saturated density and $\frac{\Delta m}{m}$ is the relative mass variation at equilibrium with the relative humidity RH for the temperature T.

Accelerated carbonation

Accelerated carbonation experiments were carried out at ambient temperature (20 ± 1°C), with a theoretical Relative Humidity (RH) of 54.4%. This value was chosen to reach a maximal rate of carbonation (Papadakis et al., 1991; Verbeck, 1958) and an initial dehydration of samples at a level lower than those imposed by the osmotic technique. In order to get a good representativity of atmospheric carbonation and a reasonable carbonation duration, the partial pressure of CO_2 was fixed to 3%. In addition, it has been previously shown that using partial pressure higher than 3% induces important mineralogical changes in a cement paste (Castellote et al., 2009; Galan et al., 2013; Ashraf, 2016).

The accelerated carbonation device consists of an acrylic desiccator (16 x 28 x 28 cm) in which the relative humidity (RH) and temperature are measured with a thermohygrometer (Testo, 608-H2). A RH = 54.4% is imposed by using an oversaturated salt solution of magnesium nitrate hexahydrate, prepared from 2,049 g/L of salt (purity > 99%, Aldrich, solubility = 1,250 g/L) and MilliQ water. CO_2 gas is injected from an Ar/CO_2 mixture at 3 vol.% CO_2 and the pressure was controlled with a manometer. The addition of gas was manually done twice a day at a flowrate of 0.4 L/min during 10 to 40 min depending on experimental conditions.

Prior to the carbonation step, HCP samples were stabilized at the chosen relative humidity (54.4% RH) in a similar device (but without CO_2 gas injection system). The kinetics of the water exchange was monitored by regularly weighing HCP samples until a constant mass (mass variation between two successive weightings < 0.05% over 60 days). After being stabilized at the chosen RH value, the HCP samples were then transferred into the carbonation device. During the carbonation process, the kinetics were monitored with the same protocol. The overall variation of mass (at steady state) is used to estimate the amount of calcium carbonates precipitated in the HCP samples.

Another quantification of calcium carbonates has achieved by using specific inorganic carbon (TOC analyzer, CSH-Shimadzu TOC-V) and the mineralogical changes induced by carbonation were analyzed by X-ray Diffraction (XRD) (D5000 diffractometer, Bruker using Cu Kα line, λ = 1.54184 Å).

Tritiated water (HTO) Through-Diffusion experiments

The through-diffusion device consists of two polyvinyl chloride (PVC) 100 mL reservoirs (upstream and downstream) and a two-part PVC sample holder manufactured by Subatech. For experiments under unsaturated conditions, HCP disks were sandwiched between the two semi-permeable membranes, inserted and glued in the sample holder to prevent any leakage. The effective final diameter of the sample is 40 mm. Through-diffusion experiments were carried out at room temperature (20 ± 1°C) in a glove box under Ar atmosphere to minimize any atmospheric carbonation. Four experiments have been performed with non-carbonated and carbonated samples, in the following experimental conditions:

- Saturated ($S_w = 1$), [PEG] = 0
- Non-saturated ($S_w < 1$), [PEG] = 0.95 $kg/kg_{solution}$

HCP disks inside diffusion cells were equilibrated for one week with a weighed volume of ACW solution (with or without PEG depending on experimental conditions). Then, tritiated water (HTO, Eckert & Ziegler) was added into the upstream reservoir (2.0 ± 0.1 kBq/g_{sol}). An aliquot of solution (1 mL) was periodically taken from the downstream reservoir and replaced by an equivalent volume of solution (ACW solution with or without PEG depending on experimental conditions) in order to maintain the total volume at its initial value. Radioassay was performed using a TriCarb 3170 TR-SL (Perking Elmer) liquid scintillation analyzer. Samples for radioassay were prepared by mixing 8 mL of aliquot to 12 mL of Ionic Fluor™ (Perking Elmer) scintillation cocktail suitable for alkaline samples. The ratio $V_{aliquot}/V_{scintillator}$ was specifically optimized to take into account the effects of PEG solution on counting (demixture, quenching...).

The experimental data were modelled applying the analytical solution of the second Fick law (Crank, 1975):

$$\frac{\partial C}{\partial t} = \frac{D_e}{\alpha}\frac{\partial^2 C}{\partial x^2} = \frac{D_e}{\varepsilon_a + \rho R_d}\frac{\partial^2 C}{\partial x^2} \qquad \textit{Eq. 4}$$

where C is the HTO volumetric activity (Bq/m^3); t, the time (s); D_e, the effective diffusion coefficient (m^2/s); ε_a, the diffusion accessible porosity; ρ, the bulk dry density (kg/m^3); R_d, the sorption distribution ratio (m^3/kg); and α, the rock capacity factor (also called apparent porosity). The initial and boundary conditions are: $C(x,t) = 0$, $t = 0$; $C(x,t) = C_0$, $x = 0$, $t > 0$; $C(x,t) = 0$, $x = L$, $t > 0$ with L is the sample thickness.

In first approximation, tritiated water may be considered as a "non-reactive species" (i.e., $R_d = 0$) even so some results show that isotopic exchange with hydrated minerals takes place during the diffusion process (Tits et al., 2003). Results are expressed in term of the time evolution of i) the HTO normalized cumulative activity, which is the ratio of the total diffused HTO activity (Bq) to the volume activity in the upstream reservoir (Bq/m^3) and

ii) the normalized flux which is the ratio of the instantaneous flux ($Bq/m^2/s$) to the volume activity of the upstream compartment (Bq/m^3). The analysis of those data allowed the extraction of values for D_e and α (here ε_α).

Results and discussion

Characterization of carbonated samples

<u>*Carbonation process and quantification of calcium carbonate*</u>

The kinetics of carbonation is presented Figure 1.

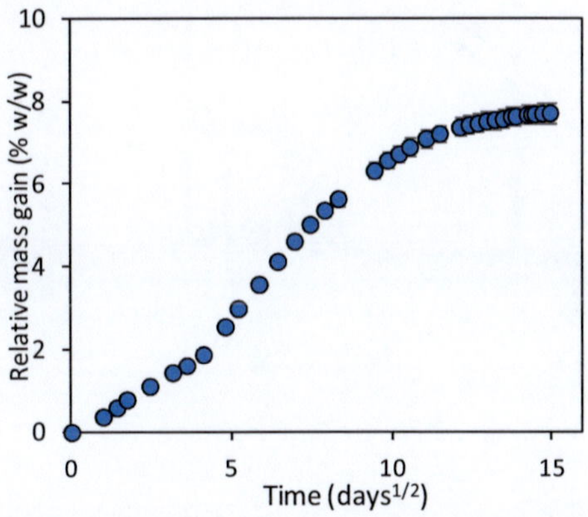

Figure 1: *Kinetics of carbonation of HCP disks (m_{sample} = 9.6567 g, T = 20 ± 1°C, RH = 55 ± 2%, P_{CO2} 3%).*

The kinetics of carbonation is a slow process and 220 days are necessary to achieve a steady state. The relative mass gain, associated with CO_2 uptake, was 7.7 ± 0.2 w/w.%. This result is in agreement with Auroy (2015) who obtained a mass gain of 8% for the same conditions (275 days, HR = 55%, P_{CO2} 3%) with a similar material (HCP CEM V/A). By comparison, Castellote et al. (2009) obtained a mass increase of 12% with an Ordinary Portland Cement paste (w/c = 0.5) carbonated at 3% of CO_2 (100 days, HR = 65%), which naturally contains a much higher amount of Portlandite (likely to be carbonated) than CEM V paste.

The amount of calcium carbonate precipitated during the carbonation process was 77 ± 2 g per kg of HCP sample. This value can be directly compared to Inorganic Carbon (IC) analysis (Table 2). Reference value is taken from samples stored under air ($P_{CO2} = 10^{-3.5}$ atm).

Table 2: *Quantification of calcium carbonate ($CaCO_3$) by IC analysis.*

	$CaCO_3$ (g/kg_{sample})	
	IC	**Mass monitoring**
Reference	11 ± 0.2	-
Carbonated sample	102 ± 15	77 ± 2

Within its relatively large uncertainty, the IC analysis confirmed the data obtained by mass monitoring and the steady state of carbonation. The two methods are consistent and give the same value of calcium carbonate precipitated during the carbonation process (77 ± 2 g of $CaCO_3$ per kg of sample). The reference value (11 ± 0.2 g of $CaCO_3$ per kg of sample) corresponds to the sample prior to the carbonation process.

Mineralogical analysis

The XRD diffractograms acquired for carbonated samples and a comparison to the reference (samples stored under air, $P_{CO2} = 10^{-3.5}$ atm) are shown in Figure 2.

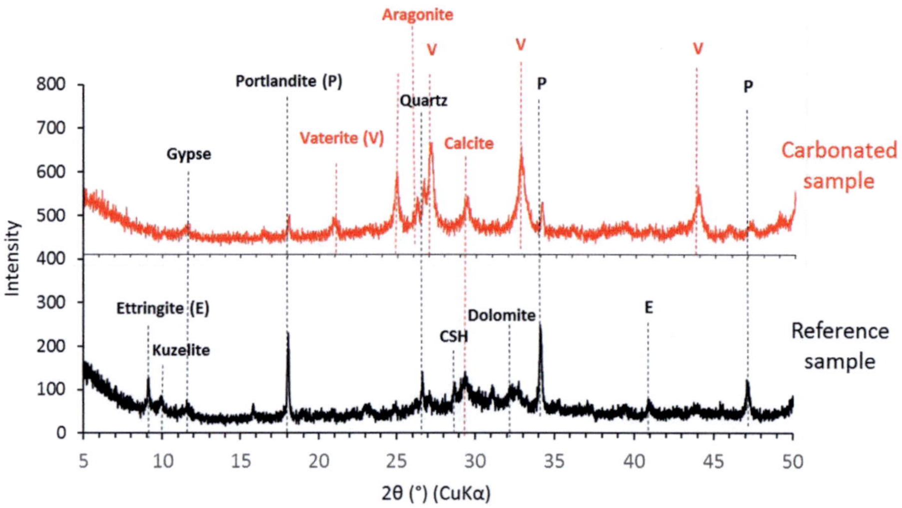

Figure 2: *X-ray diagrams acquired for carbonated HCP CEM V sample (C-S-H: Calcium silicate hydrate; E: Ettringite; P: Portlandite; V: Vaterite).*

The main mineralogical phases constituting the HCP (CEM V) reference sample are: ettringite (2θ ≈ 9°, 16°, 41°), kuzelite (AFm phase, 2θ ≈ 10°), portlandite (2θ ≈ 18°, 34°, 47°), dolomite (2θ ≈ 31°), quartz (2θ ≈ 27°), calcite (2θ ≈ 29.5°) and calcium silicate hydrate (CSH) (2θ ≈ 18 - 35°).

Mineralogical changes induced by carbonation were highlighted by XRD analysis. Indeed, the carbonation process leads to the dissolution of phases mainly portlandite and C-S-H and the precipitation of calcium carbonates; calcite but also metastable forms such as vaterite and aragonite. Peaks corresponding to ettringite, kuzelite and C-S-H disappear, while those corresponding to portlandite strongly decrease. The presence of residual portlandite after carbonation may be attributed to calcium carbonate formation around portlandite crystals, which inhibits their dissolution (Swenson and Sereda, 1968). The peaks corresponding to calcium carbonate precipitation appear such as to suggest formation of vaterite (2θ ≈ 21°, 25°, 27°, 33°, 44°) and of aragonite (2θ ≈ 26°).

2.2 Porosity and degree of saturation

The results of porosity measurements, compared to literature values are presented in Table 3.

Table 3: *Water porosity of carbonated HCP CEM V sample. [1] Auroy et al. (2015); [2] Drouet et al. (2015).*

	$\Phi_{60°C}$ (%)	**$\Phi_{80°C}$ (%)**	**$\Phi_{105°C}$ (%)**
Reference	**33.5 ± 0.1**	**35.7 ± 0.2**	**38.7 ± 0.2**
		36.9 [1]	38.2 [1]
		38.9 [2]	39.1 [2]
Carbonated	**20.6 ± 0.3**	**23.7 ± 0.3**	**26.3 ± 0.3**
		27.6 [1]	29.5 [1]

Measured water porosity values are consistent with literature. As expected, porosity increases with the drying temperatures ($\Phi_{60°C} < \Phi_{80°C} < \Phi_{105°C}$). A high drying temperature tends to overestimate the amount of free water and consequently the total porosity. Indeed, it is well known that cement mineral phases (such as ettringite) are not stable above 60°C. So, the values of water porosity obtained at the lowest temperature, $\Phi_{60°C}$ (reference sample) = 33.5% and $\Phi_{60°C}$ (carbonated sample) = 20.6% were chosen for diffusion calculation. The precipitation of calcium carbonate led to a decrease of water porosity by about 33%. This decrease of porosity is in agreement with results obtained by Auroy et al. (2015) and Ngala and Page (1997) that being 32% and 33% (OPC paste, w/c 0.5) respectively. The reduction of total porosity is mainly related to the carbonation of portlandite and C-S-H, leading to the increase of the solid phase volume.

The values of degree of saturation (S_w) determined on cement samples according to the conditions of through-diffusion experiments are presented in Table 4.

Table 4: *Degree of saturation imposed by the osmotic technique for non-carbonated and carbonated samples.*

	Degree of saturation (S_w)	
	[PEG] = 0, Ψ = 0 MPa	**[PEG] = 0.95 kg/kg, Ψ = 9 MPa**
Reference	1	0.81
Carbonated	1	0.79

A degree of saturation of about 80% was obtained for non-carbonated and carbonated samples by applying a potential suction of about 9 MPa.

Through-Diffusion experiments with HTO

Non-carbonated samples

The evolutions of the through-diffusion experiments with HTO in non-carbonated samples in full ($S_w = 1$) and partially ($S_w = 0.8$) saturated conditions are presented in Figure 3 and Figure 4 respectively. The predictions were obtained based on the experimental porosity ($\varepsilon_\alpha = 33.5\%$) and diffusion coefficients on a similar HCP CEM V sample estimated by N. Macé ($D_e = 2 \cdot 10^{-13}$ m²/s for $S_w = 1$, personal communication) and Savoye et al. (2017) ($D_e = 4 \cdot 10^{-13}$ m²/s for $S_w = 0.85$).

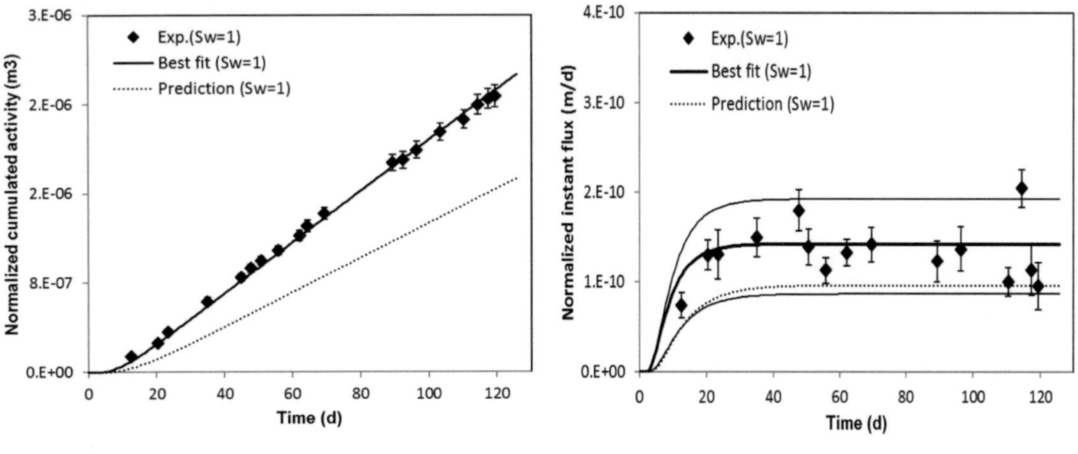

Figure 3: *Normalized cumulative activity (left) and instant fluxes (right) for HTO through-diffusion in saturated non-carbonated samples.*

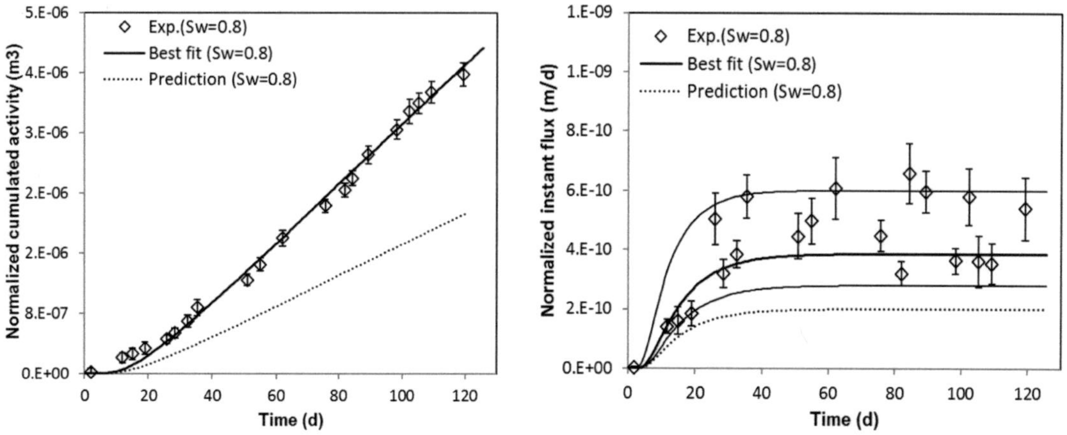

Figure 4: *Normalized cumulative activity (left) and instant fluxes (right) for HTO through-diffusion in partially saturated non-carbonated samples.*

For fully saturated conditions (S_w = 1), the best fit of the experimental points, with α = 33.5%, give D_e(HTO) = $3 \pm 1 \cdot 10^{-13}$ m^2/s. Those values are in the expected range compared to the prediction.

For partially saturated conditions (S_w = 0.8), the best fit of the experimental points give α = 28% and D_e(HTO) = $9 \pm 3 \cdot 10^{-13}$ m^2/s. The diffusion coefficients, D_e, are in the same range for the two conditions of saturation even if a tendency of high D_e value at high suction potential (S_w = 0.8) is indicated. However, according to the literature (Bourbon, 2013; Savoye et al., 2017), we expected a decrease of the diffusion coefficient values for partially saturated conditions. As a possible assumption, the duration of the hydric equilibrium phase has been considered to explain these results. One week is maybe not sufficient for the complete equilibrium to be achieved before starting HTO through-diffusion experiment. Next experiments will be then investigated with a longer equilibration time (one month) in order to confirm or invalidate the initial results.

<u>*Carbonated samples*</u>

The evolutions of the HTO through-diffusion experiments in carbonated samples at full (S_w = 1) and partially (S_w = 0.8) saturated conditions are presented in Figure 5.

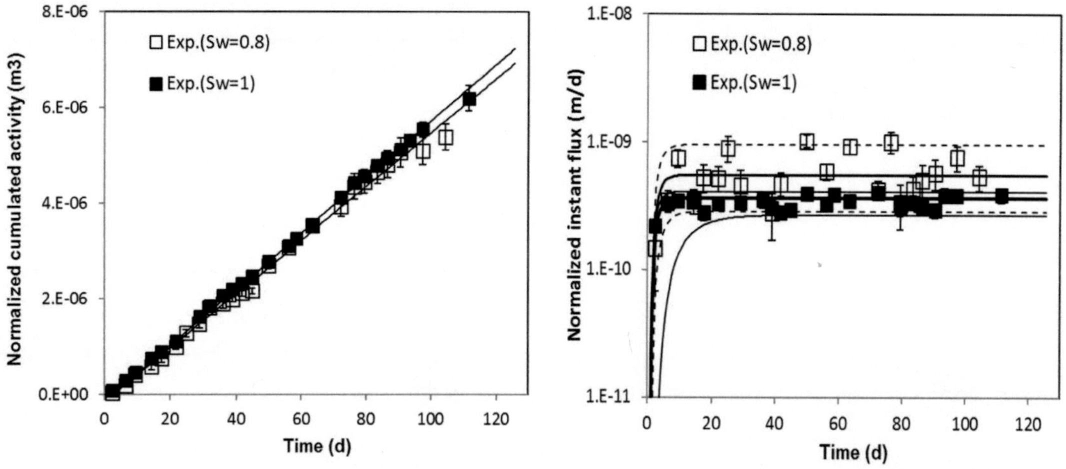

Figure 5: *Normalized cumulative activity (left) and instant fluxes (right) for HTO through-diffusion in partially and full saturated carbonated samples.*

For full saturated conditions (S_w = 1), the best fit of the experimental points, with α = 20.6%, give D_e(HTO) = 9 ± 1·10^{-13} m^2/s. For partially saturated conditions (S_w = 0.8), the best fit of the experimental points give α = 18% and D_e(HTO) = 14 ± 5·10^{-13} m^2/s. This means that whatever the saturation conditions are, the determined D_e(HTO) values are in the same range (within experimental uncertainties). As previously explained, a much longer equilibration time will be investigated for the next experiments in order to confirm or invalidate these initial results.

Conclusions and Future work

This progress report aimed at describing preparation and characterization of the carbonated samples and validating the osmotic technique on HTO through-diffusion in cement materials (HCP CEM V).

Carbonated samples were obtained by the accelerated carbonation device developed at Subatech with the following conditions: RH 55%, P_{CO2} 3%. The amount of calcium carbonate precipitated during the carbonation process was quantified by mass monitoring and inorganic measurement. The resulting data tend, whatever the method used, to confirm the results of literature, indicating a stabilized state of carbonation. An overall value of 77 ± 2 g of $CaCO_3$ per kg of HCP sample was obtained. The mineralogical changes induced by carbonation (dissolution of portlandite and C-S-H) were highlighted by XRD analysis and 3 types of calcium carbonate precipitates were identified: calcite, vaterite and aragonite. The next step will be to quantify the mineral phases (portlandite and calcium carbonate) by thermogravimetric analysis.

Diffusion parameters (porosity α, diffusion coefficient D_e) for HTO in the partially saturated samples were investigated using modified through-diffusion cells, in which suction is generated by an osmosis process. The preliminary results show that, within the experimental uncertainty, HTO diffusion rates in the fully saturated samples (S_w = 1) are within the same order of magnitude as in partially saturated samples (S_w = 0.8). The precipitation of calcium carbonate led to a decrease of water porosity (from 28% to 18%) but no effect of this reduction of porosity was clearly evidenced with respect to HTO diffusion. For both type of samples (non-carbonated and carbonated), in the next experiments, the time for ensuring a complete hydric equilibrium of the sample will be extended (up to one month) before starting through-diffusion experiments in order to confirm or invalidate these initial results in partially saturated conditions.

The schedule for the next months will focus on:

- Next HTO through-diffusion experiments for validating the assumption of the influence of the equilibration
- Inorganic carbon-14 (sodium bicarbonate) through-diffusion experiments, which can start after completion of the HTO through-diffusion experiments
- Application of a quantitative digital autoradiography of β particle emission for analyzing diffusion profiles (Sardini et al., 2016)
- Thermal analysis (TGA) on non-carbonated and carbonated samples
- Preparation and characterization of HCP samples at different water degradation stage (pH < 13.3)

Finally, 6 experiments (2 diffusion cells per condition) are planned for the following experimental conditions (Table 5).

Table 5: *Planned through-diffusion experiments.*

	Full saturated ($S_w = 1$) [PEG] = 0	**Partially saturated** ($S_w = 0.8$) [PEG] = 0.95 $kg/kg_{solution}$
Non-carbonated	1	2
Carbonated	3	4
Leached sample (pH < 13.3)	5	6

Acknowledgement

The research leading to these results has received funding from the European Union's Horizon 2020 Reasearch and Training Programme of the European Atomic Energy Community (EURATOM) (H2020-NFRP-2014/2015) under grant agreement n° 662147 (CEBAMA).

References

Ashraf, W. (2016). Carbonation of cement-based materials: Challenges and opportunities. Construction and Building Materials, 120, 558-570.

Auroy, M. (2015). Impact of the carbonation on the water transport properties in cementitous material. PhD thesis, Paris Est University.

Auroy, M., Poyet, S., Le Bescop, P., Torrenti, J-M., Charpentier, T., Moskura, M., Bourbon, X. (2015). Impact of carbonation on unsaturated water transport properties of cement-based materials. Cement and Concrete Research, 74, 44-58.

Bourbon, X. (2013). Référentiel des matériaux d'un stockage de déchets de haute activité et de moyenne activité à vie longue - Tome 2: les matériaux cimentaires. Rapport Andra, CG.RP. ASCM.12.0014.

Castellote, M., Fernandez, L., Andrade, C., Alonso, C. (2009). Chemical changes and phase analysis of OPC pastes carbonated at different CO_2 concentrations. Materials and Structures, 42, 515-525.

Cole, W.F. and Kroone, B. (1959). Carbonate minerals in hydrated portland cement. Nature, 184, B.A.57.

Crank, J. (1975). The Mathematics of Diffusion. University, Second Editions, Clarendron press.

Delage, P., Howat, M.D., Cui, Y.J. (1998). The relationship between suction and swelling properties in a heavily compacted unsaturated clay. Engineering Geology, 50, 31-48.

Drouet, E., Poyet, S., Torrenti, J.-M. (2015). Temperature influence on water transport in hardened cement pastes. Cement and Concrete Research, 76, 37-50.

Galan, I., Andrade, C., Castellote, M. (2013). Natural and accelerated CO_2 binding kinetics in cement paste at different relative humidities. Cement and Concrete Research, 49, 21-28.

Lagerwerff, J.V., Ogata, G., Eagle, H.E. (1961). Control of osmotic pressure of culture solutions with polyethylene glycol. Science, 133, 1486-1487.

Landesman, C. and Grambow, B. (2016). Diffusion properties of inorganic ^{14}C species (dissolved and gaseous) through unsaturated hardened cement paste: Influence of water saturation (SUBATECH). CEBAMA Project Deliverable, D2.03: WP2: State of the Art Report.

Macé, N., Nerfie, P., Coreau, N., Thory, E., Le Bescop, P., Touze, G. (2015). Préparation et caractérisations des PCH CEM V/A ROMBAS dans le cadre du GL CTEC. Report CEA, DPC/SECR/ST/2015/057 indice A.

Ngala, V.T. and Page, C.L. (1997). Effects of carbonation on pore structure and diffusional properties of hydrated cement pastes. Cement Concrete Research, 27, 995-1007.

Papadakis, V.G., Vayenas, C.G., Fardis, M.G. (1991). Fundamental Modeling and Experimental Investigation of Concrete Carbonation. Materials Journal, 88, 363-373.

Pham, Q.T., Vales, F., Malinsky, L., Nguyen Minh, D., Gharbi, H. (2007). Effects of desaturation-resaturation on mudstone. Physics and Chemistry of the Earth, 32, 646-655.

Savoye, S., Page, J., Puente, C., Imbert, C., Coelho, D. (2010). New experimental approach for studying diffusion through an intact and unsaturated medium: a case study with Callovo-Oxfordian argillite. Environmental Sciences of Technology, 44, 3608-3704.

Savoye, S., Macé, N., Lefèvre, S., Rajyaguru, A., Spir, G., Robinet, J.-Ch. (2017). How mobile is tritiated water through unsaturated cement-based materials? Migration conference, September 2017, Barcelona, Spain.

Sardini, P., Angileri, A., Descostes, M., Duval, S., Oger, T., Patrier, P., Rividi, N., Siitari-Kauppi, M., Toubon, H., Donnard, J. (2016). Quantitative autoradiography of alpha particle emission in geo-materials using the Beaver™ system. Nuclear Instruments and Methods in Physics Research A, 833, 15-22.

Swenson, E.G. and Sereda, P.J. (1968). Mechanism of the carbonation shrinkage of lime and hydrated cement. Journal of Applied Chemistry, 18, 111-117.

Tits, J., Jakob, A., Wieland, E., Spieler, P. (2003). Diffusion of tritiated water and $^{22}Na^{+}$ through non-degraded hardened cement pastes. Journal of contaminant Hydrology, 61, 45-62.

Verbeck, G.J. (1958). Carbonation of hydrated Portland Cement. ASTM Special Publication, 205, 17-36.

Villain, G., Thiery, M., Platret, G. (2007). Measurement methods of carbonation profiles in concrete: Thermogravimetry, chemical analysis and gamma densimetry. Cement and Concrete Research, 37, 1182-1192.

Zur, B. (1966). Osmotic control the matrix soil water potential. Soil Scientist, 102, 394-398.

Study of radium uptake by cementitious materials relevant for LILW disposal in the Czech Republic

Barbora Drtinová[1]*, Jana Kittnerová[1], Dušan Vopálka[1]

[1] Department of Nuclear Chemistry, Czech Technical University in Prague (CZ)

* Corresponding author: barbora.drtinova@fjfi.cvut.cz

Abstract

The contribution deals with the uptake of radium on three cementitious materials: a hardened cement paste and two types of concrete (both actually used in the storages of radioactive waste in the Czech Republic); all crushed and sieved on fraction ≤ 0.4 mm. The cementitious materials were analyzed by X-ray diffraction and used to the study of equilibrium sorption of the ^{223}Ra (used as an analogue of ^{226}Ra) under various conditions.

Sorption experiments were carried out in a carrier-free arrangement, i.e., nearly in the absence of change of solution composition caused by uptake of Ra sorption via ion exchange. Varying conditions included the temperature, the solid-to-liquid ratio m/V, and the composition of the starting solution which was either Portlandite water (saturated $Ca(OH)_2$) or synthetic cement pore water CPW.

Equilibrium sorption experiments were evaluated using a distribution coefficient K_d. When changing the liquid phase from Portlandite water to CPW, K_d increased for all materials studied. The temperature rise from 22 to 80°C in the experiment exposed an unexpected trend for two materials in the dependence of K_d on m/V.

Introduction

Radium is a nuclide that needs to be studied for its importance for the safety assessment of repository spent fuel. It is necessary to examine its mobility, which is influenced, among other things, by the retardation via sorption on materials of engineered barriers, including cement materials. In this work, the Ra uptake is characterized by the distribution coefficient between liquid and solid phases, K_d.

However, instead ^{226}Ra the study was performed with the isotope ^{223}Ra with a half-life of 11.43 days, decaying to ^{219}Rn with a half-life of 4 s, which does not interfere with the measurement of ^{223}Ra. Due to the short half-life of generated radon, safety of work is easier to ensure compared to the work with ^{226}Ra and ^{222}Rn. The ^{223}Ra is obtained from an ^{227}Ac/ ^{223}Ra generator (Kozempel et al., 2015).

For a repository for intermediate level radioactive waste, the concentrations of radium in water are affected by interaction with engineering barriers as described in the work of Berner (2002). According to the calculations performed there, the maximum concentration of radium in the cement pore water (pH 12.55, 25°C) would be given by a solubility limit that was assessed to be about $1 \cdot 10^{-5}$ mol/L. Our working concentrations were from $4.7 \cdot 10^{-13}$ to $1.8 \cdot 10^{-12}$ mol/L in the carrier-free configuration, so, no precipitation should occur.

The interaction of radionuclide with the cementitious materials depends on the type of radionuclide and also on the type of cement. The uptake of radionuclides (electrostatic sorption, chemisorption and adsorption) and substitution (e.g., Sr substitutes for Ca) can be considered as the most important process influencing the radionuclide retardation in the presence of the cementitious materials. Other forms are precipitation, complexation, and incorporation (Evans, 2008).

Among the few works that dealt with the sorption of Ra on cementitious materials (here specifically on C-S-H and HCP) belongs the work of Tits et al. (2006). The C-S-H phase was prepared by these authors with various $CaO:SiO_2$ ratios. The sorption proceeded from a non-alkaline medium of ultrapure water and an alkaline synthetic Artificial Cement Pore Water (ACW). HCP was prepared from CEM I-type cement. In batch experiments with HCP a rapid initial sorption was observed that, for the degraded HCP (obtained by leaching in ultrapure water), reached an equilibrium characterized by K_d = 140 L/kg, whereas for fresh HCP, K_d = 260 L/kg was determined after one day and then the K_d slowly increased up to 400 L/kg after sixty days. For C-S-H phases, component in cementitious materials most likely responsible for sorption of radium, fast sorption kinetics and equilibrium after one day was observed. The $CaO:SiO_2$ ratio affected the distribution coefficient so that with a higher ratio, the distribution coefficient decreased. Also the pH of the starting solution had an effect: with ACW the pH was 13.3, while in an environment exposed to ultrapure water the pH was only in the range 11.5 - 12.5, which would result in less sorption in the latter case. The distribution coefficients for the pure C-S-H phases were determined in the order of 10^2 - 10^4 L/kg.

As the conclusion of the cited study it can be mentioned that radium sorption on the C-S-H phase is a linear and reversible process that is probably governed by the ion exchange of Ra^{2+} - Ca^{2+}.

Preparation of cement pastes

Three cementitious materials prepared in the form of blocks of hydrated raw cement in a highly humid environment were used for the experiments:

- *Hardened Cement Paste (HCP)* - type of cement CEM II / A-S 42.5 R, hydration six months;
- *Concrete A* - type of cement CEM I 42.5 with fine and rough aggregates and fly ash, hydration one month, utilized in ÚJV Řež a.s. to stabilize the solidified solid waste container;
- *Concrete B* - type of cement CEM III B/32.5 with fine and rough aggregates, hydration one month, concrete used to fill chambers in intermediate radioactive waste repository Richard.

For experiments the prepared hydrated materials were crushed and sieved for fraction ≤ 0.4 mm. It should be noted that all studied materials contain Sr. The exchangeable amount of Sr was determined to be equal to 4.14 mmol/kg for HCP. For concrete A, the Sr content is about 60% compared to HCP, and in concrete B about 30%.

Characteristics of cement pastes

All three cementitious materials were studied by X-ray diffraction (Rigaku Mini Flex 600). The XRD spectrum of HCP is presented in Figure 1, the spectrum for Concrete A and Concrete B on Figure 2. The comparison of measured spectrum with the database ICDD PDF-2 (version 2013) of the measuring system enabled to identify four crystalline mineral phases in HCP, namely calcite $CaCO_3$, portlandite $Ca(OH)_2$, hydrotalcite

$Mg_6Al_2CO_3(OH)_{16}\cdot 4(H_2O)$ and ettringite $Ca_6Al_2(SO_4)_3(OH)_{12}\cdot 26H_2O$, and two phases in concretes, namely portlandite $Ca(OH)_2$ and quartz SiO_2.

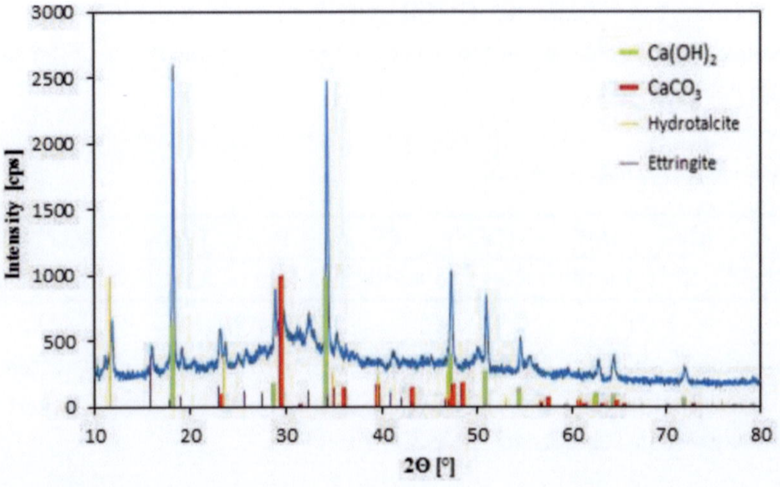

Figure 1: XRD spectrum of the HCP.

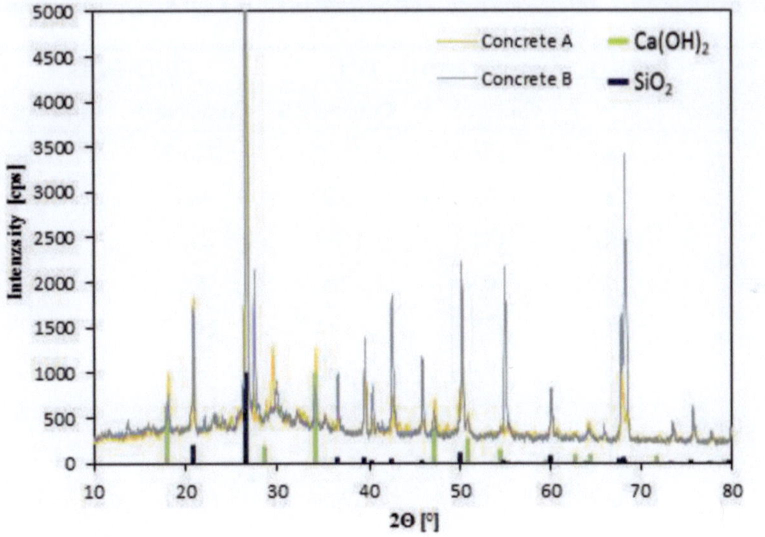

Figure 2: XRD spectrum of the Concrete A and Concrete B.

Sorption experiments with ^{223}Ra

The sorption experiment addressed both kinetic and equilibrium properties. Parameters influence of which was studied included the composition of the liquid phase, the solid phase (corrected for humidity - moisture determination in samples after 3 days of drying at 105°C), and the temperature. The concentration of ^{223}Ra could not be changed in this study purposefully because of a carrier-free arrangement. Activities ranging from about 400 to 1,500 Bq were used in 2 mL of solution, corresponding to initial concentration of ^{223}Ra between $4.7\cdot 10^{-13}$ and $1.8\cdot 10^{-12}$ mol/L. This concentration is many times smaller than the concentration of Ra assumed in the storage. The observed loss of ^{223}Ra due to sorption to walls of ampules in the absence of solid phase was about 7.75%.

Studying the kinetics of sorption for all cementitious materials, it was verified that equilibrium was established within one day. This finding enabled us to determine distribution coefficients K_d for different experimental conditions and sorbing materials. The main experimental temperatures were 22 and 80°C. Values of K_d were obtained both for Portlandite water (saturated $Ca(OH)_2$) and synthetic water (CPW), composition of which represents a deeper groundwater circulating in the fissures of the crystalline rocks of the Bohemian Massif (see Table 1), in equilibrium with $Ca(OH)_2$.

Table 1: *Composition of synthetic granite groundwater.*

Ion	Na^+	K^+	Mg^{2+}	Ca^{2+}	Cl^-	F^-	HCO_3^-	SO_4^{2-}	NO_3^-	PO_4^{3-}
c (mol/L)	$7.1\cdot10^{-4}$	$5.1\cdot10^{-5}$	$3.5\cdot10^{-4}$	$8.8\cdot10^{-4}$	$8.5\cdot10^{-5}$	$2.1\cdot10^{-5}$	$2.7\cdot10^{-3}$	$1.9\cdot10^{-4}$	$3.2\cdot10^{-5}$	$4.2\cdot10^{-5}$

The obtained distribution coefficients, K_d, were compared for different materials, different temperatures and two liquid phases, in dependence on the phase ratio (Figure 3 and Figure 4). Average values of K_d with estimates of standard deviations are given in Table 2 and Table 3.

The equilibrium value of K_d depends on mass / volume ratio proving that the sorption isotherm is non-linear.

Table 2: *Distribution coefficients K_d (L/kg) of used cementitious materials in $Ca(OH)_2$ for two temperatures.*

m:V \ Material		$Ca(OH)_2$, 22°C		$Ca(OH)_2$, 80°C	
		Concrete A	Concrete B	Concrete A	Concrete B
1:10	K_d	131.0	95.9	243.3	72.2
	s	0.2	5.2	0.8	3.7
1:60	K_d	181.1	179.0	89.9	83.6
	s	0.9	7.6	1.1	2.2
1:100	K_d	225.8	224.0	116.7	122.7
	s	9.4	7.6	9.6	3.9
1:600	K_d	377.2	495.6	161.1	258.5
	s	11.1	51.0	3.7	6.7
1:1,000	K_d	403.2	573.6	333.9	438.4
	s	25.1	24.1	1.0	18.0

Table 3: *Distribution coefficients K_d (L/kg) of HCP for different temperatures and liquid phases.*

m:V \ Temperature		HCP, $Ca(OH)_2$			HCP, CPW		
		22°C	50°C	80°C	22°C	50°C	80°C
1:10	K_d	72.5	128.8	470.6	102.5	139.6	589.1
	s	4.5	8.1	3.0	1.2	1.2	6.7
1:60	K_d	91.0	108.2	176.0	129.9	116.9	251.8
	s	1.6	0.1	8.0	5.4	3.1	0.9
1:100	K_d	114.8	126.2	178.2	158.4	135.8	210.6
	s	6.4	0.6	0.4	0.4	0.4	9.3

m:V	Temperature	HCP, Ca(OH)2 22°C	HCP, Ca(OH)2 50°C	HCP, Ca(OH)2 80°C	HCP, CPW 22°C	HCP, CPW 50°C	HCP, CPW 80°C
1:600	K_d	246.8	--	174.0	479.4	--	162.5
	s	1.8	--	16.3	24.4	--	13.9
1:1,000	K_d	477.6	218.6	265.7	763.5	276.9	228.1
	s	3.7	10.8	4.0	14.2	33.3	26.0

Regardless of the conditions at 22°C, K_d values increased significantly with the increasing phase ratio m/V, especially at the ratio of 1:1,000.

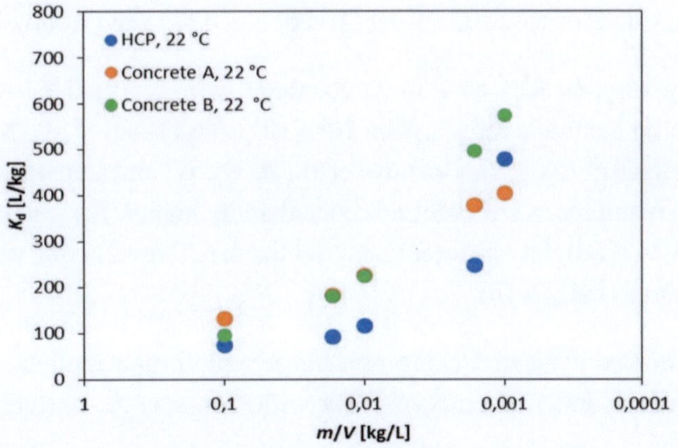

Figure 3: *Dependence of determined K_d values describing ^{223}Ra uptake on cementitious materials on the phase ratio m/V in $Ca(OH)_2$ at 22°C.*

For HCP and Concrete A, a certain trend was observed at 80°C: with the increasing phase ratio the K_d firstly decreased and then increased. It could correspond to Sr content in studied materials. From results of 7-day leaching in demineralized water by m/V equal to 1:3 ($7.97 \cdot 10^{-5}$ mol/L in HCP leachate, $4.64 \cdot 10^{-5}$ mol/L and $2.59 \cdot 10^{-5}$ mol/L in leachates of Concrete A and Concrete B) it was concluded that the highest content of Sr is in HCP.

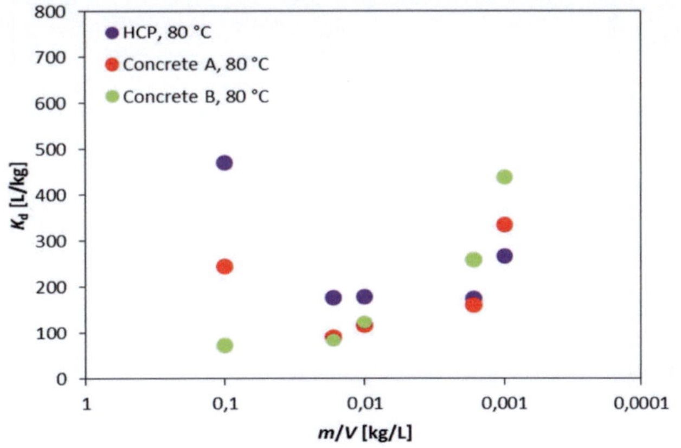

Figure 4: *Dependence of determined K_d values describing ^{223}Ra uptake on cementitious materials on the phase ratio m/V in $Ca(OH)_2$ at 80°C.*

The distribution coefficients of all materials at high phase ratios showed a decrease with temperature, with no uniform shift at lower m/V ratios.

Due to the observed trend in the experiment at 80°C, which is described above (Figure 4, Table 2 and Table 3), sorption of ^{223}Ra on HCP at 50°C was also studied. At this temperature, only a slight decrease of K_d was observed at smaller m/V ratios.

For experiments performed in the CPW environment, a K_d increase over the $Ca(OH)_2$ medium is evident.

In a study of Tits et al. (2006), the distribution coefficient of ^{226}Ra (with initial concentration about 10^{-8} mol/L) on HCP (type CEM I) was set at 140 L/kg for degraded HCP (phase ratio 1:200) at pH 12.5 and about 400 L/kg for fresh HCP (1:500) at pH 13.3. The CEM I type of cement is contained in our Concrete A sample, which at this m/V ratio produced a K_d values of about 235 L/kg (1:100) and 427 L/kg (1:600), both at 22°C.

Information about radium sorption may as well be compared with reported K_d values for bentonites and/or soils. K_d for radium sorption on bentonite ranges from 10^2 - 10^4 L/kg (Tachi et al., 2001), for soil K_d values can assume values of tens to hundreds of L/kg (Vandenhove et al., 2009). When comparing the distribution coefficient of Ra and Sr on cementitious materials, K_d of radium is considerably higher. E.g., in our previous research, K_d for strontium on HCP of type CEM II (slightly different material than HCP used in this work) was determined within the range 12 - 22 L/kg (Drtinová et al., 2016).

The obtained results were also evaluated using non-linear equilibrium models, namely the Langmuir and Freundlich isotherms. The interaction isotherms of ^{223}Ra with Concrete A, $Ca(OH)_2$ at 22°C is presented on Figure 5 as an example.

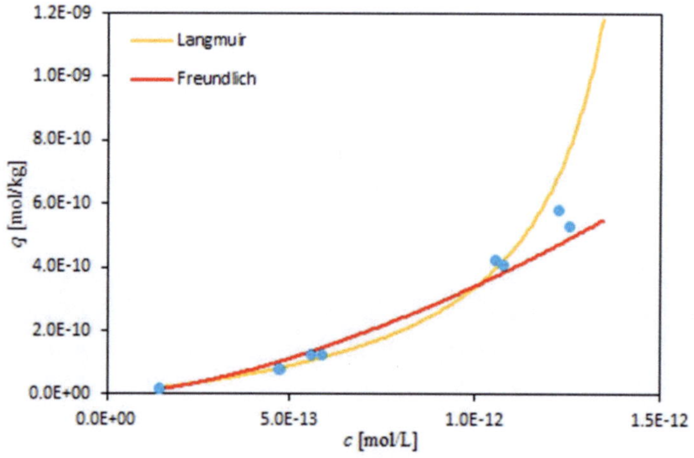

Figure 5: *Evaluation of a set of interaction experiments of ^{223}Ra with Concrete A in Portlandite water, at 22°C, with Langmuir and Freundlich isotherms.*

Conclusions and Future work

A XRD analysis of HCP and two concretes showed a significant presence of calcite accompanied by portlandite in concrete samples. In HCP four crystalline mineral phases were found: calcite, portlandite, hydrotalcite and ettringite.

Studying the kinetics of sorption proved that sorption proceeds quickly and equilibrium is reached within one day for all studied materials. During sorption experiments at equilibrium the change of the liquid phase from the saturated $Ca(OH)_2$ to the synthetic cement water CPW resulted in a distribution coefficient K_d increase. The increase in the temperature at which the experiment was performed from 22 to 80°C had an interesting effect on the behavior of the HCP and Concrete A materials. The K_d decreases firstly and then increases with increasing phase ratio. When verifying this trend at 50°C for HCP material, only a slight decrease with smaller phase ratios was observed.

Based on the data obtained, it would be appropriate to verify the sorption properties of both concrete materials in CPW. Another suitable completion of the study would be the addition of similar experiments at some other temperatures in the range of 50 - 80°C and to perform the same experiments with strontium, the assumed analogue of the radium.

Acknowledgement

The research leading to these results has received funding from the European Union's Horizon 2020 Reasearch and Training Programme of the European Atomic Energy Community (EURATOM) (H2020-NFRP-2014/2015) under grant agreement n° 662147 (CEBAMA).

This contribution is a partially result of Radioactive Waste Repository Authority project "Research support for Safety Evaluation of Deep Geological Repository".

References

Berner, U. (2002). Project Opalinus Clay. Radionuclide concentration limits in the cementitious near-field of an ILW Repository. PSI Report, 02-26.

Drtinová, B., Kittnerová, J., Vopálka, D. (2016). Characterization of hydrated cement paste (CEM II) by selected instrumental methods and a study of ^{85}Sr uptake. Proceedings of the 1st CEBAMA Annual Workshop. KIT Scientific Publishing, KIT-SR 7734.

Evans, N.D.M. (2008). Binding mechanisms of radionuclides to cement. Cement and Concrete Research, 38, 543-553.

Kozempel, J., Vlk, M., Málková, E., Bajzíková, A., Bárta, J., Santos-Oliveira, R., Malta Rossi, A. (2015). Prospective carriers of ^{223}Ra for targeted alpha particle therapy. Journal of Radioanalytical and Nuclear Chemistry, 304, 443-447.

Tachi, Y., Shibutani, T., Sato, H., Yui, M. (2001). Experimental and modeling studies on sorption and diffusion of radium in bentonite. Journal of Contaminant Hydrology, 47, 171-186.

Tits, J., Iijima, K., Wieland, E., Kamei, G. (2006). The uptake of radium by calcium silicate hydrates and hardened cement paste. Radiochimica Acta, 94, 637-643.

Vandenhove, H., Gil-García, C., Rigol, A., Vidal, M. (2009). New best estimates for radionuclide solid-liquid distribution coefficients in soils. Part 2. Naturally occurring radionuclides. Journal of Environmental Radioactivity, 100, 697-703.

Effect of redox conditions on sulfur and selenium binding in AFm phases

Latina Nedyalkova[1,2*], Barbara Lothenbach[2], Jan Tits[1], Erich Wieland[1], Urs Mäder[3]

[1] Paul Scherrer Institute, Laboratory for Waste Management (CH)
[2] Empa, Laboratory for Concrete & Construction Chemistry (CH)
[3] Institute of Geological Sciences, University of Bern (CH)

* Corresponding author: latina.nedyalkova@psi.ch

Abstract

The incorporation of Se and S in AFm phases was investigated. Se(VI)-, Se(IV)-, S(VI)-, S(IV)- and S(II)-AFm phases were synthesized accounting for a range of redox conditions representative of a cement-based L/ILW repository. The AFm phases were characterized by TGA, FTIR and X-ray diffraction. From bulk chemical analysis of the liquid phase and the pH values, the solubility products were determined. Furthermore, the effect of the equilibration time (3 months *vs.* 6 months) and pH (higher pH of ~ 13 *vs.* lower pH of ~ 12) on the stability of the phases was examined.

The X-ray spectra revealed crystalline phases with rhombohedral and monoclinic structure. For the phases with rhombohedral structure different interlayer distances (hkl 003 and 006) were observed, whereas no shift of the reflection referring to the main layer (hkl 110) could be detected. The TGA analyses showed typical water loss patterns for AFm phases giving total water contents of 11 H_2O molecules for the S(IV)- and the Se(IV)-AFm, 12 H_2O molecules for the S(VI)-AFm and 13 H_2O molecules for the S(II)- and Se(VI)-AFm. The variations in the crystal symmetry and the basal spacing seem to be a function of the size of the interlayer atoms and/or the number of water molecules. These observations suggest an intercalation of the Se and S anions in the AFm interlayers.

Introduction

One concept for radioactive waste management foresees the disposal of low- and intermediate level nuclear waste (L/ILW) in deep geological repositories. Cementitious materials are used for the solidification of the waste and for the construction of the engineered barrier system whose function is among others to retard the mobility of the radionuclides in the repository near-field. Under the alkaline and reducing conditions prevailing in such repositories many common near- and far-field minerals carry negatively charged surfaces which efficiently retard cationic radionuclides. Radionuclides with anionic speciation, such as selenium, however, are expected to be only weakly retarded. Potential anionic exclusion could decrease the effective diffusion coefficient of anions compared to diffusion coefficient of cations. Nevertheless, due to its long half-life, selenium is defined as an important dose-determining radionuclide in L/ILW repositories (Nagra, 2002).

Safety assessment so far did not take fully into consideration the potential uptake by some positively charged anion exchangers present in the cementitious near-field matrix, such as AFm(Al_2O_3-Fe_2O_3-mono) phases. AFm phases belong to the layered double hydroxides (LDH) and form during the hydration of cement. They have a lamellar structure with a positively charged main layer, $[Ca_4(Al,Fe)_2(OH)_{12}]^+$, and a negatively charged interlayer, $[X \cdot nH_2O]^{2-}$, where X denotes two singly charged or a doubly charged anion. The main layers consist of sheets of

$Ca(OH)_6$ octahedra with every third Ca^{2+} site being occupied by Al^{3+} and/or Fe^{3+}, generating thus a positive net charge. Different charge balancing anions and different numbers of water molecules can be present in the interlayer. The anions typically found in AFm phases are OH^-, SO_4^{2-}, Cl^- and CO_3^{2-} forming hydroxyl-AFm, monosulfate, Friedel's salt, hemi- and monocarbonate, respectively (Evans and Slade, 2006).

Under the alkaline conditions (10 < pH 13.5) in a cement-based repository Se will be present in anionic form and may exchange for the common interlayer anions. The aqueous speciation is dominated by SeO_4^{2-} (Se(VI), selenate) under oxidizing conditions and SeO_3^{2-} (Se(IV), selenite ion), HSe^- (Se(-II), selenide) and a series of polyselenides (Se_x^{2-}, x = 2, 3, 4) under reducing conditions (Olin et al., 2005) (Figure 1). In addition, reduced anionic species of S, namely SO_3^{2-} (S(IV), sulfite), $S_2O_3^{2-}$ (S(II), thiosulfate) and HS^- (S(-II), sulfide), may also be present as competitive anions for the AFm interlayer sites.

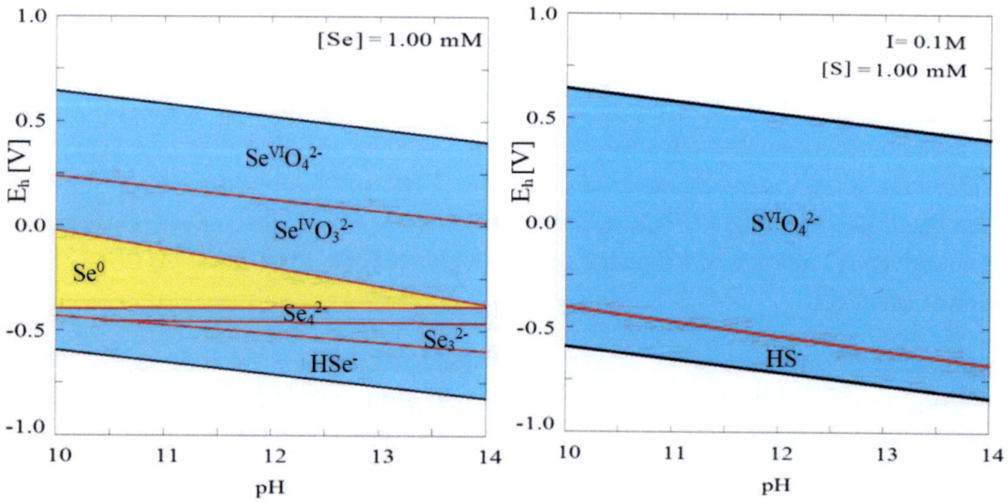

Figure 1: *Predominance diagram for selenium (left) and sulfur (right) in the pH and Eh regions relevant for the cementitious near field of a L/ILW repository; the reduced sulfur species SO_3^{2-} and $S_2O_3^{2-}$ are thermodynamically metastable relative to the main species SO_4^{2-} and HS^- and are therefore not visible on a predominance diagram. Calculations were performed using the code Medusa (Puigdomenech, 1983) and thermodynamic data from the NEA thermodynamic database.*

As AFm phases have the potential to bind Se in their structure and may play an important role in the immobilization of this hazardous element, a thorough understanding of the incorporation of selenium and sulfur in AFm phases is required. The binding of S(VI) in AFm phases and its structure are very well investigated. The S(VI)-AFm has a rhombohedral structure belonging to the $R\bar{3}$ space group (Allmann, 1977). The S(IV)-AFm phase has been synthesized and characterized before (Motzet and Pöllmann, 1999) but no solubility measurements or thermodynamic data for it are available. Data for the more reduced sulfur species S(II) and S(-II) are completely missing, although their existence has been suggested (Vernet, 1982). Similarly for Se: several studies deal with the Se(VI) and Se(IV) uptake by AFm phases (e.g., Baur and Johnson, 2003a; Bonhoure et al., 2006; Mace et al., 2007), while data for Se(-II) are still missing.

The objective of this study is to further investigate the incorporation of selenium and sulfur in AFm phases and work towards the construction of a thermodynamic model describing the uptake. For this purpose, pure Se(VI)-, Se(IV)-, S(VI)-, S(IV)-, S(II)-AFm phases were synthesized, accounting for a range of possible redox conditions in a cement-based repository in deep geological formations. The AFm phases were characterized by various techniques (XRD, TGA, FTIR, Raman, aqueous phase analysis) and their solubility products calculated.

Results and discussion

Characteristics of the synthesized AFm phases

Five different AFm phases were synthesized: SO_4-, SeO_4-, SO_3-, SeO_3- and S_2O_3-AFm, using C_3A, CaO and Na_2-salts as starting materials (pH ~ 13). After three months of equilibration time, the samples were filtered and dried in a desiccator over a saturated NaOH solution (~ 10% relative humidity) at room temperature for five to six weeks.

<u>TGA</u>

The TGA-DTG curves show a typical water loss pattern for AFm with total mass losses of ~ 34.5 wt.% for the SO_4-AFm, between 33.0 and 33.2 wt.% for the SO_3- and S_2O_3-AFm, ~ 32.3 wt.% for SeO_4-AFm and ~ 31.2 wt.% for SeO_3-AFm (Figure 2). The water removal occurs in several steps: up to 250°C the loosely bound water in the interlayer is lost, while at temperatures between 250 - 500°C water in the main layer is lost. The total amount of water estimated is ~ 11 H_2O for the SO_3- and SeO_3-AFm phases, ~ 12 H_2O for the SO_4-AFm, and ~ 13 H_2O for the S_2O_3- and SeO_4-AFm phases. All synthesized AFm phases contain ~ 6 H_2O in the main layer.

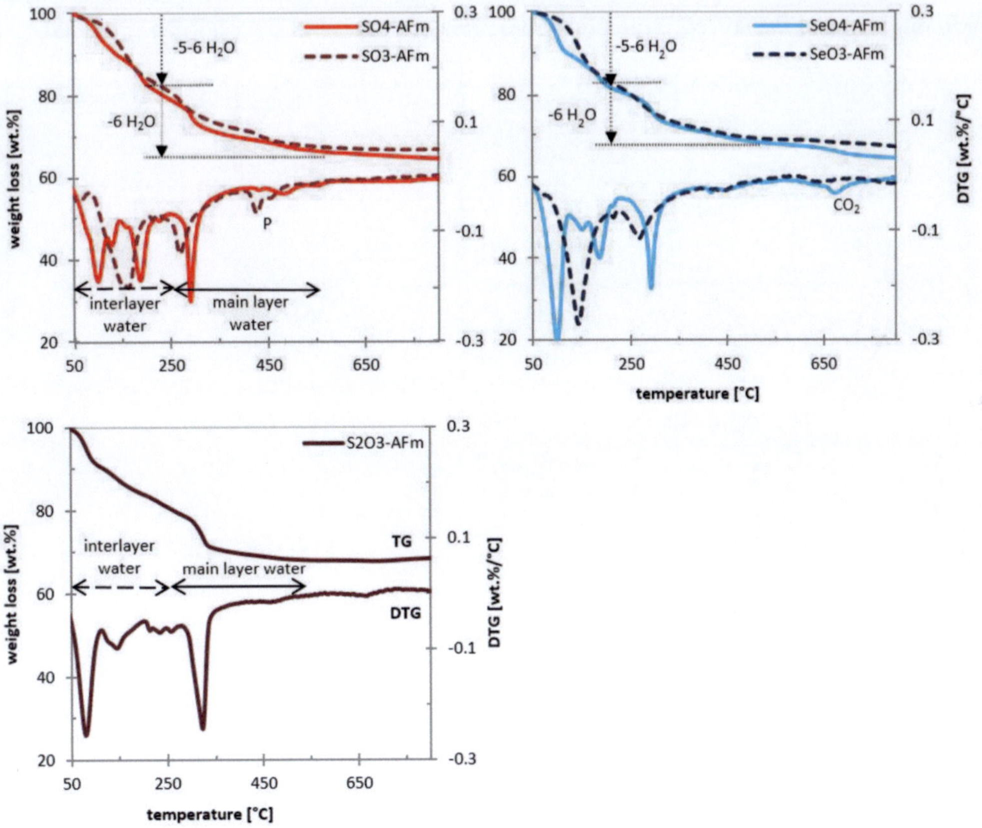

Figure 2: *TGA and DTG curves of the synthesized AFm phases (pH ~ 13) after 3 months equilibration time and drying over a saturated NaOH solution (r.h. ~ 10%). P: portlandite.*

<u>XRD</u>

The XRD-patterns show that well-crystalline AFm phases have formed (Figure 3). Comparison of all synthesized samples, however, indicates that the SeO_3- and the S_2O_3-AFm are somewhat less crystalline with less clearly defined peaks. Impurities such as ettringite (Et) in the SO_4-AFm, portlandite (P) in the SeO_4- and SeO_3-AFm, hemicarbonate in the S_2O_3- and SeO_4-AFm, as well as some katoite (K) in the SeO_3-AFm were also detected. Evidence for the rhombohedral structure of the SO_4-AFm comes from the main layer peak (hkl 110) at 31° 2θ. In addition, its two main diffraction peaks are at 9.9° and 19.9° 2θ corresponding to Miller indices (003) and (006) and d-values of 8.93 Å and 4.47 Å, respectively (Allmann, 1977). Depending on the exposure temperature and relative humidity, several hydration states of SO_4-AFm can exist - Ms9, Ms10.5, Ms12 and Ms14 (with the index number giving the water content of the phase in moles), each characterized by different interlayer distances (Baquerizo et al., 2015). Although the main hydration state observed here is Ms12, the additional peaks left of the (hkl 003) and (hkl 006) reflexions indicate also the presence of minor Ms14 which is most probably the result of insufficient drying.

The SO_3- and SeO_3-AFm phases exhibit the same main layer peak (hkl 110) at 2θ ~ 31° indicating also a rhombohedral structure. The SO_3-AFm is characterized by higher 2θ values and hence smaller basal spacings (hkl 003 and 006), consistent with the presence of less water in the interlayer space and smaller anionic size. Although the XRD pattern of the S_2O_3-AFm appears to be similar to those of the SO_4- and the SO_3-AFm, it has an additional diffraction peak at ~ 11° 2θ which cannot be explained by a rhombohedral symmetry and is more consistent to a monoclinic structure. Similarly, the numerous diffraction peaks of the SeO_4-AFm pattern indicate a lower, also monoclinic symmetry. The crystal structure analyses for the SO_3-, S_2O_3-, SeO_4- and SeO_3-AFm phases are still ongoing and the correct structure symmetries of the phases are yet to be specified.

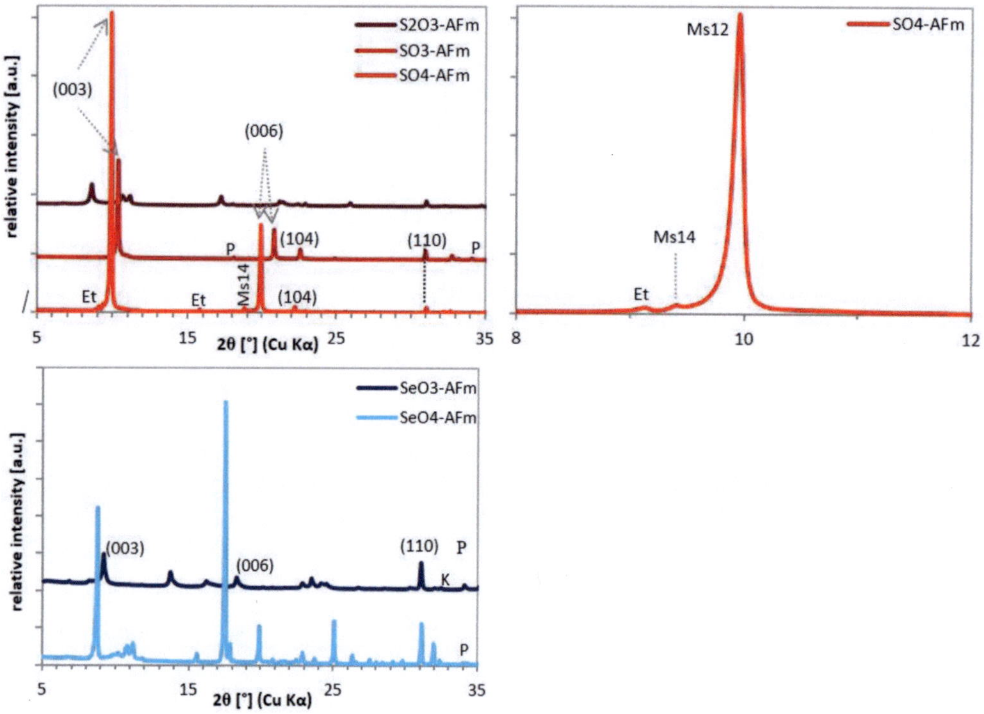

Figure 3: *XRD-pattern of the synthesized AFm phases (pH ~ 13) after 3 months equilibration time and drying over a saturated NaOH solution (r.h. ~ 10%); the image on the top right shows the (003)-peak area of the SO_4-AFm in details. Et: ettringite; Ms: monosulfate; K: katoite.*

FTIR/Raman spectra

The IR-spectra (Figure 4 and Table 1) of the synthesized samples indicate two different types of hydrous components. The broad band in the 3,600 - 3,000 cm^{-1} region as well as the absorption band at approximately

1,650 - 1,600 cm^{-1} is caused by H_2O from the interlayer, while an absorption at ~ 3,670 - 3,650 cm^{-1} is characteristic of a single hydroxyl group. Generally, sharp peaks between 3,600 and 3,000 cm^{-1} indicate more highly coordinated interlayer water (Dilnesa et al., 2011). The IR-spectra show better ordered water in the SO_4-, SeO_4- and S_2O_3-AFm phases; whereas in the SO_3- and the SeO_3-AFm less ordering can be observed. All samples show minor CO_2-contamination, visible by a band at 1,450 - 1,250 cm^{-1}. IR measurements were also carried out on the Na_2-X salts used for the synthesis in order to verify the peak positions of the main anions.

Raman spectra are plotted against the IR spectra on Figure 4 for comparison. Both spectra show similar patterns. The band position for $Al(OH)_6$ appears at ~ 532 cm^{-1}; the S-O and Se-O vibrations occur between ~ 970 - 980 cm^{-1} for SO_3 and SO_4, and between ~ 793 - 832 cm^{-1} for SeO_3 and SeO_4. Small bands at ~ 1,085 cm^{-1} belong to weakly bound carbonate due to CO_2 contamination. The hydrogen bond network vibrations can be seen in the range ~ 2,800 - 4,000 cm^{-1}. A broad band at the beginning of this range can be attributed to H_2O, while a sharper signal at the end (~ 3,688 cm^{-1}) results from the presence of OH groups (Renaudin et al., 2007).

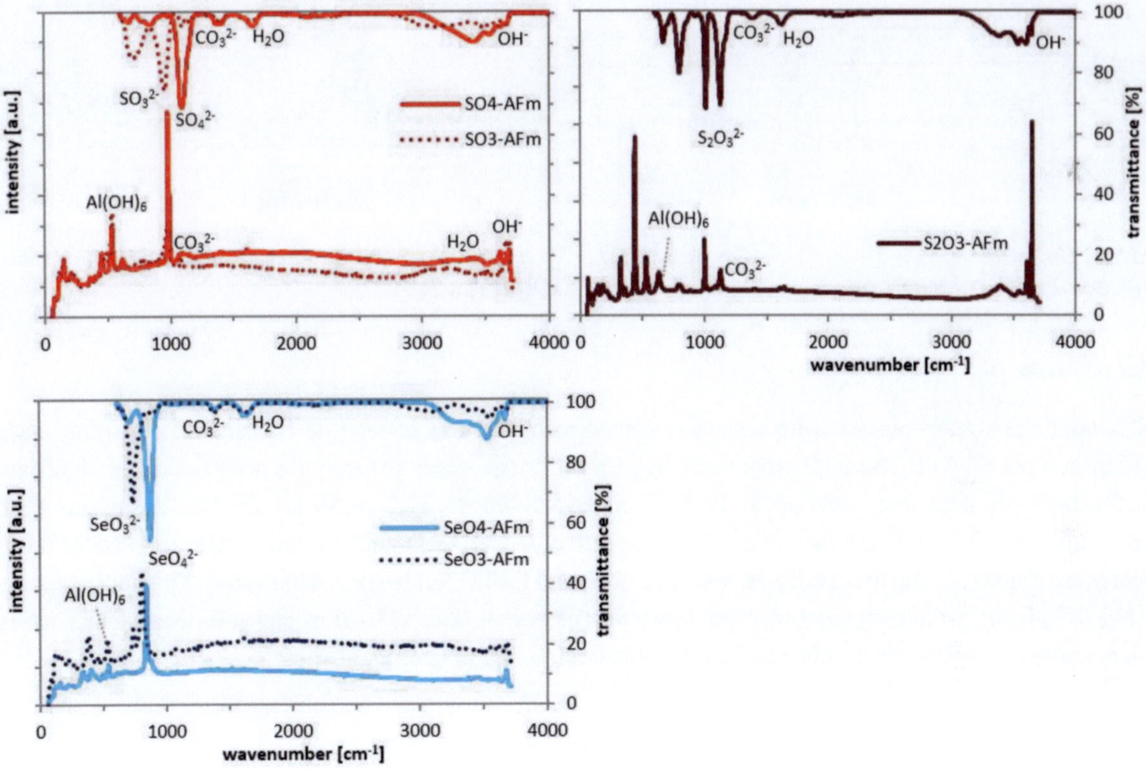

Figure 4: FTIR (top) and Raman (bottom) spectra of the synthesized AFm phases (pH ~ 13) after 3 months equilibration time and drying over a saturated NaOH solution (r.h. ~ 10%).

Table 1: Vibrations in the IR spectra of the synthesized AFm-phases (Dilnesa et al., 2011).

Wave number [cm^{-1}]	Vibration	Wave number [cm^{-1}]	Vibration
480 - 960	Al-O, Ca-OH	1,250 - 1,450	CO_2
~ 735	SeO_3^{2-}	1,600 - 1,650	v_2-H_2O
~ 874	SeO_4^{2-}	3,000 - 3,600	H_2O interlayer
~ 955	v_3-SO_3^{2-}	3,650 - 3,670	OH main layer
~ 1,105	v_1-SO_4^{2-}		

Effect of equilibration time

The five AFm phases were also analysed after longer equilibration times, i.e., after 6 months. The XRD-patterns of the 6 month samples (Figure 5) show no notable mineralogical changes; some of the AFm phases, however, display sharper peaks and higher intensities (in particular S_2O_3-AFm) indicating better ordering and/or an increase in the number of layers stacked. The SeO_4-AFm has an additional peak next to the (003)-reflex (MSe10) indicating the presence of an SeO_4-AFm phase containing 10 H_2O molecules, which is the result of longer drying.

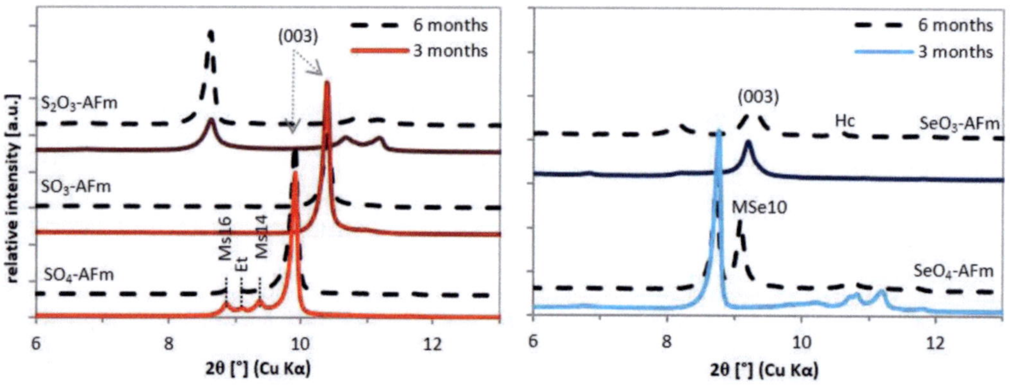

Figure 5: *XRD-pattern of the synthesized AFm phases (pH ~ 13) after 3 and 6 months (dotted line) equilibration time and drying over a saturated NaOH-solution (r.h. ~ 10%). MSe10: SeO_4-AFm with 10 H_2O.*

Effect of lower pH

SO_4- and SeO_3-AFm phases were also synthesized using Ca-salt instead of Na-salts as a starting material resulting in a pH of ~ 12. The total amount of water loss for the lower pH samples does not differ significantly from the high pH ones and sums up to 12 H_2O corresponding to 34.2 wt.% for the SO_4-AFm and 11 H_2O corresponding to 33.2 wt.% for the SeO_3-AFm, respectively. The XRD-patterns are similar to those of the high pH samples (Figure 6). The main difference can be observed for the SeO_3-AFm which shows higher intensities of the (hkl 003)-band, which suggests increased stacking of layers. The SO_4-AFm has additional, small intensity peaks near the (hkl 003)-reflex indicating the presence of hemicarbonate impurities.

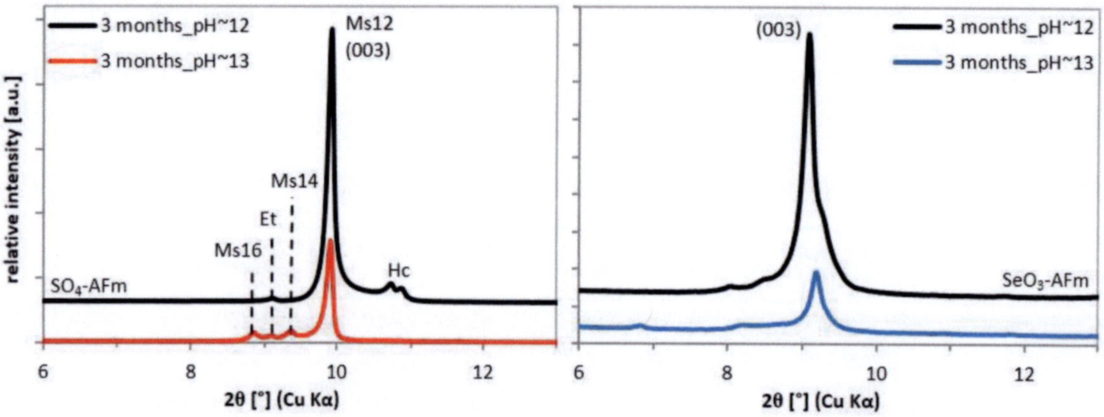

Figure 6: *XRD-pattern for the SO_4- and SeO_3-AFm phases comparing the effect of pH. All samples were dried over a saturated NaOH-solution (r.h. ~ 10%).*

Solubility products

The measured concentrations of calcium, sodium, aluminium, sulfur and selenium in the liquid phase in equilibrium with the solids were used to calculate the solubility products using GEMS (Kulik et al., 2013), including a full speciation calculation. The activity coefficients of the aqueous species were calculated with the built-in extended Debye-Hückel equation. The results are plotted in Figure 7 together with available published data. The solubility products refer to Ca^{2+}, AlO_2^-, OH^-, H_2O, $S_2O_3^{2-}$, SO_3^{2-}, SO_4^{2-}, SeO_4^{2-} and SeO_3^{2-}, respectively and 4 Ca within the main layer (e.g., $K_{s0}(SO_4\text{-AFm}) = \{Ca^{2+}\}^4 \{AlO_2^-\}^2 \{SO_4^{2-}\} \{OH^-\}^4 \{H_2O\}^{12}$ for C_4AsH_{14}). The obtained results for the SO_4- and the SeO_4-AFm are in good agreement with the literature data (Matschei et al., 2007; Baur and Johnson, 2003b; Zhang, 2000). The slightly lower solubility products of the SO_4- and SeO_4-AFm in comparison to the solubility products of the SO_3- and SeO_3-AFm indicate that SO_4- and SeO_4-AFm tend to be more stable than SO_3- and SeO_3-AFm phases.

The calculated solubility products after 6 months are comparable to those after 3 months, although some scatter is observed. The solubility products obtained for the SeO_3-AFm phase at pH ~ 12 are comparable to the results obtained in the presence of 0.2 M Na and pH ~ 13. The apparent higher solubility products obtained for the SO_4-AFm at pH 13 could be due to the presence of a solid solution between hydroxide and sulfate as discussed in detail in Matschei et al. (2007). A part of the OH^- is incorporated with the SO_4^{2-} in the interlayer of the AFm which leads to apparent higher solubility products at high pH values, if referred to SO_4-AFm only.

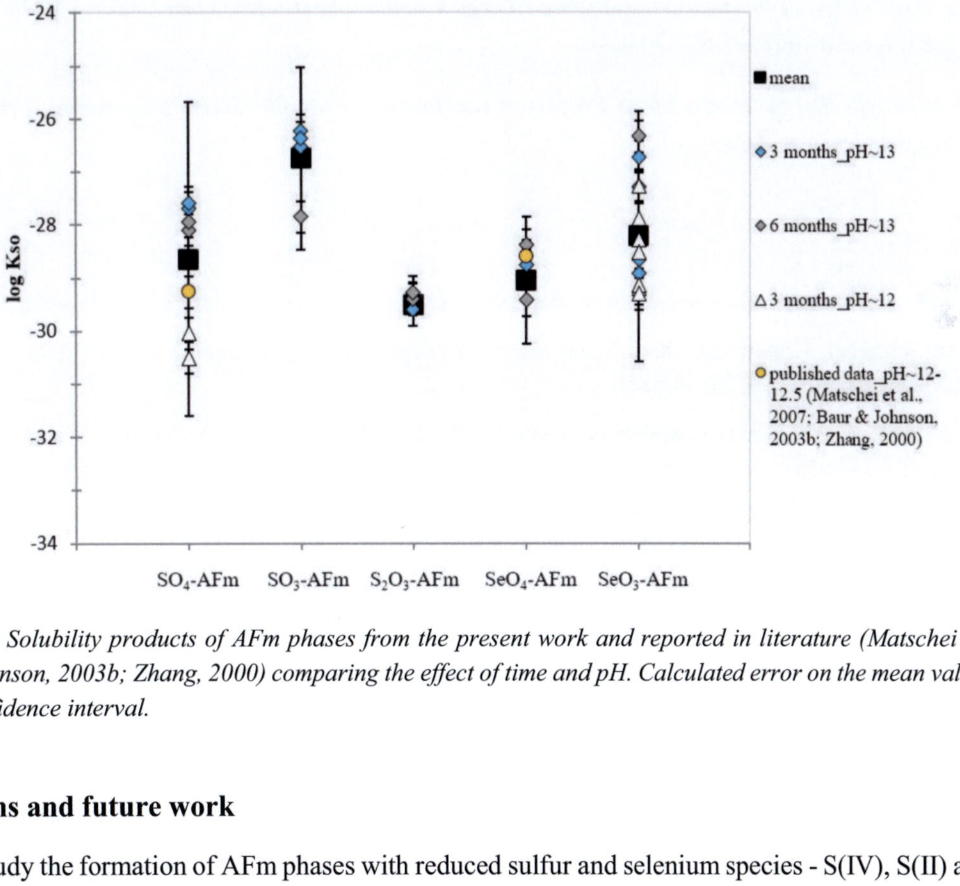

***Figure** 7: Solubility products of AFm phases from the present work and reported in literature (Matschei et al., 2007; Baur and Johnson, 2003b; Zhang, 2000) comparing the effect of time and pH. Calculated error on the mean values indicates the 95% confidence interval.*

Conclusions and future work

In this study the formation of AFm phases with reduced sulfur and selenium species - S(IV), S(II) and Se(IV) - was investigated. All synthesized products show a crystalline structure with rhombohedral or monoclinic symmetry. The S(IV)- and the Se(IV)-AFm have an identical position of the main layer reflection (hkl 110) as the S(VI)-AFm, indicating also a rhombohedral structure. Different interlayer distances (hkl 003 and 006) for these two phases are observed as a function of the size of the interlayer anions and/or the number of water molecules present in the interlayer. The S(II)- and the Se(VI)-AFm show a lower symmetry consistent with a monoclinic structure. TGA displays typical water loss pattern of AFm phases with a water content of 11 H_2O for the SO_3- and the SeO_3-AFm, 12 H_2O molecules for the SO_4-AFm, and 13 H_2O for the S_2O_3- and the SeO_4-AFm.

Longer equilibration times lead to more stacking of layers resulting in higher intensities of the reflection related to the basal spacing on a XRD spectrum.

Future objectives are:

- Synthesis and characterization of binary AFm solid solutions containing various ratios of Se(VI), Se(IV), I(-I), and X^{n-} with $X^{n-} = SO_4^{2-}, SO_3^{2-}, S_2O_3^{2-}, CO_3^{2-}, OH^-$
- Synthesis and characterization of S(-II)-AFm (HS^--AFm) and Se(-II)-AFm (HSe^--AFm)
- Construction of thermodynamic models describing the above-mentioned solid solutions using GEMS, taking into account the observed uncertainties.

Acknowledgement

The research leading to these results has received funding from the European Union's Horizon 2020 Research and Training Programme of the European Atomic Energy Community (EURATOM) (H2020-NFRP-2014/2015) under grant agreement n° 662147 (CEBAMA).

The authors would like to acknowledge Professor Guillaume Renaudin (SIGMA-Clermont, France) for his work on the crystal structure analyses.

References

Allmann, R. (1977). Refinement of the hybrid layer structure. Neues Jahrbuch für Mineralogie-Monatshefte, 3, 136-144.

Baquerizo, L.G., Matschei, T., Scrivener, K.L., Saeidpour, M., Wadsö, L. (2015). Hydration states of AFm cement phases. Cement and Concrete Research, 73, 143-157.

Baur, I. and Johnson, C.A. (2003a). Sorption of selenite and selenate to cement minerals. Environmental Science & Technology, 37, 3442-3447.

Baur, I. and Johnson, C.A. (2003b). The solubility of selenate-AFt ($3CaO{\cdot}Al_2O_3{\cdot}3CaSeO_3{\cdot}37.5H_2O$) and selenate-AFm ($3CaO{\cdot}Al_2O_3{\cdot}CaSeO_4{\cdot}xH_2O$). Cement and Concrete Research, 33, 1741-1748.

Bonhoure, I., Baur, I., Wieland, E., Johnson, C.A., Scheidegger, A.M. (2006). Uptake of Se(IV/VI) oxyanions by hardened cement paste and cement minerals: An X-ray absorption spectroscopy study. Cement and Concrete Research, 36, 91-98.

Dilnesa, B.Z., Lothenbach, B., Le Saout, G., Mesbah, A. Filinchuk, Y. Wichser, A., Wieland, E. (2011). Iron in carbonate containing AFm phases. Cement and Concrete Research, 41, 311-323.

Evans, D.G. and Slade, R.C.T. (2006). Structural aspects of layered double hydroxides. *In:* Duan, X. and Evans, D.G. (Eds.) Layered Double Hydroxides. Springer-Verlag, 119, 1-87.

Kulik, D.A., Wagner, T., Dmytrieva, S.V., Kosakowski, G., Hingerl, F.F., Chudnenko, K.V., Berner, U. (2013). GEM-Selektor geochemical modeling package: revised algorithm and GEMS3K numerical kernel for coupled simulation codes. Computational Geosciences, 17, 1-24.

Mace, N., Landesman, C., Pointeau, I., Grambow, B., Giffaut, E. (2007). Characterisation of thermally altered cement pastes. Influence on selenite sorption. Advances in Cement Research, 19, 157-165.

Matschei, T., Lothenbach, B., Glasser, F.P. (2007). The AFm phase in Portland cement. Cement and Concrete Research, 37, 118-130.

Motzet, H. and Pöllmann, H. (1999). Synthesis and characterization of sulfite-containing AFm phases in the system $CaO-Al_2O_3-SO_2-H_2O$. Cement and Concrete Research, 29, 1005-1011.

NAGRA (2002). Project Opalinus Clay. Safety report. Demonstration of disposal feasibility for spent fuel, vitrified high-level waste and long-lived intermediate level waste. Nagra Technical Report, NTB 02-05.

Olin, Å., Noläng, B., Osadchii, E.G., Öhman, L.-O., Rosén, E. (2005). Chemical thermodynamics of Selenium. Elsevier.

Puigdomenech, I. (1983). INPUT, SED, and PREDOM: Computer programs drawing equilibrium diagrams. TRITA-OOK-3010. Royal Institute of Technology (KTH), Dept. Inorg. Chemistry, Stockholm, Sweden.

Renaudin, G., Segni, R., Mentel, D., Nedelec, J.M., Leroux, F., Taviot-Gueho, C. (2007). A Raman study of the sulfated cement hydrates: ettringite and monosulfoaluminate. Journal of Advanced Concrete Technology, 5, 299-312.

Vernet, C. (1982). Comportement de l'ion S^{2-} au cours de l'hydratation des ciments riches en laitier (CLK). Formation de solutions solides de S^{2-} dans les aluminates hydrates hexagonaux. Silicates Industriels, 3, 85-90.

Zhang, M. (2000). Incorporation of oxyanionic B, Cr, Mo, and Se into hydrocalumite and ettringite: Application to cementitious systems. PhD Thesis at the University of Waterloo, Ontario, Canada.

Radionuclide retention in cementitious systems and single mineral phases

Matthew Isaacs[1*], Steve Lange[2], Guido Deissmann[2], Dirk Bosbach[2], Emily Rastrick[1], Mónica Felipe-Sotelo[1], David Read[1,3]

[1] University of Surrey (UK)
[2] Forschungszentrum Jülich GmbH (DE)
[3] National Physical Laboratory (UK)

* Corresponding author: m.isaacs@surrey.ac.uk

Abstract

This paper describes the characterisation of cement materials and equilibrated waters being used throughout the experimental programme forming the University of Surrey contribution to the CEBAMA project. The methodology for the solubility determination of long-lived fission and decay products (Ra, Tc, I, Se, Cl), by over- and under-saturation measurements, in a representative number of cementitious systems is also discussed. A series of batch sorption experiments is being carried out with five different cement formulations used either for conditioning and packaging of radioactive waste, construction or as backfill material, as well as with specific mineral phases present in the cements (CSH, AFm, AFt). These studies are complemented by through-diffusion experiments using established protocols. Details of the solid phase characterisation techniques are described below. Initial results from batch sorption tests of I^-, IO_3^-, SeO_3^{2-}, and SeO_4^{2-} onto single mineral phases and a CEM I hardened cement paste indicate the presence of the anions tested caused a degradation of the AFm phase to acicular crystals. The incorporation of iodine onto these crystals is shown via SEM/EDX.

Introduction

Cementitious materials are used widely in radioactive waste management, for example in the solidification of low and intermediate level wastes, or as construction and barrier materials in underground and surface repositories. The retention of radionuclides in cements is controlled by solubility phenomena, diffusion, adsorption and/or incorporation into solids, including the formation of solid solutions. Within the framework of CEBAMA WP2, we are studying the solubility of selected radionuclides in conditions relevant to a geological disposal facility (GDF). The uptake of long-lived fission and decay products (Ra, Tc, I, Se, Cl) in cementitious materials, the diffusion of radionuclides through hardened cement paste and the radionuclide distribution between and within various cement phases on the micro-scale are also being investigated, using advanced micro-analytical and spectroscopic tools. The objective of these investigations is to enhance mechanistic understanding of their uptake and retention and to assess the relevance of chemical alteration processes, such as carbonation, in aged concrete. In this context, a bottom-up approach is being pursued using synthesised cement phases (model phases) on the one hand and hardened cement pastes with different compositions on the other.

Numerous studies have been carried out investigating: 1) the solubility of the radionuclides in solution under alkaline conditions analogous to those expected in a GDF (Baston et al., 1997; Heath et al., 1998; Felipe-Sotelo et al., 2014), 2) batch sorption studies with cement admixtures (Glasser et al., 1985; Beaudoin et al., 1990; Solem-Tishmack et al., 1995; Bonhoure et al., 2006; Tits et al., 2006; Mace et al., 2007; Tanabe et al., 2010) and

3) diffusion studies in cement using different experimental protocols (under equilibrium conditions, i.e., in-, out-, or through-diffusion, or non-equilibrium, i.e., leaching; Atkinson and Nickerson, 1984; Tallent et al., 1987; Brodda and Mingxia, 1988; Gilliam et al., 1990; Sarott et al., 1992; Mattigod et al., 2001; Chida and Sugiyama, 2008; Felipe-Sotelo et al., 2014; van Es et al., 2015). A full review of these topics can be found in the CEBAMA WP2 state of the art report (Lange et al., 2016). Several of the radionuclides addressed in this study can exhibit a range of oxidation states under anticipated GDF conditions (Se, Tc, I). Selenium will be considered as both selenite and selenate, whereas iodine will be considered as iodide and iodate; these are believed to be the dominant species likely under GDF conditions (Thoenen et al., 2014). Technetium will be considered as Tc(IV).

Methodology

Materials

The solids of interest for this work fall into two categories: hardened cement pastes (HCP) and single mineral phases. Five cement blends are included in the work plan; namely CEM I, a ground granulated blast furnace slag: ordinary Portland cement blend (GGBS:OPC), a pulverised fuel ash blend (PFA:OPC), a backfill material (NRVB, Nirex Reference Vault Backfill) and a CEBAMA reference cement blend. The compositions used for the formation of the hardened cement pastes are detailed in Table 1. To make each cement blend the powders were weighed out in accordance to the ratios in Table 1 and mixed in a powder mixer for 2 hours.

Table 1: *Powder ratios of cement to be studied within the scope of this work.*

Cement Blend	OPC	PFA	GGBS	Hydrated Lime	Lime Flour	Silica Fume
CEM I	1	-	-	-	-	-
PFA:OPC	1	3	-	-	-	-
GGBS:OPC	1	-	9	-	-	-
NRVB	1	-	-	0.38	1.1	-
CEBAMA Reference mix	1	-	0.62	-	-	0.87

The bulk composition of the individual components in the cement powders was assessed by X-ray fluorescence and the results are summarised in Table 2.

Table 2: *XRF data providing the bulk composition of the powders used to form the cement blends.*

Cement Powder	CaO (%)	SiO_2 (%)	Al_2O_3 (%)	SO_3 (%)	MgO (%)	Fe_2O_3 (%)	K_2O (%)	P_2O_5 (%)	TiO_2 (%)	MnO (%)
CEM I	66.27	17.86	4.78	3.98	2.75	2.65	1.52	<LOD	0.2	
PFA	3.86	48.66	25.97	1.82	1.27	12.36	5.05	0.18	0.84	
GGBS	42.93	31.92	11.40	4.10	6.76	0.45	1.01	-	0.41	0.59
Hydrated Lime	97.75	1.37	0.35	-	0.20	-	0.33	-	-	-
Lime Flour	96.57	1.70	0.71	0.18	0.29	0.16	0.39	-	-	-
Silica Fume	1.46	93.10	1.44	0.47	0.88	0.91	1.73	-	-	-
CEM I (VTT)	67.72	17.60	3.42	3.81	0.6	5.17	1.3	-	0.17	0.21
GGBS (VTT)	43.13	32.3	9.85	3.68	7.40	0.74	1.20	-	1.36	0.34

The individual phases studied include: calcium silicate hydrate (CSH), ettringite, monosulphate and hydrotalcite. These phases have been selected as they comprise the major components of a CEM I cement, excluding portlandite; no significant interaction of the latter with radionuclides is expected. CSH is produced according to a direct synthesis method from Atkins et al. (1992) performed by mixing CaO with either silicic acid or silica fume in distilled, degassed water. Ca/Si ratios of between 0.9 and 1.4 are obtained using this method, with the lower ratios being applicable to an aged cement. Ettringite is synthesised using a method from Atkins et al. (1991) in which a slurry of water containing CaO is prepared at 5°C (w/s ratio = 10) and added to a solution containing $Al_2(SO_4)_3 \cdot 16H_2O$. This is stirred for 24 hours followed by aging at 25°C. Excess gypsum is removed by repeatedly re-dispersing in water. The ettringite formed is used to synthesise monosulphate via a reaction with tricalcium aluminate (C_3A) over a 2 week period. Hydrotalcite is prepared via solid state reaction (Long et al., 2014), in which aluminium sulphate, magnesium sulphate and sodium carbonate are ground along with polyethylene glycol-400, and stored at 80°C for 3 hours to give a phase pure product after washing with deionised water, anhydrous ethanol and drying. Phase purity is being confirmed via X-ray diffraction (XRD).

Cement Equilibrated Water Preparation

To prepare the cement equilibrated water for use in the solubility and batch sorption experiments 50 g of the powders for each respective cement blend detailed in Table 1 was weighed and deionised water (1 L) added under a N_2 atmosphere. The suspension was agitated daily to prevent sedimentation. After a 28 day period the solid was removed via filtration and the waters were analysed by ion chromatography (Table 3).

Table 3: *Compositional data of the equilibrated waters provided by ion chromatography.*

Cement Blend	Cl (ppm)	SO_4 (ppm)	NO_3 (ppm)	Na (ppm)	K (ppm)	Mg (ppm)	Ca (ppm)	pH
CEM I	18.0	36.9	18.1	267.6	453.3	-	1,388.4	12.82
PFA:OPC	14.4	31.0	16.3	129.8	195.2	39.0	675.8	12.83
GGBS:OPC	16.9	34.7	17.6	68.6	70.2	-	299.6	12.80
NRVB	16.0	33.6	17.2	172.5	224.7	66.3	1,374.3	12.90
CEBAMA Reference mix	24.0	50.3	20.4	189.7	202.3	137.0	1,046.6	12.69

Solubility

Solubility studies are being performed using under-saturation and over-saturation methodologies (Felipe-Sotelo et al., 2017) using water that has been equilibrated with cement for a minimum of 28 days. All experiments are being performed under N_2. Differences between the results of over- and under-saturation tests can arise from several sources. Incipient precipitation in the case of the former may generate colloidal particles that pass filtration, leading to over-estimation of solubility. In order to minimise this effect, two filtration procedures are being tested; ultrafiltration with 30 kDa MWCO (molecular weight cut off) regenerated cellulosed and with fine (0.22 μm) PES (polyester sulphone). Conversely, the dissolution kinetics of solids used in under-saturation measurements may be sufficiently slow that equilibrium is not reached, even on longer timescales (~ 1 year in this work), biasing the data to lower values. Finally, there may be compositional or textural differences in the solids formed by each method. In order to address the latter, structural techniques can be employed to highlight any changes in particle size or crystal form that could have a bearing on phase solubility. The concentration of elements in solution is being measured by inductively-coupled plasma optical emission spectroscopy (ICP-OES,

Se) and liquid scintillation counting (LSC, ^{79}Se, ^{99}Tc, ^{226}Ra). XRD is utilised to determine solubility limiting phases.

Sorption / Desorption

Sample preparation is carried out in a glove box (N_2 or Ar atmosphere) to avoid carbonation. The element of interest is added from a stock solution to an equilibrated water solution, to give an appropriate initial concentration of $5 \cdot 10^{-3}$ mol/dm^3 for Se (Bonhoure et al., 2006) or $1.5 \cdot 10^{-8}$ mol/dm^3 for ^{226}Ra (Tits et al., 2006). A liquid to solid ratio of 100:1 is used in the batch studies and the tests run for a month with daily sampling in the first week followed by weekly sampling for the remainder of the experiment. Where applicable, radioactive tracers are added. Samples are allowed to reach equilibrium and aliquots of the supernatant are taken and analysed by ICP-OES (Se, I, Cl) or LSC (^{36}Cl, ^{79}Se, ^{99}Tc, ^{129}I, ^{226}Ra). Residual solids are filtered and analysed by XRD, energy dispersive X-ray spectroscopy, scanning electron microscopy and extended X-ray absorption fine structure (EXAFS) to provide insight to the distribution of radionuclides on the phases present. Desorption studies are performed by the addition of cement / phase-equilibrated water to the solids. Samples are again allowed to reach equilibrium and the concentration in the supernatant is monitored.

These results should allow the determination of uptake and release kinetics, as well as the partitioning of the radionuclide between solid and solution, quantified in terms of a distribution ratio (R_d, Eq. 1):

$$R_d = \frac{c_s}{c_{aq}} = \left(\frac{c_{tot}-c_{aq}}{c_{aq}}\right) \cdot \left(\frac{V}{m}\right) \qquad \textbf{\textit{Eq. 1}}$$

where c_s and c_{aq} denote analyte concentrations measured on the solid (mol/kg) and in solution (mol/dm^3), respectively at equilibrium. c_{tot} denotes the total analyte concentration added, V and m denote the volume of the suspension (m^3) and mass of solid phase (kg), respectively.

Through-diffusion

Through-diffusion experiments are performed in the manner described by Felipe-Sotelo et al. (2014, 2016 and 2017), Hinchliff et al. (2016) and van Es et al. (2015). Cylindrical blocks of hardened cement paste (40 mm diameter and a 45 - 50 mm height) are drilled longitudinally to create a well for the addition of radioisotopes. The well is drilled centrally and to a depth of 25 - 30 mm. The radionuclide of interest is spiked into the central well, which is then sealed and the block placed in equilibrated water. The movement of the radionuclide through the block is monitored by measuring the equilibrated water surrounding the block for the presence of the radionuclide. After either breakthrough of a radionuclide to constant concentration, or a specified maximum time (~ 1 year), the block is removed from the equilibrated water, sawn axially and the distribution of radionuclides measured by digital autoradiography. The movement of radionuclides is determined via a 2D model, of which the details will be finalised after consultation with WP3 partners AMPHOS 21.

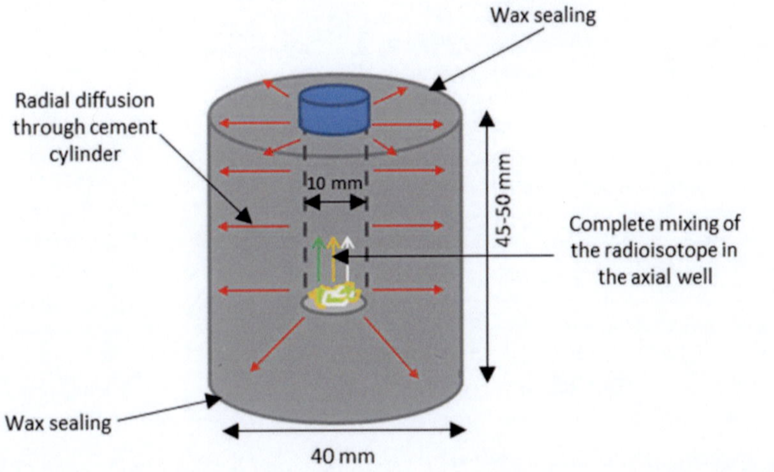

Figure 1: *Schematic of the radial diffusion experiments.*

Results to date

Batch sorption

Preliminary results from the batch sorption experiments using individual phases and a CEM I hardened cement paste have been obtained. For I^- and IO_3^- it was possible to calculate distribution ratios of the analytes in the different systems by measuring the change in concentration of iodine in solution. In the batch sorption tests using selenites and selenates, precipitation of discrete selenium phases was observed when the solids were analysed by SEM (Figure 2b). The experimental procedure has been amended, lowering the selenium concentration so that meaningful adsorption results can be obtained. There are clear indications of interactions with solid phases, notably the surface of the CSH (Figure 2a).

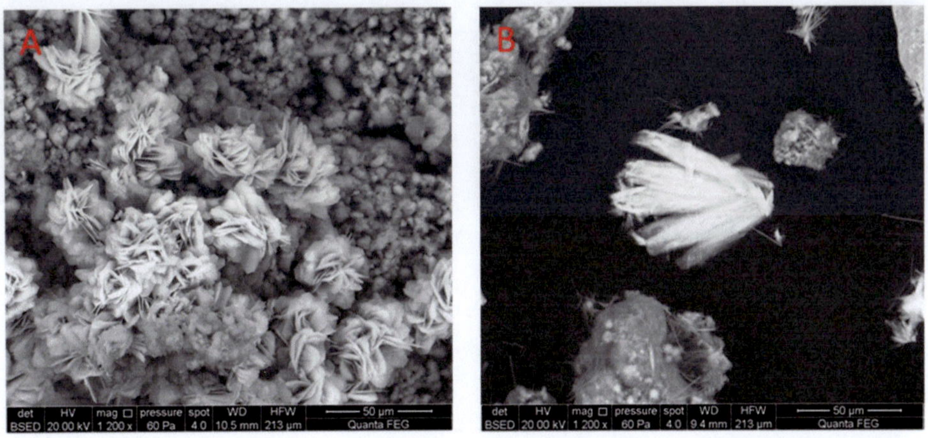

Figure 2: *A: SeO_3^{2-} precipitation to CSH surface and B: selenium-containing precipitate present within the CEM I solids.*

Within the iodine systems, iodate was shown to adsorb very quickly to a range of the solids tested including CEM I, AFt and AFm. Distribution ratios against time for two of these systems are shown in Figure 3 and display fast initial uptake. Further data points will be required to increase sampling during the initial period. There is limited evidence of iodide adsorption onto the phases from analysis of the liquids and longer experimental runs will be required to determine the extent of interaction occurring.

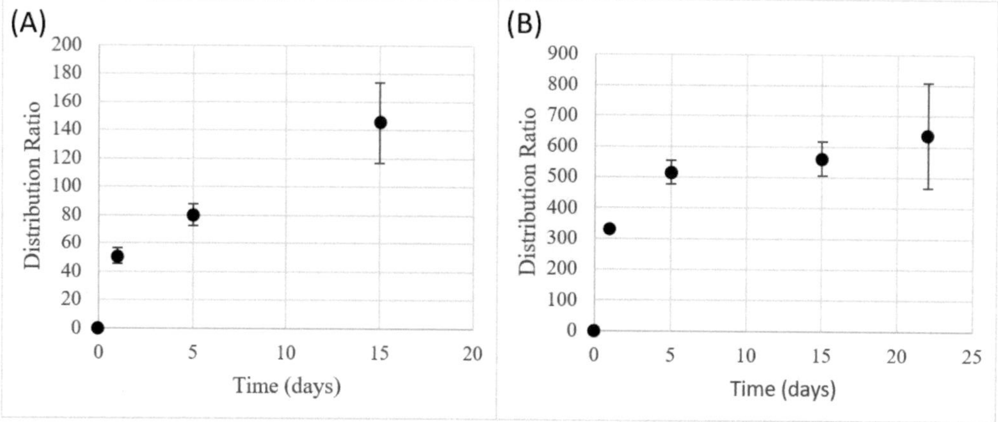

Figure 3: *A: Distribution ratio of IO_3^- on a CEM I HCP versus days and B: Distribution ratio of IO_3^- on AFt versus days.*

Whilst only limited evidence of iodide sorption was found from the concentration present in the liquid, interesting needle-like structures were found in the CEM I paste that had been treated with I^-. These thin structures (Figure 4) appear to have varying amounts of iodine associated with different regions along the axis. EDX data (Table 3) show that the brighter parts of the crystals contain greater amounts of iodine and the centre of the structure appears iodine free. This suggests that the crystal incorporates iodine as it grows. The crystals are large enough to perform crystallographic analyses.

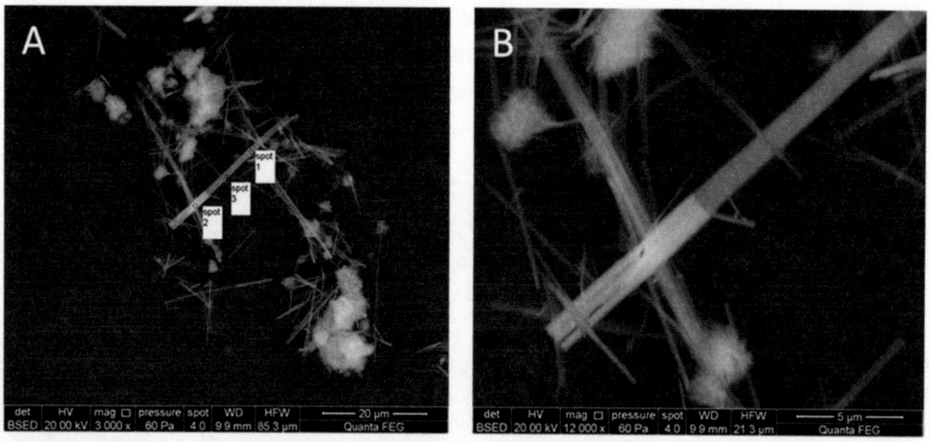

Figure 4: *I^- A: Location of EDX analysis of the crystals and B: incorporation to needle-like phases found within a CEM I paste.*

Table 3: *EDX data showing the elemental abundance at three spots on a single crystal from a CEM I paste after the adsorption of I^-.*

Element	Spot 1 (wt.%)	Spot 2 (wt.%)	Spot 3 (wt.%)
O	59.83	55.56	66.74
Al	5.86	5.42	6
S	9.72	6.12	9.09
Ca	23.33	21.84	18.17
I	1.26	11.05	0
Total	100	100	100

Another observation was that, in the presence of all of the anions tested, the phase pure AFm appeared to degrade from a planar hexagonal structure to an ill-defined aggregate of acicular crystals (Figure 5). The mechanism of this transformation is unclear and more work is required to determine the reasons why this morphological change is occurring.

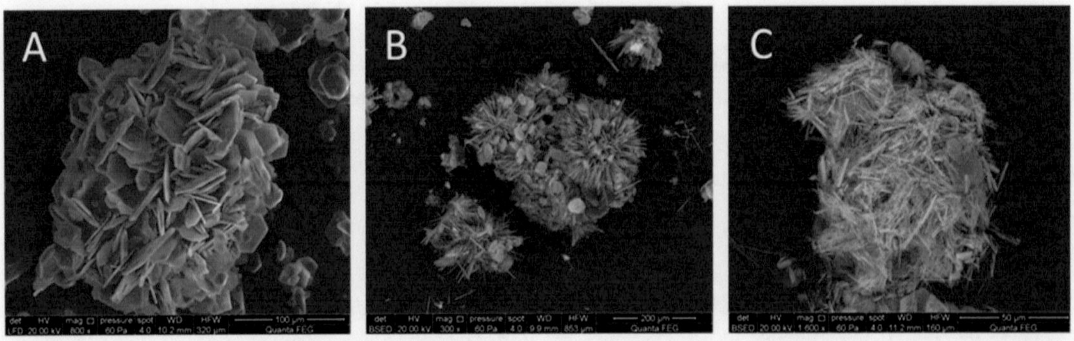

Figure 5: *A: Phase pure AFm, B: AFm after coprecipitation with* SeO_3^{2-} *and C: AFm after batch sorption experiments with* IO_3^-.

Conclusions and Future work

The work conducted so far has provided a foundation for the rest of the project. Characterisation of the cements and equilibrated waters has been carried out and a series of preliminary batch sorption experiments has been performed identifying systems of interest for further study. The solubility, through-diffusion and additional batch sorption experiments are all underway.

Acknowledgement

The research has received funding from the European Union's Horizon 2020 Research and Training Programme of the European Atomic Energy Community (EURATOM) (H2020-NFRP-2014/2015) under grant agreement no. 662147 (CEBAMA).

References

Atkins, M., Macphee, D., Kindness, A., Glasser, F.P. (1991). Solubility properties of ternary and quaternary compounds in the CaO-Al_2O_3-SO_3-H_2O system. Cement and Concrete Research, 21, 991-998.

Atkins, M., Glasser, F.P., Kindness, A. (1992). Cement hydrate phase: Solubility at 25°C. Cement and Concrete Research, 22, 241-246.

Atkinson, A. and Nickerson, A.K. (1984). The diffusion of ions through water-saturated cement. Journal of Materials Science, 19, 3068-3078.

Baston, G.M.N., Brownsword, M., Smith, A.J., Smith-Briggs, J.L. (1997). Further Near-field Solubility Studies. Nirex Report, NSS/R257.

Beaudoin, J.J., Ramachandran, V.S., Feldman, R.F. (1990). Interaction of chloride and CSH. Cement and Concrete Research, 20, 875-883.

Bonhoure, I., Baur, I., Wieland, E., Johnson, C.A., Scheidegger, A.M. (2006). Uptake of Se(IV/VI) oxyanions by hardened cement paste and cement minerals: An X-ray absorption spectroscopy study. Cement and Concrete Research, 36, 91-98.

Brodda, B.-G. and Mingxia, X. (1988). Leaching of chlorine, cesium, strontium and technetium from cement-fixed intermediate level liquid waste. MRS Proceedings, 127, 481-488.

Chida, T. and Sugiyama, D. (2008). Diffusion behavior of organic carbon and iodine in low-heat Portland cement containing fly ash. MRS Proceedings, 1124, 1124-Q10-15.

Felipe-Sotelo, M., Hinchliff, J., Drury, D., Evans, N.D.M., Williams, S., Read, D. (2014). Radial diffusion of radiocaesium and radioiodide through cementitious backfill. Physics and Chemistry of the Earth, Parts A/B/C, 70-71, 60-70.

Felipe-Sotelo, M., Hinchliff, J., Field, L., Milodowski, A., Preedy, O., Read, D. (2017). Retardation of uranium and thorium by a cementitious backfill developed for nuclear waste disposal. Chemosphere, 179, 127-138.

Gilliam, T.M., Spence, R.D., Bostick, W.D., Shoemaker, J.L. (1990). Solidification/stabilization of technetium in cement-based grouts. Journal of Hazardous Materials, 24, 189-197.

Glasser, F.P., Rahman, A.A., Macphee, D., Angus, M.J., Atkins, M. (1985). Immobilization of radioactive waste in cement-based matrices. Department of the Environment Report, DOE-RW-85.063.

Heath, T., Pilkington, N.J., Tweed, C., Williams, S. (1998). Radionuclide solubility at high pH. *In:* Baker, S., Heath, T.G. McCrohon, R. The Chemistry of Deep Disposal of Radioactive Waste. Proceedings of the Royal Society of Chemistry/Nirex Symposium. Nirex Report, S/98/008.

Hinchliff, J., Felipe-Sotelo, M., Evans, N., Read, D. (2016). Solubility constraints affecting the migration of selenium through the cementitious backfill of a geological disposal facility. Journal of Hazardous materials, 305, 21-29.

Long, Q., Xia, Y., Liao, S., Li, Y., Wu, W., Huang, Y. (2014). Facile synthesis of hydrotalcite and its thermal decomposition kinetics mechanism study with masterplots method. Thermochimica Acta, 579, 50-55.

Mace, N., Landesman, C., Pointeau, I., Grambow, B., Giffaut, E. (2007). Characterisation of thermally altered cement pastes. Influence on selenite sorption. Advances in Cement Research, 19, 157-165.

Mattigod, S.V., Whyatt, G.A., Serne, R., Martin, P.F., Schwab, K.E., Wood, M.I. (2001). Diffusion and leaching of selected radionuclides (iodine-129, technetium-99 and uranium) through Category 3 waste encasement concrete and soil fill material. PNNL Report, PNNL-13639.

Solem-Tishmack, J.K., McCarthy, G.J., Docktor, B., Eylands, K.E., Thompson, J.S., Hassett, D.J. (1995). High-calcium coal combustion by-products: Engineering properties, ettringite formation, and potential application in solidification and stabilization of selenium and boron. Cement and Concrete Research, 25, 658-670.

Sarott, F.A., Bradbury, M.H., Pandolfo, P., Spieler, P. (1992). Diffusion and adsorption studies on hardened cement paste and the effect of carbonation on diffusion rates. Cement and Concrete Research, 22, 439-444.

Tallent, O.K., McDaniel, E.W., Del Cul, G.D., Dodson, K.E., Trotter, D.R. (1987). Immobilization of technetium and nitrate in cement-based materials. MRS Proceedings, 112, 23-32.

Tanabe, H., Sakuragi, T., Yamaguchi, K., Sato, T., Owada, H. (2010). Development of new waste forms to immobilize iodine-129 released from a spent fuel reprocessing plant. Advances in Science and Technology, 73, 158-170.

Thoenen, T., Hummel, W., Berner, U., Curti, E. (2014). The PSI/Nagra Chemical Thermodynamic Databse 12/07. PSI Report, 14-04.

Tits, J., Kamei, G., Iijima, K., Wieland, E. (2006). The uptake of radium by calcium silicate hydrates and hardened cement paste. Radiochimica Acta, 38, 637-643.

van Es, E., Hinchliff, J., Felipe-Sotelo, M., Milodowski, A.E., Field, L.P., Evans, N.D.M., Read, D. (2015). Retention of chlorine-36 by a cementitious backfill. Mineralogical Magazine, 79, 1297-1305.

Modelling of radionuclides migration in the low pH cement / clay interface

Vanessa Montoya[1], Naila Ait Mouheb[1], Thorsten Schäfer[1,2]

[1] KIT-INE, Karlsruhe Institute of Technology, Institute for Nuclear Waste Disposal (DE)
[2] Friedrich-Schiller-Universität Jena, Institute of Geosciences (DE)

* Corresponding author: vanessa.montoya@kit.edu

Abstract

Reactive transport modelling activities described in this work are focused on the definition of the conceptual model and the selection of the chemical and transport parameters to be applied on laboratory through diffusion experiments performed in KIT-INE (see contribution of Ait Mouheb et al., *this proceedings*). The model includes different coupled processes which are thought to play a role in a through diffusion experiment of HTO, $^{36}Cl^-$, $^{129}I^-$ and Be across the interface between bentonite porewater and low pH cement. One of these processes is the impact of porosity changes due to dissolution / precipitation reactions.

Introduction

Current deep geological repository concepts are based on the confinement of radioactive waste over a long period of time by multiple barriers. Many of the concepts developed internationally use concrete and clay as confinement barriers (ANDRA, 2005; ENRESA, 1995). During the prolonged post disposal period, cementitious materials can undergo alterations through evolution of the chemical and physical properties of the repository. One important process that could reduce the durability of a concrete barrier is the leaching / degradation of the solid due to the clay porewater in direct contact with the cementitious material.

In order to minimize the interaction between the "classical" high-pH cement materials and the bentonite, low-pH cements were developed within the nuclear waste disposal context in the late 90's. The main feature of these cements is the less alkaline pore solution (pH ~ 11) compared to what is observed in "classical" cements and the absence of portlandite as hydrated solid phase. The Ca/Si ratio in the hydrated low-pH cement paste is less than 1.0 and C-S-H phases are the main solids present.

The understanding of the evolution of the interface low pH cements / bentonite over geological time scales and the impact on radionuclide migration requires a detailed knowledge of a series of highly complex coupled processes. The conceptual understanding of the different processes and their integration in numerical models will increase the reliability and predictive capabilities of the simulations. By using small-scale laboratory experiments, under well-defined boundary conditions, reactive transport modelling can be used for their interpretation and provide valuable information for the repository design and predict radionuclide migration. Studies concerning the evolution of low-pH cementitious materials in clayey environment are scarce and the knowledge of their physicochemical evolution is very limited (Dauzeres et al., 2016). To the best of our knowledge the impact of these processes in radionuclide migration has never been studied.

This work includes modelling studies of interface processes between a low pH cement (50% CEM I 52.5N + 50 silica fume) and a MX-80 bentonite porewater and assesses the specific impact on HTO, $^{36}Cl^-$, $^{129}I^-$ and Be migration. The modelling work supports the interpretation of the experimental results performed in the CEBAMA project and helps to identify possible missing parameters and to the process understanding. The processes considered are diffusion transport in saturated media under isothermal conditions, sorption reactions, and both local equilibrium and kinetically controlled mineral dissolution-precipitation reactions. Update of the effective diffusion coefficient as a function of porosity changes is also considered. The reactive transport code where the model is implemented is PHREEQC v.3 (Parkhurst and Appelo, 2013).

Conceptual model and selected parameters

The reactive transport model consists of a one dimensional (1D) fully water saturated isothermal (298 K) problem representing a laboratory through diffusion experiment of HTO, Be, $^{36}Cl^-$, and $^{129}I^-$ across the interface bentonite porewater / low pH cement. Geometrical and transport parameters including the discretization of the system (water and solid domains) have been initially implemented (see Figure 1). The mesh size and the time steps have been selected to ensure a satisfactory compromise between computation time and sufficient spatial resolution of the expected geochemical and transport processes, especially at the interface between the bentonite porewater and the low pH cement hydrated phases. A constant concentration and closed boundary condition have been imposed on the extremities of the upstream and downstream reservoir, respectively in order to reproduce the boundary conditions imposed in the experiments.

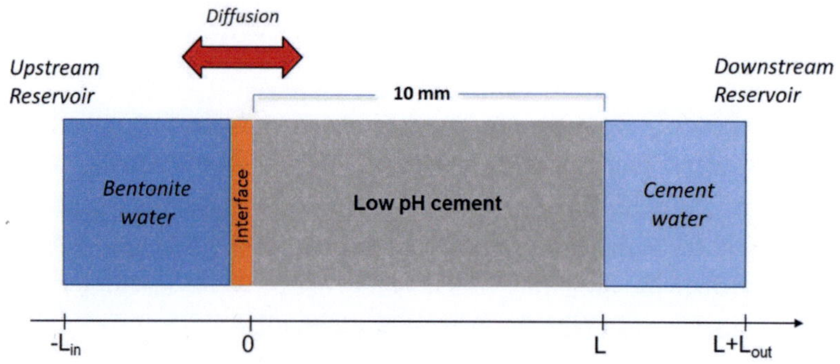

Figure 1: *Schematic representation of the through diffusion experiments.*

Considering the very low permeability of cement (10^{-10} - 10^{-14} m/s), mass transport will be considered diffusion-driven only, following Fick's law (Eq. 1).

$$J = -D\frac{\partial C}{\partial x} \qquad \textit{Eq. 1}$$

where J is the substance flux (kg/m^2·s); $\frac{\partial C}{\partial x}$ is the concentration gradient (kg/m^4); and D is the diffusion coefficient (m^2/s). In the absence of experimental data available concerning the effective diffusion in low-pH cements, a value of 10^{-10} m^2/s was tentatively selected as initial value taking into account the properties of the studied cement (Ait Mouheb et al., 2017). The effective diffusion coefficient evolution was then related to the porosity change according to Archie's law (Eq. 2) assuming n = 1.

$$D = D_w(porosity)^n \qquad \textbf{Eq. 2}$$

Initial porosity of 35% has been determined experimentally by mercury intrusion porosimetry (MIP). Additionally, porosity changes due to mineral precipitation/dissolution and feedback on the effective diffusion coefficient are also taken into account in the model considering the molar volumes of the different solids formed or dissolved. Electrostatic surface interactions are not included, although they may influence the transport of the anionic tracers $^{36}Cl^-$, $^{129}I^-$ and $Be(OH)_3^-$ (Chagneau et al., 2015).

The system studied is implemented in the code PHREEQC v.3 (Parkhurst and Appelo, 2013) which can take into account geochemical and physical parameters variations due to mineralogical evolutions as a function of time. The initial mineralogical composition of the cement hydrated phases considered is representative of a full hydrated low-pH cement (pH ~ 11.0) with a 93 wt.% being formed by C-S-H phases with a Ca/Si ratio of 0.8. Mineralogical composition has been determined experimentally (Ait Mouheb et al., 2017) with a combination of different techniques (X-ray diffraction (XRD), thermogravimetric - differential thermal analysis (TG-DTA), ^{29}Si and ^{27}Al Magic Angle Spinning Nuclear Magnetic Resonance (^{29}Si and ^{27}Al MAS NMR) and Scanning Electron Microscopy - Energy Dispersive X-ray spectroscopy (SEM-EDX) At this moment, it has not been possible to experimentally identify the solid phase containing iron and for this reason we have assumed that iron phases are in the form of Fe-ettringite. The initial porewater composition of the low-pH cement is defined in equilibrium with the hydrated solid phases present in the system and compares well with the measured concentrations. The porewater composition of the clay is representative of the MX-80 bentonite described in the literature (Wersin et al., 2003) and has been synthetized and measured in the laboratory.

Chemical reactions at equilibrium and kinetically controlled have been simulated using the thermodynamic database Cemdata07 available in PHREEQC format (Lothenbach and Winnefeld, 2006). Debye-Hückel equation, valid for the ionic strength of the studied system (< 0.3 M) is preferred to save computational time as it has been observed that calculations performed with SIT (Specific ion Interaction Theory) equation needs more computational effort. Cemdata07 includes hydrates commonly encountered in Portland cement systems in the temperature range 0 - 100°C. Rate equations of precipitation/dissolution of secondary / primary phases are provided directly in the input files of PHREEQC. Kinetic parameters for C-S-H phases and ettringite have been selected from Marty et al. (2015) and Baur et al. (2004).

In the present status, sorption of the tracers in the cement matrix is not included in the model. Sorption reactions of $^{36}Cl^-$, $^{129}I^-$ and Be tracers into the low pH cement matrix will be considered as a thermodynamic mechanistic sorption model to be implemented in PHREEQC. Observations described in the recent review of Ochs et al. (2016) will be considered to select the most appropriate data, as well as data generated in the Cebama project, consisting of batch experiments of the studied radionuclides on the same solid materials (Ait Mouheb et al., 2017).

Results and discussion

The simulations were carried out for different times of interaction (14 min, 5 and 30 days) predicting that the low-pH cement in contact with the bentonite pore water will have a degraded area of ~ 2 mm after one month of alteration. No significantly changes can be appreciated in the cement solid after 5 days (Figure 2). The alteration of the low-pH cement paste is mainly due to the partial dissolution of C-S-H phases controlled by kinetic rates resulting in an increase of the porosity due to the difference of the molar volumes between the initial solids and the final products of the alteration process in the interface. Carbonation is linked to the precipitation of calcite and thus the decalcification of C-S-H phases (Jenni et al., 2014). Magnesium enrichment in the decalcification area has been reported in the literature (Jenni et al., 2014; Dauzères et al., 2016). In our system, brucite is undersaturated and attempts to model the magnesium perturbation using available thermodynamic data suggested the formation of hydrotalcite. However, the recent determination of solubility data for M-S-H phases will make it possible to account for the potential formation of M-S-H phases (Dauzères et al., 2016) and will be considered in the future. Finally, no attempts have been done to model the iron evolution and the diffusion of HTO in the system will reach steady-state after 5 days of interaction.

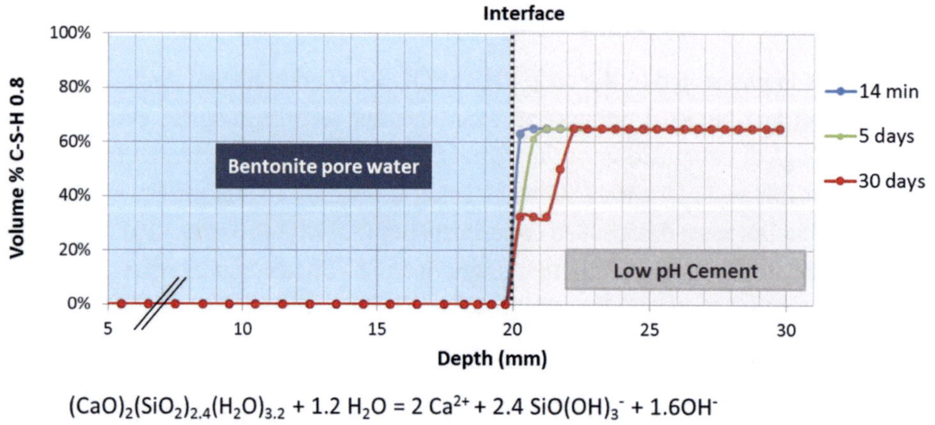

Figure 2: *Simulation of the low-pH cement alteration at different periods of time. The main process identified is the partial dissolution of C-S-H phases resulting in an increase in porosity at few millimeters close to the interface bentonite / cement.*

Conclusion and Future work

We have developed and implemented a 1D model to simulate the interaction between a low pH cement and bentonite porewater. The model considered a low-pH cement paste (pH ~ 11.0) synthesized in the CEBAMA project with a 93 wt.% of the initial mineralogy being C-S-H phases with a Ca:Si ratio of 0.8. The simulations predict the formation of a degraded area in the cement phase of ~ 2 mm after one month of alteration. The alteration of the low-pH cement paste is mainly due to the partial kinetic dissolution of C-S-H phases resulting in an increase of the porosity. The carbonation of the cement is linked to the precipitation of calcite and thus the additional decalcification of the C-S-H phases.

The model will be improved including the iron and M-S-H solid phases in the system and the sorption properties of the cement to the selected radionuclides and toxic elements. Additionally, the model will be implemented in the iCP interface (Nardi et al., 2014) and compared with the results obtained with PHREQC.

Acknowledgement

The research leading to these results has received funding from the European Union's Horizon 2020 Research and Training Programme of the European Atomic Energy Community (EURATOM) (H2020-NFRP-2014/2015) under grant agreement n° 662147 (CEBAMA).

Additionally, the authors would like to express their appreciation to Volker Metz for valuable comments.

References

Ait Mouheb, N., Montoya, V., Schild, D., Soballa, E., Adam, C., Geyer, F., Schäfer, T. (2017). Characterization and sorption properties of low pH cements. Proceedings of the 2nd CEBAMA Annual Workshop (*this document*).

ANDRA (2005). Andra Research on the Geological Disposal of Highlevel Long-lived Radioactive Waste - Results and Perspectives. June 2005.

Baur, I., Keller, P., Mavrocordatos, D., Wehrli, B., Johnson, C.A. (2004). Dissolution-precipitation behaviour of ettringite, monosulfate, and calcium silicate hydrate. Cement and Concrete Research, 34, 341-348.

Cau dit Coumes, C., Courtois, S., Nectoux, D., Leclercq, S., Bourbon, X. (2006). Formulating a low-alkalinity, high-resistance and low-heat concrete for radioactive waste repositories. Cement and Concrete Research, 36, 2152-2163.

Chagneau, A., Tournassat, C., Steefel, C.I., Bour, I.C., Gaboreau, S., Esteve, I., Kupcik, T., Claret, F., Schäfer, T. (2015). Complete Restriction of $^{36}Cl^{-}$ Diffusion by Celestite Precipitation in Densely Compacted Illite. Environmental Science & Technology Letters, 2, 139-143.

Dauzeres, A., Achiedo, G., Nied, D., Bernard, E., Alahrache, S., Lothenbach, B. (2016). Magnesium perturbation in low-pH concretes placed in clayey environment-solid characterizations and modelling. Cement and Concrete Research, 79, 137-150.

ENRESA (1995). Almacenamiento geológico profundo de residuos radiactivos de alta actividad (AGP). Diseños conceptuales genéricos. Publicación Técnica ENRESA, 11/95.

Jenni, A., Mäder, U., Lerouge, C., Gaboreau, S., Schwyn, B. (2014). *In-situ* interaction between different concretes and Opalinus Clay. Physics and Chemistry of the Earth, Parts A/B/C, 70-71, 71-83.

Lothenbach, B. and Winnefeld, F. (2006). Thermodynamic modelling of the hydration of Portland cement. Cement and Concrete Research, 36-2, 209-226.

Marty, N., Claret, F., Lassin, A., Tremosa, J., Blanc, P., Madé, B., Giffaut, E., Cochepin, B., Tournassat, C. (2015). A database of dissolution and precipitation rates for clay-rocks minerals. Applied Geochemistry, 55, 108-118.

Nardi, A., Idiart, A., Trinchero, P., de Vries, L.M., Molinero, J. (2014). Interface COMSOL-PHREEQC (iCP), an efficient numerical framework for the solution of coupled multiphysics and geochemistry. Computers & Geosciences, 69, 10-21.

Ochs, M., Mallants, D., Wang, L. (2016). Radionuclide and Metal Sorption on Cement and Concrete. Springer International Publishing.

Wersin, P. (2003). Geochemical modelling of bentonite porewater in high-level waste repositories. Journal of Contaminant Hydrology, 61, 405-422.

Chemo-mechanical modelling of calcium leaching experiments in cementitious materials

Andrés Idiart[1*], Emilie Coene[1], Marcelo Laviña[1]

[1] Amphos 21 Consulting S.L. (ES)

* Corresponding author: andres.idiart@amphos21.com

Abstract

In order to assess the long-term integrity and performance of cementitious barriers for their use in underground nuclear waste disposal, the evolution of their main physical properties over time needs to be well understood. These properties can be altered by chemical reactions as a result of interaction with clay or groundwater. The transport properties need to be estimated, as well as the mechanical long-term behaviour, to ensure the mechanical stability of the repository. This contribution presents the results of the modelling work aimed at simulating calcium leaching experiments on cement paste and concrete samples. It is well-known that this degradation process leads to a more porous microstructure, negatively impacting the transport and mechanical concrete properties. To study these coupled processes, a reactive transport model is coupled to a micromechanical model in order to simulate a set of existing accelerated calcium leaching experiments. The impact of calcium leaching on the mechanical and transport properties is explicitly considered in the model. Simulation results are compared to recent experimental data obtained from accelerated calcium leaching experiments, showing how chemically-induced damage impacts the physical properties of cement paste and concrete.

Introduction

Degradation of cementitious materials due to mineral dissolution and precipitation processes can significantly impact their physical properties and thus impair their barrier functions. One of the most important functions of cementitious materials to be used for disposal of radioactive waste is to limit and retard the ingress of deleterious species to the waste packages and the release of radionuclides in case of failure. Transport properties such as the diffusion coefficient and the permeability govern these processes. In addition, the mechanical stability and long-term performance of the cementitious barriers also needs to be assessed to guarantee a correct functioning.

One of the main degradation processes expected to play a role in deep geological repositories is calcium leaching from the solid matrix of cementitious materials due to chemical gradients between the barriers and the surrounding environment. Due to the fact that this process is very slow, calcium leaching experiments are typically performed using a more aggressive leaching solution than expected under repository conditions, e.g., using deionized water or different concentrations of ammonium nitrate (Adenot and Buil, 1992; Faucon et al., 1996).

The effect of calcium leaching on concrete properties is a result of the dissolution of a sequence of cement hydrates (portlandite, C-S-H gels, ettringite, etc.) leading to a more porous microstructure. Consequently, the diffusion coefficient and permeability can significantly increase (e.g., Camps, 2008). Moreover, the mechanical properties such as the Young's modulus and the tensile and compressive strengths can decrease substantially (Gérard et al., 1998; Nguyen et al., 2007; Heukamp et al., 2005).

Brief description of the experimental data

Recently, a set of calcium leaching experiments using OPC hardened cement paste and concrete specimens has been conducted at Chalmers University of Technology in Sweden (Babaahmadi, 2015; Babaahmadi et al., 2015). In their setup, a 0.3 M ammonium nitrate solution is used coupled to electrical migration (with a constant current of 0.25 A) through cement paste and concrete samples. In this way, the leaching process is substantially accelerated, enabling the use of relatively large specimens which can then be used for macroscopic characterization tests (permeability, diffusivity and mechanical properties).

The characterization of the experiments was performed over the reference (i.e., intact) and aged (i.e., leached) states of the same samples. It included microstructural aspects such as porosity and pore size distribution by MIP and gas sorption, as well as macroscopic mechanical (tensile and compressive strength, elastic modulus), hydraulic (gas and water permeability, water sorptivity) and (reactive) transport properties (diffusion coefficients of cations and chloride, sorption properties, qualitative mineral phase assemblages, and profiles of the Ca/Si ratio in the solid phase). Table 1 summarizes the characterization techniques. The accelerated leaching experiments were conducted until reaching a point where portlandite has been entirely depleted from the sample, while only partial degradation of the C-S-H gels was observed (change in Ca/Si ratios using LA-ICP-MS). All the specimens were casted with a diameter of 50 mm and a length of 75 mm, except the concrete specimens used for measuring the tensile strength. The cement used is an OPC CEM I 42.5N MH/SR3/LA mixed with different water-to-cement ratios close to 0.5.

Table 1: *Characterization techniques used in the experiments by Babaahmadi et al. (2015).*

	Chemical properties				**Transport properties**						**Mechanical properties**			
Specimen type	IC & PT	XRD	LA-ICP-MS	SEM-EDX	MIP & Gas Sorption	Freezable Water	Diffusion Cell Tests	Diffusion Adsorption Test	Gas Permeability	Water Permeability	Tensile Strength	Compressive Strength	Elastic Modulus	Resonant Frequency
hcp	✓	✓	✓	✓	✓	✓	✓	✓	✗	✗	✗	✓	✗	✓
Mortar	✓	✓	✓	✓	✓	✓	✗	✗	✗	✗	✗	✓	✗	✓
Concrete	✓	✓	✓	✓	✓	✗	✗	✗	✗	✓	✗	✓	✓	✓
	✓	✓	✓	✓	✓	✗	✗	✗	✓	✗	✓	✗	✗	✗

hcp = hardened cement paste; PT = Potentiometric Titration

Multi-scale chemo-mechanical model

A multi-scale chemo-mechanical model is implemented in iCP (interface Comsol-Phreeqc, Nardi et al., 2014) at the continuum scale, where the cementitious material is regarded as homogeneous. The model encompasses a classical reactive transport framework based on (1) solute transport and (2) chemical reactions, with (3) a non-linear mechanical model of the cementitious system, and (4) a multi-scale micromechanical model. The latter is used to describe the coupling between changes in the microstructure of cement paste and concrete (due to chemically-driven degradation) and the mechanical properties. The micromechanical model computes the elastic constants as a function of the volume fractions of all the minerals in concrete and is used to calculate the isotropic chemical damage, dc (between 0 and 1). The non-linear mechanical constitutive law is based on Mazars' damage model (Mazars, 1986), regularized following an implicit gradient formulation (Simone, 2007), and is used to calculate the isotropic mechanical damage, dm (between 0 and 1). Both damage variables are scalar and coupling of chemical and mechanical damage is considered to be multiplicative (see e.g., Gérard et al., 1998). For these experiments, solute transport is limited to Fickian diffusion (i.e., using a single diffusion coefficient for every species to maintain electroneutrality of the pore solution), although both the diffusion coefficient and permeability are updated as a function of porosity in the model. Aqueous speciation and mineral dissolution / precipitation are calculated at chemical equilibrium using the PhreeqC format of the thermodynamic database CEMDATA07 (Lothenbach et al., 2008; Jacques, 2009). Changes in porosity due to chemical reactions are explicitly considered as well as the impact on transport properties.

The model developed in this work does not intend to reproduce the accelerated experiments in terms of the time needed to reach an aged sample. The goal has been to determine, by means of reactive transport modelling, the leaching acceleration factor obtained with the experiment when compared to the leaching of the same samples subjected to groundwater interaction. Furthermore, the outcome of the model in terms of chemical and physical properties of the aged samples can be compared to the experimental data to gain insight into the mechanisms of chemical degradation when using the accelerated leaching test.

Given that the experiments were accelerated by means of an imposed electrical current, a proper representation of this system should take into account the electrochemical migration effect, i.e., the Nernst-Planck equations, and adapted boundary conditions (e.g., Liu et al., 2012). This is not standard in currently available reactive transport codes and is not addressed in this study.

The multi-scale micro-mechanical model proposed by Stora et al. (2009) and Bary et al. (2014) is implemented in iCP. It uses homogenization theory to describe the effective properties at multiple scales based on structural information at a lower scale and the volume fractions of the different constituents. Four different levels are defined, ranging from a few microns for low and high density C-S-H structures to the cm-scale for concrete (Table 2). Three homogenization schemes are used to better adapt the modelled microstructure at a given scale to experimental observations.

Table 2: *Summary of multi-scale micromechanical model based on Stora et al. (2009).*

Level	Homogenized domain	RVE size	Microstructure	Homogenization scheme	Components
I	High Density C-S-H (HD)	10 µm	matrix-inclusion type	Interaction Direct Derivative (IDD)	Porous C-S-H gel and other hydrates
I'	Low Density C-S-H (LD)	10 µm	matrix-inclusion type	Interaction Direct Derivative (IDD)	Porous C-S-H gel, capillary pores and other hydrates
II	Hardened cement paste (hcp)	100 µm	doubled coated spheres	Generalized Self-Consistent Scheme (GSCS)	Unhydrated clinker, HD & LD C-S-H
III	Mortar	mm-scale	doubled coated spheres	Generalized Self-Consistent Scheme (GSCS)	hcp, ITZ and sand particles
IV	Concrete	cm-scale	matrix-inclusion type	Mori Tanaka (MT)	mortar and larger aggregates

RVE: representative volume element; ITZ: interfacial transition zone

Input data and results

Diffusion-driven leaching of cement paste (initial effective diffusivity of $1.3 \cdot 10^{-11}$ m²/s) samples is simulated in a 2D axisymmetric setup using a FE mesh with triangular elements and a maximum element size of 2.5 mm. The mineralogical, porewater, and boundary groundwater compositions are depicted in Table 3. The evolution of the resulting Ca/Si ratio of the sample is compared to the one obtained experimentally in Figure 1. The depletion of portlandite as a function of time and the evolution of the physical properties over time at a point in the modelled domain are presented in Figure 2. Portlandite is completely depleted from the sample after 11,000 years in the model, while in the experiments it took only ~ 53 days (Babaahmadi et al., 2015), that is ~ 75,000 times slower than the accelerated leaching test, where ammonium nitrate and electrical migration are considered. It may be observed that the porosity increases substantially due to the leaching process even though with this groundwater composition significant calcite formation is expected in the long-term.

Table 3: *Composition of the cement paste, porewater and groundwater used in the simulations.*

Component	Groundwater (mol/L)	Porewater (mol/L)	Cement phases	Volume fraction (-)
pH	8.64	12.975	Portlandite	0.176
Al	$1.21 \cdot 10^{-6}$	$5.54 \cdot 10^{-7}$	C-S-H (jennite)	0.314
C	$6.91 \cdot 10^{-4}$	$1.00 \cdot 10^{-5}$	Ettringite	0.029
Ca	$5.26 \cdot 10^{-4}$	$3.68 \cdot 10^{-3}$	Hydrogarnet (Si)	0.050
Cl	$4.54 \cdot 10^{-3}$	$5.55 \cdot 10^{-5}$	Hydrotalcite (OH)	0.012
K	$7.60 \cdot 10^{-5}$	$8.84 \cdot 10^{-2}$	Thaumasite	0.039
Mg	$1.48 \cdot 10^{-4}$	$2.53 \cdot 10^{-8}$	Unhydrated clinker*	0.014
Na	$4.79 \cdot 10^{-3}$	$2.70 \cdot 10^{-2}$	Initial porosity	0.366
S(VI)	$3.73 \cdot 10^{-4}$	$3.29 \cdot 10^{-3}$		
Si	$1.42 \cdot 10^{-4}$	$4.52 \cdot 10^{-5}$		

*considered unreactive

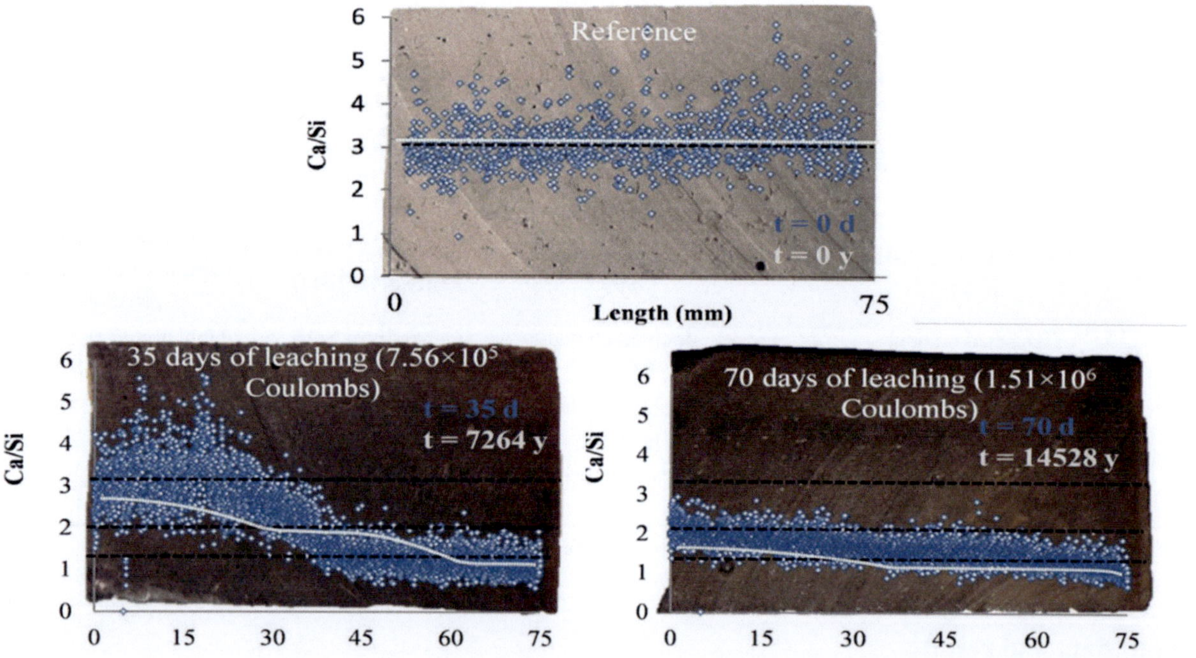

Figure 1: *Calcium-to-Silicon ratio profiles in the cement paste specimens obtained at different times from laboratory experiments (blue dots, Babaahmadi, 2015) and the chemo-mechanical model (grey lines). Times in blue and light grey correspond to experimental data and model results, respectively. Horizontal dimension is given in mm.*

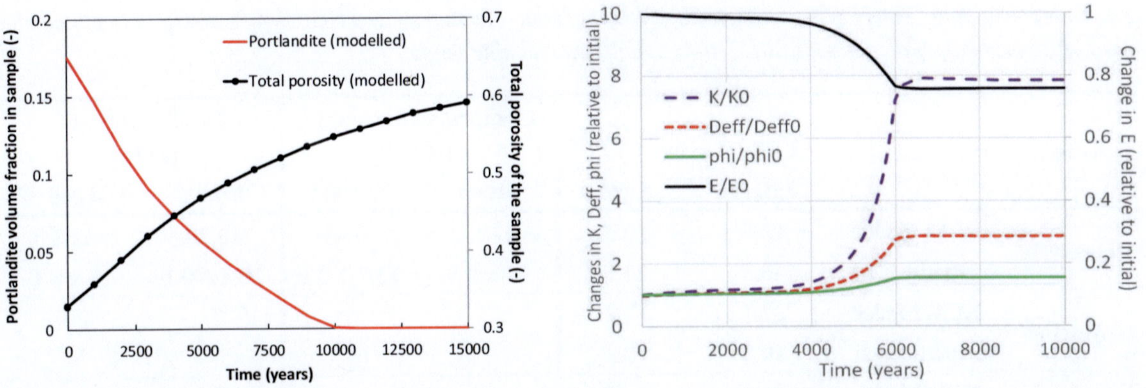

Figure 2: Total portlandite volume fraction and total porosity in the modelled sample (left) and physical properties at 16 mm from leaching surface (right) as a function of time.

The same simulation has been performed for a concrete sample with an initial effective diffusivity of $2.8 \cdot 10^{-12}$ m^2/s. The concrete consists of the hardened cement paste described in Table 3 with 34.6 vol.% of large aggregates and 34.7 vol.% of sand. Figure 3 displays the time evolution of porosity and mineral volume fractions and of the Young's modulus at the centre of the sample obtained with the chemo-mechanical model. The mechanical properties of the concrete are greatly affected by portlandite and ettringite dissolution and the consequent increase in porosity. In a second stage of degradation, C-S-H is gradually decalcified (from jennite-like to tobermorite-like C-S-H), leading to a further increase in porosity and reduction of the Young's modulus. The initial and aged Young's modulus are compared to the experimental values in Table 4. Furthermore, the aged-to-intact ratio of effective diffusivity obtained for concrete is 4.05, whereas the experimental value is 3.94.

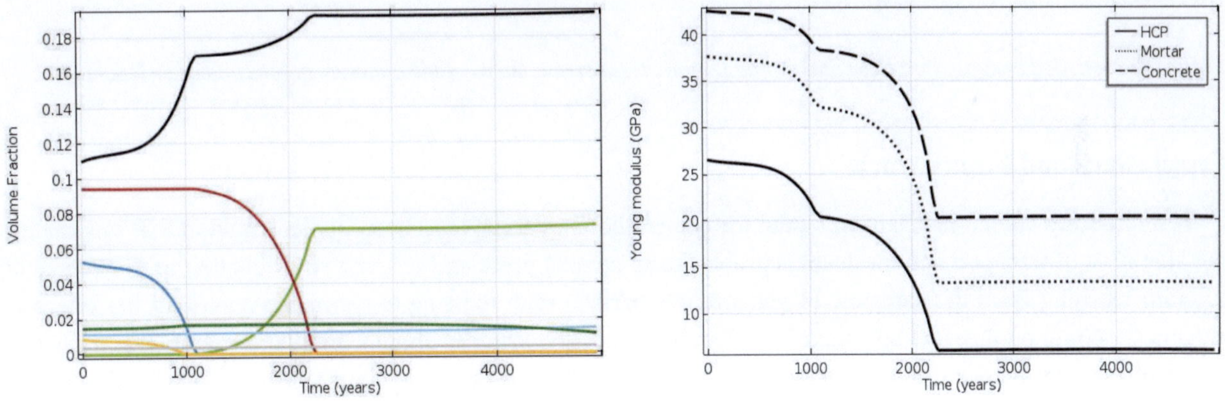

Figure 3: Time evolution at the centre of the concrete specimen according to the chemo-mechanical model: (left) porosity and mineral volume fractions and (right) Young's modulus (GPa) of the concrete sample (at the concrete, mortar, and HCP levels).

Finally, the regularized mechanical damage model of Mazars has been used to determine the compressive and tensile strength of intact and aged concrete specimens. The model parameters are fitted with the initial compressive and tensile strengths of concrete measured by Babaahmadi (2015). These are: $\varepsilon_0 = 1.24 \cdot 10^{-4}$, $A_t = 1.36$, $B_t = 2{,}135$, $A_c = 0.95$, and $B_c = 2{,}200$. The compressive and tensile stress-strain curves are shown in Figure 4. Comparison with experimental data is presented in Table 4 in terms of strength.

Table 4: *Young's modulus, compressive and tensile strength of concrete, mortar and HCP before and after leaching obtained by means of experiments (Babaahmadi, 2015) and the calcium leaching model.*

Sample		Young's modulus (GPa)		Compressive strength (MPa)		Tensile strength (MPa)	
		Intact	**Aged**	**Intact**	**Aged**	**Intact**	**Aged**
Concrete	HCM Model	42.6	20.3	43.17	16.98	10.20	4.37
	Experimental	46.5	24.1	43 ± 0.3	12.7 ± 0.1	10.2 ± 0.1	2.8 ± 0.1
Mortar	HCM Model	37.7	13.2				
	Experimental	30.4	15.2				
HCP	HCM Model	26.6	5.95				
	Experimental	11.4	6.24				

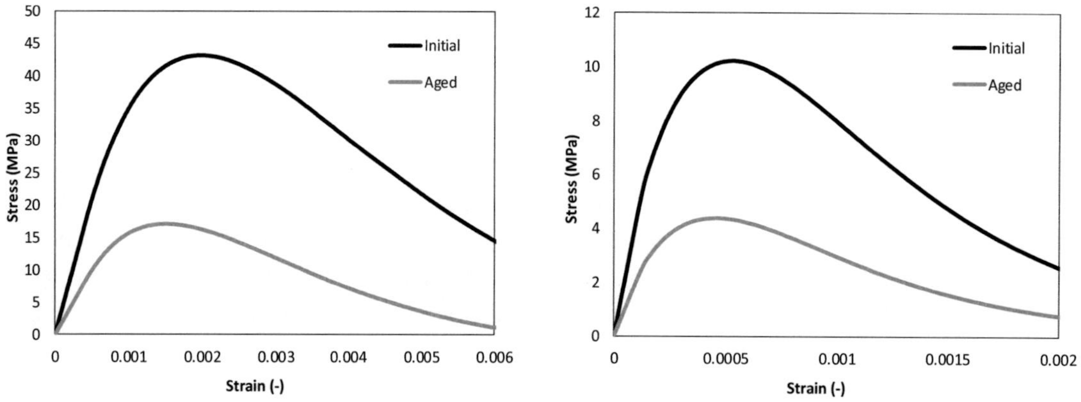

Figure 4: *Modelled compression (left) and tensile (right) stress-strain curve of the concrete before and after leaching.*

Conclusions and Future work

A multi-scale chemo-mechanical model for cementitious materials has been implemented in iCP to simulate the alteration of transport and mechanical properties of cement paste and concrete as a function of the degree of calcium leaching from the samples. In general, comparison with existing experiments regarding the effect of calcium leaching on physical and chemical properties of tested samples show a good agreement. The measured effect of calcium leaching on increased diffusivity of cement paste is also correctly captured by the model. Furthermore, the extension of homogenization methods to predict the transport properties such as diffusivity will also be explored (see for instance Stora et al., 2009) as well as the implementation of Nernst-Planck-Poisson equations to simulate electrochemical migration.

Acknowledgements

The research leading to these results has received funding from the European Atomic Energy Community's (Euratom) Horizon 2020 Programme (NFRP-2014/2015) under grant agreement 662147 – Cebama. Financial support from SKB is also gratefully acknowledged.

References

Adenot, F. and Buil, M. (1992). Modelling of the corrosion of the cement paste by deionized water. Cem. Conc. Res., 22, 489-496.

Babaahmadi, A. (2015). Durability of Cementitious Materials in Long-Term Contact with Water. PhD Thesis. Chalmers University of Technology, Gothenburg, Sweden.

Babaahmadi, A., Tang, L., Abbas, Z., Mårtensson, P. (2015). Physical and Mechanical Properties of Cementitious Specimens exposed to an electrochemically derived accelerated leaching of calcium. Int. J. Conc. Struct. Mater., 9(3), 295-306.

Bary, B., Leterrier, N., Deville, E., Le Bescop, P. (2014). Coupled chemo-transport-mechanical modelling and numerical simulation of external sulfate attack in mortar. Cem. Conc. Comp., 49, 70-83.

Camps, G. (2008). Etude des interactions chemo-mécaniques pour la simulation du cycle de vie d'un élément de stockage en béton. PhD Thesis. Paul Sabatier Univ., Toulouse, France.

Faucon, P., Le Bescop, P., Adenot, F., Bonville, P., Jacquinot, J.F., Pineau, F., Felix, B. (1996). Leaching of cement: Study of the surface layer. Cem. Conc. Res., 26, 1707-1715.

Gérard, B., Pijaudier-Cabot, G., Laborderie, C. (1998). Coupling diffusion-damage modelling and the implications on failure due to strain localisation. Int. J. Solids Struct., 35, 4107-4120.

Heukamp, F.H., Ulm, F.-J., Germaine, J.T. (2005). Does calcium leaching increase ductility of cementitious materials? Evidence from direct tensile tests. J. Mater. Civ. Eng., 17, 307-312.

Jacques, D. (2009). Benchmarking of the cement model and detrimental chemical reactions including temperature dependent parameters. Project near surface disposal of category A waste at Dessel. NIRAS-MP5-03 DATA-LT(NF) Version 1. NIROND-TR 2008-30 E.

Liu, Q.F., Li, L.Y., Easterbrook, D., Yang, J. (2012). Multi-phase modelling of ionic transport in concrete when subjected to an externally applied electric field. Engng. Struct., 42, 201-213.

Lothenbach, B., Le Saout, G., Gallucci, E., Scrivener, K. (2008). Influence of limestone on the hydration of Portland cements. Cem. Conc. Res., 38, 848-860.

Mazars, J. (1986). A description of micro- and macroscale damage of concrete structures. Eng. Fract. Mech., 25(5-6), 729-737.

Nardi, A., Idiart, A., Trinchero, P., de Vries, L.M., Molinero, J. (2014). Interface COMSOL-PHREEQC (iCP), an efficient numerical framework for the solution of coupled multiphysics and geochemistry. Comput. Geosci., 69, 10-21.

Nguyen, V.H., Nedjar, B., Torrenti, J.M. (2007). Chemo-mechanical coupling behaviour of leached concrete: Part II: Modelling. Nucl. Eng. Des., 237, 2090-2097.

Simone, A. (2007). Explicit and implicit gradient-enhanced damage models. Revue Européenne de Génie Civil, 11(7-8), 1023-1044.

Stora, E., Bary, B., He, Q.-C., Deville, E., Montarnal, P. (2009). Modelling and simulations of the chemo-mechanical behaviour of leached cement-based materials: Leaching process and induced loss of stiffness. Cem. Conc. Res., 39, 763-772.

Spectral induced polarization measurements of low-pH concrete

Johan Alexander Huisman[1], Egon Zimmermann[1], Stéphane Gaboreau[2], Philippe Leroy[2*], Francis Claret[2], Christophe Tournassat[2], Johannes Lützenkirchen[3]

[1] Forschungszentrum Jülich (DE)
[2] BRGM (FR)
[3] KIT (DE)

* Corresponding author: p.leroy@brgm.fr

Abstract

Low-pH concretes are foreseen to be used in nuclear waste disposal underground. Their confining properties depend on their interactions with the host-rock. Evolution of mineralogy, porosity, pore size distribution and connectivity can be monitored in situ using geophysical methods such as Induced Polarization (IP). This electrical method consists of injecting an alternating current and measuring the resulting voltage in the porous medium of interest. The measured complex resistivity depends on the water chemical composition but also on the geometry and morphology of the pore network. In this study, Spectral Induced Polarization (SIP) measurements on low-pH concrete and cement paste are presented. These measurements will be interpreted in terms of mineralogy and pore size distribution. The SIP measurements of the cementitious samples perturbed by a CO_2 enriched solution over a period of one year and their interpretation in terms of changes of mineralogy and microstructure will be presented in the next report.

Introduction

The low-pH concrete ($10 \leq pH \leq 11$) is a material under investigation for the storage of high-level nuclear wastes (HLW) in deep geological formations because of its low permeability, high chemical stability (towards dissolution / precipitation reactions) and less impact on the performance of bentonite compared to medium and high-pH concretes (Calvo et al., 2010). Despite its relatively high chemical stability, changes of the mineralogy and microstructure are observed when the material is in contact with the clay materials and steel reinforcement (Berner et al., 2013; Calvo et al., 2013). These changes may modify the porosity, permeability and the confining properties of the concrete (Berner et al., 2013).

Hydraulic and chemical analyses of concretes are local, invasive and may be time-consuming and expensive. Geophysical methods and in particular electrical methods are non-intrusive and may provide precious information on the evolution of the mineralogy and microstructure of low-pH concretes in situ. Amongst electrical methods, there are the Self-Potential (SP), Electrical Resistivity Tomography (ERT), and Induced Polarization (IP) methods (Binley et al., 2015). IP is an active method involving the injection of an alternating current into the porous medium in order to displace the mobile charges in bulk water and at the mineral / water interface. During IP, the voltage resulting from conduction and polarization currents is measured (Kemna et al., 2012). IP is an extension of the classical resistivity method that includes the imaginary resistivity. Spectral Induced Polarization (SIP) is referred to induced polarization measurements in the frequency domain when a sinusoidal current in the mHz to kHz frequency range is injected in the porous medium.

We propose, for the CEBAMA project, spectral induced polarization measurements of low-pH concrete and cement paste at equilibrium and in contact with an enriched CO_2 solution. These measurements are carried out during a period of one year at the IBG 3 institute of the Forschungszentrum Jülich to monitor chemical, mineralogical and microstructural changes at the concrete and cement/water interface. We expect to observe the geophysical signature of dissolution and precipitation reactions associated with leaching and carbonation in concrete and cement paste, and their effect on the pore size distribution and connectivity.

Spectral Induced Polarization (SIP) system

The complex conductivity of low-pH concrete and cement paste samples is currently investigated (frequency range from 10 mHz to 10 kHz):

- water saturated low-pH concrete (containing ~ 70% by mass of coarse carbonates and silica aggregates, ~ 30% by mass of alpha C-S-H ($0.75 \leq$ Ca/Si ratio ≤ 1) (Haas and Nonat, 2015), ~ 12% porosity);
- saturated cement paste (containing mostly alpha C-S-H, ~ 40% porosity).

The SIP measurements are carried out using a novel sample holder design developed within the CEBAMA project and the ZEL-SIP-04 system built at Forschungszentrum Jülich GmbH including a function generator (Agilent 33120A), a measurement amplifier, an ADC card, and a PC-based measurement control system (Figure 1) (Zimmermann et al., 2008).

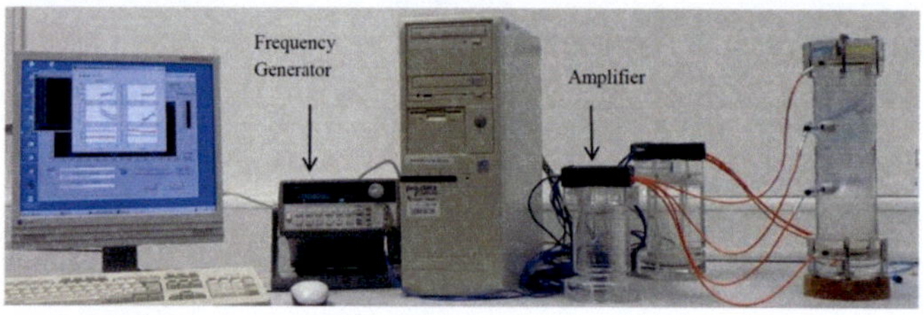

Figure 1: *Photo of the ZEL-SIP-04 system for SIP measurements.*

The dimensions of the cylindrical samples are:

- 5 cm diameter and 20 cm length for the concrete (Figure 2),
- 5 cm diameter and 10 cm length for the shorter cement paste.

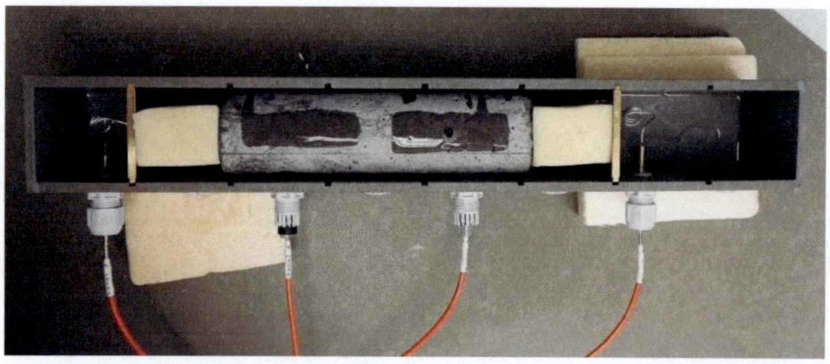

Figure 2: *Sample holder for SIP measurements on concrete and the four-electrode set-up for current (outer electrodes) and electrical potential (inner electrodes) measurements.*

Test measurements on a reference electrical circuit

Preliminary measurements indicated the high resistivity of both the concrete and cement paste (CEBAMA reference mix designs provided by VTT). Measured impedance was in the range of 1 MΩ, a range where the SIP measurement system has not been properly tested before. Therefore, initial measurements were performed on a reference electrical circuit that mimics the properties of the measurement set-up for the concrete and cement paste samples. A sample resistance of 909 kΩ with a nominal accuracy of 1% was selected and realistic contact impedances for both the current and potential electrodes were used. The uncorrected SIP measurement in Figure 3 (red line) shows that the measured impedance magnitude corresponds well with the expected resistance of 909 kΩ. The zoom provided in the right column shows that measured impedance falls within the reported accuracy of the resistance used within the reference electrical circuit. The measured phase shift (between injected current and resulting voltage) starts to deviate from zero at approximately 100 Hz. This is due to measurement errors associated with the contact impedance of the potential electrodes and remaining leaking currents in the SIP system and experimental set-up (Huisman et al., 2016). At 1 kHz, the measurement error amounts to 30 mrad. After correction for the contact impedance of the potential electrodes using the method outlined in Huisman et al. (2016), this error is reduced to 10 mrad (Figure 3). The close-up provided on the right of Figure 3 indicates that the measurement error for the phase shift is approximately 1 mrad for frequencies below 100 Hz. Of course, these errors are much higher than previously reported errors for the SIP measurement system. This is simply related to the challenges associated with making accurate measurements on highly resistive samples. This test measurement will be used in the following to evaluate the significance of the SIP measurements on cement paste and concrete samples.

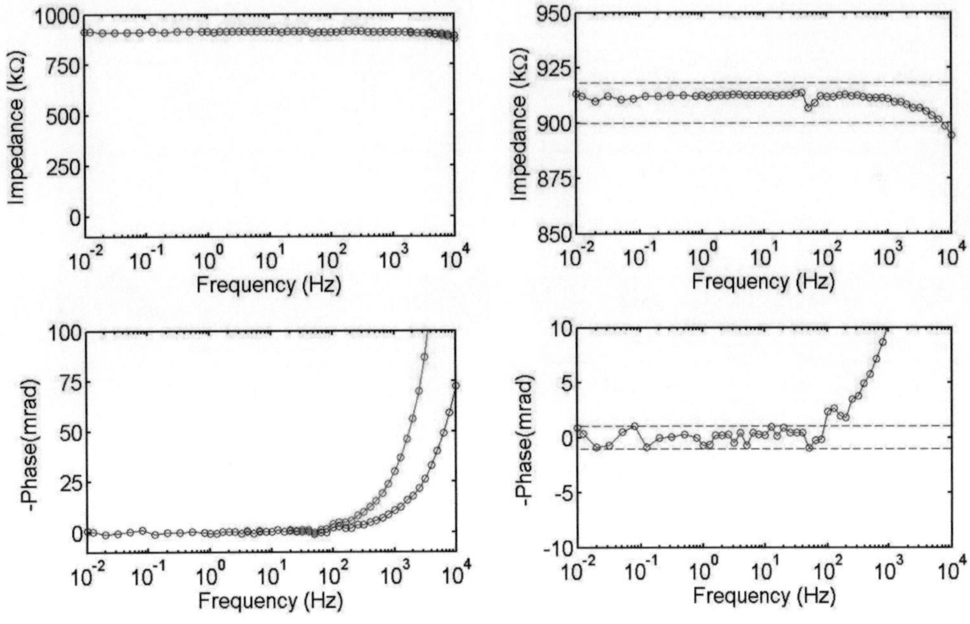

Figure 3: Uncorrected (red) and corrected (blue) SIP measurements on a reference electrical circuit with a sample resistance of 909 kΩ and realistic contact impedances for the current and potential electrodes. Ideally, the magnitude of the measured impedance should be 909 kΩ with zero phase shift. The right column provides a close-up of the data provided in the left column. It additionally indicates the 1% accuracy of the used resistance in the reference electrical circuit and a phase shift accuracy of 1 mrad for frequencies below 100 Hz.

SIP measurements on cement paste and concrete at equilibrium

SIP measurements on cement paste and concrete specimens were made on four different days (end of January 2017). The specimens were removed from the sample holder after each measurement, and placed back in the equilibrium water provided with the samples by BRGM. The results of these measurements are provided in Figure 4. It can be seen that the four repeat measurements agreed very well, in particular for the measured phase shift below 1 kHz. The magnitude of the measured impedance is somewhat more variable, which is at least partly related to small differences in the position of the potential electrodes between measurements that affect the geometrical factor used to obtain complex resistivity from the measured impedance. The measured phase shift above 1 kHz is also more variable, and this is attributed to poorly understood electrode polarization processes at the potential electrodes. A comparison with the SIP measurements on the reference electrical circuit shows that the observed phase shifts for cement paste and concrete are considerably above the estimated measurement error. Polarization of C-S-H is probably responsible for these measured phase shifts that are particularly high (≥ 100 mrad) for frequencies > 100 Hz. This result is not surprising because of the nanometric pores of C-S-H (Gaboreau et al., 2012) that may be responsible for the measured high relaxation frequencies (small relaxation times).

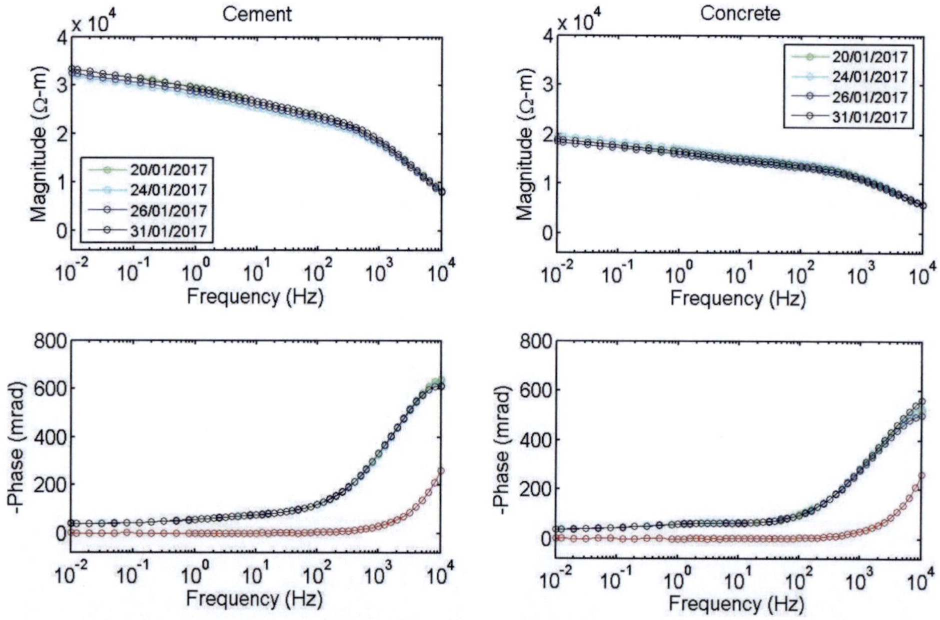

Figure 4: *Magnitude (top) and phase shift (bottom) of four repeat SIP measurements on cement (left) and concrete (right). The red lines indicate the approximated measurement error obtained from the SIP measurement on a reference electrical circuit presented in Figure 3.*

The SIP measurements indicate that the magnitude of the electrical resistivity is higher for cement than for concrete, despite the much higher porosity of the cement. Therefore, the concrete is more conductive than the cement paste. This is an unexpected result that needs explanation. The higher ion concentration and/or diffusivity in the concrete than in the cement paste may explain this observation (Larbi et al., 2016). Another explanation may be the increase of the surface conductivity of the C-S-H with the compaction.

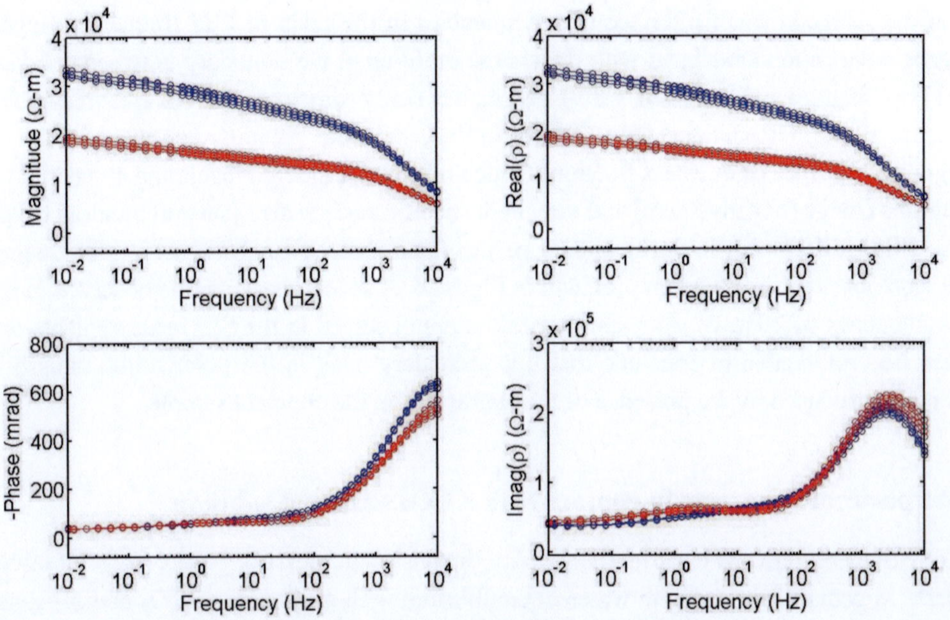

Figure 5: *(left) Comparison of magnitude (top) and phase shift (bottom) of four repeated SIP measurements on cement (blue) and concrete (red). (right) Comparison of real part (top) and imaginary part (bottom) of the complex electrical resistivity for four repeat SIP measurements on cement (blue) and concrete (red).*

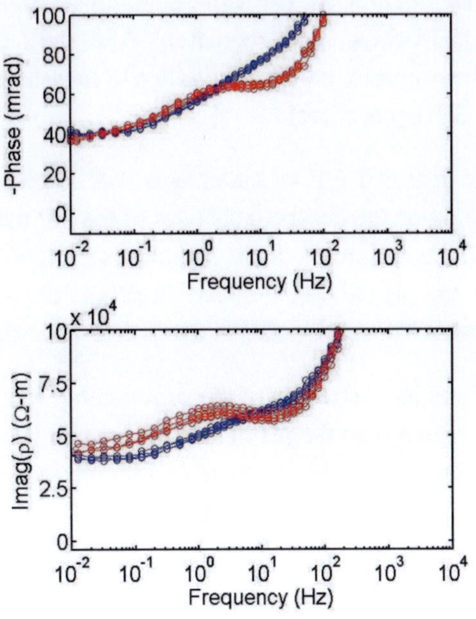

Figure 6: *Zoom of the results presented in Figure 5 for the phase shift and imaginary resistivity (blue: cement paste; red: concrete).*

The measured phase shift for cement and concrete is quite similar, in particular for frequencies below 100 Hz (Figure 5). Since the interpretation of the measured phase shifts is partly complicated by changes in both conduction and polarization (phase shift ≅ polarization / conduction), Figure 5 also provides a comparison between the real (conduction) and imaginary (polarization) part of the complex resistivity. A peak in the imaginary part of the electrical resistivity is observed for a frequency of 3 kHz. The polarization of the concrete is also slightly higher than the polarization of the cement paste. Polarization is due to the polarization of the electrical double layer at the mineral / water interface (for frequencies in the mHz to kHz frequency range) and to the Maxwell-Wagner polarization associated with the charge build-up at the boundary between solid and water for frequencies ≥ kHz (Chelidze and Gueguen, 1999). Again, it is really surprising that concrete may polarize slightly more than cement paste that contains only C-S-H. Indeed, concrete contains a large fraction of silica and carbonates aggregates (~ 70% of the mass fraction), which may be considered insulating compared to C-S-H that carry a high surface charge (negative here) and very high specific surface area (several hundred m^2/g) (Labbez et al., 2006). In addition, differences in the imaginary part of the resistivity can be observed for frequencies below 1 kHz. This is more apparent in the zoom provided in Figure 6. A second peak can be observed in the imaginary resistivity at a frequency of 2 Hz. Since a similar peak is not observed in the SIP measurements on the cement paste specimen, it is reasonable to conclude that this secondary peak in the polarization is associated with a polarization length introduced by the presence of the aggregates in the concrete sample.

SIP of cement paste and concrete in contact with a CO_2 enriched solution

A new set of SIP measurements is currently ongoing. It consists in measuring the voltage of low-pH concrete and cement paste in contact with a dilute water in equilibrium with a mixture of CO_2 and N_2 gases (1% CO_2, 99% N_2). A salinity of 10^{-2} mol/L (M) NaCl has been chosen for the starting condition but one should keep in mind that the concrete dissolution will impose its salinity to the solution. Such a salinity of 10^{-2} M corresponds to the averaged ionic strength of clay rocks in the nuclear waste storage of Mol in Belgium (Beaucaire et al., 2000). The low-pH concrete and cement paste are brought into a cylindrical box of dimensions 30 cm long and 10 cm diameter and are in contact with the saline solution. The saline solution will modify the chemistry and mineralogy of the concrete and cement paste (e.g., leaching and carbonation). After the surrounding water in the box reaches equilibrium with the concrete or cement paste, monitoring with SIP measurements will be performed during a short period (e.g., few days of daily SIP experiment).

The SIP experiment will last one year and SIP measurements will be performed at different times since the beginning of the experiment, depending on the observed changes of the SIP measurement. Post-mortem analyses (mineralogy, petrography, etc…) will be performed on the samples at BRGM. For each SIP measurement, water electrical conductivity, temperature, and pH will be measured directly in the box or by extracting water from the box. Additional chemical measurements can also be carried out to determine the water chemistry.

We expect an increase of the pH and ionic strength of the surrounding water in the box because the pH of the concrete or cement paste is initially higher than the pH of the surrounding water and of the resulting dissolution of some minerals (leaching) in the concrete or cement paste (at the interface between concrete or cement paste and external water). We also expect carbonation at the surface of the samples. These phenomena may modify their measured complex resistivity.

Conclusions and Future work

Spectral induced polarization (SIP) measurements on low-pH concrete and cement paste (in the frequency range 10 mHz to 10 kHz) are currently being performed to monitor the evolution of their mineralogy and microstructure. The concrete and cement paste are very resistive (measured impedance in the range of 1 MΩ) and exhibit high phase shifts (magnitude > 40 mrad) between injected current and measured voltage that are not due to experimental artifacts. Nanoporous and highly charged C-S-H may contribute to the high phase shifts observed for frequencies > 100 Hz and C-S-H and silicates and carbonates aggregates to the lower phase shifts for frequencies < 100 Hz. SIP measurements on the cementitious materials in contact with a CO_2 enriched solution are currently being carried out and the first results of this experiment will be presented soon. We expect to observe the electrical response associated with the leaching and carbonation of the cementitious materials over the duration of the SIP experiment (one year) and notably the evolution of the porosity and pore size distribution. These observations will help the reactive transport models for predicting the long term performance of concretes for the storage of nuclear wastes. SIP measurements will be modelled by combining an electrostatic surface complexation model with an induced polarization model based on membrane polarization (Bucker and Hordt, 2013; Elakneswaran et al., 2009). Mineralogical and petrographic measurements will also be used to constrain the SIP model.

Acknowledgement

The research leading to these results has received funding from the European Union's European Atomic Energy Community's (Euratom) Horizon 2020 Programme (NFRP-2014/2015) under grant agreement, 662147 – Cebama.

References

Beaucaire, C., Pitsch, H., Toulhoat, P., Motellier, S., Louvat, D. (2000). Regional fluid characterisation and modelling of water-rock equilibria in the Boom clay Formation and in the Rupelian aquifer at Mol, Belgium. Appl. Geochem., 15, 667-686.

Berner, U., Kulik, D.A., Kosakowski, G. (2013). Geochemical impact of a low-pH cement liner on the near field of a repository for spent fuel and high-level radioactive waste. Phys. Chem. Earth., Parts A/B/C, 64, 46-56.

Binley, A., Hubbard, S.S., Huisman, J.A., Revil, A., Robinson, D.A., Singha, K., Slater, L.D. (2015). The emergence of hydrogeophysics for improved understanding of subsurface processes over multiple scales. Water Resour. Res., 51, 3837-3866.

Bucker, M. and Hordt, A. (2013). Analytical modelling of membrane polarization with explicit parametrization of pore radii and the electrical double layer. Geophys. J. Int., 194, 804-813.

Calvo, J.L.G., Hidalgo, A., Alonso, C., Luco, L.F. (2010). Development of low-pH cementitious materials for HLRW repositories Resistance against ground waters aggression. Cem. Concr. Res., 40, 1290-1297.

Calvo, J.L.G., Moreno, M.S., Alonso, M.C.A., Lopez, A.H., Olmo, J.G. (2013). Study of the Microstructure Evolution of Low-pH Cements Based on Ordinary Portland Cement (OPC) by Mid- and Near-Infrared Spectroscopy, and Their Influence on Corrosion of Steel Reinforcement. Materials, 6, 2508-2521.

Chelidze, T.L. and Gueguen, Y. (1999). Electrical spectroscopy of porous rocks: a review - I. Theoretical models. Geophys. J. Int., 137, 1-15.

Elakneswaran, Y., Nawa, T., Kurumisawa, K. (2009). Electrokinetic potential of hydrated cement in relation to adsorption of chlorides. Cem. Concr. Res., 39, 340-344.

Gaboreau, S., Claret, F., Crouzet, C., Giffaut, E., Tournassat, C. (2012). Caesium uptake by Callovian-Oxfordian clayrock under alkaline perturbation. Appl. Geochem., 27, 1194-1201.

Haas, J. and Nonat, A. (2015). From C-S-H to C-A-S-H: Experimental study and thermodynamic modelling. Cem. Concr. Res., 68, 124-138.

Huisman, J.A., Zimmermann, E., Esser, O., Haegel, F.-H., Treichel, A., Vereecken, H. (2016). Evaluation of a novel correction procedure to remove electrode impedance effects from broadband SIP measurements. J. Appl. Geophys., 135, 466-473.

Kemna, A., Binley, A., Cassiani, G., Niederleithinger, E., Revil, A., Slater, L., Williams, K.H., Orozco, A.F., Haegel, F.H., Hordt, A., Kruschwitz, S., Leroux, V., Titov, K., Zimmermann, E. (2012). An overview of the spectral induced polarization method for near-surface applications. Near Surf. Geophys., 10, 453-468.

Labbez, C., Jonsson, B., Pochard, I., Nonat, A., Cabane, B. (2006). Surface charge density and electrokinetic potential of highly charged minerals: Experiments and Monte Carlo simulations on calcium silicate hydrate. J. Phys. Chem. B, 110, 9219-9230.

Larbi, B., Dridi, W., Dangla, P., Le Bescop, P. (2016). Link between microstructure and tritiated water diffusivity in mortars: Impact of aggregates. Cem. Concr. Res., 82, 92-99.

Leroy, P., Li, S., Jougnot, D., Revil, A., Wu, Y. (2017). Modeling the evolution of complex conductivity during calcite precipitation on glass beads. Geophys. J. Int., 209, 123-140.

Li, S., Leroy, P., Heberling, F., Devau, N., Jougnot, D., Chiaberge, C. (2016). Influence of surface conductivity on the apparent zeta potential of calcite. J. Colloid Interface Sci., 468, 262-275.

Zimmermann, E., Kemna, A., Berwix, J., Glaas, W., Munch, H.M., Huisman, J.A. (2008). A high-accuracy impedance spectrometer for measuring sediments with low polarizability. Meas. Sci. Technol., 19, 105603.

Implementation of crystallization and precipitation mechanisms in pore-scale models based on the Lattice-Boltzmann Method

Stephan Rohmen[1*], Andrés Idiart[2], Guido Deissmann[1], Dirk Bosbach[1]

[1] Forschungszentrum Jülich GmbH, Institute of Energy and Climate Research (IEK-6): Nuclear Waste Management and Reactor Safety, Jülich (DE)
[2] Amphos 21 Consulting S.L. (ES)

* Corresponding author: st.rohmen@fz-juelich.de

Abstract

Pore-scale models represent an appealing approach for obtaining a more accurate and mechanistic description of physical and chemical processes in heterogeneous porous media such as cement-based materials. For this purpose, a coupled code, which is able to calculate reactive transport processes in porous media on the pore scale using the Lattice-Boltzmann approach, is developed. The aim of this coupled code is to simulate chemically-driven alteration and degradation processes and solute transport in cement-based materials. Geometrical changes of the pore structure due to dissolution and precipitation processes at the microscopic scale can have significant effects on the macroscopic hydrodynamic properties of porous media. In this context, the current development status and the techniques used to simulate precipitation processes on the pore-scale in a realistic way are outlined. A model consisting of a gypsum containing pore leading to ettringite precipitation on monosulfoaluminate is depicted to illustrate the effect of the recent development iteration.

Introduction

Cement-based materials have been widely adopted as engineered barrier materials in nuclear waste repositories both for their radionuclide fixation and immobilization properties as well as for their low permeability and diffusivity. Cementitious materials generally consist of various hydrated phases such as nanocrystalline calcium-silicate-hydrates (C-S-H), portlandite, and aluminate / ferrate compounds (AFm/AFt). The composition of a specific cementitious material and its microstructure (e.g., pore size distribution and pore structure) depend on the cement types and the mixing and curing processes. The macroscopic physical properties of these materials strongly depend on the pore structure and the (micro)structural and chemical changes resulting from long-term alteration / degradation processes (Taylor, 1997).

In the frame of the CEBAMA project, a pore-scale modelling framework for the simulation of concrete degradation processes is developed. Compared to continuum-scale models, pore-scale models can be used to obtain a more accurate and mechanistic description of physical and chemical processes in heterogeneous porous media such as cement-based materials. Processes of interest to be addressed with this approach within CEBAMA comprise chemical degradation processes due to calcium leaching and the carbonation of cementitious materials and its impact on physical and transport properties, such as porosity, pore-size distribution, permeability and diffusivity.

The code under development is based on the coupling of a transport code based on the Lattice-Boltzmann-Method (Palabos; FlowKit, 2011) with a geochemical code (PhreeqC) and is called *iPP* (interface Palabos PhreeqC). The Lattice-Boltzmann method (LBM) is derived from Boltzmann's kinetic theory and statistical mechanics by simulating the microscopic behavior of particles in a fluid in a discrete way (space, time and velocity vectors are discretized). By choosing proper discretization steps and equilibrium functions (in terms of LBM collision step), mesoscopic and macroscopic properties as density and flux can be recovered. Thus, as in finite element methods (FEM) the LBM approximates the solution of the Navier-Stokes equation for advective-diffusive transport processes, but derived from kinetic laws instead of continuum (e.g., momentum conservation). One advantage of using LBM is the explicit description of geometry and the possibility to update it on the fly, e.g., without rather complicated automatic remeshing steps as in FEM (cf., Mohamad, 2011; Succi, 2001).

The coupling code *iPP* follows the operator splitting approach, i.e., flow and solute transport are calculated using the LBM at a given time step, followed by the calculation of chemical reactions in PhreeqcRM (cf., Parkhurst and Wissmeier, 2015). The latter calculates various solution properties including aqueous species concentrations, pH, pe and saturation indices for solid phases. Moreover, the abundance of the various solid phases and their composition are obtained and the porosity values are updated after each time step.

Precipitation and crystallization algorithm

Geometrical changes of the pore structure due to dissolution and precipitation processes at the microscopic scale can have significant effects on the macroscopic hydrodynamic properties of porous media (Raoof et al., 2013). The impact of these geometrical changes in the pore network on transport properties depend in particular on the spatial location of the precipitated solids. Hence, the spatial distribution of newly formed precipitates within the modelling domain has to be addressed in the simulations in an appropriate way. In the literature this is often referred to as nucleation theory. Several nucleation theories have been proposed (e.g., Kashchiev and van Rosmalen, 2003; Kashchiev, 2000; Prieto, 2014).

To mimic the nucleation mechanism for heterogeneous precipitation processes an overgrowth algorithm is implemented in the framework of *iPP*. This is achieved by defining a set of rules and constraints for the precipitation processes, i.e., where precipitation is possible and under which conditions. A simple variant of this approach was presented by (Huber et al., 2014) and referred to as phase field technique. With this approach, solids can only precipitate and dissolve at the grain surfaces. However, this assumption can sometimes oversimplify the system since it is known that precipitation can also occur in the free pore space (homogeneous nucleation), if a critical supersaturation of a solid is exceeded (e.g., Putnis et al., 1995; Scherer, 2004).

In PhreeqC the amount of a precipitating solid is calculated using a (user defined) target saturation index. Translating the phase field approach of Huber et al. (2014) into the terminology of geochemistry, the target saturation index (SI_T) at grain surfaces is always zero (equilibrium condition). At a distance of more than one grid cell from the grain surface into the pore space no precipitation is possible allowing for supersaturation of the respective solid (non-equilibrium condition). In the case of the implementation of Huber et al. (2014), this is translated into a sufficiently large target saturation index in PhreeqC terminology so that supersaturation (and thus precipitation) is never achieved.

In order to allow for homogeneous nucleation effects in our modelling approach, an extension of the precipitation algorithm is proposed here. To this end, the distance to the grain surfaces needs to be considered, which requires information on the local neighborhood to a grain surface in the modelling domain. This information can be retrieved via a signed distance field (SDF) providing the Euclidean distances to grain surfaces for all grid node positions. By convention, positive values refer to "outside the grain" positions (i.e., "in pore space"), while negative values are representing "inside the grains" positions. In *iPP* the SDF is implemented by an algorithm called 8SSEDT (Leymarie and Levine, 1992). The SDF is updated after each iteration step that involves changes in pore geometry.

In this way, the target saturation index (SI_T) used in the geochemical calculations in PhreeqC is calculated via transfer functions (which can be defined arbitrarily) using the information on distance to grain surfaces. For illustrating the generality of the approach, an exponential function of the distance to the grain surface, d (µm), was implemented here. In this case, the following rules are applied (Eq. 1 to Eq. 3):

$$SI_T(d) = 0 \; for \; d \leq 0 \qquad \textbf{\textit{Eq.1}}$$

$$SI_T(d) = SI_{cr} \; for \; d > d_{th} \qquad \textbf{\textit{Eq.2}}$$

$$SI_T(d) = (1 - e^{k \cdot d}) \cdot SI_{cr} \; for \; 0 < d \leq d_{th} \qquad \textbf{\textit{Eq.3}}$$

The k factor in the exponent can be used to fine tune the slope of the exponential function. In the present model it was determined from the threshold distance d_{th}. With this relation, 99% of the critical SI (SI_{cr}) is reached at the given threshold distance, i.e.,:

$$k = ln\,(0.01)/d_{th} \qquad \textbf{\textit{Eq.4}}$$

Figure 1 shows a comparison of three different approaches of defining the target saturation indices as a function of the grain boundary distance, including the above described exponential function. Moreover, a flat transformation implementing a simple approach at which always equilibrium conditions ($SI_T = 0$) are assumed, and the PhreeqC terminology analogue to the step function used by Huber et al. (2014) is depicted.

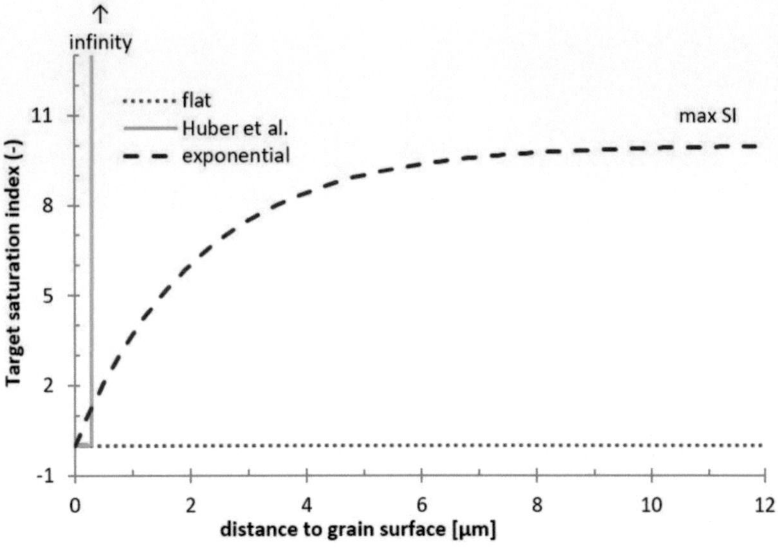

Figure 1: *Comparison of three different target saturation index transfer functions. For the function of Huber et al. (2014) a discretization step of 0.333 µm was used. For the exponential function a threshold distance of 10 µm and a k value of ln(0.01)/10 µm = -0.461 µm*$^{-1}$ *was adopted.*

Local saturation index and solute concentration values are inverse proportional to the applied target saturation index values in case of identical total concentrations. Applying a gradual transfer function leads to the effect that the solute concentration values in the free pore volume can be higher than at the grain surface. On this basis the concentration gradient in the pore water is more likely towards the grain surface. Therefore, using a gradual function instead of a simple "flat" or threshold step function has the advantage that there is always a driving force towards the grain surfaces in potentially leading to the formation of grain overgrowth, since the required target saturation index is lower towards the grain surface. Thus, minor amounts of solids potentially dispersed in the free pore space, due to exceedance of the critical supersaturation SI_{cr} previously, will be transferred towards grain surfaces. Thus the precipitate will not remain in the free pore space resulting in to a more compact aggregation of solids.

Ettringite precipitation as pore overgrowth

To illustrate the effect of the new precipitation algorithm implemented in our code, the formation of ettringite due to reaction of sulfate ions with AFm (mechanism usually referred to as sulfate attack) was simulated using a simple pore geometry. The simulated system consists of a pore filled with pure water in a monosulfoaluminate (AFm) domain and a spherical gypsum inclusion within the center of the pore water (see Figure 2, left). The dissolving gypsum is the source for sulfate ions which diffuse towards the surface of the AFm phase, leading to the formation of ettringite. The pore was designed not to be fully sphero-symmetrical to obtain different supersaturation of ettringite. The size of the system was set to 100 x 100 µm and discretized in 300 x 300 Lattice-Boltzmann nodes, leading to a spatial cell resolution of 0.333 µm. Assuming a diffusion coefficient of $1.0 \cdot 10^{-9}$ m^2/s for the dissolved ions in the pore solution and setting the LBM relaxation time to 1.0, one iteration step is equivalent to a time step of $1.85 \cdot 10^{-5}$ s. The simulations were run for 5,400 iteration steps corresponding to 0.1 s in physical time units.

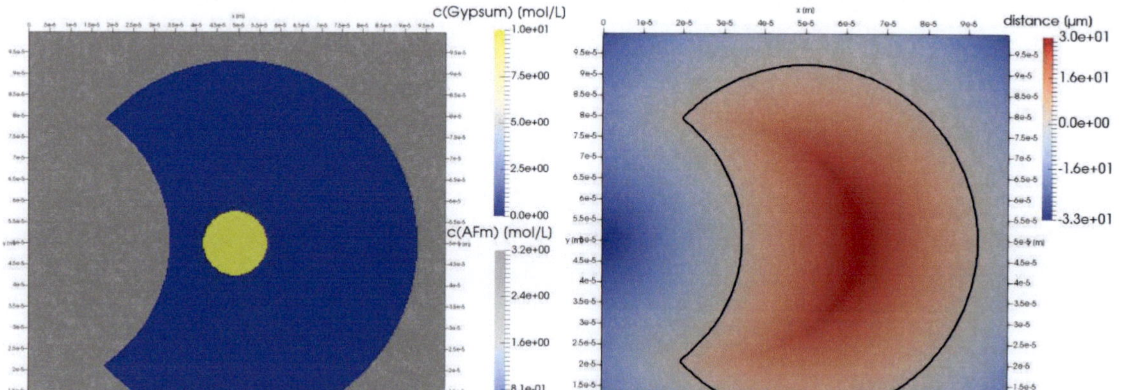

Figure 2: *Left: Initial pore geometry consisting of AFm (grey), pore water (blue) and gypsum (yellow) in the center of the pore space. Right: Initial distance field of the AFm / Ettringite pore model. Pore surface is indicated as black line.*

Two different distance transfer functions were used for comparison, namely a "flat" (always equilibrium condition) and the exponential function described above. Here, the critical saturation index SI_T for the exponential function was set to 10, while the distance threshold d_{th} was set to 10 µm. The initial distance field is shown in Figure 2 (right).

The results of simulations using two different transfer functions are shown in Figure 3 and Figure 4. In both simulations at the elapsed time the concentration gradients in the pore water exist, i.e., the system is not in equilibrium. Therefore, there are still sulfate, aluminum and calcium source terms present. The simulation using the flat (always equilibrium case, $SI_T = 0$) transfer function shows a broad precipitation zone with ettringite dispersed in the entire pore volume (Figure 3). Only at the closer grain surface there is a higher amount of precipitation due to higher local concentrations of sulfate reaching the AFm phase. Since there is no driving force in the saturation index depending on the distance to the grain surface the precipitated ettringite will not be transferred towards the pore walls and form any overgrowth. In Figure 4 the results for the exponential transfer function are depicted. In this case ettringite precipitates as an overgrowth on the surface of the AFm phase. On the left hand side, the precipitated amount and the thickness of the ettringite overgrowth is higher, due to the higher local sulfate concentration. The thickness of this ettringite layer is governed by the slope (k value) of the exponential function. In the example simulation, a relatively high value of 10 μm for the distance threshold d_{th} was used. Depending on the precipitating solids this value has to be adjusted accordingly.

The total amount of ettringite precipitated was 4.78 fmol and 3.92 fmol for the flat case and the exponential configuration, respectively. Using a molar volume of ettringite of 707 cm^3/mol, the porosity is reduced by 0.143% and 0.118%, respectively. Thus, the porosity reduction in case of the flat transfer function is 20% higher compared to the exponential function model.

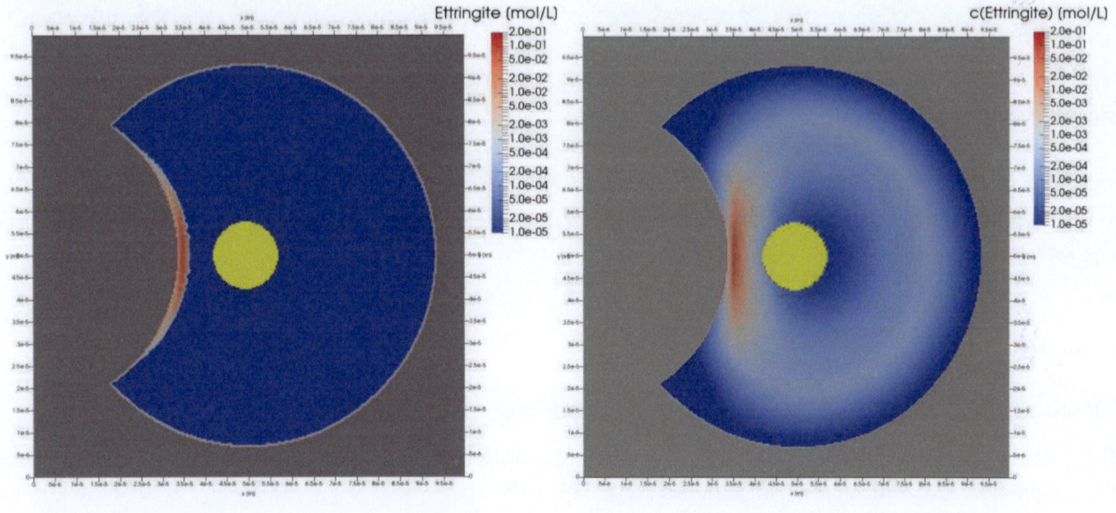

Figure 3: *Ettringite concentration after 0.1 s runtime using the flat (left) and exponential (right) target saturation transfer function (log color scale).The residual gypsum and AFm are indicated in yellow and grey color, respectively.*

Conclusions and Future work

A coupling code for pore-scale simulations of reactive transport processes based on the Lattice-Boltzmann method, named *iPP*, has been developed. Preliminary verification and benchmarking of the code against analytical solutions provided evidence that the tool is appropriate for the simulation of complex advective and diffusive solute transport processes (Rohmen et al., 2017). Recently, the code was extended to support the simulation of heterogeneous precipitation processes leading to the formation of grain overgrowth via distance field transformation. This demonstrates the general ability that different kind of precipitation processes can be simulated with this technique, using different transfer functions. At this point it is emphasized that a mathematical formalism is developed which allows controlling the locality of precipitation. The type of transfer function to use is currently under development but can be chosen arbitrarily depending on the demands of the model.

In contrast to the approach of Huber et al. (2014), where precipitation is completely prevented aside of pore surface domains, the presented scheme allows precipitation of a limited amount of solids in free pore space as well. The existence of homogenous nucleation phenomena is underlined by other authors (e.g., Poonoosamy et al., 2016). Open questions remain with respect to the parametrization of the functions describing the distance-dependent target saturation indices. Experimental data for more common minerals such as barite are available and underline the importance of capturing those effects (Putnis et al., 1995). However, at present the critical supersaturation values and distance threshold for hydrated cement phases are uncertain. Following the nucleation theory describing homogeneous nucleation, it is thought that the critical supersaturation values can be obtained on a theoretical basis. Conceptually, taking into account geometrical properties such as surface curvature and gravitational forces would enable the simulation of crystal growth with respect to morphology.

In future development steps the simulated precipitation patterns will be compared to results obtained by other simulation codes and experimental data produced within CEBAMA, addressing in particular the CEBAMA WP1 experiments performed by USFD and SURREY. In the scope of these experiments data about the microstructural evolution in cementitious materials leached by various types of water (clay water, granite water, seawater) will be obtained, which will aid to derive a better understanding of the mechanisms which have to be implemented. Until these experimental data are available, it is intended to benchmark the code development against appropriate experiments on the degradation of cementitious materials and other heterogeneous materials. Further input for determining the transfer function parameters (in particular addressing cement phases) will be derived from theoretical nucleation models and/or by reviewing of existing experimental data in the available literature.

Acknowledgement

The research leading to these results has received funding from the European Union's Horizon 2020 Research and Training Programme of the European Atomic Energy Community (EURATOM) (H2020-NFRP-2014/2015) under grant agreement n° 662147 (CEBAMA).

References

FlowKit (2011). Palabos user guide. *http://www.palabos.org/documentation/userguide*

Huber, C., Shafei, B., Parmigiani, A. (2014). A new pore-scale model for linear and non-linear heterogeneous dissolution and precipitation. Geochimica et Cosmochimica Acta, 124, 109-130.

Kashchiev, D. (2000). Nucleation: basic theory with applications. Butterworth Heinemann.

Kashchiev, D. and van Rosmalen, G.M. (2003). Review: Nucleation in solutions revisited. Crystal Research and Technology, 38, 555-574.

Leymarie, F. and Levine, M.D. (1992). Fast raster scan distance propagation on the discrete rectangular lattice. CVGIP: Image Understanding, 55, 84-94.

Mohamad, A.A. (2011). Lattice Boltzmann Method. Springer.

Parkhurst, D.L. and Wissmeier, L. (2015). PhreeqcRM: A reaction module for transport simulators based on the geochemical model PHREEQC. Advances in Water Resources, 83, 176-189.

Poonoosamy, J., Curti, E., Kosakowski, G., Grolimund, D., Van Loon, L.R. (2016). Barite precipitation following celestite dissolution in a porous medium: A SEM/BSE and μ-XRD/XRF study. Geochimica et Cosmochimica Acta, 182, 131-144.

Prieto, M. (2014). Nucleation and supersaturation in porous media (revisited). Mineralogical Magazine, 78, 1437-1447.

Putnis, A., Prieto, M., Fernandez-Diaz, L. (1995). Fluid supersaturation and crystallization in porous media. Geological Magazine, 132, 1-13.

Raoof, A., Nick, H.M., Hassanizadeh, S.M., Spiers, C.J. (2013). PoreFlow: A complex pore-network model for simulation of reactive transport in variably saturated porous media. Computers and Geosciences, 61, 160-174.

Rohmen, S., Shafei, B., Idiart, A., Deissmann, G., Bosbach, D. (2017). Pore-scale reactive transport modelling of cementitius systems: Concept development based on the Lattice-Boltzmann method. Proceedings of the 1st CEBAMA Annual Workshop. KIT Scientific Publishing, KIT-SR 7734.

Scherer, G.W. (2004). Stress from crystallization of salt. Cement and Concrete Research, 34, 1613-1624.

Succi, S. (2001). The Lattice Boltzmann Equation for fluid dynamics and beyond. Oxford.

Taylor, H.F.W. (1997). Cement chemistry. Thomas Telford.

H-M-C coupling analysis considering several scenarious of long-term alteration in cement-bentonite system

Shinya Tachibana[1*], Shinji Ito[1], Atsushi Iizuka[1], Hitoshi Owada[2], Daisuke Hayashi[2]

[1] Kobe University (JP)
[2] Radioactive Waste Management Funding and Research Center (JP)

* Corresponding author: stachi@people.kobe-u.ac.jp

Abstract

Two-dimensional Hydro-Mechanical-Chemical (H-M-C) coupling analyses with assumptions of several scenarios of long-term alteration of montmorillonite were performed to estimate changes in the mechanical and hydraulic characteristics of clay-based buffer in the barrier system. This paper mainly presents the numerical results of a series of H-M-C coupling analyses focusing on the long-term changes in stress distribution and permeability generated in the bentonite buffer. A newly developed constitutive model to describe the mechanical response of the bentonite buffer considering the effects of volume change of clay minerals induced by the alteration or the dissolution of montmorillonite is also explained briefly.

Introduction

Bentonite or bentonite mixture is anticipated as a mechanical and hydraulic buffer achieving the swelling pressure and the low permeability to maintain the sealing ability in the concept of TRU (TRans-Uranic waste) geological disposal in Japan. The saturation of the near-field will allow for highly coupled and non-linear reactions between clay minerals and alkaline solutions produced by the cement degradation. This could result in various types of clay alteration, which might have a detrimental effect locally on the long-term performance of the barriers, such as swelling pressure, hydraulic conductivity, gas permeability and so forth (Sellin and Leupin, 2013). The dissolution of clay minerals including montmorillonite, the main constituent of bentonite buffer material, and the subsequent formation of secondary minerals such as analcime, heulandite and clinoptirolite are possible reactions depending on temperature, groundwater composition and pH under hyperalkaline conditions (e.g., Pusch et al., 2010; Fujii et al., 2010; Satoh et al., 2013; Sawaguchi et al., 2016). Thus, it is generally recognised that more attention needs to be given to the coupled and complex effects of chemical and hydraulic / mechanical interactions to evaluate the long-term performance of the engineered barrier. This paper addresses the H-M-C coupling analysis by using the existing geochemical and hydraulic/mechanical codes.

Methodology of coupling analysis

The geochemical code used in this study is PHREEQC-TRANS (Parkhurst and Appelo, 1999; JAEA and FEPC, 2005b) for the purpose of calculating the changes in montmorillonite content at every nodal point in an object domain over time. Geochemical data obtained in this way are transferred as input data to the hydraulic / mechanical code DACSAR-BA being developed within the authors' research group. This means that the quantities in geochemical field are unidirectionally transferred to hydraulic / mechanical field in successive steps and that the geochemical field is determined a priori, independently of the hydraulic / mechanical quantities, e.g., volume change of mass, groundwater flow and confining pressure. Although the bidirectional data transfer is favourable whether by weak or by strong coupling, a weak coupling requires some sort of iteration as discussed by Ishii et al. (2013) and a strong coupling requires the reformulation of equations to be solved on matrix level (Hameyer et al., 1999). This study allows such an ad hoc solution procedure to reduce the operation burden in exchange for high accuracy, while a more efficient interface which couples two standalone simulation programs has been developed and implemented recently (i.e., Nardi et al., 2014). It should however be noted that DACSAR-BA is a fully coupled finite element code which deals with the problem describing the equations of equilibrium and continuity condition where the changes in average density of soil particles and volume change of solid part are considered.

Scenarios of clay alteration

First, the changes in montmorillonite content at every location in each time step are calculated using the code PHREEQC-TRANS. Then, the history of average density of soil particles ρ_s and ratio of volume change of solid part α (= $V_s/V_{s,ini}$; V_s represents the volume of solid part), which will be perturbations in the hydraulic / mechanical field, are locally determined in accordance with the scenario of clay alteration after Morimoto (2014). This study assumes two scenarios of clay alteration: the formation of clinoptirolite zeolite as a secondary mineral and the dissolution of montmorillonite. In the former scenario, the quantities ρ_s (Eq. 2) and α (Eq. 1) are calculated as follows:

$$\alpha = 1 + \sum_{k=\mathrm{m,si}} \left(\eta_{\mathrm{cp}} - 1\right)\frac{\rho_{\mathrm{s},ini}}{\rho_k}\alpha_{k,ini}\beta$$

Eq. 1

$$\rho_{\mathrm{s}} = \frac{\rho_{\mathrm{s},ini}}{\alpha}\left[1 + \sum_{k=\mathrm{m,si}}\left(\eta_{\mathrm{cp}}\frac{\rho_{\mathrm{cp}}}{\rho_k} - 1\right)\alpha_{k,ini}\beta\right]$$

Eq. 2

where, $\beta = 1 - \alpha_m/\alpha_{m,ini}$ is the alteration ratio, α_m the montmorillonite content, α_{si} the silicon dioxide content and the quantity with subscript *ini* stands for its initial value. ρ_m, ρ_{si} and ρ_{cp} denote the density of montmorillonite, silicon dioxide and clinoptirolite, respectively. η_{cp} denotes the ratio of the volume of reaction product (i.e., clinoptirolite) to that of reactants (i.e., montmorillonite and silicon dioxide) per one mole reaction of montmorillonite. On the other hand, the latter scenario for dissolution of montmorillonite gives the following conditions (Eq. 3 and Eq. 4):

$$\alpha = 1 - \frac{\rho_{\mathrm{s},ini}}{\rho_{\mathrm{m}}}\alpha_{\mathrm{m},ini}\beta$$

Eq. 3

$$\rho_{\mathrm{s}} = \frac{\rho_{\mathrm{s},ini}}{\alpha}\left(1 - \alpha_{\mathrm{m},ini}\beta\right)$$

Eq. 4

Table 1: *Input data for the calculation of α and ρ_s.*

Input data	Notation	Value	Unit
Density of montmorillonite	ρ_{mon}	2.74	Mg/m^3
Density of silicon dioxide	ρ_{si}	2.65	Mg/m^3
Density of clinoptirolite	ρ_{cp}	2.24	Mg/m^3
Initial content of montmorillonite	$\alpha_{m,ini}$	0.34	-
Initial content of silicon dioxide	$\alpha_{si,ini}$	0.57	-
Reaction parameter for clinoptirolite	η_{cp}	3.07	-

The dissolution scenario assumes that there is no formation of secondary minerals. Input data used in the analysis for the chemical reactions is based on the thermodynamic data on Arthur et al. (2005) and on data summarized in Table 1.

Constitutive model for bentonite

The elasto-plastic constitutive model applied in the analysis is based on the Cam-clay model for a saturated soil but allows the void ratio change induced not only by changes in the effective stress but also by the loss / generation of solid volume due to chemical reactions. Under the condition that the solid volume in a soil mass changes, the volumetric strain ε_v can be defined as Eq. 5.

$$\varepsilon_v = \frac{V_0 - V}{V_0} = 1 - \frac{1+e}{1+e_0}\alpha$$

Eq. 5

Figure 1 shows the relationship between void ratio *e*, volumetric strain ε_v and effective mean stress p' under isotropic stress conditions. If the volume of solid part remains unchanged (i.e., $\dot{\alpha} = 0$), the loading after yielding leads to contraction or the decrease of void ratio along the normally consolidation line having the slope λ in e - ln p'. On the other hand, unloading leads to swelling or the increase of void ratio along the swelling line with the slope κ in a similar manner of Cam-clay model. In addition, the model considers the void ratio change due to the change in volume of particles even when the effective means stress is constant. The path on e - ε_v plane with a change in solid volume under constant effective mean stress condition is simply assumed as e = $k_\alpha \varepsilon_v$ with a material constant k_α. Thus, the compression and swelling indices, λ and κ, for any condition can be respectively determined as Eq. 6 and Eq. 7.

$$\lambda = \frac{k_\alpha + (1+e_0)}{k_\alpha \alpha + (1+e_0)}\lambda_0$$

Eq. 6

$$\kappa = \frac{k_\alpha + (1+e_0)}{k_\alpha \alpha + (1+e_0)}\kappa_0$$

Eq. 7

where, λ_0 and κ_0 are the compression and swelling indices at α = 1. Taking such an effect of the change in volume of particles on mechanical properties into account, the Cam-clay model is modified and applied in the analysis.

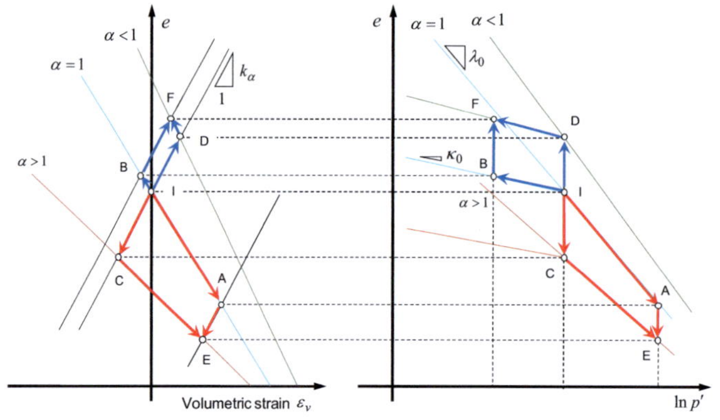

Figure 1: *Schematic diagram for the constitutive modelling.*

Analysis conditions and parameters

Figure 2 shows the model domain of analysis object, which is a part of the engineered barrier consisting of the bentonite buffer and mortar, together with initial and boundary conditions. The system is assumed to be initially saturated. The advection and the chemical reaction are dealt with in the PHREEQC-TRANS simulations, while the deformation and the flow of water are analysed in DACSAR-BA simulations. As for the boundary condition in the geochemical analysis, no solute flux condition is imposed on all boundaries except for the left end. The left end is assumed to be directly in contact with groundwater ignoring the existence of concrete invert and is set to be a specified concentration boundary. The chemical compositions input in geochemical analysis are determined referring to the evidence material 4 - 5 in JAEA report (2005) in general. The groundwater is assumed to be a fresh reducing high pH (FRHP) and its chemical composition is determined as shown in Table 2, also referring to Yui et al. (1999). On the other hand, as for the boundary conditions in the hydraulic / mechanical analysis, all boundaries except for the right end are set to be permeable with the total head of 300 m assuming that the engineered barrier locates at a depth of 300 m below the groundwater level. Displacement in in/out-of-plane directions is not allowed as a geometric boundary condition. A cement / mortar layer is assumed to consist of the ordinary Portland cement (OPC). Its chemical composition is set as shown in Table 3. Table 4 summarises the initial specification of the cement / mortar layer. The bentonite buffer is assumed to consist of a mixture of a Japanese Na type bentonite (Kunigel V1) and 30 wt.% silica. The initial average dry density of bentonite buffer was set to be 1.6 Mg/m^3, while two patterns of its distribution were considered: homogeneous and heterogeneous cases. In the heterogeneous case, the bentonite buffer consists of three layers having the dry density of 1.4, 1.6 and 1.8 Mg/m^3 from the bottom. The initial particle density of bentonite buffer is set to be 2.67 Mg/m^3. Table 5 and Table 6 summarise the initial mineral composition and the exchangeable cation composition of bentonite, respectively. The initial mineral contents of cement / mortar and bentonite buffer are given in Table 7. The effective diffusion coefficient for cement/mortar and bentonite buffer are calculated as Eq. 8.

$$D_e = D^* \phi^n; \quad \begin{cases} n = 3.05 \text{ for cement/mortar} \\ n = 2.22 f_s^{0.13} + 1.0 \text{ for bentonite buffer} \end{cases} \qquad \textbf{\textit{Eq. 8}}$$

where, D* denotes a free water diffusion coefficient of $2.27 \cdot 10^{-9}$ m^2/s, f_s is a montmorillonite content, and ϕ is a porosity. The dissolution rate of montmorillonite is calculated based on a formulation in JAEA (2005a and 2005b).

In hydraulic / mechanical analysis, the bentonite buffer is modelled by the elasto-plastic constitutive model described above, while a linear elastic model is considered for the waste / mortar domain. Material parameters used in the hydraulic / mechanical analysis are summarised in Table 8. The initial stress state is assumed to be isotropic and the initial value of effective mean stress is set to be 0.17 MPa so that the loosest part in bentonite buffer satisfies the normally consolidated state. Four cases of simulation were conducted assuming two scenarios of chemical alteration against two distributions of initial dry density, as summarised in Table 9.

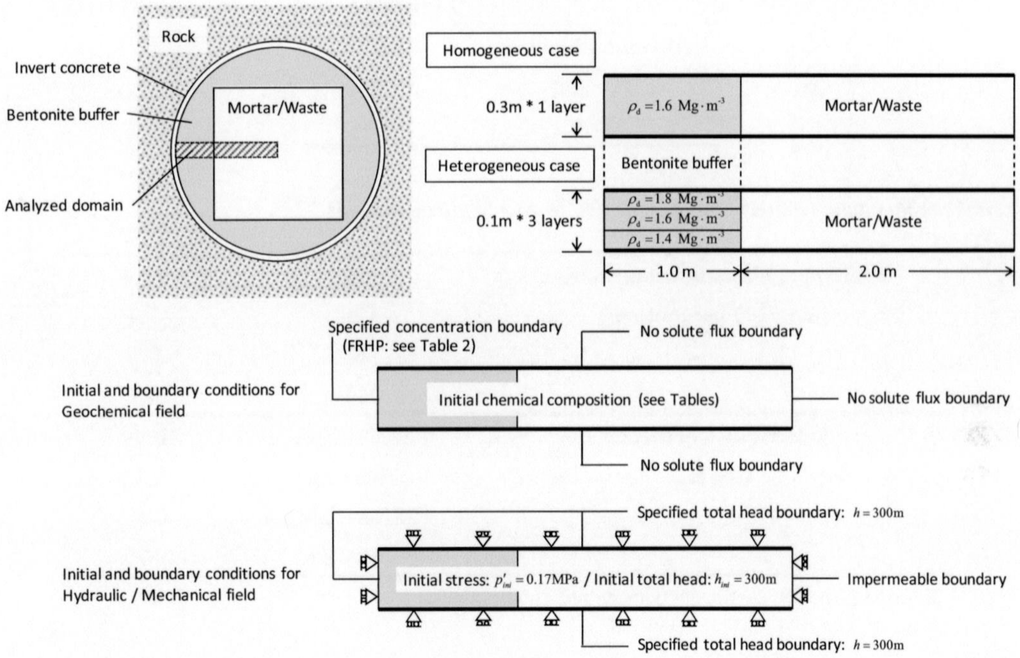

Figure 2: *Analysed geometry with initial and boundary conditions.*

Table 2: *Chemical composition of the groundwater in geochemical analysis.*

pH	Na	Ca	K	Mg	Al	C	S	Cl	Si
8.5	$3.4 \cdot 10^{-3}$	$1.1 \cdot 10^{-4}$	$6.2 \cdot 10^{-5}$	$5.0 \cdot 10^{-5}$	$3.4 \cdot 10^{-7}$	$3.5 \cdot 10^{-3}$	$1.1 \cdot 10^{-4}$	$1.5 \cdot 10^{-5}$	$3.4 \cdot 10^{-4}$

(unit: mol/L)

Table 3: *Chemical composition of the cement (OPC) in geochemical analysis.*

SiO_2	Al_2O_3	Fe_2O_3	CaO	MgO	SO_3	Na_2O	K_2O
21.5	5.2	2.9	64.0	1.5	2.0	0.3	0.5

(mass%)

Table 4: *Initial specification of cement/mortar layer.*

Quantity	Value	Unit
Porosity	0.19	-
Water cement ratio	55.0	%
Unit content of cement	483.0	kg/m^3
Unit content of water	266.0	kg/m^3

Quantity	Value	Unit
Unit content of fine and coarse aggregates	1,449.0	kg/m^3

***Table 5:** Initial mineral composition of Kunigel V1 in geochemical analysis.*

Initial mineral	Mass content (mass%)
Montmorillonite	48.0
Chalcedony	38.0
Calcite	2.4

***Table 6:** Exchangeable cation composition of bentonite in geochemical analysis.*

Exchangeable cation composition	NaZ	51.4
(meq/100 bentonite-g)	CaZ_2	7.4
	KZ	0.6
	MgZ_2	0.7
Selectivity Coefficient	2ZNa - Z_2Ca	0.69
(log K)	ZNa - ZK	0.42
	2ZNa - Z_2Mg	0.67
	ZNa - ZH	0.70

K: Gaines-Thomas coefficient (Gaines and Thomas, 1953)

***Table 7:** Initial mineral content in geochemical analysis.*

Mineral	Mineral content (mol/L_{water})				Molar volume (L/mol)
	Mortar / Cement	Bentonite buffer			
		1.4 Mg/m^3	1.6 Mg/m^3	1.8 Mg/m^3	
Portlandite	9.98	-	-	-	0.0332
Ettringite	0.212	-	-	-	0.726
Brucite	0.946	-	-	-	0.0244
CSH (Ca/Si = 1.8)	16.4	-	-	-	0.0445
Na_2O	0.123	-	-	-	-
K_2O	0.135	-	-	-	-
Calcite	-	0.536	0.727	1.01	0.0369
Quartz	-	13.0	17.7	24.5	0.0227
Na-montmorillonite	-	2.31	3.13	4.34	0.134
Ca-montmorillonite	-	0.332	0.451	0.624	0.134
K-montmorillonite	-	0.0269	0.0366	0.0506	0.136
Mg-montmorillonite	-	0.0314	0.0427	0.0591	0.133

***Table 8:** Material parameters for the hydraulic / mechanical coupling analysis.*

Material	Parameter	Notation	Value	Unit
Bentonite buffer	Compression index	λ_0	0.14	-
	Swelling index	κ_0	0.02	-
	Critical state parameter	M	0.5	-
	Poisson ratio	ν'	0.45	-
	Chemical / mechanical parameter	k_α	1.0	-
Waste / Mortar	Lame's constants	$\tilde{\lambda}$	495.8	GPa
		$\tilde{\mu}$	743.8	GPa

Table 9: *Analysis Cases.*

Case	Initial dry density distribution	Chemical scenario
Case 1	Homogeneous	Formation of clinoptirolite
Case 2	Homogeneous	Dissolution of montmorillonite
Case 3	Heterogeneous	Formation of clinoptirolite
Case 4	Heterogeneous	Dissolution of montmorillonite

Numerical results

Figure 3 shows the effective mean stress distributed in bentonite buffer after 100,000 years and the chronological changes in normal stress distributions acting on the boundaries. The trend differs remarkably between scenarios of chemical alteration: the formation of clinoptirolite results in stress generation, whereas the dissolution of montmorillonite causes a stress reduction. Differences in the initial distribution of dry density seem to give little impact to the stress distribution, while the horizontal stress distribution is somewhat affected. Figure 4 shows the changes in hydraulic conductivity which is calculated from the void ratio and the montmorillonite content by applying the permeability model proposed in Kobayashi et al. (2011) (see Appendix A). It can be seen obviously that the formation of clinoptirolite causes a further lowering of permeability induced by the decrease in void ratio, while the dissolution of bentonite allows the increase of hydraulic conductivity which may result in the malfunction of bentonite barrier.

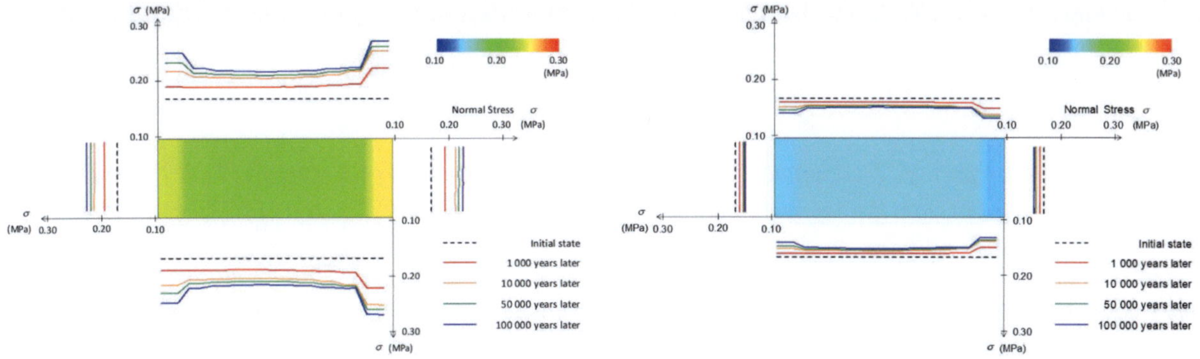

Case 1 *Case 2* *(Cont. next page)*

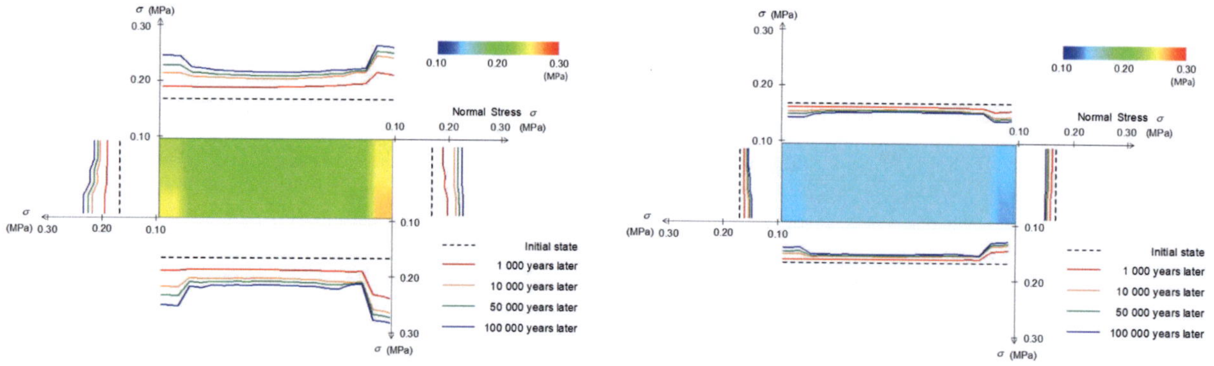

Case 3 *Case 4*

Figure 3: *Stress distribution in bentonite buffer.*

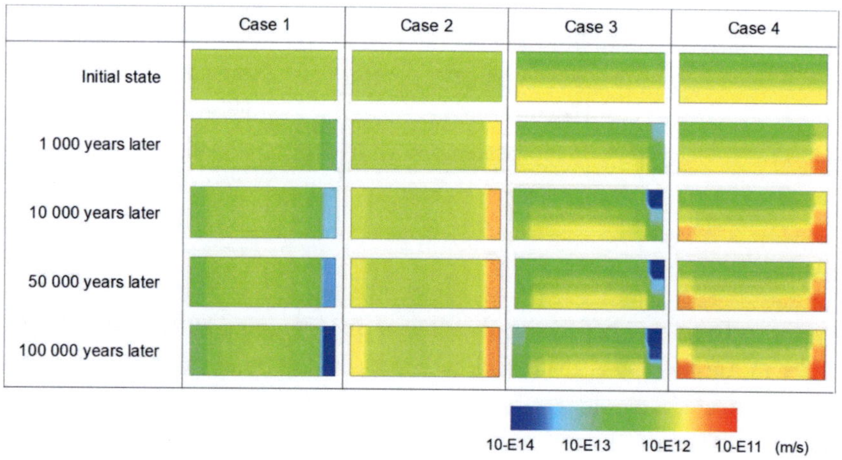

Figure 4: *Distribution of hydraulic conductivity in bentonite buffer.*

Conclusions

The scenarios of chemical reaction will affect both mechanical and hydraulic performances of the clay-based buffer. More systematic coupled analyses need to be developed. To predict the long term performance of the EBS, it will be important to model the mechanical and hydraulic properties including the cement-clay interaction.

Acknowledgement

This research is a part of "Development of the Treatment and Disposal Techniques for TRU Waste Disposal" (FY2017) under a grant from the Japanese Ministry of Economy, Trade and Industry (METI).

The research leading to these results has received funding from the European Union's Horizon 2020 Research and Training Programme of the European Atomic Energy Community (EURATOM) (H2020-NFRP-2014/2015) under grant agreement n° 662147 (CEBAMA).

Appendix A

The permeability model proposed by Kobayashi et al., (2011) was applied in this study to evaluate the hydraulic conductivity of bentonite buffer. They distinguished inter-layer voids (inner voids) of montmorillonite from other voids (outer voids) and formulated the overall hydraulic conductivity by the following equation:

$$\frac{1}{k}=\frac{1}{k_{in}}+\frac{1}{k_{ex}} \qquad \textbf{\textit{Eq. A1}}$$

where, k is a hydraulic conductivity, k_{in} a hydraulic conductivity for inner voids, and k_{ex} a hydraulic conductivity for outer voids. Using Kozeny-Carman's equation, conductivities are expressed as follows:

$$k_{in}=\frac{1}{C}\frac{\rho g}{\mu}\frac{1}{S_{vin}^{2}}\frac{e_{in}^{3}}{1+e_{in}} \qquad \textbf{\textit{Eq. A2}}$$

$$k_{ex}=\frac{1}{C}\frac{\rho g}{\mu}\frac{1}{S_{vex}^{2}}\frac{e_{ex}^{3}}{1+e_{ex}} \qquad \textbf{\textit{Eq. A3}}$$

where, C is a shape function (= 5.0), μ a coefficient of water viscosity (= 0.001 Pa·s), e_{in} an inner void ratio, e_{ex} an outer void ratio, S_{vin} a specific surface area of inner voids, and S_{vex} a specific surface area of outer voids. These quantities are defined as follows:

$$S_{vin}=S_{v}\left(\frac{e_{in}}{e}\right)^{\frac{2}{3}}, \qquad S_{vex}=S_{v}\left(\frac{e_{ex}}{e}\right)^{\frac{2}{3}} \qquad \textbf{\textit{Eq. A4, Eq. A5}}$$

$$e_{in}=w^{*}G_{s}, \qquad e_{ex}=e-e_{in} \qquad \textbf{\textit{Eq. A6, Eq. A7}}$$

where, S_v is a specific surface area (= $36\cdot10^{6}\,w^{*}\rho$ m^2/m^3), and w^* a two-layer swelling water content given as:

$$w^{*}=\frac{\alpha_{\mathrm{mon}}/100}{0.052+10^{-5.42}\rho_{d}^{12.1}} \qquad \textbf{\textit{Eq. A8}}$$

References

Arthur, R.C., Sasamoto, H., Oda, C., Honda, A., Shibata, M., Yoshida, Y., Yui, M. (1999). Development of thermodynamic databases for hyperalkaline, Argillaceous systems. JNC Technical Report, JNC TN 8400 2005-010.

Japan Atomic Energy Agency (JAEA) and Federation of Electric Power Companies of Japan (FEPC) (2005a). Second progress report on research and development for TRU waste disposal in Japan. JNC TY1400 2005-013 / FEPC TRU-TR1-2005-02. Evidence material 4-2 *(in Japanese).*

Japan Atomic Energy Agency (JAEA) and Federation of Electric Power Companies of Japan (FEPC) (2005b). Second progress report on research and development for TRU waste disposal in Japan. JNC TY1400 2005-013 / FEPC TRU-TR1-2005-02. Evidence material 4-5 *(in Japanese).*

Fujii, N., Arcilla, C.-A., Yamakawa, M., Pascua, C., Namiki, K., Sato, T., Shikazono, N., Alexander, W.-R. (2010). Natural analogue of bentonite reaction under hyperalkaline conditions: overview of ongoing work at the Zambales ophiolite, Philippines. Proceedings of the 13th International Conference on Environmental Remediation and Radioactive Waste Mangement (ICEM2010).

Gaines, G.L. and Thomas, H.C. (1953). Adsorption studies on clay minerals. II. A formulation of the thermodynamics of exchange adsorption. Journal of Chemical Physics, 21(4), 714-718.

Hameyer, K., Driesen, J., De Gersem, H., Belmans, R. (1999). The classification of coupled field problems. IEEE Transactions on Magnetics, 35(3), 1618-1621.

Ishii, T., Yahagi, R., Owada, H., Kobayashi, I., Takazawa, M., Yamaguchi, K., Takayama, Y., Tsutsumi, S., Iizuka, A. (2013). Coupled chemical-hydraulic-mechanical modelling of long-term alteration of bentonite. Clay Minerals, 48, 331-341.

Kobayashi, I., Owada, H., Ishii, T. (2011). Hydraulic/mechanical modelling of smectitic materials for HMC analytical evaluation of the long term performance of TRU geological repository. Proceedings of the 14th International Conference on Environmental Remediation and Radioactive Waste Management (ICEM2011).

Morimoto, K. (2014). Modelling of mechanical properties of bentonite buffer materials in consideration with the chemical alteration. Graduation thesis, Kobe University *(in Japanese).*

Nardi, A., Idiart, A., Trinchero, P., de Vries, L.M., Molineto, J. (2014). Interface COMSOL-PHREEQC (iCP), and efficient numerical framework for the solution of coupled Multiphysics and geochemistry. Computer and Geoscience, 69, 10-21.

Parkhurst, D.L. and Appelo, C.A.J. (1999). User's guide to PHREEQC (version 2) - a computer program for speciation, batch-reaction, one-dimensional transport, and inverse geochemical calculations. U.S. Geological Survey Water-Resources Investigations Report, 99-4259.

Pusch, R., Kasbohm, J., Thao, H.T.M. (2010). Chemical stability of montmorillonite buffer clay under repository - like conditions - A synthesis of relevant experimental data. Applied Clay Science, 47, 113-119.

Satoh, H., Ishii, T., Owada, H. (2013). Dissolution of compacted montmorillonite at hyperalkaline pH and 70°C: *in-situ* VSI and *ex-situ* AFM measurements. Clay Minerals, 48, 285-294.

Sawaguchi, T., Tsukada, M., Yamaguchi, T., Mukai, M. (2016). Effects of OH^- activity and temperature on the dissolution rate of compacted montmorillonite under highly alkaline conditions. Clay Minerals, 51, 267-278.

Sellin, P. and Leupin, O.X. (2013). The use of clay as an engineered barrier in radioactive-waste management - a review. Clays and Clay Minerals, 61(3), 477-498.

Yui, M., Sasamoto, H., Arthur, R. (1999). Groundwater evolution modeling for the second progress performance assessment (PA) report. JNC Technical Report, JNC TN 8400 99-030.

Discretization errors and porosity feedback for the long-term reactive transport model of the interactions of concrete, compacted bentonite and clay in a HLW repository in clay

Javier Samper[1*], Alba Mon[1], Jesús Fernández[1], Luis Montenegro[1]

[1] Escuela de Ingenieros de Caminos. Universidad de A Coruña (ES)

* Corresponding author: j.samper@udc.es

Abstract

Carbon steel, compacted bentonite and concrete will come in contact with each other in the near field of deep geological repositories (DGR) for high-level radioactive waste (HLW) in clay formations. The bentonite barrier could be affected by the corrosion products at the canister-bentonite interface and the hyper-alkaline conditions caused by the degradation of concrete at the bentonite-concrete interface. Additionally, the host clay formation could also be affected by the hyper-alkaline conditions at the concrete-clay interface. Here we present a non-isothermal multicomponent reactive transport model of the long-term (1 Ma) interactions of compacted bentonite with the corrosion products of a carbon-steel canister and the concrete liner of the engineered barrier of a high-level radioactive waste repository in clay. The degradation of the concrete liner leads to the precipitation of secondary minerals and a reduction of the porosity of concrete, bentonite and the clay formation at their interfaces with the concrete liner. The reduction of porosity becomes especially relevant after $t = 10^4$ years. The zones affected by pore clogging at the concrete-clay interface after 1 My are approximately equal to 3.3 cm thick. The hyper-alkaline front ($pH > 8.5$) spreads 2.5 cm into the clay formation after 1 My.

The thickness of the zones affected by a reduction of porosity and high pH computed with a coarse grid are significantly larger than those computed with a fine grid. Therefore, the finite element grid should be properly optimized to prevent numerical errors in the prediction of pore clogging and high-pH zones. The numerical results show that the porosity feedback starts to be relevant after $t > 10^4$ years. By neglecting the porosity feedback one may overestimate the thickness of the pore clogging zones for $t = 2.5 \cdot 10^4$ years. The relevance of porosity feedback could increase for larger times.

Introduction

Concrete and cement are used for mechanical support of the emplacement tunnel, and as tunnel seals and plugs in radioactive waste repositories. Here we present a non-isothermal multicomponent reactive transport model of the long-term (1 My) geochemical interactions of the concrete liner with the compacted bentonite of the engineered barrier and the clay host rock in a HLW repository according to the Spanish reference concept. As indicated by Mon et al. (2017), this non-isothermal reactive transport model greatly extends the scope of the model of Yang et al. (2008) by accounting for canister corrosion, concrete degradation, surface complexation on three types of clay mineral sorption sites, iron sorption by surface complexation, kinetic smectite dissolution and the precipitation of secondary clay minerals. We also present the analysis of the discretization errors and the feedback effect of the changes in porosity caused by the mineral dissolution / precipitation reactions.

Reactive transport model

The model corresponds to a radioactive waste repository in clay according to the Spanish Reference Concept (ENRESA, 2004). Cylindrical carbon steel canisters with 0.9 m diameter are emplaced in horizontal galleries and embedded into a 0.75 m thick bentonite buffer. A 0.3 m thick concrete support is located between the bentonite buffer and the clay formation (Figure 1). The concrete is made of ordinary Portland CEM-I cement (Hidalgo et al., 2003; Sánchez et al., 2006b). The geochemical conceptual model was taken from ENRESA (2004). The porosity of the concrete is equal to 0.085. The initial concrete composition is 0.1 vol.% of calcite, 1 vol.% of brucite, 16.5 vol.% of portlandite, 36.6 vol.% of tobermorite and 37.3 vol.% of quartz (Mon et al., 2017).

Heat and solute transport processes are assumed to have axial symmetry with respect to the axis of the galleries. Therefore, the model is discretized with a one-dimensional axisymmetric grid which accounts for the canister (8 elements), the bentonite (31 elements), the concrete (31 elements) and the clay formation (78 elements) (Figure 1). The grid size is constant in the canister with $\Delta r = 1.25$ cm. The first element in the bentonite is 1 cm long. The rest of them are 2.5 cm long. The elements in the concrete range from 0.83 to 2.5 cm. In the clay formation the elements are small near the concrete interface ($\Delta r = 0.83$ cm) and increase with radial distance. The numerical simulations were performed for a time horizon of 1 My. This is the time span used for the performance assessment of the Spanish radioactive waste repository in clay (ENRESA, 2004). An implicit Euler scheme was used for time integration. The model is performed with a sequential iterative approach. The time steps during the first 10^4 years follow a geometric sequence so that any time step after the first is obtained by multiplying the preceding time step by a constant greater than 1. Later, from 10^4 to 10^6 years the time step is equal to 100 years. Moreover, the time steps are automatically reduced if needed to achieve the convergence of the iterative process. Similar to Yang et al. (2008) and Mon et al. (2017), the simulation starts (t = 0) when the bentonite buffer is fully saturated. Therefore, there is no water flow through the system. The numerical model accounts for the thermal transient stage and the cooling of the radioactive waste (ENRESA, 2004). The numerical simulations were performed for a time horizon of 1 My. Diffusive parameters as well as the initial porewater chemical composition of the bentonite, the concrete and the clay were taken from Yang et al. (2008). The model accounts for canister corrosion, aqueous complexation, acid / base, redox reactions, mineral dissolution / precipitation, surface complexation of Fe^{2+} and H^+ on three types of sorption sites (S^SOH, $S^{W1}OH$ and $S^{W2}OH$) and cation exchange reactions of Ca^{2+}, Mg^{2+}, Na^+, K^+ and Fe^{2+}. The chemical system is defined in terms of the following 13 primary species: H_2O, H^+, Ca^{2+}, Mg^{2+}, Na^+, K^+, Cl^-, SO_4^{2-}, HCO_3^-, $SiO_2(aq)$, $O_2(aq)$, Fe^{2+} and $AlOH_4^-$. The model considers 16 minerals (brucite, calcite, gyrolite, tobermorite, quartz, dolomite, portlandite, sepiolite, gypsum, ettringite, Fe(s), magnetite, siderite, goethite, smectite and analcime) and 58 aqueous complexes which were identified from speciation runs performed with EQ3/6 (Wolery, 1992). Secondary mineral phases are allowed to precipitate anywhere in the model domain. Cation exchange reactions are modelled with the Gaines-Thomas convention (Gaines and Thomas, 1953). The triple sorption site model of Bradbury and Baeyens (1997) is used to model surface complexation reactions. All the reactions except for canister corrosion and smectite dissolution are assumed at chemical equilibrium. The corrosion rate is constant and equal to 2 μm/year (Samper et al., 2016). The model does not account for the passivation effect of the high pH. Smectite dissolution was simulated by assuming the chemical formulation of the FEBEX-montmorillonite derived by Fernández et al. (2009) and by using the following kinetic rate law (Eq. 1) (Sánchez et al., 2006b):

$$r_m = -k_m a_{OH}^n e^{-\frac{Ea}{RT}} \qquad \textbf{\textit{Eq. 1}}$$

where r_m is the dissolution / precipitation rate (mol/m²/s), k_m is the kinetic rate constant ($2\cdot10^{-13}$ mol/m²/s) at 25°C, E_a is the activation energy (22.7 kJ/mol), R is the gas constant, T is the temperature (K), a_{OH}^n is a catalytic term which depends on the activity of the OH^- and n is an exponent which is equal to 0.5. The smectite specific surface area, σ, is assumed equal to 100 dm^2 per liter of water (Fernández et al., 2009). This surface area is several orders of magnitude smaller than the total surface area.

The numerical model accounts for the thermal transient stage during the cooling of the radioactive waste. Figure 2 shows the time evolution of the temperatures prescribed at the canister-bentonite interface (r = 0.45 m), the bentonite-concrete interface (r = 1.2 m), the concrete-clay interface (r = 1.5 m) and the outer boundary (r = 25 m). These temperatures were taken from ENRESA (2004). The temperature in the canister is about 98°C at t = 10 years. Then, it decreases to 45°C at t = 10^4 years and reaches 19.7°C at 1 My. The maximum temperature at the bentonite-concrete interface is 75.8°C at t = 10 years. The maximum temperature at r = 25 m is 42°C at 10^3 years.

The model was performed with $CORE^{2D}$ V4 (Samper et al., 2009 and 2011). It is a code for transient saturated and unsaturated water flow, heat transport and multicomponent reactive solute transport under both local chemical equilibrium and kinetic conditions in heterogeneous and anisotropic media. The chemical formulation is based on the ion association theory and uses an extended version of Debye-Hückel equation (B-dot) for the activity coefficients of aqueous species. $CORE^{2D}$ V4 relies on the *com* thermodynamic database of EQ3/6 (Wolery, 1992).

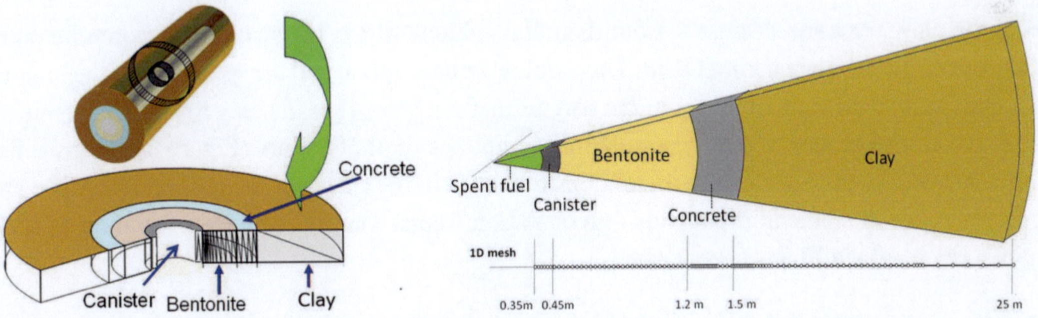

Figure 1: *Scheme of the multibarrier system of a HLW repository in clay according to the Spanish Reference Concept (Yang et al., 2008) and 1-D finite element grid which accounts for the canister, the bentonite barrier, the concrete liner and the clay formation (Mon et al., 2017).*

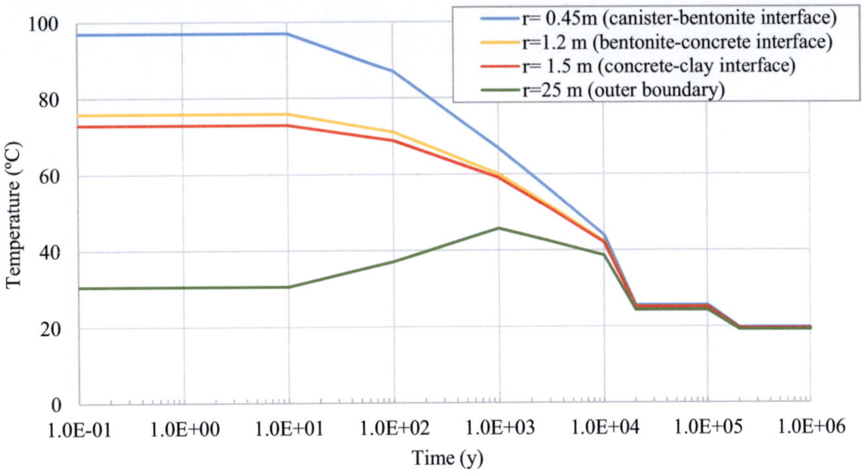

Figure 2: *Time evolution of the prescribed temperatures at the canister-bentonite interface (r = 0.45 m), the bentonite-concrete interface (r = 1.2 m), the concrete-clay interface (r = 1.5 m) and outer boundary (r = 25 m).*

Model results

Solute concentrations and pH of porewaters in the system are mostly affected by: 1) Solute diffusion from (into) the bentonite into (from) the concrete and 2) Mineral dissolution / precipitation caused by the spreading of the hyper-alkaline conditions from the concrete into the bentonite buffer and the clay formation. Figure 2 shows the computed pH at the canister-bentonite interface (r = 0.45 m), the bentonite near the concrete interface (r = 1.125 m), the concrete midpoint (r = 1.35 m) and the clay near the concrete interface (r = 1.525 m). The pH in the concrete increases until $8 \cdot 10^4$ years while portlandite dissolves and the temperature decreases. Then, the pH is buffered at 12.4 until portlandite is exhausted. Sepiolite and brucite precipitate in the concrete after 10^5 years and later also in the bentonite. The precipitation of these mineral phases leads to a decrease in pH.

The pH in the bentonite is affected by the hyper-alkaline front and canister corrosion. Near the concrete interface, pH increases because of the hyper-alkaline conditions from the concrete liner. The pH at the canister-bentonite interface increases until $8 \cdot 10^4$ years due to the canister corrosion (pH = 12.4). The pH in the bentonite increases radially from the canister outwards and radially inwards from the concrete. These pH fronts spread and lead to a uniform pH equal to 12.4 at t = $8 \cdot 10^4$ years. Once portlandite is exhausted (t = $8 \cdot 10^4$ years) and while tobermorite is dissolving until t = $1.2 \cdot 10^5$ years, the pH remains around 12.4. The pH starts to decrease at t = 10^5 years and decreases even faster when tobermorite is exhausted and sepiolite, calcite and brucite precipitate.

The pH in the clay formation decreases from its initial value until t = 10^2 years when dolomite precipitates slightly as a result of the transient thermal field. Then, dolomite dissolution and the spreading of the high-pH front from the concrete lead to an increase in pH in the clay formation. The pH in the clay near the concrete interface (r = 1.525 m) increases because of the hyper-alkaline conditions until 10^4 years (Figure 3). Later, it decreases when sepiolite and brucite precipitate. The oscillations of the pH in the clay near the concrete interface are due to the strong interplays of the mineral dissolution / precipitation fronts. The accurate representation of such fronts would require a more refined finite element mesh.

The pH front, considering a threshold value of 8.5 higher than the initial pH of the clay, migrates 2.5 cm into the clay formation after 1 My.

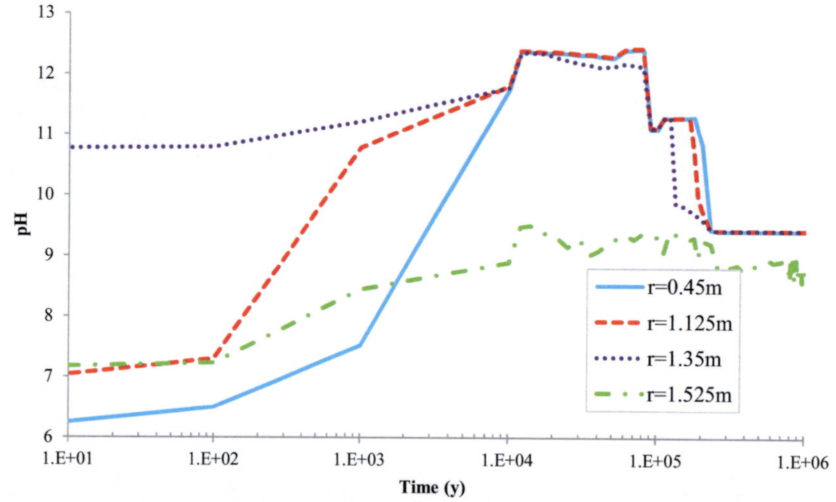

Figure 3: *Time evolution of the computed pH at the canister-bentonite interface (r = 0.45 m), the bentonite near the concrete interface (r = 1.125 m), the concrete midpoint (r = 1.35 m) and the clay near the concrete interface (r = 1.525 m).*

Figure 4 shows the time evolution of the concentration of the cumulative mineral precipitation / dissolution in the bentonite near the concrete interface (r = 1.125 m), the concrete midpoint (r = 1.35 m) and the clay near the concrete interface (r = 1.525 m). In the concrete midpoint (r = 1.35 m), portlandite dissolves and is exhausted at $t = 2\cdot10^4$ years. Tobermorite starts dissolving at $t = 9\cdot10^4$ years once portlandite is exhausted. The dissolution of tobermorite leads to the precipitation of calcite, gypsum, sepiolite and brucite. It should be noticed that the cumulative amount of precipitated gypsum in the concrete is much greater than in the bentonite.

Brucite precipitates in the bentonite near the concrete interface (r = 1.125 m) from $t = 10^2$ to $t = 2\cdot10^5$ years. Later, it dissolves. Sepiolite precipitates at $t = 2\cdot10^5$ years. Its precipitation rate increases when the dissolution of tobermorite increases the concentration of dissolved silica. Quartz dissolves until $t = 10^3$ years. Gypsum precipitates after $t = 10^5$ years when the temperature decreases. Calcite precipitates in the bentonite barrier near the concrete interface. The dissolution rate of smectite is largest near the concrete interface where the pH is highest. Analcime precipitation (not shown here) is clearly linked to the dissolution of smectite.

The increase in temperature causes the dissolution of quartz and dolomite and the precipitation of calcite at the clay near the concrete interface (r = 1.525 m). Therefore, the concentrations of some minerals do not start from zero. Brucite and gypsum precipitate. Sepiolite and calcite also precipitate, but at larger rates.

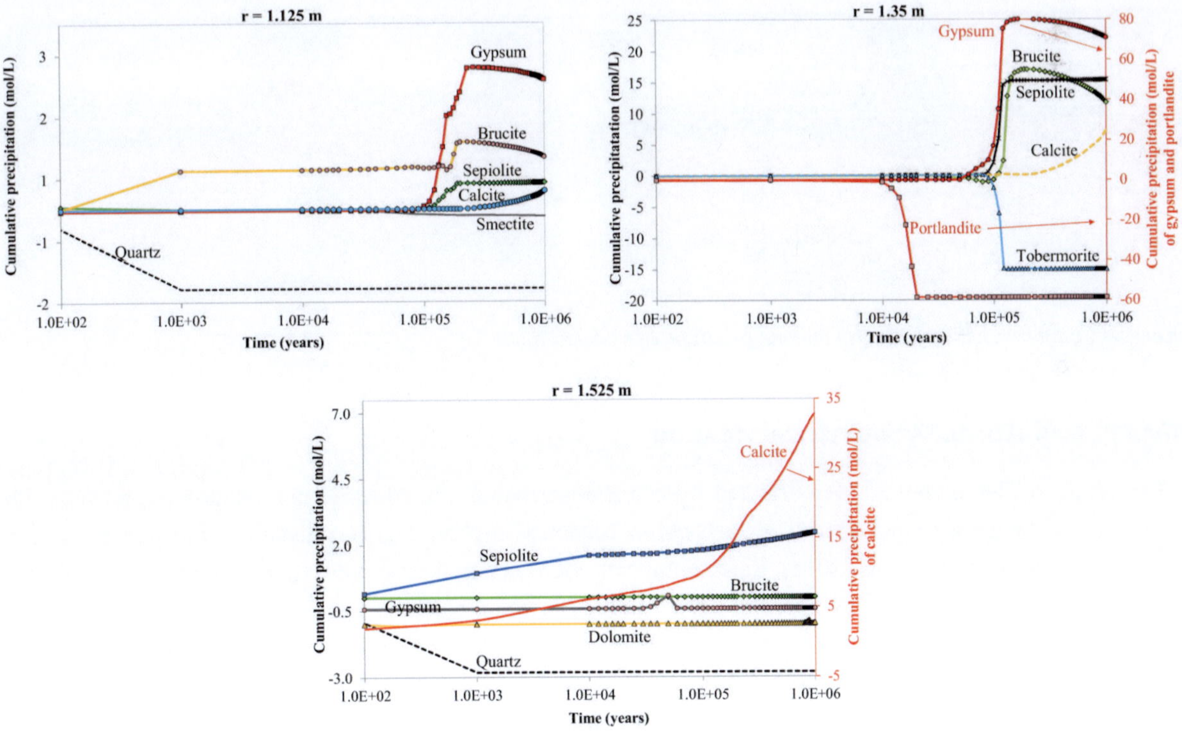

Figure 4: *Time evolution of the computed cumulative mineral dissolution / precipitation in the bentonite buffer near the concrete interface (r = 1.125 m) (top left), at the midpoint of the concrete liner (r = 1.35 m) (top right) and in the clay formation near the concrete interface (r = 1.525 m) (bottom) (positive for precipitation and negative for dissolution).*

Figure 5 shows the computed pH and mineral volume fractions at $t = 10^3$, 10^4, 10^5 and 10^6 years. It can be seen that the volume fraction of tobermorite in the concrete decreases with time and increases in the bentonite until $t = 10^5$ years. The portlandite volume fraction in the concrete decreases. The precipitation of gypsum refills some of the voids left by the dissolution of portlandite. Tobermorite dissolution causes the precipitation of sepiolite.

Sepiolite, gypsum and brucite occupy the volume fraction previously occupied by portlandite and tobermorite at t = 1 My.

Magnetite precipitation progresses in the canister and in the bentonite near the canister interface as the canister corrodes. The volume fraction of quartz in the bentonite and in the clay decreases.

The volume fraction of dolomite in the clay near the concrete interface decreases with time when calcite and sepiolite precipitate. Dolomite and brucite precipitate at 1My in this interface and therefore the porosity of the clay decreases.

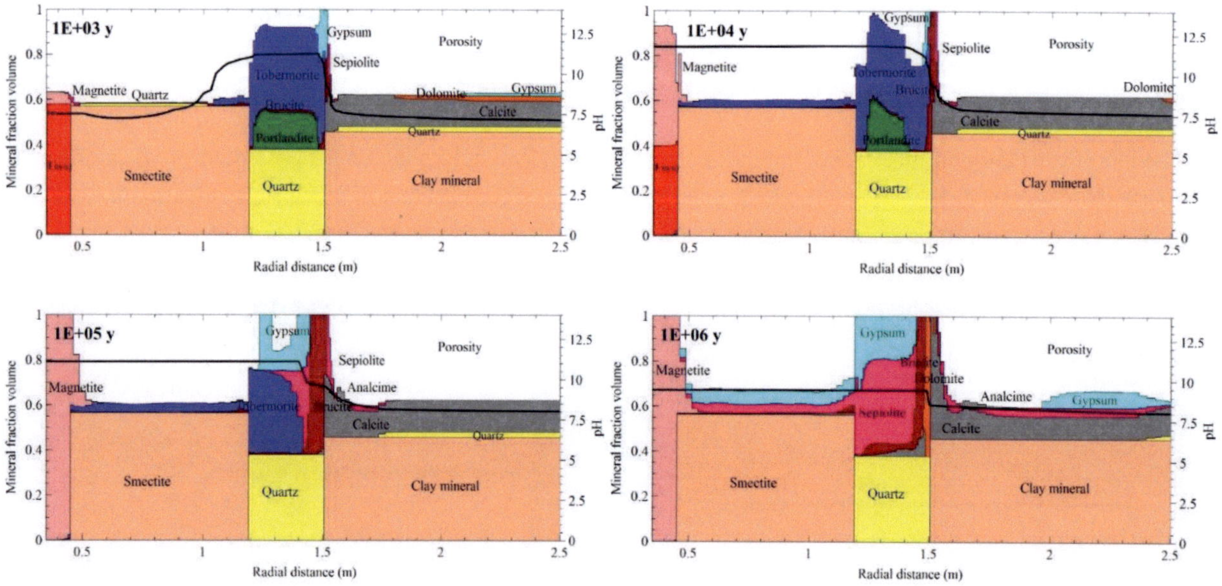

Figure 5: *Computed pH and mineral volume fractions at selected times.*

Effect of grid size on the numerical solution

The zones of bentonite and clay affected by canister corrosion and concrete degradation are narrow. Their quantification requires proper spatial discretization schemes to minimize the discretization numerical errors (Marty et al., 2009; Hayek et al., 2011). Since computer codes generally use consistent numerical methods, these errors can be decreased simply by appropriately refining the finite element grid. Three grids with 148, 108 and 66 elements (fine, intermediate and coarse grids, respectively) were used to evaluate the discretization errors in the computed pH, the mineral volume fractions, the changes in porosity and the pore clogging. The results presented in the previous section *Model results* correspond to the fine grid.

The mineral volume fractions and the pH computed with the coarse grid show significant differences compared to those computed with the intermediate and fine grids which coincide for the most part (Mon et al., 2017). The differences are especially substantial for tobermorite and portlandite in the concrete and sepiolite and gypsum in the clay near the concrete interface for $t < 10^4$ years. These differences occur also in the bentonite for $t > 10^4$ years (see Mon et al., 2017). The spreading of the high pH front in the clay formation with the coarse grid is larger than that of the fine grid. Figure 6 shows the changes in porosity computed with the fine, intermediate and coarse grids. The porosities computed with the fine and intermediate grids are always similar. The zones of bentonite and clay affected by a reduction of porosity computed with the coarse grid are significantly wider than those computed with the intermediate and fine grids. The numerical errors of the coarse grid are largest at t = 1 My.

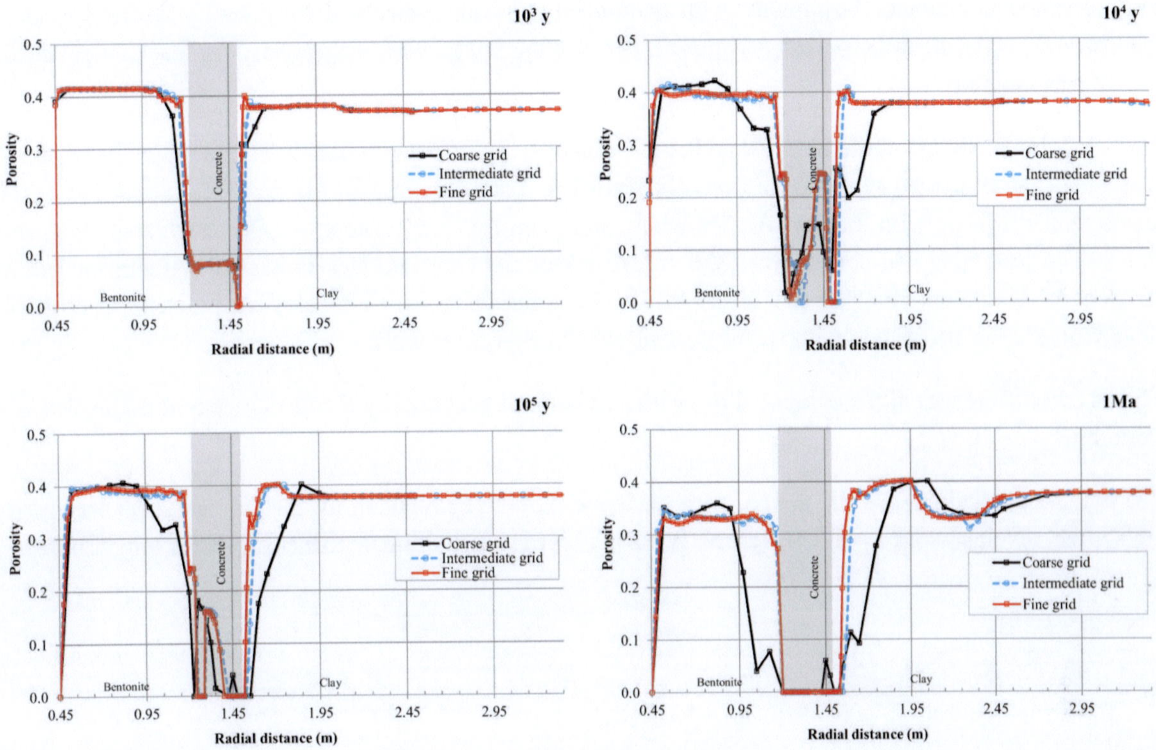

Figure 6: *Radial distribution of the porosity at t = 10^3, 10^4, 10^5 and 10^6 years computed with the fine, intermediate and coarse grids. Porosity changes due to mineral dissolution / precipitation reactions.*

Porosity feedback

In the simulations without porosity feedback, the porosity remains constant in the transport and chemical equations. The changes in porosity that could be caused by mineral dissolution / precipitation reactions are calculated from model results. The feedback effect of the porosity caused by dissolution / precipitation reaction has been tested with a sensitivity run. This sensitivity run has been performed with an updated version of CORE^{2D} which has been developed within the framework of the CEBAMA project. The updated version of the code takes into account the porosity changes due to mineral dissolution/precipitation reactions and the feedback on transport parameters. The porosity is updated each time step according to the computed changes in the mineral volume fractions. The porosity is not allowed to be less than a prescribed threshold porosity of 10^{-4}. The changes in the permeability are calculated with the Kozeny-Carman equation (Carman, 1937) while the pore diffusion coefficients and the solute tortuosities are updated based on Archie's law (Archie, 1942) with an exponent equal to $^4/_3$. Other flow and transport parameters such as nodal volumes, water velocities, mineral specific surface areas and nodal dissolved concentrations are also updated every time step to account for the changes in the porosities. The permeability has not been updated in the sensitivity run because there is no water flow.

Here we report the results for t = $2.5 \cdot 10^4$ years. Figure 7 shows the computed porosity and the pH with and without the porosity feedback at $2.5 \cdot 10^4$ years. The major differences in mineral precipitation and pH occur in the concrete and at the concrete-clay and canister-bentonite interfaces. Portlandite dissolution in the run with the porosity feedback is slightly faster than that computed without the porosity feedback. The results with the porosity feedback indicate less gypsum, brucite and sepiolite precipitation and more calcite precipitation in the concrete and clay near the concrete interface. The differences in the computed porosity in the concrete are due to portlandite dissolution.

The computed reduction of the porosity in the bentonite-canister interface with the porosity feedback is smaller than that computed without the porosity feedback, because the model with the porosity feedback calculates less magnetite precipitation.

Pore clogging thickness in the concrete and at the concrete-clay interface considering the porosity feedback is smaller than that computed without the porosity feedback. Pore clogging in the concrete at $2.5 \cdot 10^4$ years is predicted at r = 1.34 m with the porosity feedback, and from r = 1.24 m to r = 1.28 m without the porosity feedback. The pore clogging computed in the concrete near the clay interface is 2.5 cm without the porosity feedback and 0.83 cm with the porosity feedback. The clogging thickness in the clay near the concrete interface is 0.83 cm and 1.67 cm with and without the porosity feedback, respectively.

The largest differences in the computed pH with and without the porosity feedback occur at both sides of the concrete-liner.

The numerical results show that the porosity feedback starts to be relevant for $t > 10^4$ years. We have shown that this effect is relevant for $t = 2.5 \cdot 10^4$ years. Work is in progress to evaluate these effects for longer times.

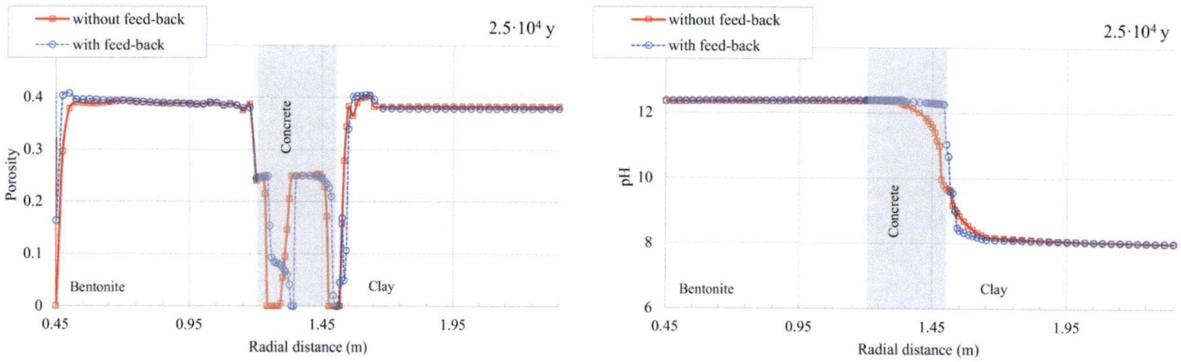

Figure 7: *Radial distribution of the porosity (left) and pH (right) at* $t = 2.5 \cdot 10^4$ *years computed with and without the porosity feedback.*

Comparison with other works

Similar to Shao et al. (2013) who found that pore clogging occurs in the clay formation after $2 \cdot 10^4$ years, our model predicts pore clogging in the clay formation at about $t = 10^4$ years. Our results share the features reported by Kosakowski and Berner (2013) such as narrow alteration zones with pore clogging. Similar to Berner et al. (2013), our predictions display that smectite dissolution is very small. Berner et al. (2013) predict that the pH front spreads about 10 cm in the clay formation after $3 \cdot 10^4$ years. The pH front in our model reaches around 4.17 cm after 10^4 years. The front of high pH migrates 2.5 cm after 1 My.

Conclusions and future work

A non-isothermal multicomponent reactive transport model of the long-term (1 My) interactions of the concrete with compacted bentonite and the clay host rock in a high-level radioactive waste repository in clay has been presented. Concrete degradation leads to the precipitation of secondary minerals and the reduction of the porosity of the bentonite and the clay formation at their interfaces with the concrete liner. The decrease of the porosity is especially important for $t > 10^4$ years. The numerical results show that magnetite is the main corrosion product. The Eh in the bentonite near the canister decreases from the initial value of -0.078 V to -6.7 V after $5 \cdot 10^4$ years (Samper et al., 2016). Its precipitation reduces the bentonite porosity near the canister. The zones affected by pore clogging at the canister-bentonite, and concrete-clay interfaces at 1 My are equal to 1 and 3.3 cm thick, respectively. The pH in the bentonite after 1 My is uniform and equal to 9.42. At that time, the hyper-alkaline front (pH > 8.5) migrates 2.5 cm into the clay formation.

The pH and porosities computed with fine and coarse grids show similar trend but the zones of concrete, bentonite and clay affected by a reduction of porosity and the spreading of the high pH front in the clay computed with coarse grids are significantly wider than those computed with finer grids are. The major differences of the porosity computed with and without the porosity feedback are located in the concrete, and at the concrete-clay and canister-bentonite interfaces. The thickness of pore clogging in the concrete and in the concrete-clay interface computed with the model with the porosity feedback is smaller than that computed without the porosity feedback.

The results of the long-term geochemical predictions presented here could be improved by: 1) Updating the smectite kinetic law and reactive surface area; 2) Reviewing and updating the concrete and bentonite mineral phases to include additional C-S-H, C-A-S-H, magnesium phases and secondary clay minerals such as zeolites; 3) Accounting for the changes in CEC caused by smectite dissolution; 4) Accounting for the passivation effect of the high pH on canister corrosion as indicated by Kursten and Druyts (2015); and 5) Accounting for the porosity feedback for 1 My.

Acknowledgements

The research leading to this work has received funding from the CEBAMA Project of the European Atomic Energy Community's (Euratom) Horizon 2020 Programme (NFRP-2014/2015) under grant agreement # 662147.

This work was partly funded by ENRESA (Spain) and the Spanish Ministry of Economy and Competitiveness (Project CGL2016-78281) with support from the FEDER funds. The second author enjoyed a research contract from University of A Coruña and the third author had a Contract from the FPI Program of the Spanish Ministry of Economy and Competitiveness.

We thank Andrés Idiart (AMPHOS 21), Georg Kosakowski (PSI), Jaime Cuevas (UAM) and the EUG reviewers for their comments and suggestions which have contributed to improve this paper. We also thank Erika Neeft (COVRA) for her comments and suggestions on canister corrosion.

References

Archie, G. (1942). The electrical resistivity log as an aid in determining some reservoir characteristics. Pet. Trans. AIME, 146, 54-62.

Berner, U., Kulik, D.A., Kosakowski, G. (2013). Geochemical impact of a low-pH cement liner on the near field of a repository for spent fuel and high-level radioactive waste. Phys. Chem. Earth., 64, 46-56.

Bradbury, M.H. and Baeyens, B. (1997). A mechanistic description of Ni and Zn sorption on Na-montmorillonite. Part II: Modelling. J. Contam. Hydrol., 27, 223-248.

Carman, P.C. (1937). Fluid through granular beds. Trans. Inst. Chem. Eng., 15, 150-166.

Gaines, G.I. and Thomas, H.C. (1953). Adsorption studies on clay minerals II. A formulation of the thermodynamics of exchange adsorption. J. Chem. Phys., 21, 714-718.

ENRESA (2004). Evaluación del comportamiento y de la seguridad de un almacén geológico profundo de residuos radiactivos en arcilla. ENRESA Report, 49-1PP-M-A1-01 *(In Spanish).*

Fernández, A., Cuevas, J., Mäder, U.K. (2009). Modelling concrete interaction with a bentonite barrier. Eur. J. Mineral., 21, 177-191.

Hayek, M., Kosakowski, G., Churakov, S. (2011). Exact analytical solutions for a diffusion problem coupled with a precipitation-dissolution reaction and feedback of porosity change. Water Resour. Res., 47, W07545.

Hidalgo, A., Alonso, C., Andrade. C., Fernández, L. (2003). ECOCLAY II. Review of concrete formulations actually defined for radioactive repositories waste. WP2. IETCC (CSIC). Interim report.

Kosakowski, G. and Berner, U. (2013). The evolution of clay rock/cement interfaces in a cementitious repository for low- and intermediate level radioactive waste. Phys. Chem. Earth., 64, 65-86.

Kursten, B. and Druyts, F. (2015). Assessment of the uniform corrosion behaviour of carbon steel radioactive waste packages with respect to the disposal concept in the geological Dutch Boom Clay formation. Technical Report, OPERA-PU-SCK513.

Marty, N.C.M., Tournassat, C., Burnol, A., Giffaut, E., Gaucher, E.C. (2009). Influence of reaction kinetics and mesh refinement on the numerical modelling of concrete/clay interactions. J. Hydrol., 364, 58-72.

Mon, A., Samper, J., Montenegro, L., Naves, A., Fernández, J. (2017). Long-term non-isothermal reactive transport model of compacted bentonite, concrete and corrosion products in a HLW repository in clay. J. Contam. Hydrol., 197, 1-16.

Shao, H., Kosakowski, G., Berner, U., Kulik, D.A., Mäder, U., Kolditz, O. (2013). Reactive transport modeling of the clogging process at Maqarin natural analogue site. Phys. Chem. Earth., 64, 21-31.

Samper, J., Xu, T., Yang, C. (2009). A sequential partly iterative approach for multicomponent reactive transport with $CORE^{2D}$. Comput. Geosci., 13, 301-316.

Samper, J., Yang, C., Zheng, L., Montenegro, L., Xu, T., Dai, Z., Zhang, G., Lu, C., Moreira, S. (2011). Chapter 7: $CORE^{2D}$ V4: A code for water flow, heat and solute transport, geochemical reactions, and microbial processes. *In:* Zhang, F., Yeh, G.-T., Parker, C., Shi, X. (Ed) Groundwater Reactive Transport Models. Bentham Science Publishers.

Samper, J., Naves, A., Montenegro, L., Mon, A. (2016). Reactive transport modelling of the long-term interactions of corrosion products and compacted bentonite in a HLW repository in granite: uncertainties and relevance for performance assessment. Appl. Geochem., 67, 42-51.

Sánchez, L., Cuevas, J., Ramírez, S., Ruiz De León, D., Fernández, R., Vigil Dela Villa, R., Leguey, S. (2006a). Reaction kinetics of FEBEX bentonite in hyperalkaline conditions resembling the cement-bentonite interface. Appl. Clay Sci., 33, 125-141.

Sánchez, L., Cuevas, J., Vigil Dela Villa, R., Ramírez, S., Fernández, R., Leguey, S. (2006b). The alkaline reaction of FEBEX bentonite: a contribution to the study of the performance of bentonite/concrete engineered barrier systems. J. Iberian Geol., 32(2), 151-174.

Wolery, T.J. (1992). EQ3/6, a software package for geochemical modeling of aqueous systems: package overview and installation guide (version 7.0). PNNL Technical Report, UCRL-MA-110662-Pt 1.

Yang, C., Samper, J., Montenegro, L. (2008). A coupled non-isothermal reactive transport model for long-term geochemical evolution of a HLW repository in clay. Environ. Geol., 53, 1627-1638.

Modelling of interactions between hydrated OPC and Czech bentonite under in-situ conditions and by heating

Tomáš Rosendorf [1,2*], Dušan Vopálka [1], Radek Červinka [2]

[1] Department of Nuclear Chemistry, Czech Technical University in Prague (CZ)
[2] Fuel Cycle Chemistry Department, ÚJV Řež a.s. (CZ)

* Corresponding author: tomas.rosendorf@fjfi.cvut.cz

Abstract

This study is a prior step towards the modelling of real interactions between hydrated OPC and groundwater from Josef Underground Facility (GW Josef) with or without addition of Czech bentonite under *in-situ* conditions (10°C) and by heating (95°C) performed in cooperation with Czech laboratories (UJV and CTU-CEG) as a part of WP1. In this contribution, bentonite dissolution and formation of new phases in model cement-leach aqueous solution has been studied by geochemical calculations in PhreeqC. Compositions of the Czech commercial Bentonite 75 (denoted as B75) and the hydrated OPC (CEM II A-S 42.5R) taken into consideration by modelling were based on experimental XDR analyses. The lack of OPC in identified phases (no S-phases, no amorphous Na/K-compounds, no quantitative information) led to a simplification of modelled systems. In the systems without bentonite (hydrated OPC contacted with GW Josef) initial concentration of dominant cations Na/K-species has been raised, with the aim to adjust ionic strength and pH, respecting real analysis of cement leachates in liquid phases performed for nine months of interaction. To improve this model, further XRF analysis of hydrated OPC is needed, as the prediction of the C/S ratio is necessary. Influences of phase ratio $m_{B75} : m_{OPC} : m_{GW}$ and temperature on dissolution, forming phases or composition of pore water were studied. Despite the simplification of the model, the results (e.g., formation of C-A-S-H and phases based on $CaCO_3$) on the boundary of cement and bentonite are in an agreement with experiments.

Introduction

Among aims of the Czech laboratories (UJV and CTU-CEG) in WP1 are experimental studies of long-term (9, 18, 27 months) interactions between hydrated OPC cylinders and underground water (GW Josef) or with bentonite suspension (Czech bentonite B75 with GW Josef) at defined temperatures of 10°C (*in-situ* conditions of Josef Underground Facility) and 95°C (heating to accelerate alteration processes) in experimental vessels. The set of experimental setups is schematically described as follows:

- Vessel 4: hydrated OPC and bentonite suspension at 95°C,
- Vessel 9: hydrated OPC and bentonite suspension at 10°C,
- Vessel 11: hydrated OPC and GW Josef at 95°C,
- Vessel 14: hydrated OPC and GW Josef at 10°C,
- Vessel 17: bentonite suspension at 10°C, see Figure 1 and Večerník et al. (2016).

The goal of the modelling to be performed in CTU-DNC (Department of Nuclear Chemistry) within CEBAMA project is an attempt to predict the alterations of hydrated OPC or Czech bentonite (or both) as a result of an interaction with groundwater. In addition, the model results should be compared with experimental data obtained by the long-term interactions of samples of hydrated cement paste and bentonite with groundwater, e.g., dissolution of bentonite or hydrated OPC, precipitation on the boundary of OPC / bentonite, and formation of new phases in all experimental vessels. Obtained data could be used in future safety assessments of deep geological repositories.

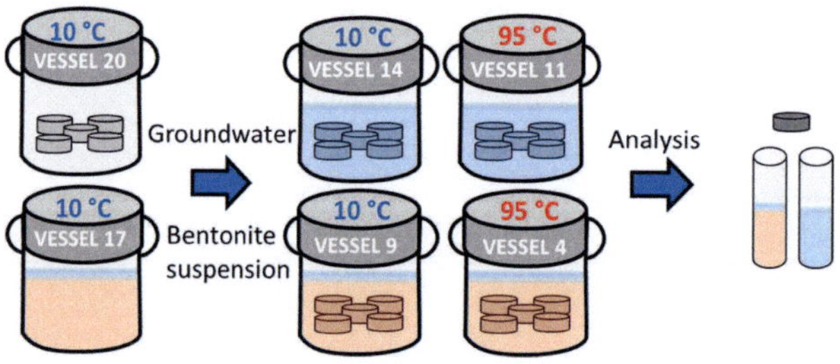

Figure 1: *Laboratory procedures: Interactions in the system: OPC thin plates / groundwater and bentonite suspension.*

Methods and conditions

For this study, PhreeqC version 3.3.11 (Parkhurst and Appelo, 2013) has been used with the ThermoChimie v. 9.0b thermodynamic database (Giffaut et al., 2014), that includes the Specific Ion Interaction Theory (SIT) parameters (Ciavatta, 1980; Guggenheim and Turgeon, 1955). Simulations have been based on batch calculations of the equilibrium state.

The composition of underground water GW Josef is described in Večerník et al. (2016). The GW Josef was equilibrated with atmospheric CO_2 (partial pressure: $10^{-3.39}$ atm) before modelled interaction with solid phases.

The bentonite has been modelled as a solid solution of Na-/K-/Mg-/Ca-montmorillonites, quartz, calcite, illite and kaolinite. Its composition is based on the results of XRD analysis and the cation distribution in the montmorillonite corresponding to the cation representation (using data obtained by Cu(II)-trien method) of the bentonite B75 (see Table 1).

Table 1: *Composition of ideal solid solutions (B75 and OPC) in weight ratio (w) and their molecular weights (M) in ThermoChimie v.9.0b SIT database.*

Bentonite phases: based on XRD	M (g/mol)	w (g/kg)	OPC major phases: estimation	M (g/mol)	w (g/kg)
Me-Montmorillonite (Me-Y· n H_2O)		550	Portlandite	74	400
→ $Na_{0.34}Y$	436	175	Ettringite	1,255	250
→ $K_{0.34}Y$	408	16	Calcite ($CaCO_3$ phases)	100	250
→ $Mg_{0.17}Y$	447	134	Amorphous + minor phases	--	100
→ $Ca_{0.17}Y$	446	225			
K-illite	393	20			
Quartz (S-phases)	60	80			
Kaolinite	258	15			
Calcite ($CaCO_3$ phases)	100	35			
Amorphous + minor phases	--	300			

The hydrated OPC has been modelled in a simplified way as an ideal solid solution of portlandite, ettringite and calcite as main phases. The XRD analysis did not reveal S-phases (that are part of C-S-H) and any Na/K-rich compounds. Solubility of alkalis is not controlled by any mineral phase; thus, they were considered as cationic species in the liquid phase influencing significantly ionic strength and pH. The lack of identified phases led to a simplification of modelled systems, in which we preferred the prediction of reactions, results of which could be compared with experimental results. Estimated weight ratio of portlandite : ettringite : calcite was set to 400 : 250 : 250 g/kg, see Table 1. To adjust ionic strength and pH for models in Vessel 11 (respectively in Vessel 14), the concentrations of dominant cations Na and K in GW Josef were set to 89 (resp. 73) and 485 (resp. 490) mg/L, that were estimated in cement leachates in these vessels.

As the first step of the modelling, phase ratio was set to: 375 g dry hardened OPC, 1 kg B75 and 5 kg GW Josef for the arrangement like real abundances of mentioned phases in Vessel 4 and Vessel 9 (related to Figure 1), respectively 375 g dry hardened OPC and 5 kg GW Josef (enriched by NaOH and KOH) for Vessel 11 and Vessel 14.

However, alteration in the whole volume of cementitious blocks cannot be expected in duration of the project, mainly in Vessel 4 and Vessel 9 (system of unstirred reactors): on the boundary of OPC and bentonite suspension, it can be expected that there will be the dissolution of cementitious phases small in comparison with the bentonite, and that the weight ratio of GW Josef : bentonite will be smaller than 5:1 due to the deposition of solids on the bottom of the vessel and by the water evaporation which was during the experiment duration compensated by adding of water every second week. Therefore, different phase ratios have been set for each model, as shown in Table 2.

The pe value was fixed to the measured experimental value that was close to unity in all vessels.

Table 2: *Conditions of model experiments: T [°C], weight ratio OPC : bentonite : GW Josef.*

	VESSEL 4 / VESSEL 9					VES. 11	VES. 14	VES. 17
T [°C]	95 / 10					95	10	10
identification:	4_A / 9_A	4_B / 9_B	4_C / 9_C	4_D / 9_D	4_E / 9_E	11_A	14_A	17_A
weight ratio [-] (OPC:B75:GW)	1:2.7:13.3	1:33:100	1:30:75	1:200:600	1:200:500	1:0:13.3	1:0:13.3	0:1:5

Results

A preliminary character of results must be underlined. An important uncertainty was caused by insufficient characterization of hardened OPC (e.g., absence of C-S-H phases, especially SiO_2, in XRD analysis that gave only qualitative information). Additional analysis (XRF, or thermogravimetric methods) can help to better estimate the ratios between main cement phases (CaO:SiO_2:H_2O-bounded) and can improve the model.

The modeling results of solid phase weight ratios in the system of hardened OPC, bentonite (B75) and GW Josef are shown in Figure 2 and Figure 3. The results show modelled dissolution of quartz, illite, kaolinite from bentonite, portlandite, in some cases ettringite (models 4_A/B/…/E, 9_D and 9_E, 11_A) from hardened OPC and conversion of the Mg-montmorillonite to that of "Ca" form. In addition, we can expect the formation of new phases (saturation index $SI \geq 0$) or increasing amount of the phases at the boundary of hardened OPC / bentonite based on $CaCO_3$ (calcite, aragonite), C(-A)-S-H phases, in some cases also ettringite (models 9_A/B/C), hydrotalcite and brucite (models 11_A, 14_A). The model proved formation of C(-A)-S-H phases or carbonate phases that was observed in real (see Figure 4).

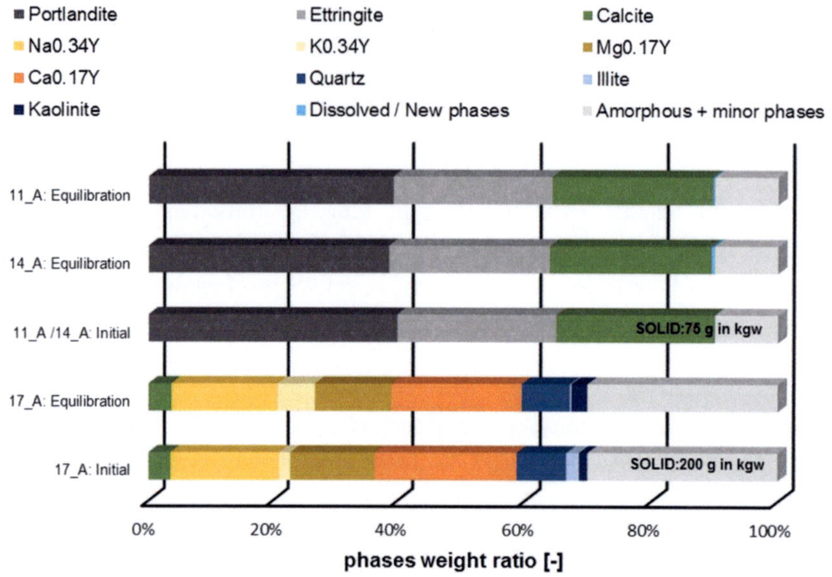

Figure 2: *Solid phase weight ratio in the vessels with OPC and GW Josef (Vessel 11 and Vessel 14) or bentonite (B75) and GW Josef (Vessel 17) before and after model equilibration of systems (for identification of models see Table 2).*

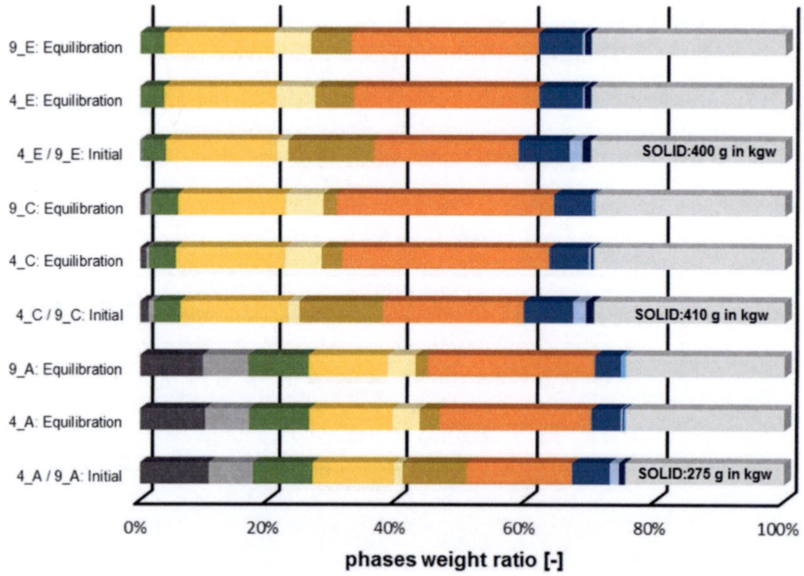

Figure 3: *Model of solid phase weight ratio in Vessel 4 and Vessel 9 (chosen series A, C, E, see Table 2) with OPC, bentonite (B75) and GW Josef before and after equilibration of systems (for a legend see Figure 2).*

Figure 4: *Formation of new phases (white areas) on the boundary of hardened cement with bentonite (Vessel 4) after 9 months of interaction.*

Composition of liquid phases presented in Table 3 demonstrated the agreement and/or differences between experimental and modelled data for concentrations of selected species. Results showed that the prediction of concentrations of cations and anions in solution is difficult, as phases that were not identified could strongly determine the composition of the pore solutions.

Table 3: *Composition of liquid phases in all models (identification see Table 2) and experimental data. IS (ionic strength), N.A. (not analysed), grey colouring (experimental analysis), red colouring (decrease in concentration), green colouring (increase in concentration) and yellow colouring (no significant change in concentration).*

	T (°C)	pH [(1)]	IS (mol/L)	Na	K	Mg	Ca	Si	Al	SO_4^{2-}	CO_3^{2-} + HCO_3^{2-}
							(mg/L)				
GW Josef: Input	25	7.6	-	18	2	18	87	N.A.	N.A.	106	206
17_A: Equilibration	10	8.6	0.005	35	20	4	21	1	0	106	40
VESSEL 17: 9 months	25	8.7	-	397	18	5	4	N.A.	N.A.	150	879
14_A: Equilibration	10	12.5	0.057	73[(2)]	490[(2)]	18	615	0	0	106	1
VESSEL 14: 9 months	10	12.4	-	73	490	0	40	N.A.	N.A.	41	0
11_A: Equilibration	95	12.4	0.037	89[(2)]	485[(2)]	18	395	0	10	157	2
VESSEL 11: 9 months	95	12.6	-	89	485	0	272	N.A.	N.A.	285	0
4_A: Equilibration	95	12.1	0.031	2	4	0	554	140	5	138	1
9_A: Equilibration	10	12.3	0.070	39	21	0	937	527	0	0	1
4_B: Equilibration	95	11.7	0.027	2	4	0	425	99	2	512	1
9_B: Equilibration	10	11.9	0.031	29	17	0	396	304	0	1	1
4_C: Equilibration	95	11.7	0.028	2	4	0	444	104	2	529	1
9_C: Equilibration	10	11.9	0.034	30	17	0	439	338	0	1	1
4_D: Equilibration	95	10.4	0.009	2	4	0	116	19	0	202	2
9_D: Equilibration	10	9.5	0.008	25	12	0	81	2	0	202	2
4_E: Equilibration	95	10.5	0.009	2	4	0	125	20	0	221	2
9_E: Equilibration	10	9.6	0.009	26	12	0	88	2	0	221	2

(1) pH values were recalculated on temperature T = 25°C to compare with experimental data that were measured at 25°C

(2) Initial model solution of GW Josef was enriched by NaOH and KOH with the aim to respect experimental concentration values of Na and K

Results from the modelled interactions in Vessel 4 and Vessel 9 (hardened OPC-B75-GW Josef) are hardly comparable with composition of squeezed bentonite pore water (Vessel 17), because the layer of reacted bentonite on the boundary with hardened OPC, in which the reaction could occur, is after nine months thin and it is not possible to obtain a sufficient volume of such samples for analyses not influenced by the unreacted layer of B75.

Under alkaline conditions, a more significant dissolution of the S-phases than that of the A-phases in modelled bentonite was observed. The absence of S-phases in the model of hardened OPC was significant. It would be worthwhile to assess better the real ratios between hydrated OPC-phases based on $CaO:SiO_2:Al_2O_3:MgO:SO_3(:CO_2)$, by methods mentioned above and by analysis of silica in the pore water.

Analysis of squeezed pore water from Vessel 17 showed that measured concentration of sodium was approximately ten times higher than that in the model. There were also more bicarbonates and sulfates, which probably originate from phases not taken into consideration yet, e.g., $NaHCO_3$ or Na_2SO_4.

Conclusions and Future work

Regarding to results of XRD analysis of hardened OPC and bentonite 75 in the Czech cooperation in WP1, the evolution of hardened OPC and bentonite mixtures was modelled, by a simplified description of solid phases, in the presence of groundwater GW Josef. We found that the absence of unidentified Na-/K-phases, S-phases (in C-S-H) or other amorphous phases in experimental characterisation of samples is significant for the geochemistry in the whole system (e.g., composition of pore water, ionic strength, forming of new phases under alkaline conditions, etc.). In the following study, it would be worthwhile to predict the ratios between OPC-phases, based on subsequent instrumental analysis, e.g., XRF and thermogravimetric methods, with the aim to improve model description of hydrated cement.

Despite of this fact, we can conclude that dissolution of quartz, illite, kaolinite, portlandite, and under certain condition of also ettringite, are the predominant mechanisms on the boundary of bentonite and hydrated OPC. From reached results it was concluded the formation of some new phases (e.g., C-A-S-H) or the increasing abundance of some phases on the boundary hydrated cement - bentonite, some of which was experimentally confirmed (e.g., that based on $CaCO_3$). The prediction of the increase of the ettringite mass or forming of hydrotalcite and brucite seems also to be important.

In the next phase of the project, we also plan to evaluate and model diffusion experiments of ^{3}H and ^{36}Cl on samples of hydrated affected by aging and heating that will be performed within WP1 in a UJV and CTU-CEG cooperation.

Acknowledgement

The research leading to these results has received funding from the European Union's Horizon 2020 Research and Training Programme of the European Atomic Energy Community (EURATOM) (H2020-NFRP-2014/2015) under grant agreement n° 662147 (CEBAMA).

References

Ciavatta, L. (1980). The specific interaction theory in the evaluating ionic equilibria. Ann. Chim. (Rome), 70, 551-562.

Guggenheim, E.A. and Turgeon, J.C. (1955). Specific interaction of ions. Trans. Faraday Soc., 51, 747-761.

Giffaut, E., Grivé, M., Blanc, P., Vieillard, P., Colàs, E., Gailhanou, H., Gaboreau, S., Marty, N., Madé, B., Duro, L. (2014). Andra thermodynamic database for performance assessment: ThermoChimie. Appl. Geochem., 49, 225-236.

Parkhurst, D.L. and Appelo, C.A.J. (2013). Description of input and examples for PHREEQC Version 3 - A computer program for speciation, batch-reaction, one-dimensional transport, and inverse geochemical calculations. U.S. Geological Survey Techniques and Methods, book 6, chapter A43.

Večerník, P., Hausmannová, L., Červinka, R., Vašíček, R., Roll, M., Hloušek, J., Havlová, V. (2016). Interaction between cement and Czech bentonite under temperature load and in in-situ conditions: an overview of experimental program. Proceedings of the 1st CEBAMA Annual Workshop. KIT Scientific Publishing, KIT-SR 7734.

Compositional parameters for solid solution C-S-H and the applicability to thermodynamic modelling

Tapio Vehmas[1*], Aku Itälä[1]

[1] VTT, Technical Research Centre of Finland Ltd. (FI)

* Corresponding author: tapio.vehmas@vtt.fi

Abstract

Cementitious materials behaviour must be predictable through the lifetime of nuclear waste repository to ensure the functionality of the engineered barrier system. Thermodynamic modelling is a useful tool to predict the behaviour of the cementitious materials in extremely long time span. Together with the experimental results, modelling provides potential scenarios of cementitious materials behaviour during the lifetime of the repository. Largest uncertainties that are related to the thermodynamic modelling of the cementitious materials are related to calcium-silicate-hydrates, the main reaction product of cementitious materials. The goal of the current study was to define modelling parameters for calcium-silicate-hydrates that are credible and transparent. Ion activity product and Gibbs formation energies were calculated from various solution compositions. Calculated results enable determination of solubility products and formation energies with varying silicate chain lengths and CaO/SiO_2-ratios. Dimeric silicate chain lengths were arbitrarily selected for further modelling. Calcium-silicate-hydrates were modelled as a solid solution and the results were compared to the experimentally measured pH, calcium and silicon concentrations. Modelled results were in good agreement with the measured results. In the future, leaching of calcium-silicate-hydrates will be modelled and compared with the experimental results. Silicate chain length and solid solution end-members effects on the modelling will be further investigated.

Introduction

Posiva Oy is constructing one of the world's first long-term nuclear waste repositories in Finland. The safety of the nuclear waste repository is ensured with a combination of natural and engineered barriers. Final nuclear waste disposal will take place in deep underground repository in Olkiluoto bedrock at a depth of 400 - 450 meters underground. The natural barrier consists of the surrounding Olkiluoto bedrock and its inherent isolating properties. The engineered barrier system consists of water- and gas-tight sealed copper canisters with a cast iron insert and bentonite-based buffer and backfill. Concrete plugs are used for closure and sealing of the repository and the deposition tunnels (Palomäki et al., 2013; Keto et al., 2010). Besides concrete plugs, cementitious materials are used in shotcrete for tunnel wall rock supports, rock-bolting grouts and injection grouts for fracture sealing. Only some of the cementitious materials will remain in the repository after final closure. Shotcrete is removed before the final closure with an estimated efficiency of 95% (Koskinen, 2013).

Although the service-life of concrete structures is only the operational lifetime of the repository, some of the concrete structures will remain in the closed repository and potentially interact with the engineered barrier system. Within the first hundred years, nuclear waste is deposited into the repository which is permanently sealed. After sealing, the repository reaches the radioactivity level of rich natural uranium deposits within 250,000 years.

Cementitious materials behaviour must be known for such time spans to ensure the functionality of the engineered barrier system. Time span for 250,000 years is very challenging for methods of experimental research, especially in cementitious materials. The main reaction product of cementitious materials is calcium-silicate-hydrate (C-S-H), which is essentially an amorphous solid-solution of calcium and silicates. Prediction of the behaviour of amorphous solid-solutions in extremely long time span in transient environment is a task that cannot be solved solely with experimental methods.

Thermodynamic modelling is a useful tool to predict the behaviour of the cementitious materials. If the modelling parameters are accurate, the composition of cementitious materials through the lifetime of nuclear waste repository can be predicted. Together with the experimental results, modelling provides potential scenarios of the cementitious materials behaviour in the repository. Largest uncertainties related to the thermodynamic modelling of the cementitious materials are related to the models that describe the behaviour of C-S-H. Various models C-S-H have been published in the last decades. The models have been built either based on empirical fittings, different solid C-S-H phases or solid solution approaches (Lothenbach and Nonat, 2015; Walker et al., 2016).

Leaching is an essential mechanism in the degradation of C-S-H in crystalline rock repository. C-S-H model should describe the leaching process accurately when the cementitious materials behaviour under repository conditions is modelled. C-S-H dissolves incongruently. C-S-H dissolution releases more calcium than the solid composition of the C-S-H predicts at CaO/SiO_2 -ratios higher than 0.9. C-S-H dissolution produces more silicon than the solid composition predicts at CaO/SiO_2 -ratios lower than 0.8. Congruent dissolution of C-S-H takes place in CaO/SiO_2 -ratios 0.8 - 0.9, where the silicon and calcium dissolves in equal ratios respect to the composition of solid C-S-H (Harris et al., 2002).

Another important feature is the low CaO/SiO_2 -ratio of cementitious materials used in the crystalline rock repository. Cementitious materials having a low CaO/SiO_2 -ratio are used in repositories to ensure compatibility with bentonite buffer and backfill. Leaching and low CaO/SiO_2 -ratio cementitious materials has not been in the focus of interest in traditional concrete technology where the modelling of cementitious materials is typically applied.

In this respect, thermodynamic modelling used to estimate the long-term behaviour of cementitious materials under repository conditions should be very transparent and credible. As the results constitute a potential part of the safety evaluation, the reliability of the results should be known and the scale of potential errors should be predictable.

The goal of the current research is to derive thermodynamic parameters for C-S-H which are accurate with respect to incongruent dissolution behaviour and describe C-S-H behaviour in various CaO/SiO_2 -ratios. The magnitude of errors, originating from the C-S-H model is also predictable.

Solubility product for C-S-H

Thermodynamic modelling is based on a number of parameters that describe the behaviour of the modelled phases. Various methods can be used to obtain these parameters. A transparent method is to calculate the solubility product for a certain phase and further calculate the thermodynamic parameters from the composition and solubility product. This method is widely used and it is very suitable for crystalline materials that dissolve congruently. As C-S-H is a solid solution that dissolves incongruently, defining the solubility product in a traditional way is not a suitable approach.

To define the solubility product of a solid solution, the active ionic species in the dissolution must be recognised. Composition of the C-S-H alters throughout the dissolution and the ionic species activities depend on the solution pH. The pH of equilibrium C-S-H can range from pH 10 to over 13, depending on the system composition and temperature. Within such pH region, calcium can exist in solution as Ca^{2+} and $CaOH^{+}$, while silicates can exist as H_4SiO_4, $H_3SiO_4^{-}$, $H_2SiO_4^{2-}$ and $HSiO_4^{3-}$. Highly charged $HSiO_4^{3-}$ are not typically accounted as active silicate species as the silicates tend to form complexes that are energetically more favourable. In general, it is possible to present the dissolution equation for C-S-H in six different compositions:

$$(CaO)_x(SiO_2)_n(H_2O)_{x+2n} \rightleftharpoons xCa^{2+} + nH_4SiO_4 + 2xOH^- \quad \textit{Eq. 1}$$

$$(CaO)_x(SiO_2)_n(H_2O)_{x+2n} \rightleftharpoons xCa^{2+} + nH_3SiO_4^- + (2x-n)OH^- + 1H_2O \quad \textit{Eq. 2}$$

$$(CaO)_x(SiO_2)_n(H_2O)_{x+2n} \rightleftharpoons xCa^{2+} + nH_2SiO_4^{2-} + (2x-2n)OH^- + 2H_2O \quad \textit{Eq. 3}$$

$$(CaO)_x(SiO_2)_n(H_2O)_{x+2n} \rightleftharpoons xCaOH^+ + nH_4SiO_4 + xOH^- \quad \textit{Eq. 4}$$

$$(CaO)_x(SiO_2)_n(H_2O)_{x+2n} \rightleftharpoons xCaOH^+ + nH_3SiO_4^- + (x-n)OH^- + 1H_2O \quad \textit{Eq. 5}$$

$$(CaO)_x(SiO_2)_n(H_2O)_{x+2n} \rightleftharpoons xCaOH^+ + nH_2SiO_4^{2-} + (x-2n)OH^- + 2H_2O \quad \textit{Eq. 6}$$

In Eq. 1 to Eq. 6, x represents the total calcium content of C-S-H and n the total silicon content. Total water content is represented as x+2n, representing molecular water and chemically bound water. The above presented equations can be summarized into a single solid solution equation accounting for the various ionic species present:

$$(CaO)_x(SiO_2)_n(H_2O)_{(x+2n)} \rightleftharpoons (x-xy)Ca^{2+} + xyCaOH^+ + (n-an-bn)H_4SiO_4$$
$$+anH_3SiO_4^- + bnH_2SiO_4^{2-} + (2x-xy-an-2bn)OH^- + (an+2bn)H_2O \quad \textit{Eq. 7}$$

In Eq. 7, distribution of silicate species in the solution are described as parameters *a* and *b*, according to Eq. 8 and Eq. 9. Distribution of calcium in the solution is described as parameter *y*, according to Eq. 10.

$$a = \frac{[H_3Si\ _4^-]}{[H_3SiO_4^-]+[H_2SiO_4^{2-}]+[H_4Si\ _4]} \quad \textit{Eq. 8}$$

$$b = \frac{[H_4Si\ _4^{2-}]}{[H_3SiO_4^-]+[H_2Si\ _4^{2-}]+[H_4Si\ _4]} \quad \textit{Eq. 9}$$

$$y = \frac{[CaOH^+]}{[Ca^{2+}]+[CaOH^+]} \quad \textit{Eq. 10}$$

In Eq. 7, the solid C-S-H is presented as general oxide formulation. Richardson and Groves (1992) presented a general formulation for C-S-H gel which is composed of isolated silicate chains with varying length and variable number of hydroxyl groups attached to silicon atoms (Eq. 11). Eq. 11 presents the large variety of potential C-S-H compositions. Two especially interesting C-S-H compositions are C-S-H structures of tobermorite and jennite (Eq. 12 and Eq. 13).

$$Ca_{x'}H_{2(n+1-x)}Si_nO_{(3n+1)} \cdot zCa(OH)_2 \cdot m(H_2O) \quad \textit{Eq. 11}$$

$$Ca_{2.5N}H_NSi_{(3N-1)}O_{(9N-2)} \cdot m(H_2O) \quad \textit{Eq. 12}$$

$$Ca_{2.5N}H_NSi_{(3N-1)}O_{(9N-2)} \cdot 2NCa(OH)_2 \cdot m(H_2O) \quad \textit{Eq. 13}$$

The equation for general C-S-H formulation by Richardson and Groves can be derived from Eq. 7 by substituting x = (x'+z). The equation for a tobermorite-based structure can be derived from Eq. 7 by using the relations x = 2.5N and n = (3N-1), where N is an integer. The equation for jennite-based structures is retrieved with the relations x = 4.5N and n = (3N-1).

The ion activity product (IAP) for the C-S-H solid solution according to Eq. 7 is presented as Eq. 14:

$$IAP_{ss}^{C-S-H} = (Ca^{2+})^{(x-y)}(CaOH^{+})^{(y)}(H_4SiO_4\ )^{(n-na-\ \ )}(H_3SiO_4^{1-})^{(na)}$$
$$(H_2SiO_4^{2-})^{(nb)}(OH^{-})^{(2x-y-na-2nb)} \quad \textit{Eq. 14}$$

The ion activity product in equation (14) is able to represent the IAP of C-S-H with various silicate chain lengths (n) and CaO/SiO_2 -ratios. If the CaO/SiO_2-ratio and silicate chain length (n) of C-S-H is known, equation (14) can be used to calculate the solubility product.

Solubility product calculations

In order to calculate ion activity products from equation (14), solution concentrations of Ca, Si and pH must be defined. Also the CaO/SiO_2 -ratio of the solid C-S-H and the polymerisation degree must be known. A Comprehensive dataset of C-S-H samples that enable the calculation of the solubility product has been published. Haas and Nonat (2015) published a dataset of solid C-S-H covering the CaO/SiO_2 ratios from 0.001 to 1.35. Glasser et al. (1987) combined results from literature that covers the CaO/SiO_2 range 0.93 to 1.63 and Chen et al. (2004) presented values ranging from 0.11 to 2.42. Calculation of ion activities from the above described datasets is presented in Figure 1, assuming various silicate chain lengths (n). Ion activities in the calculations were accounted according Davies equation.

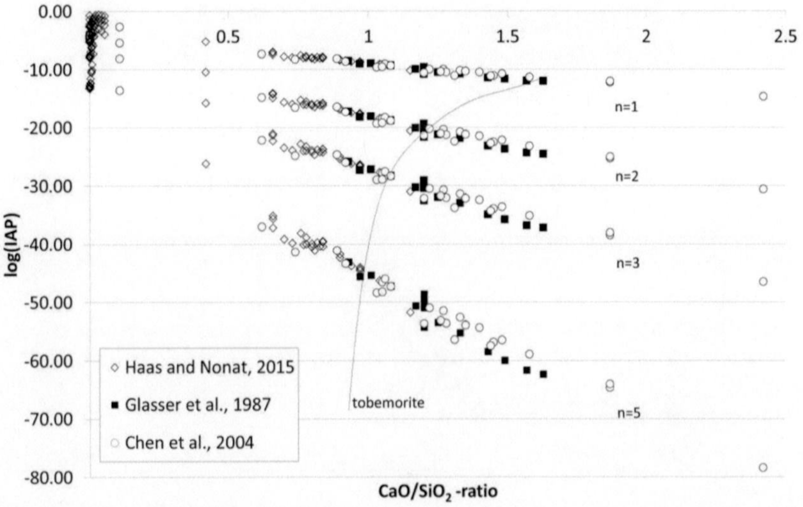

Figure 1: *Logarithm of ion activity product (IAP) as a function of CaO/SiO_2 -ratio of the solid phase. Solid black line presents the ion activity product of tobermorite. Silicate chain length of C-S-H is presented as n.*

Gibbs energy of formation

Gibbs energy of formation can be calculated from the solubility products. Results are presented in Figure 2. As the Gibbs energy of formation is calculated from ion activity product, the apparent Gibbs energy of formation presents the true energy only when ion activity product equals the solubility product.

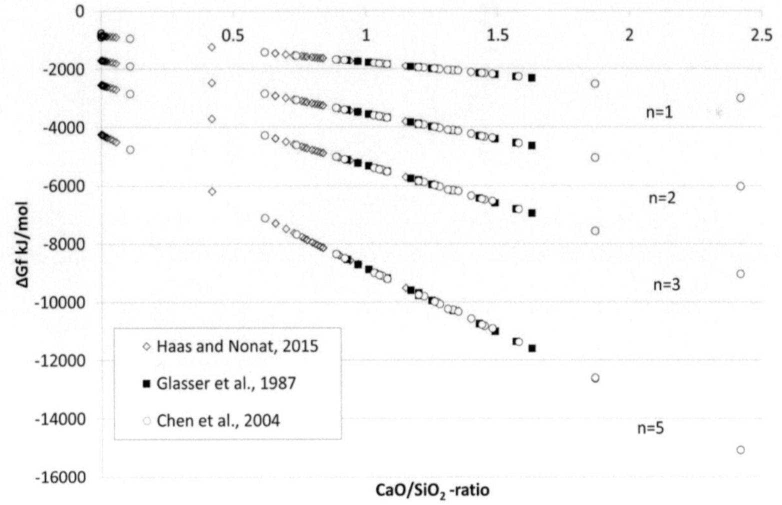

Figure 2: *Gibbs energy of formation a function of CaO/SiO_2 -ratio of the solid phase.*

Modelling Results

The model described above was tested using GEM-Selektor v3 software (Kulik et al., 2013) using a solid solution. Dimeric silicate chain length and CaO/SiO_2 -ratios of 0.66, 0.81, 0.84, 1.26, 1.33, and 2.42 were arbitrarily selected for the modelling study. The CEMDATA 14.01 (EMPA) was used together with Nagra/PSI thermodynamic database. Results are presented in Figures 3, 4 and 5. Modelled results were compared to the experimental results of Haas, Glassier and Chen (Haas and Nonat, 2015; Glasser et al., 1987; Chen et al., 2004).

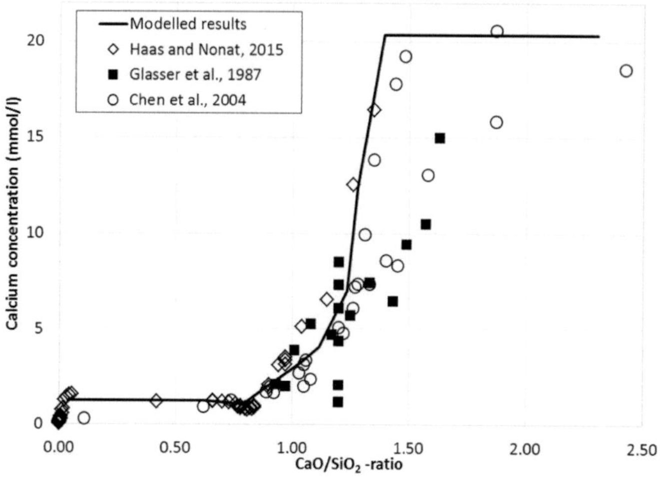

Figure 3: *Modelled calcium concentration in the pore solution as a function of CaO/SiO₂-ratio compared to experimental values.*

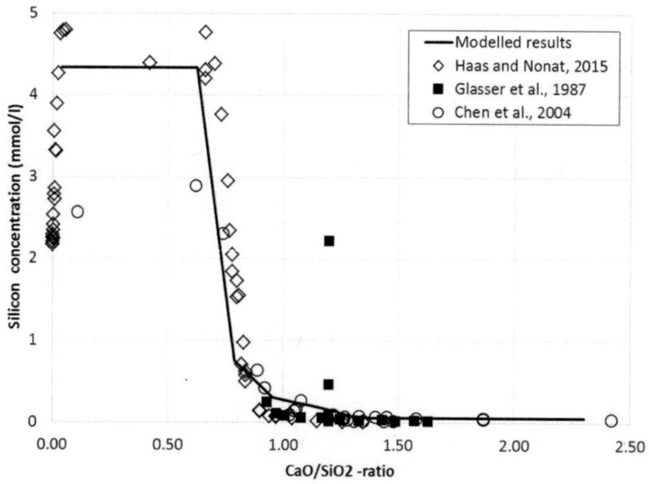

Figure 4: *Modelled silicon concentration in the pore solution as a function of CaO/SiO₂-ratio compared to experimental values.*

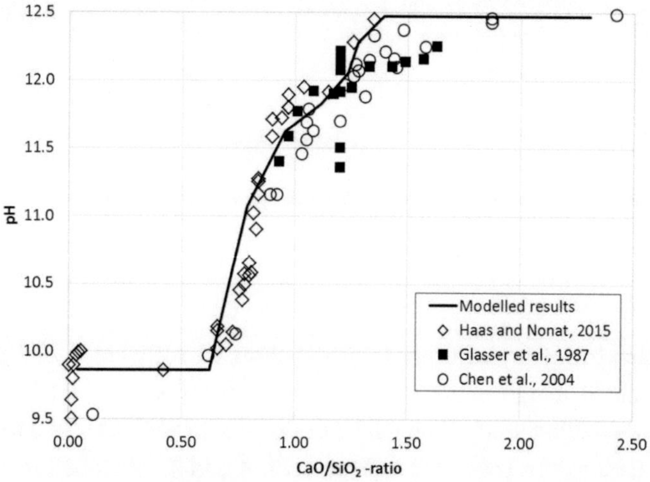

Figure 5: *Modelled pH in the pore solution as a function of CaO/SiO₂-ratio compared to experimental values.*

Conclusions and Future work

Ion activity product and Gibbs formation energies were calculated from the various solution compositions. Calculated results enable determination of solubility products and formation energies with varying silicate chain lengths and CaO/SiO_2 -ratios. Dimeric silicate chain lengths were arbitrarily selected for further modelling. Calcium-silicate-hydrates were modelled as solid solution and the results were compared to the experimentally measured pH, calcium, and silicon concentrations. Modelled results were in good agreement with the measured results. A number of similar models have been published in during last decades (Lothenbach and Nonat, 2015). Published models are also able to describe the pH, calcium and silicon concentrations.

In the future, leaching of calcium-silicate-hydrates will be modelled and compared with the VTT experimental results obtained in WP1 in the CEBAMA -project. Silicate chain length and solid solution end-members effect on the modelling will be further investigated. Generated C-S-H model enables further modifications to improve the modelling accuracy related to C-S-H leaching behaviour and low CaO/SiO_2 -ratios.

Acknowledgement

The research leading to these results has received funding from the European Union's Horizon 2020 Research and Training Programme of the European Atomic Energy Community (EURATOM) (H2020-NFRP-2014/2015) under grant agreement n° 662147 (CEBAMA).

References

Chen, J., Thomas, J., Hal, T., Jennings, H. (2004). Solubility and structure of calcium silicate hydrate. Cem. Con. Res., 34, 1499-1519.

EMPA, Cemdata 14.01 database, *www.empa.ch/web/s308/cemdata*, accessed 13.6.2017.

Glasser, F., Lachowski, E., Macphee, D. (1987). Compositional Model for Calcium Silicate Hydrate (C-S-H) Gels, Their Solubilities, and Free Energies of Formation. J. Am. Ceram. Soc., 70(7), 481-485.

Haas, J. and Nonat, A. (2015). From C-S-H to C-A-S-H: Experimental study and thermodynamic modelling. Cem. Con. Res., 68, 124-138.

Harris, A., Manning, M., Tearle, W., Tweed, C. (2002). Testing of models of the dissolution of cements - leaching of synthetic CSH gels. Cem. Con. Res., 32, 731-746.

Keto, P., Dixon, D., Jonsson, E., Börgesson, L., Hansen, J., Gunnarsson, D. (2010). Assessment of backfill design for KBS-3V Repository. Posiva Working Report, 09-115.

Koskinen, K. (2013). Effects of Cementitious Leachates on the EBS. Posiva Working Report, 2013-4.

Kulik, D.A., Wagner, T., Dmytrieva, S.V., Kosakowski, G., Hingerl, F.F., Chudnenko, K.V., Berner, U. (2013). GEM-Selektor geochemical modeling package: revised algorithm and GEMS3K numerical kernel for coupled simulation codes. Comput. Geosci., 17, 1-24.

Lothenbach, B. and Nonat, A. (2015). Calcium silicate hydrates: Solid and liquid phase compositions. Cem. Con. Res., 78, 57-70.

Palomäki, J. and Ristimäki, L. (2013). Facility Description 2012: Summary Report of the Encapsulation Plant and Disposal Facility Designs. Posiva Working Report, 2012-66.

Richardson, I. and Groves, G. (1992). Models for the composition and structure of calcium silicate hidrates (C-S-H) gel in hardened tricalcium silicate pastes. Cem. Con. Res., 22, 1001-1010.

Walker, C., Sutou, S., Oda, C., Mihara, M., Honda, A. (2016). Calcium silicate hydrate (C-S-H) gel solubility data and a discrete solid phase model at 25°C base don two binary non-ideal solid solutions. Cem. Con. Res., 79, 1-30.

Rapid development of a reactive transport code with FEniCS and reaktoro

Leonardo Hax Damiani[1,2*], Martin Glaus[1], Sergey Churakov[1,2], Georg Kosakowski [1]

[1] Paul Scherrer Institut (CH)
[2] Institute for Geological Sciences, University of Bern (CH)

* Corresponding author: leonardo.hax@psi.ch

Abstract

Reactive transport codes are used to simulate migration of reactive substances coupled with chemical reactions in various geoscience and engineering disciplines. The consideration of more and specialized process-couplings between transport and chemistry often sets limits to established codes. It is often difficult to adopt existing codes such that they can solve specific reactive transport problems accurately and efficiently. We present a problem-oriented approach by developing a new flexible reactive transport simulator that allows a rapid implementation of transport equations using the FEniCS software in combination with the cutting edge Reaktoro geochemical simulator that combines the advantages of Gibbs Energy Minimization and Law of Mass Action methods. Finally, two examples validate our model, a literature benchmark and an experiment modeling that show the diffusion of electrochemical coupled species.

Introduction

In many repository concepts for radioactive waste disposal, cement and clay materials constitute important parts of the engineered barrier system (Nagra, 2002). In a diffusive transport regime, the evolution of cement / clay interfaces is driven by geochemical gradients that cause precipitation and dissolution of minerals and alter material properties. The transport of mass and the reactions due to geochemical gradients between cement and clay could clog pores in the long-term which would hinder mass transport across the interface (Saripalli et al., 2001). The long-term stability of these materials is a topic very often investigated with the help of reactive transport codes (e.g., Berner et al., 2013; Jenni et al., 2014; Kosakowski and Berner, 2013). For materials like clays, reactive transport codes need to represent the transport of ions across various pore types confined by charged mineral surfaces, such as interlayer or inter particle pores in clays (Van Loon et al., 2007). Also, the chemically complex clay and cement mineralogy must be properly represented. Some codes solve Nernst-Planck equations but lack the possibility to model the full complexity of mineralogy as e.g., the possibility to include solid solutions. Others are not able to calculate ion transport in the presence of charged mineral surfaces.

There is an increasing need for developing reactive transport codes that are flexible to reconfigure and allow to address specific problems. This includes the possibilities to modify the transport or geochemical model effectively and the consideration of different numerical methods for solving specific reactive transport problems most efficiently or accurately.

Approach / Tools

Our motivation to develop a new simulation code (NPS) is to have additional flexibility to deal with different reactive transport problems related to the geochemical evolution of a deep geological repository. We use FEniCS (Alnaes et al., 2015) to solve the transport equations. FEniCS is a scientific tool for solving partial differential equations with several finite elements (FE) based numerical methods. The big advantage of using FeniCS, compared to other existing software packages, is the automated and efficient solution of differential equations, as the user has essentially only to define the weak form of the equation to solve and FEniCS provides the numeric infrastructure. This allows for rapid development and easy adoption of codes. Also, high-performance computing is supported by FEniCS without rewriting existing codes. With FEniCS, we explicitly define the weak form of the partial differential equation to be solved; this gives us the possibility to have full control on the PDE level and to easily change to a different numerical method.

For the chemical reactions, we use Reaktoro (Leal et al., 2015), a framework developed in C++ for modeling chemically reactive systems which can interface with the geochemical solvers PHREEQC (Parkhurst and Appelo, 2013) and GEM-Selektor (Kulik et al., 2013). Reaktoro can calculate solid solutions, different activity calculations methods, and shares several libraries with GEM-Selektor which ensures that newly implemented activity models, the equation of states and other functionalities are available in both codes. Recently a new method, the extended law of mass-action (xLMA) algorithm (Leal et al., 2016) has been implemented, which combines the LMA method with Gibbs Energy Minimization (GEM) algorithm. One disadvantage of conventional LMA methods is that the definition of the stable/unstable phases in equilibrium is not trivial. xLMA algorithm takes advantage of the GEM method to define the stable phases in equilibrium and combines it with the superior mass balance calculation method of the LMA algorithms. Through the interface to established geochemical simulators, it is possible to import complex chemical scenarios developed e.g., with the help of GEM-Selektor user interface and also use many linear and non-linear solid solution formations for minerals when calculating the equilibrium state. Besides, Reaktoro is internally coupled to an ODE solver which allows the implementation of kinetic control of mineral phase dissolution or precipitation.

To couple the transport solver (FEniCS) and the chemical solver (Reaktoro), we use the programming language Python (Python Software Foundation, *https://www.python.org/*). Python is a widely used language in applied math, science and engineering due to its transparent syntax, powerful performance and a complete set of libraries.

Diffusion Transport

We implement in our first version of the coupled code a transport equation that describes the diffusion of neutral and charged species in the presence of an electrostatic potential. The diffusive flux is described by the Fick's law as Eq. 1.

$$J_{diffusive} = -D_i \nabla c_i \qquad \textit{Eq. 1}$$

Where $J_{diffusive}$ is the molar diffusive flux [mol/m^2/s], D_i [m^2/s] is the diffusion coefficient for species i and ∇c_i is the concentration gradient with concentration c_i [mol/m^3]. The minus sign indicates that the transport is towards the lower concentration. For diffusion in porous media D_i can be interpreted as pore diffusion coefficient which includes the effect of pore space geometry in form of a tortuosity. For sake of simplicity tortuosity is not explicitly included in the derivation and diffusion coefficients should be interpreted as pore diffusion coefficients.

The electrochemical flux is described as Eq. 2.

$$J_{electrochemical} = -\frac{z_i D_i F c_i}{RT} \nabla \Phi$$ *Eq. 2*

where z_i [C] is the charge of species i, F [C/mol] is the Faraday constant, R is the gas constant [kg m^2/mol/K/s^2], c_i [mol/m^3] is the concentration for species i and $\nabla\Phi$ [V] is the electrical potential gradient. If the electrical potential is constant in the simulation domain we can assume the electroneutrality condition for the solution, i.e., no charge transport and null current method $\left(0 = F\sum_i^{N_s} z_i c_i\right)$, For this case we can simplify the equation and express the electrochemical flux based on concentration terms (Eq. 3).

$$J_{electrochemical} = -\sum_{k=1}^{Ns} D_{ik} \nabla c_k$$ *Eq. 3*

where D_{ik} are the cross diffusion coefficients (see e.g., Liu et al., 2011; Parkhurst and Appelo, 2013) given by Eq. 4.

$$D_{ik} = \frac{z_i z_k D_i D_k c_i}{\sum\limits_{k=1}^{Ns} z_k^2 D_k c_k}$$ *Eq. 4*

where N_s is the number of species in the system. With the two flux terms combined, we achieve the total flux formulation (Eq. 5).

$$J_{total} = -D_i \nabla c_i - \sum_{k=1}^{Ns} D_{ik} \nabla c_k$$ *Eq. 5*

For a porous media, we have to scale fluxes with the porosity value and combine it with the Fick´s 2nd law. In this way, we achieve a simplified Nernst-Planck equation (Eq. 6).

$$\theta \frac{\partial c_i}{\partial t} = \theta \nabla \cdot (-D_i \nabla c_i - \sum_{k=1}^{Ns} D_{ik} \nabla c_k)$$ *Eq. 6*

The equation presented above is a partial differential equation (PDE) in its strong formulation, which may not always be trivial or efficient to solve because it requires that all the solutions be differentiable and continuous. For this reason, we work with the weak formulation (Eq. 7).

$$\int_\Omega \theta \frac{\partial c_i}{\partial t} v dx = \int_\Omega \theta (-D_i \nabla c_i - \sum_{k=1}^{Ns} D_{ik} \nabla c_k) \cdot \nabla v dx$$ *Eq. 7*

where v is a test function smooth inside the domain Ω. For a detailed discussion of partial differential equations, strong and weak formulations we refer to (Gilbarg and Trudinger, 2001). To achieve the final formulation that describes the total flux, it is necessary to discretize it in time (Eq. 8).

$$\frac{\partial c_i}{\partial t} \approx \frac{c^n - c^{n-1}}{\Delta t} = F(c^{n-1})$$ *Eq. 8*

where n is the current iteration index and F is a function that depends on previous concentration. With some algebraic manipulation (Eq. 9):

$$\int_{\Omega} \theta(c_i^n) v_i dx = \int_{\Omega} \theta(c_i^{n-1} v_i - \Delta t \sum_{k=1}^{Ns} (D_i \delta_{ik} - \frac{z_i z_k D_i D_k c_i^{n-1}}{\sum_{k=1}^{Ns} z_k^2 D_k c_k^{n-1}}) \nabla c_k^{n-1} \cdot \nabla v_i) dx \qquad \textit{Eq. 9}$$

In the final equation, for simplification purposes, the diffusive part of the flux is defined by using the Kronecker symbol function $\delta_{ij} = \begin{cases} 1, & if\ i = j \\ 0, & if\ i \neq j \end{cases}$.

For solving the weak form, we use a standard Galerkin Finite Element Method (FEM) which is readily available in FEniCS. FEM is a method where the geometry is subdivided into small sub-domains, the so-called elements, which now represent the continuous domain of the problem. The division of geometry into elements decreases the complexity of the problem by placing it into small and easier to comprehend problems. The technique proposes that we substitute an infinite number of unknowns by a limited number of elements with well-defined behavior. The divisions of the domain into elements can be done with different shapes, such as linear, triangular, tetrahedral, among others, depending on the type and size of the problem. These elements are connected by nodes, and the sum of nodes and elements is the mesh. References of FEniCS and finite element formulation can be found in Alnaes et al. (2015).

Chemical reactions and transport-chemical coupling

Currently, we use the sequential non-iterative approach (SNIA), explained in details in Ingham and Pop (2005), for coupling transport and chemistry (Figure 1). After solution of the transport equations, the changed concentrations are updated at each node. The chemical solver will then calculate the equilibrium state of each node and update the concentrations for each species. In addition, porosity changes and related transport properties are calculated from the newly equilibrated mineral phase composition. According to the complexity of the spatial and temporal discretization finite element / volume (number of nodes and time step) and the chemical details needed for the model (i.e., the number of species and which possible phases are considered) the computational time needed for solving a problem can get very long.

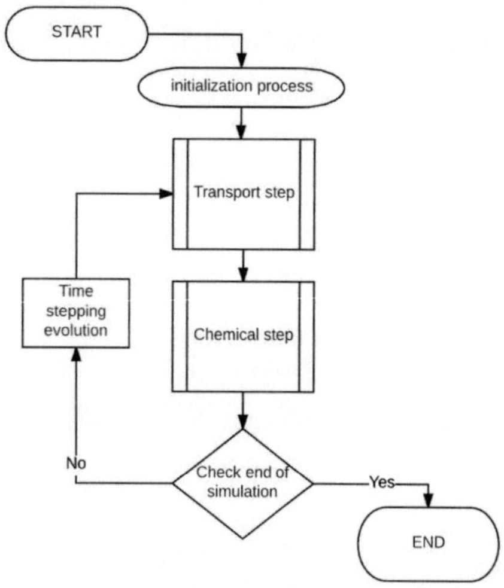

Figure 1: *Diagram showing the coupling between transport solver and chemical solver.*

Code testing and benchmarking

We verified our numerical implementation of the simplified Nernst-Planck equations by comparison with analytical solutions for salt diffusion and by reproduction of a benchmark problem from Lichtner (1994), described as case 1 in Rasouli et al. (2015). This benchmark case calculates the diffusion of HNO_3 from a low-pH solution (pH = 4, right side in Figure 2) to a higher pH solution (pH = 6, left side in Figure 2). Both solutions contain the same elevated NaCl concentrations. The expected diffusion of HNO_3 from right to left is indicated by an arrow. The chemical system is composed by H^+, NO_3^-, Na^+, and Cl^-. The initial condition is displayed in Figure 2. The left solution is imposed as Dirichlet boundary condition at x = 0.0. On the right side of the model we assigned an open boundary.

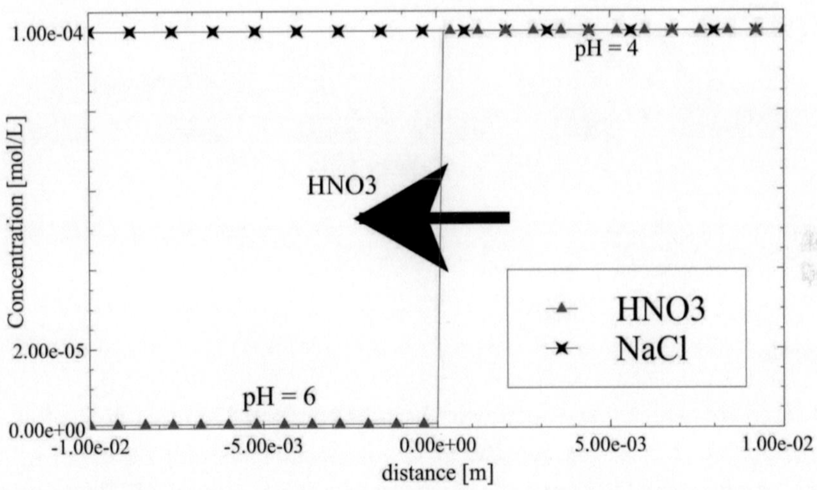

Figure 2: *Initial condition according to benchmark case 1 in Rasouli et al. (2015) with indications of expected ion fluxes due to initial and boundary conditions.*

Results from our implementation (NPS) are compared to the reference values provided by Rasouli et al. (2015) in Figure 3 after a simulation time of 1 hour. We see a very good agreement between the results which verifies our implementation in FEniCS.

Moreover, Figure 3 shows an increase in Na^+ and a decrease in Cl^- concentrations which cannot be explained with simple Fickian diffusion models, as there is initially and at the boundaries no concentration difference for NaCl. These effects can only be explained by considering additional cross-coupling terms between charged species. These terms originate from different diffusive mobilities of species and are needed to maintain a charge-balanced solution.

The observed concentration variations are sometimes called "uphill diffusion", although these are only transient phenomena. On the long term, the concentrations in the system will equilibrate to the boundary values and will be constant throughout the domain. Most reactive transport codes cannot reproduce this benchmark, as they apply Fick's law only and use the same diffusion coefficient for all transported species. The latter approach also preserves charge balance but does not take into account the effect of interactions between charged species with different diffusion coefficients.

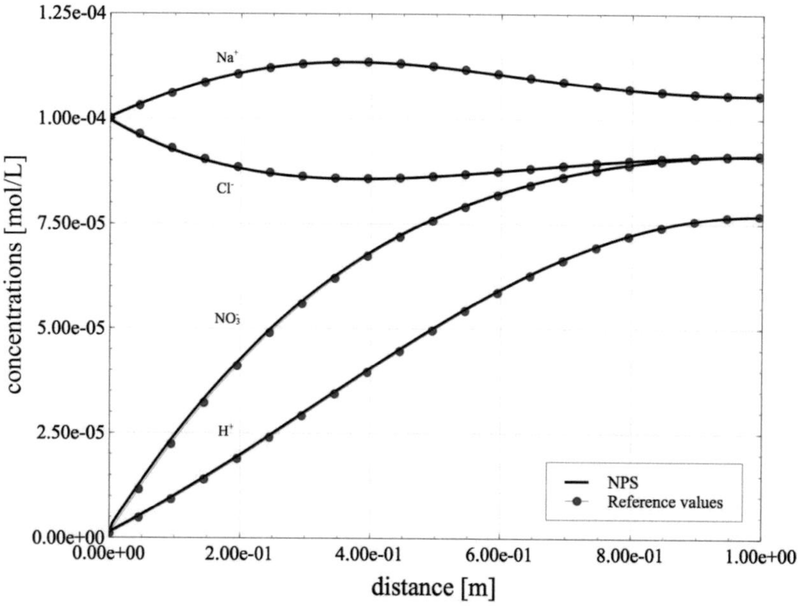

Figure 3: *Comparison between reference values from benchmark case 1 in Rasouli et al. (2015) and our results after 1 h diffusion time.*

Application example

As a first application, we modeled an experiment that was conducted to measure the flux of charged species through a porous membrane. The set up consists of two adjacent reservoirs separated by a cellulose acetate membrane. Cellulose acetate is used to represent an uncharged porous medium. HTO (tritiated water), as electro-neutral and non-sorbing tracer, was used to measure geometric transport properties (tortuosity) of the membrane. The solutions in the reservoir contain HNO_3 and a background electrolyte, in this case, KCl. The concentration of HNO_3 is different in each reservoir, imposing the concentration gradient necessary for the diffusion to occur. Background electrolyte concentrations in both reservoirs are identical. As for the benchmark example, the diffusion of HNO_3 induces an additional flux of the background electrolyte. This so-called "uphill fluxes" of the electrolyte originate from the different mobilities of the involved ions.

To properly model this experiment, a volume adjustment factor was introduced in the weak formulation (in FEniCS) to simulate the different cross-sectional area and volume sizes of the nodes representing different domains (membrane and tanks). The membrane diameter is 23 mm and the width is 0.13 mm. The domain was discretized into 29 cells, and the volume adjustment factor (V) for membrane cells is $4.1547563 \cdot 10^{-4}$ while for tanks cells it is 1.0. V was introduced in Eq. 10, and the updated weak formulation is the following:

$$\int_\Omega V\theta(c_i^n)v_i dx = \int_\Omega V\theta(c_i^{n-1}v_i - \Delta t \sum_{k=1}^{Ns}(D_i\delta_{ik} - \frac{z_i z_k D_i D_k c_i^{n-1}}{\sum_{k=1}^{Ns} z_k^2 D_k c_k^{n-1}})\nabla c_k^{n-1} \cdot \nabla v_i)dx \qquad \textbf{Eq. 10}$$

Figure 4 and Figure 5 show the experimental concentration evolution after 55 days in the left and right reservoir, respectively. HTO is a neutral species, and its breakthrough curve is used to extract an additional material dependent scaling factor for all species diffusion coefficients (tortuosity). In addition to the results for our code (NPS), PHREEQC (Parkhurst and Appelo, 2013), and Flotran (Holder et al., 2000) and, especially, HTO with OpenGeoSys-GEMS (Kosakowski and Watanabe, 2014) are included. The detailed information about the system is presented in Appendix A. For both reservoirs, we see a good agreement between the modeled and the experimental data.

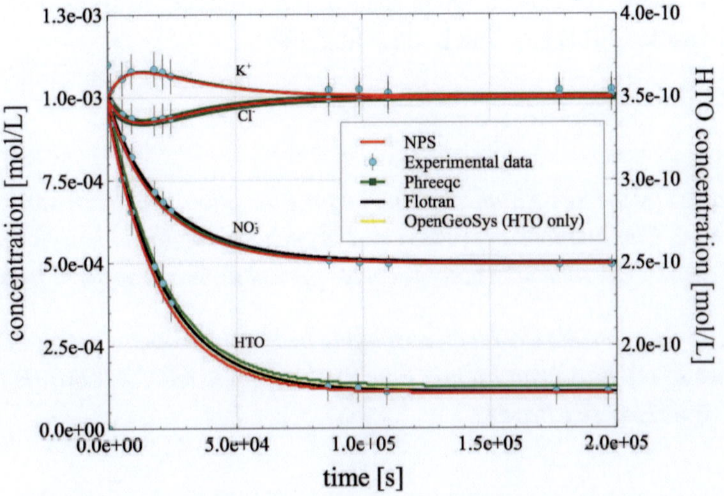

Figure 4: *Comparison of measured and modeled temporal change of concentration in the left reservoir.*

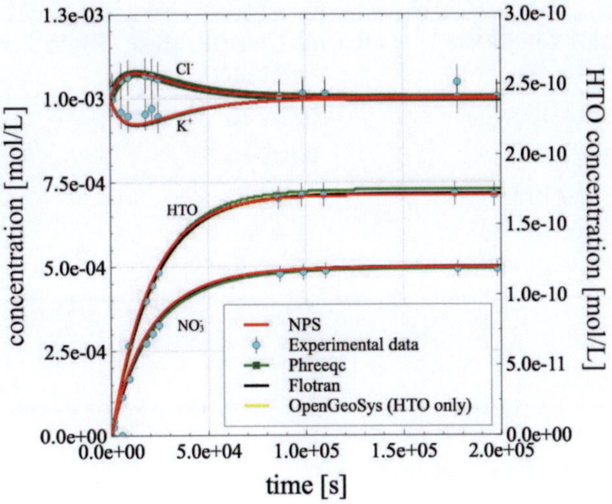

Figure 5: *Comparison of measured and modeled temporal change of concentration in the right reservoir.*

Summary and outlook

The development of a new reactive transport code based on advanced software tools gives us the possibility to set up a very flexible working modeling environment and at the same time have full control on how and with which numerical methods we solve a simulation problem. Our approach, using FEniCS and Reaktoro, allows a rapid development of an efficient tool for solving reactive transport in a results-oriented manner. We successfully tested the implementation with benchmarks and a first application.

In a next step, we would like to extend the model to consider the effects of electric double-layers at charged mineral surfaces on transport. These surface charges are responsible for the so-called surface diffusion and for anion exclusion in materials containing clay minerals. We will implement an electrostatic model in the form of a Donnan approach (Overbeek, 1956) into GEMS and Reaktoro. The Donnan model will partition the pore space into two parts with different composition and transport properties. The first volume is not influenced by surface charges and the second volume, the Donnan volume, has a different composition, as certain ions can not enter due to electrostatic repulsion. These phenomena are presented and discussed in Appelo et al. (2010), Glaus et al. (2015), Tournassat and Appelo (2011) and Van Loon et al. (2007).

Acknowledgement

This work was supported by the Swiss State Secretariat for Education, Research and Innovation (SERI) under contract number 15.0186-2. The opinions expressed and arguments employed herein do not necessarily reflect the official views of the Swiss Government. Authors receive partial financial support from Nagra.

The research leading to these results has received funding from the European Union's Horizon 2020 Reasearch and Training Programme of the European Atomic Energy Community (EURATOM) (H2020-NFRP-2014/2015) under grant agreement n° 662147 (CEBAMA).

Appendix A. Detailed information about the application example

Table A1: *Chemical composition.*

Species	Diffusion Coefficient (m^2/s)	Left Tank Concentration (mol/m^3)	Right Tank Concentration (mol/m^3)
H^+	$9.0 \cdot 10^{-9}$	$0.001 \cdot 10^{-3}$	$1.0 \cdot 10^{-3}$
NO_3^-	$1.9 \cdot 10^{-9}$	$0.001 \cdot 10^{-3}$	$1.0 \cdot 10^{-3}$
K^+	$2.0 \cdot 10^{-9}$	$1.0 \cdot 10^{-3}$	$1.0 \cdot 10^{-3}$
Cl^-	$2.0 \cdot 10^{-9}$	$1.0 \cdot 10^{-3}$	$1.0 \cdot 10^{-3}$
HTO	$2.3 \cdot 10^{-9}$	$3.44 \cdot 10^{-10}$	$1.0 \cdot 10^{-25}$

Table A2: *Spatial discretization.*

Number of cells	29
Tanks cells indexes	0 and 28
Membrane cells indexes	1 to 27
Size of a membrane cell	5.0e-6 m
Size of a tank cell	$1.0 \cdot 10^{-4}$ m (0.1 liter)
Tank adjustment factor	1.0
Membrane adjustment factor	$4.1547563 \cdot 10^{-4}$
Porosity of the membrane	0.44
Porosity of the tank	1.0

References

Alnaes, M.S., Blechta, J., Hake, J., Johansson, A., Kehlet, B., Logg, A., Richardson, C., Ring, J., Rognes, M.E., Wells, G.N. (2015). The FEniCS Project Version 1.5. Archive of Numerical Software, 3(100), 9-23.

Appelo, C.A.J., Van Loon, L.R., Wersin, P. (2010). Multicomponent diffusion of a suite of tracers (HTO, Cl, Br, I, Na, Sr, Cs) in a single sample of Opalinus Clay. Geochimica et Cosmochimica Acta, 74(4), 1201-1219.

Berner, U., Kulik, D.A., Kosakowski, G. (2013). Geochemical impact of a low-pH cement liner on the near field of a repository for spent fuel and high-level radioactive waste. Physics and Chemistry of the Earth, 64, 46-56.

Gilbarg, D. and Trudinger, N.S. (2001). Elliptic partial differential equations of second order. Springer.

Glaus, M.A., Aertsens, M., Appelo, C.A.J., Kupcik, T., Maes, N., Van Laer, L., Van Loon, L.R. (2015). Enhanced mass transfer rates for Sr^{2+}, Co^{2+} and Zn^{2+} in compacted illite caused by cation diffusion in the electrical double layer. Geochimica et Cosmochimica Acta, 165, 376-388.

Holder, A.W., Bedient, P.B., Dawson, C.N. (2000). FLOTRAN, a three-dimensional ground water model, with comparisons to analytical solutions and other models. Advances in Water Resources, 23(5), 517-530.

Ingham, D.B. and Pop, I.I. (2005). Transport phenomena in porous media. Volume III. Elsevier.

Jenni, A., Mäder, U., Lerouge, C., Gaboreau, S., Schwyn, B. (2014). *In-situ* interaction between different concretes and Opalinus Clay. Physics and Chemistry of the Earth, 70-71, 71-83.

Kosakowski, G. and Berner, U. (2013). The evolution of clay rock/cement interfaces in a cementitious repository for low- and intermediate level radioactive waste. Physics and Chemistry of the Earth, 64, 65-86.

Kosakowski, G. and Watanabe, N. (2014). OpenGeoSys-Gem: A numerical tool for calculating geochemical and porosity changes in saturated and partially saturated media. Physics and Chemistry of the Earth, 70-71, 138-149.

Kulik, D.A., Wagner, T., Dmytrieva, S.V., Kosakowski, G., Hingerl, F.F., Chudnenko, K.V., Berner, U.R. (2013). GEM-Selektor geochemical modeling package: Revised algorithm and GEMS3K numerical kernel for coupled simulation codes. Computational Geosciences, 17(1), 1-24.

Leal, A.M.M., Kulik, D.A., Kosakowski, G. (2015). Computational methods for reactive transport modeling: A Gibbs energy minimization approach for multiphase equilibrium calculations. Advances in Water Resources, 88, 231-240.

Leal, A.M.M., Kulik, D.A., Kosakowski, G., Saar, M.O. (2016). Computational methods for reactive transport modeling: A revised law of mass-action approach for multiphase equilibrium calculations. Advances in Water Resources, 96, 405-422.

Lichtner, P.C. (1994). Principles and practice of reactive transport modeling. MRS Proceedings, 353, 117-130.

Liu, C., Shang, J., Zachara, J.M. (2011). Multispecies diffusion models: A study of uranyl species diffusion. Water Resources Research, 47(12), 1-16.

Nagra (2002) Project Opalinus clay. Safety report: demonstration of disposal feasibility for spent fuel, vitrified high-level waste and long-lived intermediate-level waste (Entsorgungsnach- weiss). Nagra Technical Report, NTB 02-05.

Overbeek, J.T. (1956). The Donnan equilibrium. Progress in Biophysics and Biophysical Chemistry, 6, 57-84.

Parkhurst, D.L. and Appelo, C.A.J. (2013). Description of Input and Examples for PHREEQC Version 3 - A Computer Program for Speciation, Batch-Reaction, One-Dimensional Transport, and Inverse Geochemical Calculations. U.S. Geological Survey Techniques and Methods, book 6, chapter A43.

Rasouli, P., Steefel, C.I., Mayer, K.U., Rolle, M. (2015). Benchmarks for multicomponent diffusion and electrochemical migration. Computational Geosciences, 19(3), 523-533.

Saripalli, K.P., Meyer, P.D., Bacon, D.H., Freedman, V.L. (2001). Changes in Hydrologic Properties of Aquifer Media Due to Chemical Reactions: A Review. Critical Reviews in Environmental Science and Technology, 31(4), 311-349.

Tournassat, C. and Appelo, C.A.J. (2011). Modelling approaches for anion-exclusion in compacted Na-bentonite. Geochimica et Cosmochimica Acta, 75(13), 3698-3710.

Van Loon, L.R., Glaus, M.A., Müller, W. (2007). Anion exclusion effects in compacted bentonites: Towards a better understanding of anion diffusion. Applied Geochemistry, 22(11), 2536-2552.

POSTERS

TABLE OF POSTERS

Experiments on Interface Processes at the Cement/Callovo-Oxfordian Claystone Interface and the Impact on Physical Properties; Mechanical results from Callovo-Oxfordian Claystone

Robert Cuss[1*], Andrew Wiseall[1], Jon Harrington[1], Jean Talandier[2] and Xavier Bourbon[2]

[1] British Geological Survey, Keyworth, Nottingham, NG12 5GG, United Kingdom (*rjcu@bgs.ac.uk)

[2] Andra, 1-7, rue Jean Monnet, 92298 Chatenay-Malabry, France

1. Introduction

The weakest part of any gallery or deposition hole seal is likely to be at the interface between the sealing components and the host rock. A single seal completion may comprise a number of elements reflecting different design criteria in order to address specific engineering challenges associated with changes in geochemistry and stress. The interaction of these components with the host rock, their evolution in terms of strength/bonding, cation exchange behaviour and interfacial permeability, and the sensitivity of these properties to an evolving geochemical and physical environment will be key factors in determining the long-term seal performance. The current study aims to examine the temporal evolution of the host-rock/low-alkali cement interface in the French repository concept to changes in geochemistry, mineralogy and stress to assess their impact on the development of hydraulic permeability and strength (shear strength).

Since the previous meeting experimental emphasis has been placed on preparing the Callovo-Oxfordian claystone samples ready for casting of concrete to create the interfaces of interest. This has included re-hydration of all samples and shearing of intact samples in order to create realistic, rough, surfaces to act as representative interfaces.

2. Experimental apparatus

BGS have three bespoke shear rigs, 2 of which have been modified for the Cebama project. The Direct Shear Rig (DSR) has been used to investigate the flow characteristics of rock (Opalinus Clay, Callovo-Oxfordian Claystone). The apparatus has the following specifications:

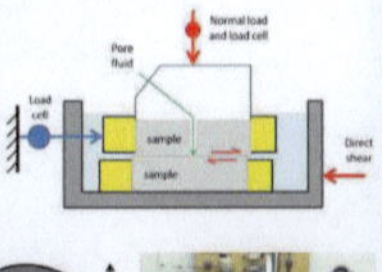

- Normal/vertical stress up to 20 MPa
- Variable shear rate of 1 mm per second to 3 month
- Vertical displacement accurate to < 60 nm
- Sample dimension 60 mm diameter, 53 mm height
- Pore pressure 0.5 - 25 MPa
- Central injection point ensuring pore fluid is delivered directly to the fracture

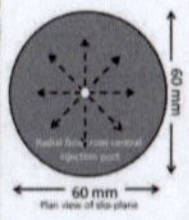

3. Test protocol

- Preserved core barrels supplied by Andra from the Meuse/Haute-Marne URL at Bure, France
- Sub-sampled into 65 mm long sections
- Machine lathed to 60 mm diameter, 53 mm height samples
- Samples vacuum packed and stored in a fridge
- Off-cut material weighed and basic geotechnical properties determined

- Samples placed into the experimental apparatus
- Initially rehydrated using synthetic pore fluid to bring samples up to full saturation at a normal stress of 6.5 MPa, representative of effective stress at repository depths
- Once swelling had stopped, samples were sheared for 2.5 days at a rate of 2 mm/day

Data recorded:

- pore-fluid flow, shear displacement, shear stress, normal stress, and vertical displacement

Giving

- initial flow rate, shear yield stress, shear modulus (G), shear peak strength, shear residual strength, dilation/contraction during shear, and change in flow during shear

- Fracture surfaces scanned using a NextEngine 3D laser scanner
- Samples vacuum packed for medium-term storage.

4. Example shear test: Cebama_01

The initial stage of testing aimed to raise the saturation of the COx back to full saturation, as shown below. Approximately 50 µm dilation was observed (approximately 0.1 % strain), suggesting that the starting saturation of the test material was high.

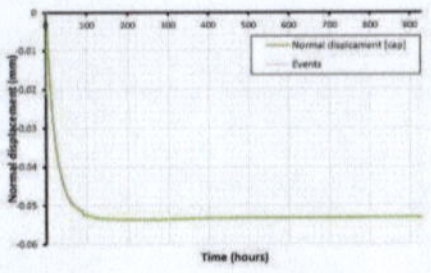

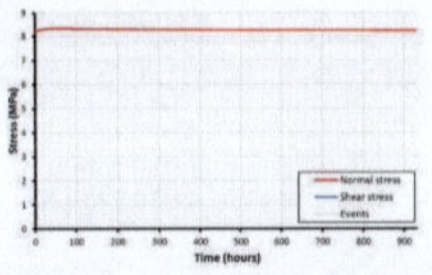

During shearing (see below) the initial increase in shear stress was non-linear. This is commonly observed in shear tests using the Direct Shear Rig. The stress-strain response can be seen to become linear, giving the shear modulus of COx (~340 MPa).

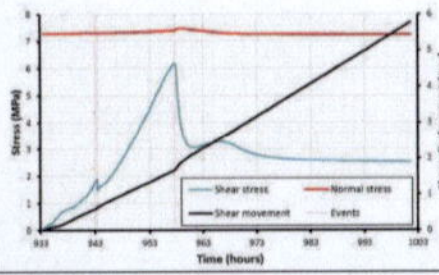

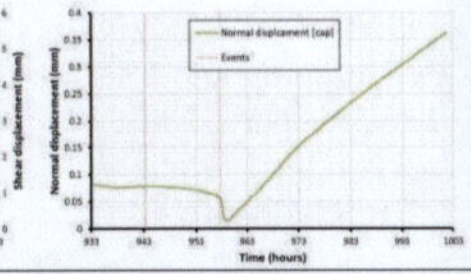

Approaching 6.0 MPa shear stress, deviation from linear stress-strain behaviour is seen at the yield shear stress, followed by classic brittle deformation with a stress drop to 3.05 MPa and dilation of approximately 60 µm. At this point the stress starts to increase again, probably as a result of the mismatch of the two formed fracture surfaces and a second low-angle peak is reached, before a second stress decrease occurs. A secondary residual stress level was achieved, but as can be seen in the vertical displacement data, one fracture surface was now riding above the other. The test was terminated at approximately 9.5 % strain. The table below summarises the data for test Cebama_01.

Test	Cebama_01		Normal stress at peak	8.251	MPa
Rock type	COx		Displacement at peak	1.668	mm
Core barrel	EST 53639		Strain at peak	2.781	%
Sample diameter	59.986	mm	Residual shear stress	3.052	MPa
Sample height	52.873	mm	Residual norm' stress	8.270	MPa
Sample volume	149.423	cc	Displ' at residual	2.120	mm
Sample weight	360.307	g	Strain at residual	3.534	%
Sample density	2.411	g/cc	Shear rate	90.740	µm/h
Rig used	DSR1		Shear rate set point	150	µl/h
Test start date	28/11/2016 15:21:18		Maximum displ'	5.767	mm
Shear start date	06/01/2017 13:01:04		Total strain	9.614	%
Shear finish date	09/01/2017 09:07:30		Average norm' stress	8.136	MPa
Shear duration	68.107	hours	Shear modulus	339.598	MPa
Total test duration	41.740	days	Intercept	-3.047	MPa
Starting yield stress	0.070	MPa	Start height	0.032	mm
Starting normal stress	8.133	MPa	Start date	28/11/2016 16:23:01	
Yield shear stress	6.042	MPa	Peak height	-0.021	mm
Yield normal stress	8.239	MPa	Peak date	09/12/2016 21:24:15	
Displacement at yield	1.624	mm	Total swelling	0.054	mm
Strain at yield	2.707	%	Swelling strain	0.102	%
Peak yield stress	6.185	MPa	Swelling duration	11.209	days

5. Post-shear observation of the fracture surface

The fracture surface created during test Cebama_01. Some clear smearing and polishing was observed, as was the disaggregation of some clay as a fault gouge had started to form.

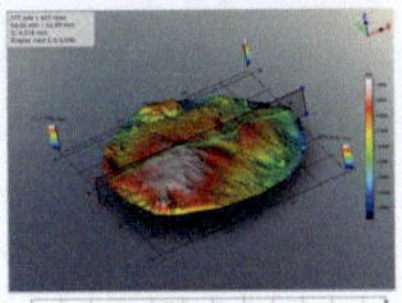

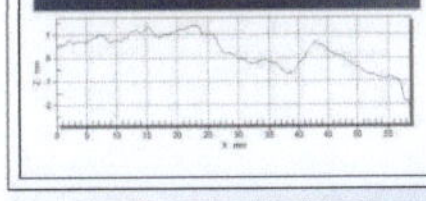

The sample was laser scanned and vacuum packed for storage. Left shows the scan result for the test sample, with the table below summarising the surface textural data achieved for the lower sample of test Cebama_01.

Parameter	Value	Description
Sa	0.697 mm	Average roughness
Sq	0.879 mm	RMS roughness
St	4.578 mm	Peak to valley height
SPaX	0.585 mm	Mean profile Pa x-axis
SPaY	0.465 mm	Mean profile Pa y-axis
SPtX	2.846 mm	Mean profile Pt x-axis
SPtY	2.37 mm	Mean profile Pt y-axis
Std	0.334°	Texture direction

6. Shear properties of COx

All sample hydration tests have been conducted, including 6 shear tests to create 12 rough interface surfaces and 8 hydration tests of planed (smooth) interface surfaces. Two varieties of Callovo-Oxfordian have been tested;

- the repository depth rock (COx) and
- the overlying high carbonate variety (USC)

USC has a considerably greater shear strength than COx. The shear apparatus had to be modified in order to achieve shear failure of the sample at about 20 MPa. Good repeatability was seen between the 3 repeat tests conducted on the two varieties, even though samples derived from more than one core barrel. Shear modulus was seen to be higher in USC compared with COx. The results for starting, yield, peak, and residual shear stress for all 6 shear experiments are shown right. USC is much stronger than COx. In COx, there is little difference between the yield and peak stresses, whilst in USC there is considerable difference. In both varieties, the residual stress is less than half the peak stress condition.

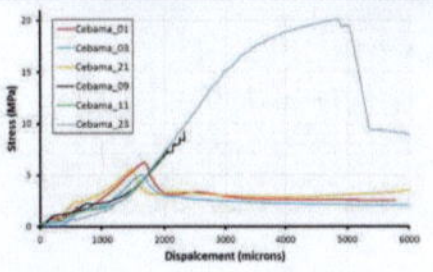

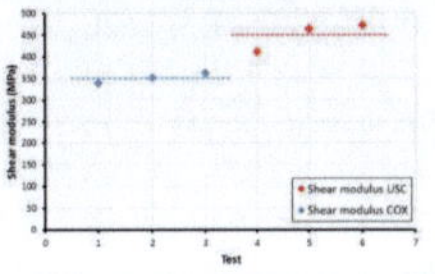

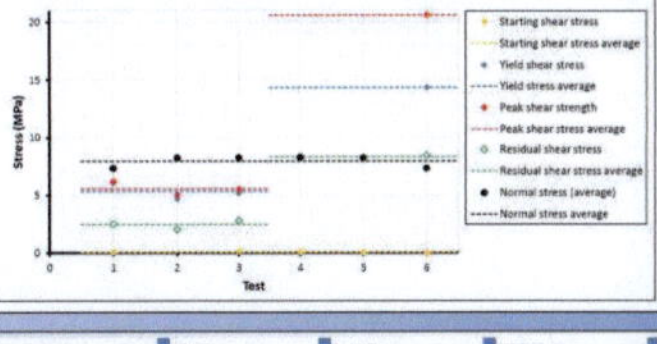

7. Current status and future activities

- All COx and USC samples rehydrated, scanned and stored
- Casting will occur in early June, followed by curing
- Shear testing conducted at repeat intervals until the end of the project
  - Determine changes in shear and flow properties as the interface ages
- Series of uniaxial compression tests will be conducted on the concrete to look at aging

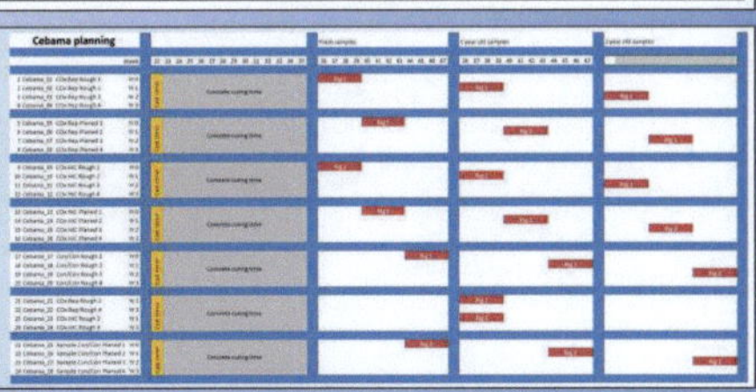

British Geological Survey
NATURAL ENVIRONMENT RESEARCH COUNCIL

The research leading to these results has received funding from the European Union's Horizon 2020 Research and Training Programme of the European Atomic Energy Community (EURATOM) (H2020-NFRP-2014/2015) under grant agreement n° 662147 (CEBAMA).

Contact:

Dr Robert Cuss
Fracture Physics Laboratory
British Geological Survey, Keyworth, Nottingham, NG12 5GG, UK
Tel: +44 (0)115 936 3486 Email: rjcu@bgs.ac.uk

Use of Neutron and X-ray Tomography based visualization of water uptake in cementitious materials for OPERA disposal

REACTOR INSTITUTE DELFT

Yasin Yigittop[1], Zhou Zhou[2], Denis Bykov[1], Andrea Sabău[1], Lambert van Eijck[2], Jan-Leen Kloosterman[1]

E-mail: d.bykov@tudelft.nl

[1]Section Nuclear Energy and Radiation Applications, Department of Radiation Science and Technology

[2]Section Neutron and Positron Methods in Materials, Department of Radiation Science and Technology

Mekelweg 15, 2629 JB Delft, The Netherlands

Introduction

The performance of concrete subjected to aggressive environments is a function, to a large extent, of the penetrability of the pore system. Three mechanisms can be used to describe transport in cementitious systems: (1) permeability, (2) diffusion and (3) absorption [1]. In the Netherlands Boom clay has been identified as a potential host formation for the disposal of nuclear waste and was studied within OPERA programme on geological disposal of radioactive waste (Onderzoeks Programma Eindberging Radioactief Afval). The composition of Boom clay pore water (BCPW) in the Netherlands has strong seawater signature with chloride concentrations corresponding to 70-96 % of chloride concentrations in seawater [2]. Oxidation of pyrite can account for the elevated sulphate and calcium concentrations measured in some of the pore waters [3]. The present work is dedicated to the study of porosity evolution and water absorption in concrete exposed to high chloride and sulphate concentrations, using X-ray and neutron imaging techniques.

Materials and Methods

Concrete samples: CEM III /B 42.5N LH/SR

Component	Receipt for 1 m³ of Aerolite FC 1200 to 1000 kg/m³	Type for OPERA	1500 kg m⁻³
Cement	360 to 400 kg	CEM III/B 42.5 LH/SR	391 kg m^{-3}
Water	140 to 160 kg	-	154 kg m^{-3}
Fine aggregate	750 to 1100 kg	Sand: 0-2 mm	955 kg m^{-3}
Foaming agent Synthetic surfactant	0.57 to 0.36 l	Foaming agent TM 80/23 Synthetic	
Water	21.3 to 13.6 l	Water	
Air	434 to 277 l	Air	0

Degradation 6 weeks [4]:
NH_4Cl solution (10 g/L)
Na_2SO_4 solutions (35 g/L)
(Seawater (ocean average):
Cl^- 0.558M, SO_4^{2-} 0.0289M)

Sulfate degradation → formation of ettringite ($Ca_6Al_2(SO_4)_3(OH)_{12}\cdot 26H_2O$), gypsum ($CaSO_4\cdot 2H_2O$) or thaumasite ($Ca_3Si(OH)_6(SO_4)(CO_3)\cdot 12H_2O$) → volumetric expansion, fracture formations

Chloride attack → formation of Friedel's salt ($3CaO\cdot Al_2O_3\cdot CaCl_2\cdot 10H_2O$) as solid solution, chloride can undergo adhesion on C-S-H surface

Experimental set-up

X-ray radiography

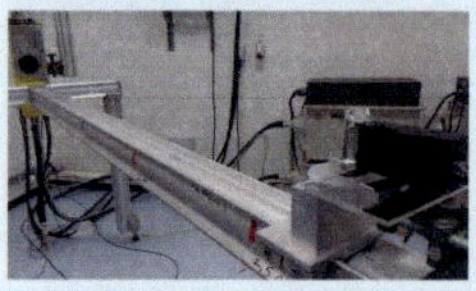

- Philips MSC 321 X-ray generator
- Medipix QTPX-262k camera
- Tracer (Cs_2CO_3) (200g/L)
- L/D: 150

Neutron tomography

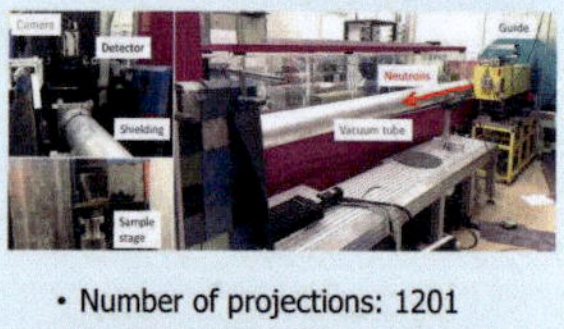

- Number of projections: 1201
- Exposure time: 60 seconds
- Scintillator thickness: 100 μm
- Reconstruction by Octopus
- Analysis by Avizo & Matlab
- Pixel size: 46 μm

X-ray radiography

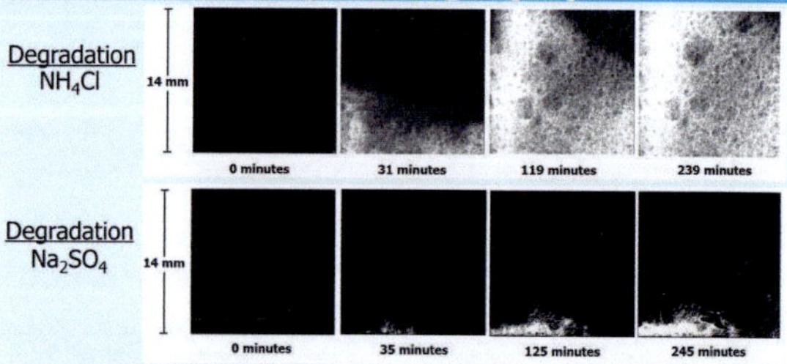

The radiographs are recorded in similar intervals, the chloride degraded sample shows higher capillary suction/rate of water uptake. This is confirmed by ASTM sorptivity measurements.

Sorptivity according to ASTM standard C1585 (absorption in mm/($s^{1/2}$).):

	Non-degraded CEM III	Degraded in Na_2SO_4	Degraded in NH_4Cl	Non-degraded CEM I
Initial rate of absorption	$4.6\cdot 10^{-3}$	$27\cdot 10^{-3}$	$38\cdot 10^{-3}$	$1.1\cdot 10^{-3}$
Secondary rate of absorption	$1.0\cdot 10^{-3}$	$7.3\cdot 10^{-3}$	$8.7\cdot 10^{-3}$	$0.3\cdot 10^{-3}$

Neutron Tomography

Avizo: void identification and categorization

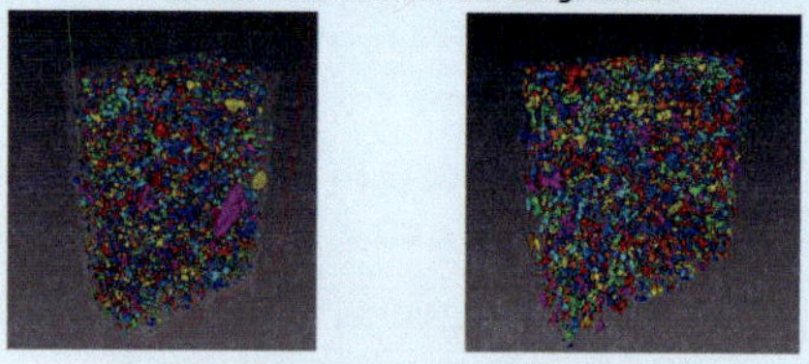

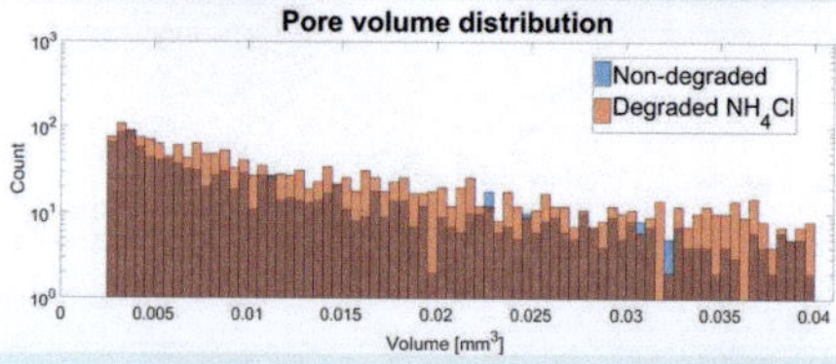

Summary

1. Chloride degraded specimens show higher impact on sorptivity.
2. Neutron imaging is suitable for pore structure reconstruction above 100 μm. To see changes below – small scales nano-tomography or scattering techniques are needed. In case of X-rays, quartz and pores are not distinguishable.
3. Boom clay pore water is expected to have impact on water absorption. In the future experiments we are planning to study the influence of synthetic BCPW, a multielement solution, on concrete. The typical pH of the BCPW is close to 8 and hence the influence on concrete may not be as significant [5].

References

1. R. Henkensiefken, J. Castro, D. Bentz et al., Cement Concrete Res., 39 (2009) 883-892.
2. T. Behrends, I. van der Veen, A. Hoving, J. Griffoen, report OPERA-PU-UTR521, 2015, 53 p.
3. T. Behrends, I. van der Veen, A. Hoving, J. Griffoen, Neth. J. Geosci., 95 (2016) 315-335.
4. C. Xiong, L. Jiang, Y. Xu et al., Constr. Build. Mater., 116 (2016) 52-62.
5. S. Seetharam, D. Jacques, report OPERA-PU-SCK514, 2015, 90p.

TUDelft

Implementation of crystallization and precipitation mechanisms in pore-scale models based on the Lattice-Boltzmann method

S. Rohmen[1*], A. Idiart[2], G. Deissmann[1], D. Bosbach[1]

[1]Institute of Energy and Climate Research (IEK-6): Nuclear Waste Management and Reactor Safety, Forschungszentrum Jülich GmbH, Jülich, Germany
[2]Amphos21 Consulting, Barcelona, Spain
*e-mail: st.rohmen@fz-juelich.de

Motivation

- Application of reactive transport simulations on chemically complex systems like alteration/degradation of cement-based materials
- Geometric changes on microscopic scale can have significant effects on macroscopic hydrodynamic properties of porous media
- Pore-scale models provide an approach for a more accurate and mechanistic description of physical/chemical processes in heterogeneous porous media

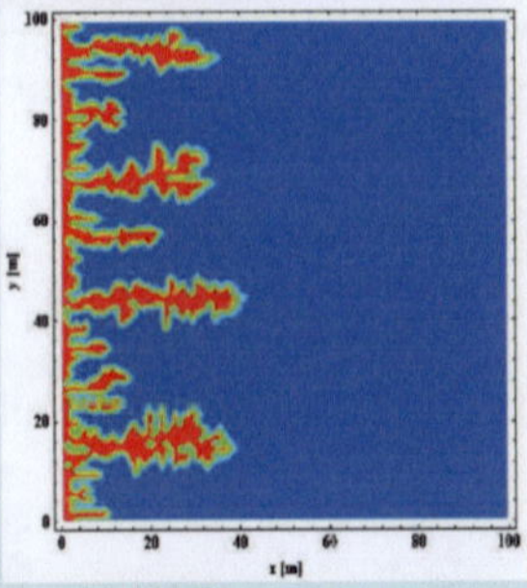

(Steefel et al., 2015)

Coupling code *iPP* (interface Palabos PhreeqC)

- Transport code: Lattice-Boltzmann code **Palabos** (Flowkit, 2011)
- Geochemical speciation code: **PhreeqcRM** (Parkhurst et al., 2015)
- Sequential non-iterative approach (SNIA)
- Iterative synchronizing between codes
- Implemented with C++ template meta-programming techniques and MPI capabilities for HPC facilities

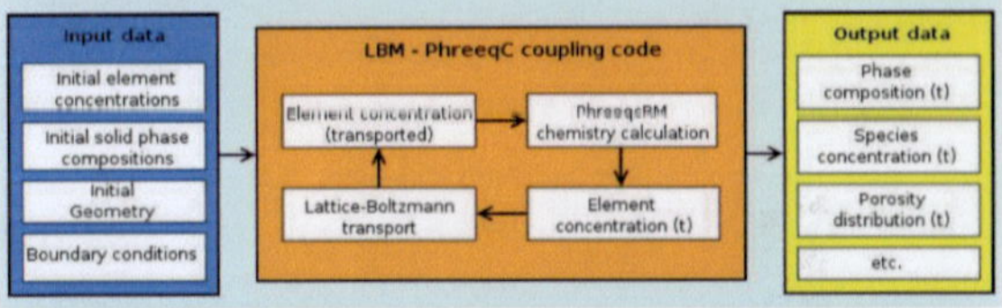

Data input/output capabilities

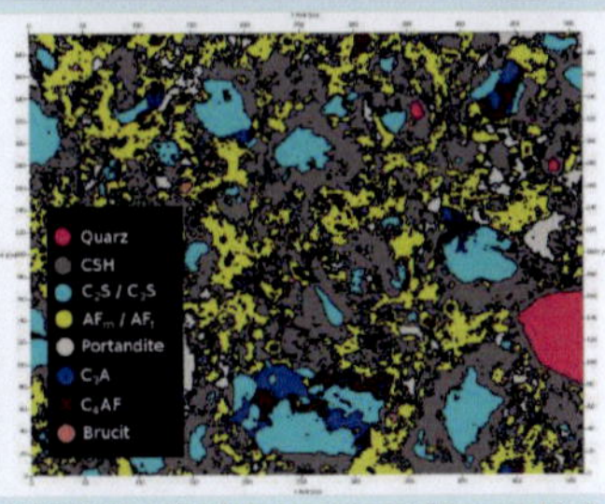

- Various data input formats are considered (incl. C++ interface)
- 2-D and 3-D geometries defined in input files (i.e. CSV, VTK, CEMHYD)
- SEM/EDX mapping techniques can provide geometrical 2-D input and phase compositions for iPP
- μCT volume data can provide 3-D pore space geometries

Benchmark against analytical solutions of diffusion problems

- Benchmark of iPP against analytical solution of diffusion problem
- Symmetrical 1-D model including non-reactive tracer with extended initial distribution
- NaCl as tracer
- Low NaCl concentration (1 mol/L) to prevent precipitation
- Simulated tracer transport (diamonds) in good agreement with analytical solution (lines)

Analytical solution: $c(x,t) = \frac{1}{2}C_0\left(erf\left(\frac{h-x}{2\sqrt{Dt}}\right) + erf\left(\frac{h+x}{2\sqrt{Dt}}\right)\right)$ (Cranc J., 1975)

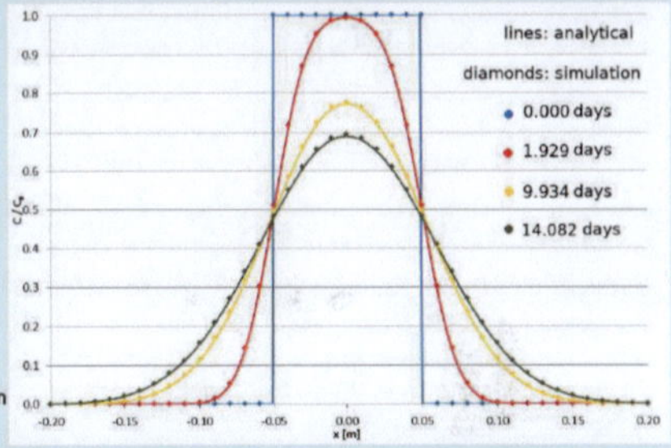

Precipitation and crystallization algorithm: Comparison of two different **S**igned **D**istance **F**ield transfer functions

- Local target saturation index (SI_T) as input for PhreeqC depending on **S**igned **D**istance **F**ield (SDF) transfer function

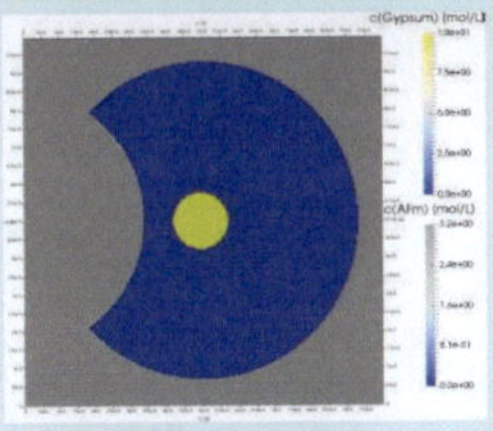

- Initial artificial pore geometry for sulphate attack example
- AFm (grey) and gypsum (yellow) forming ettringite (red)
- Initially, pore space filled with pure water

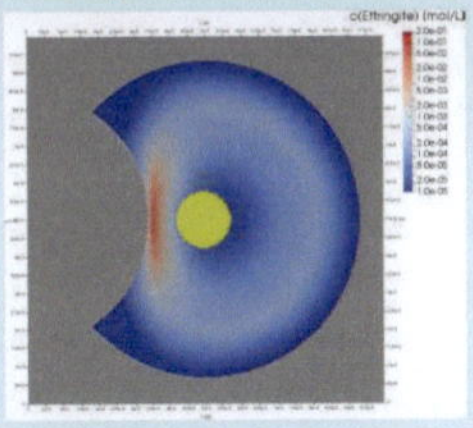

- Simple equilibrium only (flat) transfer function: $SI_T = 0$
- Broad ettringite precipitation zone
- Precipitate dispersed in whole pore volume
- No driving force for overgrowth formation

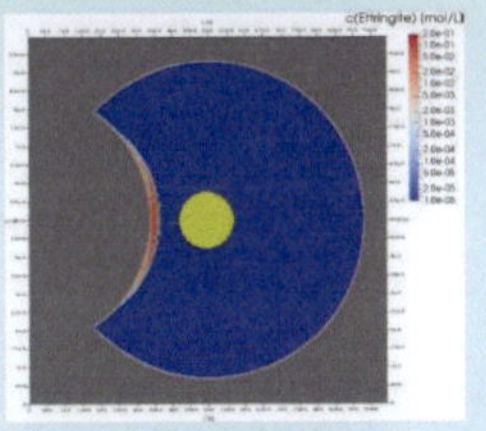

- Exponential transfer function: $SI_T = (1 - e^{k*d}) * maxSI$
- Ettringite precipitates at the grain surface
- Precipitate forms overgrowth
- Driving force for overgrowth due to exponential transfer function

Conclusion and outlook

- Development of coupling code iPP for pore-scale simulations based on Lattice-Boltzmann concept
- Data input into iPP code possible via JSON input files or abstract C++ programming interface
- Successful benchmark of solute transport against analytical solutions
- Implementation of heterogeneous and homogeneous precipitation via Signed Distance Field approach
- Parameters for distance dependent transfer functions for hydrated cement phase must be obtained
- Extension of algorithm possible to include further effects
- Validation/verification against long-term leaching experiments with various cementitious materials using different aqueous media
- Further validation of code with experimental data from CEBAMA WP1 collaboration partners when available

References

Bentz, D. P. (2005), CEMHYD3D: A three-dimensional cement hydration and microstructure development modeling package. Version 3.0
Crank, J. (1975), The mathematics of diffusion (Second Edi.), Oxford: Clarendon Press
FlowKit. (2011), Palabos user guide, http://www.palabos.org/documentation/userguide/
Parkhurst, D. L., Wissmeier, L. (2015), PhreeqcRM: A reaction module for transport simulators based on the geochemical model PHREEQC. Advances in Water Resources, 83, 176–189
Steefel C. et al. (2015), Reactive transport modeling: An essential tool and a new research approach for the Earth sciences, Earth and Planetary Science Letters, 240, 539-558

Acknowledgment

The research leading to these results has received funding from the European Union's Horizon 2020 Research and Training Programme of the European Atomic Energy Community (EURATOM) (H2020-NFRP-2014/2015) under grant agreement n° 662147 (CEBAMA).

Member of the Helmholtz Association

Cebama RWMC

THERMAL ALTERATION OF C-S-H PHASE IN CEMENTITIOUS MATERIALS

Daisuke Hayashi [1*], Masaya Ida [1], Hitoshi Owada [1], Naoki Fujii [1] and Kumi Negishi [2]
[1] Radioactive Waste Management Funding and Research Center (Japan RWMC)
[2] Taiheiyo Consultant (Japan)
* Corresponding author: d-hayashi@rwmc.or.jp

1. Introduction

◆ **In the study of geological disposal of TRU radioactive waste in Japan**

- Materials of the engineered barrier system (EBS) : Cementitious material and Bentonite
- The temperature of the EBS is kept below 80°C
- To mitigate the thermal alteration of cementitious material caused by heat generation of the waste

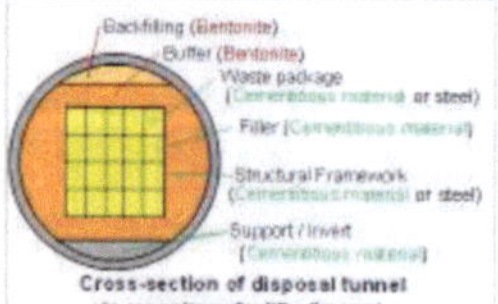

Cross-section of disposal tunnel in repository facility (image)

◆ **Previous research**

- crystalline calcium silicate hydrate (C-S-H) generate at 80 ° C or less (Cong and Kirkpatrick, 1996; Glasser and Hong, 2003; Kalousek and Roy, 1957).
- Synthetic C-S-H gel changes into crystalline C-S-H (11Å tobermorite) at 85 °C (Atkins, Glasser and Mironi, 1991)

- C-S-H gel in cementitious material may change into crystalline C-S-H at 80°C or less.
- Thereby bentonite alteration behavior by cementitious material may change. (mechanical property and nuclide migration may ..)

◆ **This study investigated** Thermal alteration of C-S-H gel in aged concrete which had been exposed to thermal effect. In addition dissolution test of aged concrete carried out.

2. Aged concrete specimens

Table 1 : Specifications of Aged Concrete.

Specimen	Exposure to thermal effect	Year of construction	Age of concrete structure (years)	Period of exposing to thermal effect (years)	Estimated temperature# (°C)
33H		1933	73	50	70-100
33N	No	1933	73	---	---
60H	Yes	1960	47	30	70-100
59H	Yes	1959	47	25	40-60
75H	Yes	1975	40	30	Around 50

#: The temperature estimated from heat conduction analysis using some actual measured temperature value.

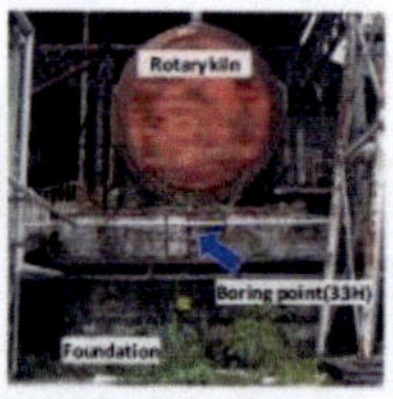

Figure 1: Boring point of specimen 33H
(foundation of cement rotary kiln)

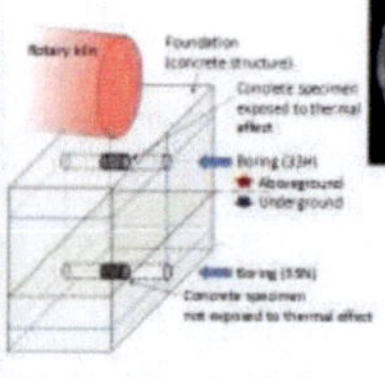
Figure 2: Sampling point of specimen 33H and 33N

Figure 3: Cross section of aged concrete (diamaters are 100mm)

- Cement type : Japanese ordinary Portland cement (OPC) (similar to European type I cement)

3. Experimental result of aged concrete

◆ **Analysis of C-S-H phase by XRD**

XRD specimens were pulverized after removal of the aggregate

- **Crystalline C-S-H was identified** (similar to tobermorite) :
  33H > 60H > 75H
  Elapsed time
  Thermal effect
- **Not identified:** 33N 59H
  Non thermal effect

✓ **Portlandite ($Ca(OH)_2$)**
Identified : 33N
Non thermal effect

Not identified: 33H 60H 59H 75H
Exposed to thermal effect

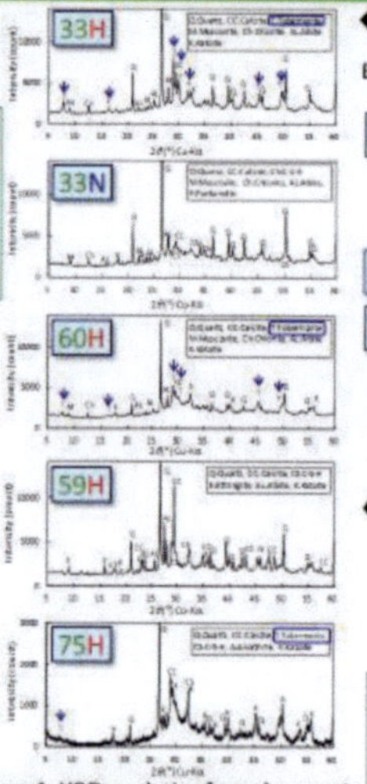

Figure 4: XRD analysis of aged concrete

◆ **Ca/Si molar ratio (C/S) of C-S-H by EPMA**

EPMA: Area analysis on Cross-section of concrete specimens(Area size: 300μm square, Pixel size: 0.5μm)

33H ➤ Average C/S was 0.84 (close to 0.83)
(0.83 is the composition of tobermorite)

60H 59H 33N ➤ Average C/S were between 1.31 and 1.36

- Decreases of C/S are related to elapsed time since construction and exposure to thermal effects.

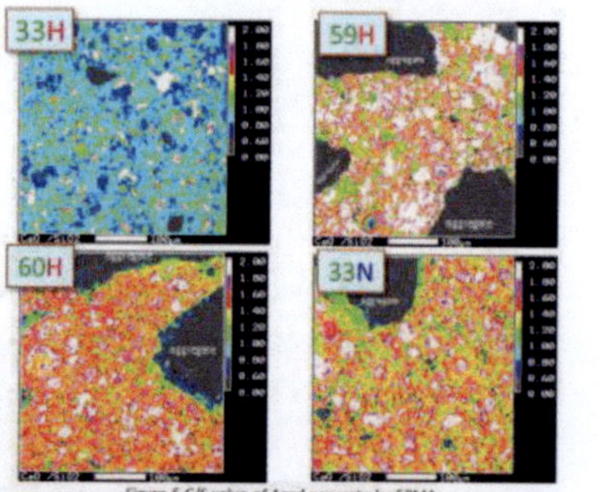

Figure 5 C/S value of Aged concrete by EPMA

◆ **C/S distribution around the aggregate**

- The cement paste (contain C-S-H) in contact with aggregate has lower C/S than the paste that is farther away from the aggregate.

- It is considered that the dissolved Si from the quartz within the aggregate was supplied to formation of low C/S value C-S-H gel.

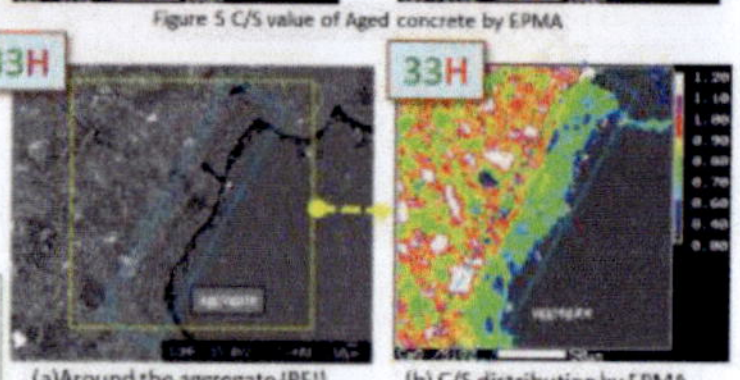

(a)Around the aggregate (BEI) (b) C/S distribution by EPMA
Figure 6 C/S distribution around the aggregate

4. Dissolution test of aged concrete

Immersed in ion-exchanged water at liquid-to-solid ratios (L/S) of 10, 200, 1000, and 2000.

◆ **thermal effects on pH**

33N Almost 12 or less
33H **11 or less**

◆ **Thermal effects on Ca dissolution**

33H < 33N
Exposed to Thermal effects / Non Thermal effects

Concrete that is exposed to thermal effects is expected to have a smaller impact on the bentonite alteration in EBS.

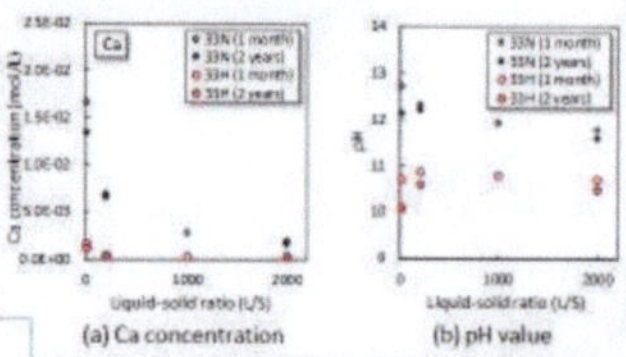

(a) Ca concentration (b) pH value
Figure 7: pH value and Ca concentration of liquid phase

5. Conclusions and Future work

- In aged concrete which has aged about from 40 to 70 years, it was observed that C-S-H gel decreased C/S and changed to crystalline C-S-H by thermal effect. The crystalline C-S-H is similar to tobermorite.
- Due to these phenomena, Ca dissolution and pH value were decreased. Therefore, concrete that is older than 50 years and has been exposed to thermal effects is expected to have a smaller impact on the bentonite alteration in EBS.
- Therefore, it is considered that these phenomena are necessary to properly consider in the long-term behaviour evaluation of EBS.
- In the future, it is necessary to clarify the condition of the phenomena (e.c. temperature) based on the results of the laboratory test.

Acknowledgement: This research is a part of the results of "Evaluation Experiments of Long-Term Performance of Engineered Barriers (FY2006, FY2007, FY2008 and FY2011)" and "Advancement of Processing and Disposal Technique for the Geological disposal of TRU Waste (FY2013 and FY2016)" under a grant from the Agency for Natural Resources and Energy (ANRE) in the Ministry of Economy, Trade and Industry (METI) of Japan. The research leading to these results has been done in cooperation with the European Union's Horizon 2020 Research and Training Programme of the European Atomic Energy Community (EURATOM) (H2020-NFRP-2014/2015) under grant agreement n° 662147 (CEBAMA).

Ce ba ma

RWMC

RWMC'S CONTRIBUTION FOR CEBAMA IN PERIOD 1

- DISSOLUTION AND GROWTH RATE OF SECONDARY MINERALS IN BENTONITIC MATERIAL -

Hitoshi Owada [1]*, Daisuke Hayashi [1], Naoki Fujii [1] and Hisao Satoh [2]
1 Radioactive Waste Management Funding and Research Center (Japan RWMC)
2 Mitsubishi Material Corp. (Japan)
* Corresponding author: owada@rwmc.or.jp

Abstract

To evaluate the influences on Hydro-Mechanical-Chemical coupling (HMC) phenomenon arisen by alkaline alteration of bentonitic material, pseudo-in-situ observation of dissolution of montmorillonite and generation of secondary zeolitic minerals was carried out by using vertical phase shift interferometory (VSI). From results of VSI observation, growth ratio of expected secondary minerals such as Phillipsite, Clinoptilolite and Analcime.

1. Introduction

Bentonite and bentonitic clay materials are expected to use as buffer and backfill in Japanese geological disposal. Water tightness and low diffusivity are dominant function of those bentonitic materials. Bentonitic materials will be altered by the reaction with alkaline cement leachate. Because hydraulic and mechanical properties of such bentonitic materials depend on the porosity and the composition of minerals, to evaluate the alkaline alteration, such as dissolution of montmorillonite and generation and growth of zeolitic secondary minerals, is an important issue of the performance assessment of EBS system. In the 2nd progress report of geological disposal of TRU waste[1], phillipsite, clinoptilolite, analcime and laumontite are expected as secondary zeolitic minerals those are generated by alkaline alteration of bentonitic materials. In this report, generation and growth ratio of such zeolitic minerals were obtained by VSI.

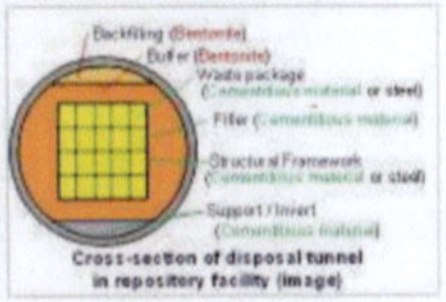

Cross-section of disposal tunnel in repository facility (image)

2. Specimen for VSI-observation

Fine grinded KunigelV1 bentonite (KV1) is used for the start material (KV1-fine). Chemical composition of KV1-fine bentonite is shown in table 1.

Table 1: chemical composition of KV1

Oxide	Content (wt. %)
SiO2	71.86
Al2O3	15.70
TiO2	0.24
Fe2O3	0.61
MnO	0.08
MgO	2.69
CaO	2.06
Na2O	1.96
K2O	0.29
Total	97.49
Loss of ignition	8.56

3. Experimental conditions

Pseudo in-situ observation of the growth of secondary minerals are carried out by using VSI equipment such as shown in Figure 1.

In ALT-VSI2.5 to 4 samples, fine clinoptilolite was added as "seed" for precipitation of secondary minerals.

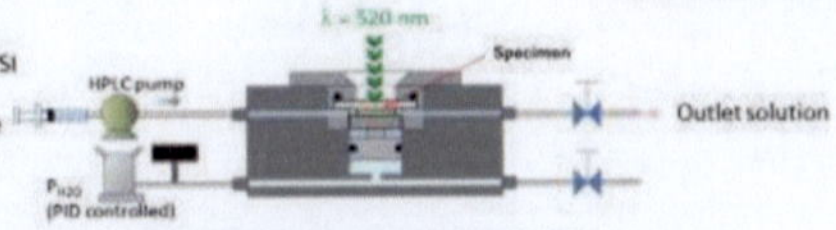

Figure 1: VSI equipment for in-situ observation

Table 2 : Test condition of VSI observation

Sample ID	Compressive pressure (MPa)	Temp (degree C)	Start Solid	Solution	Analyses
ALT-VSI0	0.4-0.8	70	KV1-fine	0.5mol/L NaOH (Region1 leachate from OPC)	ICP-AES Micro-XRD FESEM-EDS AFM
ALT-VSI1	4				
ALT-VSI2	20				
ALT-VSI2.5	1		KV1-fine and Clinoptilolite		
ALT-VSI3	4				
ALT-VSI4	20				

4. Results of VSI-observation and analyses of the outlet solution

Table 3 : Chemical composition of the outlet water of KV1

Run# ALT-VSI0	Solution	compaction P (MPa)	Temperature (°C)	pH calc	solution volume (ml)	flow rate (µl/min)	duration (min)	reaction period (h)
Kunigel-V1fine								
Run1-1	N-I	0.4	70	12.3	1.0	0.4	2247	37.5
Run1-2	N-I	0.4	70	12.3	8.2			
Run2-2	N-I	0.8	70	12.3	7.4			
Run2-3	N-I	0.8	70	12.3	26.5			
Run3	N-I	0.8	70	12.3	0.0			
Run4	N-I	0.8	70	12.3	0.0			
Run5	N-I	0.8	70	12.3	0.0			
Run6	N-I	0.8	70	12.3	0.0			
Run7	N-I	0.8	70	12.3	0.0			
Run8	N-I	0.8	70	12.3	0.0			

Run# ALT-VSI1	Solution	compaction P (MPa)	Temperature (°C)	pH calc	solution volume (ml)	flow rate (µl/min)	duration (min)	reaction period (h)
Kunigel-V1fine								
Run1	N-I	4.0	70	12.3	1.0	1.0	980	16.3
Run2-1	N-I	4.0	70	12.3	8.2	4.5	2801	46.7
Run2-2	N-I	4.0	70	12.3	7.4	7.8	3751	62.5
Run2-3	N-I	4.0	70	12.3	26.5	8.8	6752	112.5
Run3	N-I	4.0	70	12.3	0.0	0.0	10991	183.2
Run4	N-I	4.0	70	12.3	0.0	0.0	12836	213.9
Run5	N-I	4.0	70	12.3	0.0	0.0	18662	311.0
Run6	N-I	4.0	70	12.3	0.0	0.0	24698	411.6
Run7	N-I	4.0	70	12.3	0.0	0.0	30907	515.1
Run8	N-I	4.0	70	12.3	0.0	0.0	32837	547.3

Run# ALT-VSI2	Solution	compaction P (MPa)	Temperature (°C)	pH calc	solution volume (ml)	flow rate (µl/min)	duration (min)	reaction period (h)
Kunigel-V1fine								
Run2-1	N-I	20.0	70	12.3	7.5	7.9	951	15.9
Run2-2	N-I	20.0	70	12.3	9.0	6.2	2399	40.0
Run2-3	N-I	20.0	70	12.3	7.3	5.1	3826	63.8
Run2-4	N-I	20.0	70	12.3	15.0	5.4	6620	110.3
Run3	N-I	20.0	70	12.3	6.8	5.2	7939	132.3
Run4	N-I	20.0	70	12.3	0.0	0.0	10739	178.9
Run5	N-I	20.0	70	12.3	0.0	0.0	13645	227.4
Run6	N-I	20.0	70	12.3	0.0	0.0	14977	249.6
Run7	N-I	20.0	70	12.3	0.0	0.0	22402	373.7
Run8	N-I	20.0	70	12.3	0.0	0.0	27255	454.3

Table 4 : Chemical composition of the outlet water of KV1+clinoptilolite

Run# ALT-VSI2.5	Solution	compaction P (MPa)	Temperature (°C)	pH calc	solution volume (ml)	flow rate (µl/min)	duration (min)	reaction period (h)	Remarks
Kunigel-V1 fine+CLP									
Run1-1	N-I	1.0	70	12.3	3.0	1.5	2247	37.5	
Run1-2	N-I	1.0	70	12.3	6.0	3.6	3662	60.9	
Run1-3	N-I	1.0	70	12.3	14.0	2.0	6980	116.3	
Run2	N-I	1.0	70	12.3	0.0	0.0	11542	192.4	
Run3	N-I	1.0	70	12.3	0.0	0.0	18325	305.8	
Run4	N-I	1.0	70	12.3	0.0	0.0	24308	406.3	
Run5	N-I	1.0	70	12.3	0.0	0.0	26049	434.2	
Run6	N-I	1.0	70	12.3	0.0	0.0	30394	506.6	
Run7	N-I	1.0	70	12.3	0.0	0.0	31880	531.3	
Run8	N-I	1.0	70	12.3	0.0	0.0	37491	624.9	

Run# ALT-VSI3	Solution	compaction P (MPa)	Temperature (°C)	pH calc	solution volume (ml)	flow rate (µl/min)	duration (min)	reaction period (h)	Remarks
Kunigel-V1 fine+CLP									
Run1	N-I	4.0	70	12.3	1.0	1.0	980	16.3	
Run2-1	N-I	4.0	70	12.3	8.2	4.5	2801	46.7	
Run2-2	N-I	4.0	70	12.3	7.4	7.8	3751	62.5	
Run2-3	N-I	4.0	70	12.3	26.5	8.8	6752	112.5	
Run3	N-I	4.0	70	12.3	0.0	0.0	10991	183.2	
Run4	N-I	4.0	70	12.3	0.0	0.0	12836	213.9	
Run5	N-I	4.0	70	12.3	0.0	0.0	18662	311.0	
Run6	N-I	4.0	70	12.3	0.0	0.0	24698	411.6	
Run7	N-I	4.0	70	12.3	0.0	0.0	30907	515.1	[illegible]
Run8	N-I	4.0	70	12.3	0.0	0.0	32837	547.3	

Run# ALT-VSI4	Solution	compaction P (MPa)	Temperature (°C)	pH calc	solution volume (ml)	flow rate (µl/min)	duration (min)	reaction period (h)	Remarks
Kunigel-V1 fine+CLP									
Run2-1	N-I	20.0	70	12.3	7.5	7.9	951	15.9	
Run2-2	N-I	20.0	70	12.3	9.0	6.2	2399	40.0	
Run2-3	N-I	20.0	70	12.3	7.3	5.1	3826	63.8	
Run2-4	N-I	20.0	70	12.3	15.0	5.4	6620	110.3	
Run3	N-I	20.0	70	12.3	6.8	5.2	7939	132.3	
Run4	N-I	20.0	70	12.3	0.0	0.0	10735	178.9	
Run5	N-I	20.0	70	12.3	0.0	0.0	13645	227.4	
Run6	N-I	20.0	70	12.3	0.0	0.0	14977	249.6	
Run7	N-I	20.0	70	12.3	0.0	0.0	22401	373.7	
Run8	N-I	20.0	70	12.3	0.0	0.0	27255	454.3	

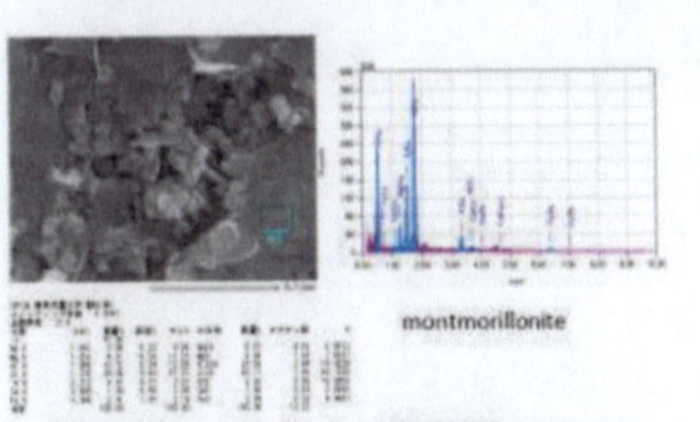

Figure 2: An example of results of FESEM-EDS measurement

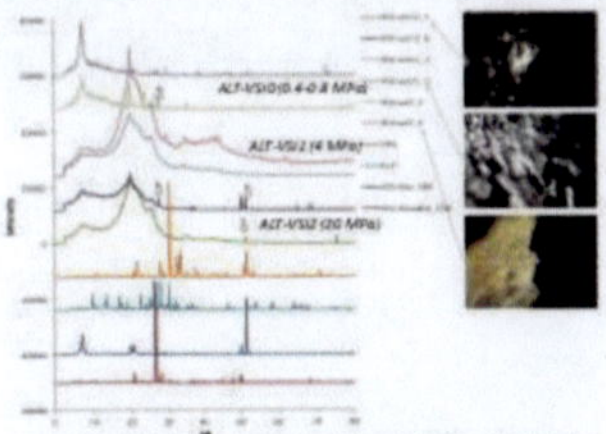

Figure 3: Results of µ-XRD measurement of KV1-samples

After the experiment in ALT-VSI0, K and Na rich phase which is expected as sanidine and growth of silica phase was observed. In ALT-VSI1 and ALT-VSI2, Al and Ca containing secondary mineral and dissolution of SiO_2 phase such as calcedony were observed. From these results, though K-containing secondary mineral such as sanidine grow under alkaline and lower compressive pressure condition, dissolution of montmorillonite and silica minerals were thought as dominant alteration process in higher pressure condition. From the results of micro-XRD analyses, clinoptilolite was precipitated as secondary mineral under high pressure (20MPa) condition (shown in Figure 3).

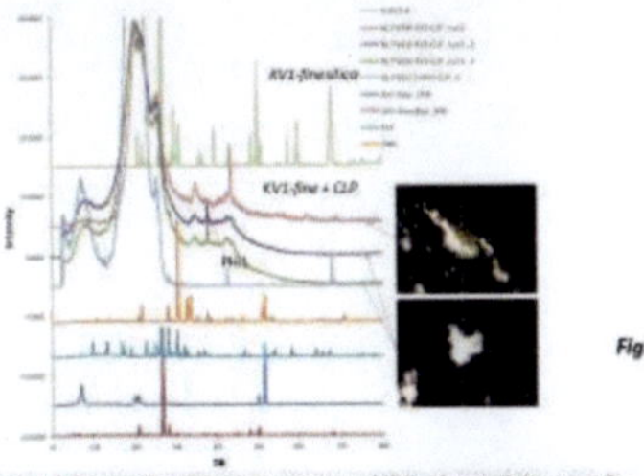

Figure 4: Results of µ-XRD measurement of KV1+clinoptilolite samples

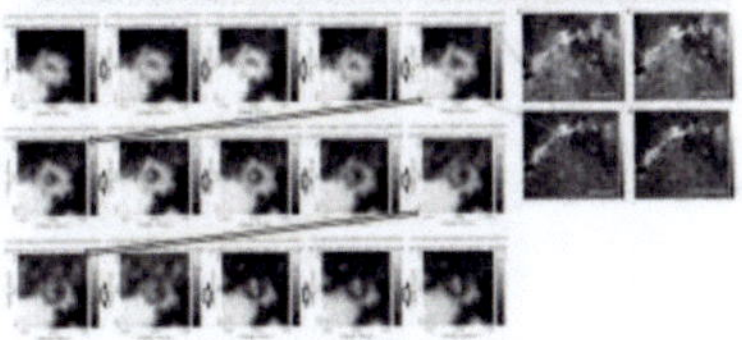

Figure 5: An example of the process of nucleation of zeolitic mineral

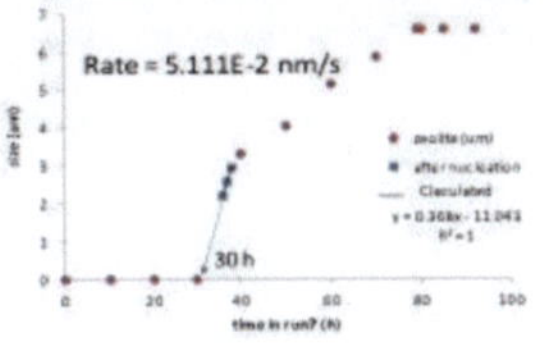

Figure 6: Initiation time and accelelation of nucleation of zeolitic mineral

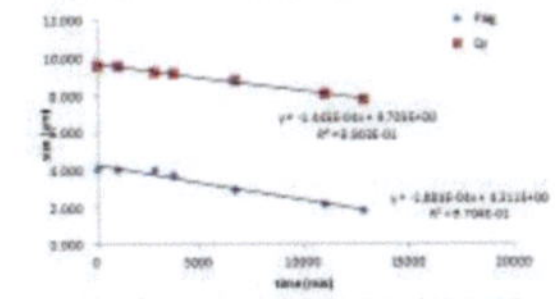

Figure 7: Dissolution rate of accessory minerals

Results of micro-XRD analyses of KV1+Clinoptilolite samples are shown in figure 4. Phillipsite was detected as secondary zeolite. Nucleation and growth was observed as shown in Figure 5. From VSI observation and the change of chemical composition of outlet solutions, initiation time and growth ratio of zeolitic mineral and dissolution ratio of accessory minerals are determined as shown in Figures 6 and 7 respectively.

Acknowledgement:
This research is a part of the results of "Evaluation Experiments of Long-Term Performance of Engineered Barriers (FY2011)" and "Advancement of Processing and Disposal Technique for the Geological disposal of TRU Waste (FY2013 and FY2016)" under a grant from the Agency for Natural Resources and Energy (ANRE) in the Ministry of Economy, Trade and Industry (METI) of Japan.
The research leading to these results has received funding from the European Union's Horizon 2020 Research and Training Programme of the European Atomic Energy Community (EURATOM) (H2020-NFRP-2014/2015) under grant agreement n° 662147 (CEBAMA).

2nd Annual Workshop of the Cebama Project, Espoo 16-19 May, 2017

H-M-C COUPLING ANALYSIS CONSIDERING SEVERAL SCENARIOS OF LONG-TERM ALTERATION IN CEMENT-BENTONITE SYSTEM

Shinya Tachibana[1*], Shinji Ito[1], Atsushi Iizuka[1], Hitoshi Owada[2] and Daisuke Hayashi[2]
1 Kobe University (Japan) * Corresponding author: stachi@people.kobe-u.ac.jp
2 Radioactive Waste Management Funding and Research Center (Japan RWMC)

INTRODUCTION

It is generally recognized that more attentions need to be given to coupling and complex effects on chemical and hydraulic/mechanical interactions to evaluate the long-term performance of the engineered barrier. The dissolution of montmorillonite and/or the subsequent formation of secondary minerals are the possible chemical reactions, which may causes the changes in the properties and performance of the bentonite buffer over a long period. Thus, this study aims at contributing the establishment of method and addressing the H-M-C coupling analysis by using the present numerical codes.

METHODROLOGY

The geochemical and hydraulic/mechanical codes were used to calculate the long-term performance of bentonite buffer: PHREEQC-TRANS and DACSAR-BA, respectively. The elasto-plastic constitutive model was newly developed so as to express the changes in mechanical properties induced by the chemical alteration and was incorporated to code DACSAR-BA.
A part of the engineered barrier consisting of the bentonite buffer and mortar was set to be as an analysis object domain.
Two scenarios of chemical alteration: formation of clinoptirolite and dissolution of montmorillonite, were considered. In addition, two kinds of initial condition: initially homogeneous and heterogeneous distributions of dry density in bentonite buffer, were taken into consideration.

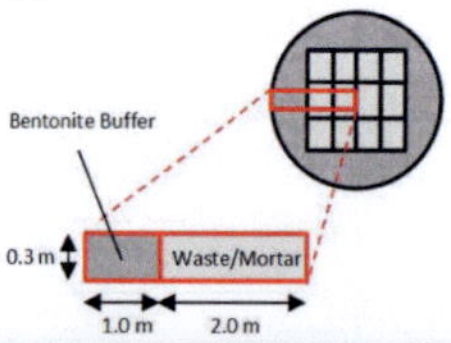

GEOCHEMICAL ANALYSIS

Changes in montmorillonite content were calculated by using code PHREEQC-TRANS ahead of H-M-C coupling analysis. These data would be converted to "Rate of volume change of solid phase, α" in respond to each scenario and be provided to the coupling analysis as input data.

DISTRIBUTION OF ALPHA AFTER 100 000 YEARS

Clinoptirolite formation causes the increases in solid volume in the vicinity of mortar/cement, while montmorillonite dissolution results in the loss of solid volume.

RESULTS OF H-M-C COUPLING ANALYSIS: PROPERTIES DURING/AFTER 100 000 YEARS

HOMOGENEOUS INTIAL CONDITION

CLINOPTIROLITE FORMATION

STRESS DISTRIBUTION

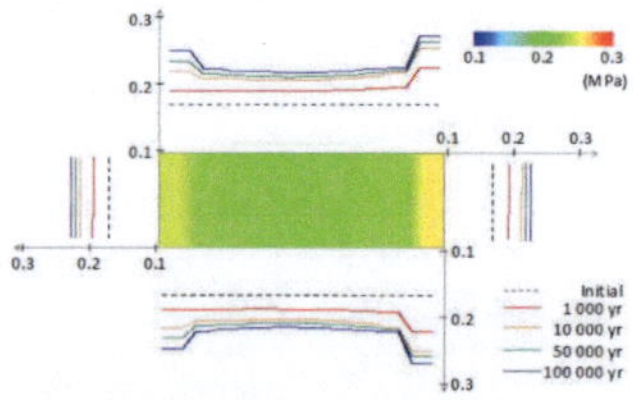

PERMEABILITY

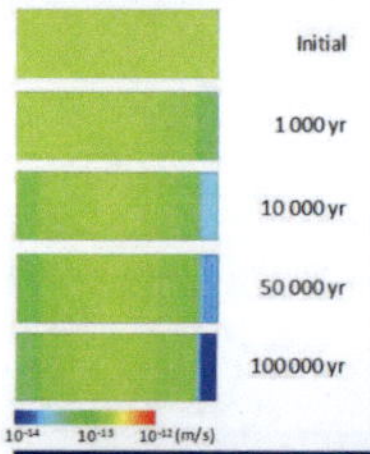

PERMEABILITY

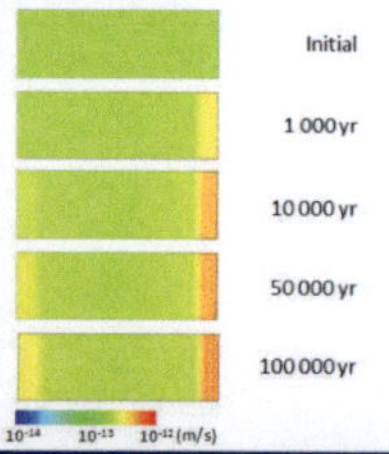

MONTMORILLONITE DISSOLUTION

STRESS DISTRIBUTION

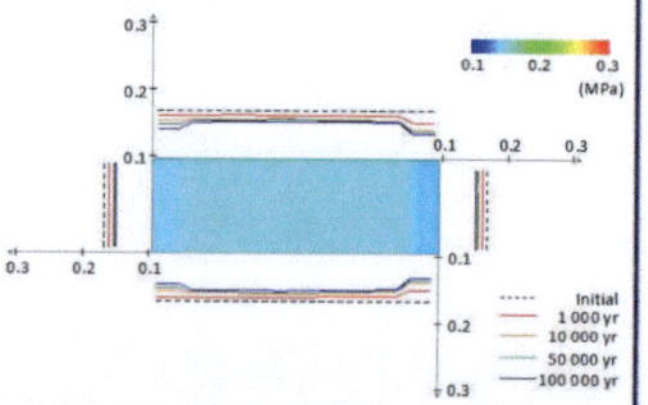

HETEROGENEOUS INITIAL CONDITION

CLINOPTIROLITE FORMATION

STRESS DISTRIBUTION

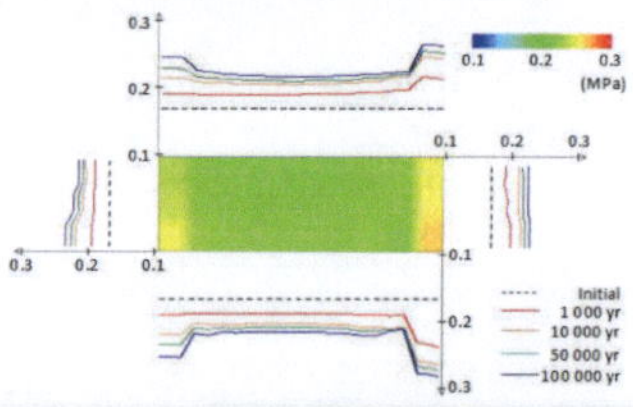

PERMEABILITY

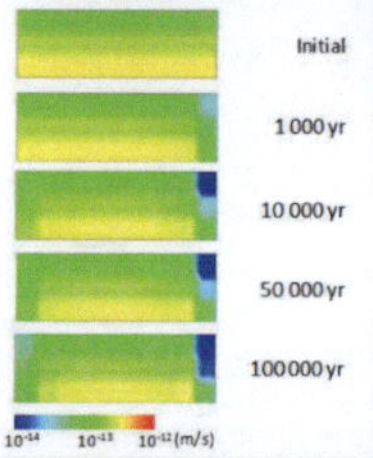

PERMEABILITY

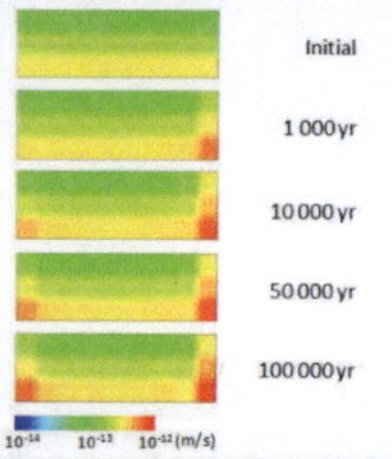

MONTMORILLONITE DISSOLUTION

STRESS DISTRIBUTION

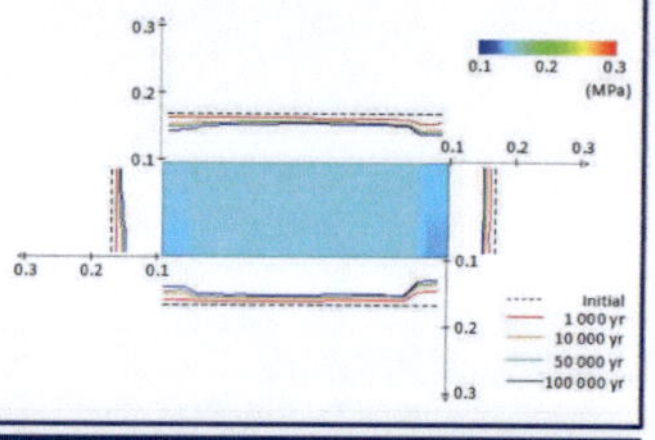

CONCLUSION

The results of H-M-C coupling analysis suggest that the long-term performance of buffer system strongly depends on the chemical alteration scenario. Considering the uncertainty of chemical reaction even locally, it remains necessary to develop the systematic analysis scheme which enable to deal with every possible scenarios and assess the long-term performance stochastically.

ACKNOWLEDGEMENT

This research is a part of the results of "Advancement of Processing and Disposal Technique for the Geological disposal of TRU Waste (FY2016)" under a grant from the Agency for Natural Resources and Energy (ANRE) in the Ministry of Economy, Trade and Industry (METI) of Japan. The research leading to these results has received funding from the European Union's Horizon 2020 Research and Training Programme of the European Atomic Energy Community (EURATOM) (H2020-NFRP-2014/2015) under grant agreement n° 662147 (CEBAMA).

The University Of Sheffield. NucleUS Immobilisation Science Laboratory midas Radioactive Waste Management HORIZON 2020 Cebama

Characterisation of UK cement backfill material and preliminary groundwater leaching experiments

Rita G. W. Vasconcelos*[1], Neil C. Hyatt[1], John L. Provis[1] and Claire L. Corkhill[1]

NucleUS Immobilisation Science Laboratory, Department of Materials Science and Engineering, University of Sheffield, Sheffield, United Kingdom
*presenting author: rgwvasconcelos1@sheffield.ac.uk

1. Introduction

Nirex Reference Vault Backfill (NRVB) is a high pH cementitious material considered for use as a backfill material in one of the conceptual scenarios for the UK geological disposal of intermediate level waste (Fig. 1). The aim of this project is to understand how the physico-chemical properties of NRVB will be affected by interactions with groundwater. We present the results of a full characterisation of NRVB, prepared using different precursor materials, at 28 days of curing, and results from preliminary leaching experiments.

(1)

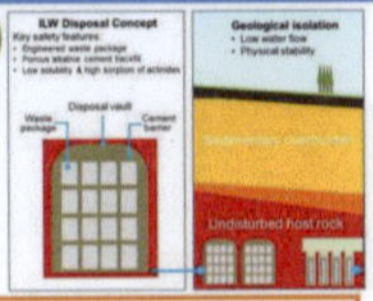

2. Characterisation

Characterisation of NRVB synthesised using different sources of materials (industrial materials or lab reagents) was performed. NRVB (Ind) was synthesised using 450 kg/m³ CEM I 52.5 N; 170 kg/m³ hydrated lime; and 495 kg/m³ limestone flour (w/s = 0.55). NRVB (Lab) used the same proportions and w/s ratio, but $Ca(OH)_2$ and $CaCO_3$ were used instead of hydrated lime and limestone flour.

(2)

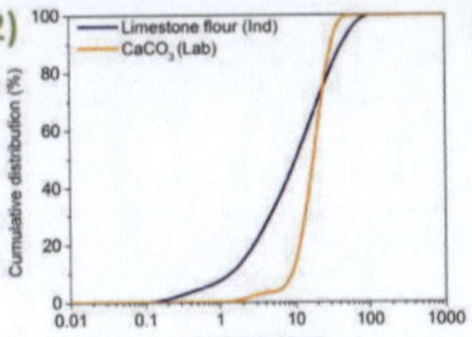

(3)

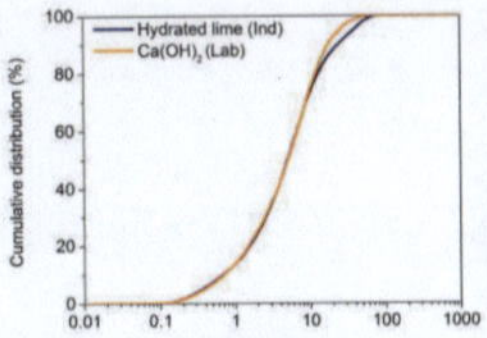

- Differences in the particle size distribution between limestone flour and $CaCO_3$ were observed (Fig. 2), while the $Ca(OH)_2$ and hydrated lime had similar particle size (Fig. 3).
- This corresponded to a higher surface area for the limestone flour (5.2 ± 0.2 m²/g), compared with (3.7 ± 0.2 m²/g) for $CaCO_3$

Table 1. Workability and Vicat results

	Workability (mm)	Vicat Setting time (hours)
NRVB (Lab)	56.5 ± 0.8	Initial – 5.3 Final – 7.7
NRVB (Ind)	68.4 ± 1.7	Initial – 5.5 Final – 7.3

After mixing, the workability and setting time of the paste were measured, and isothermal calorimetry was performed at 20 °C to understand the hydration reaction of the two cements (Fig. 4).

- NRVB (Ind) evidenced a more intense sulfate depletion reaction.
- This was associated with the formation of more monocarboaluminate (Fig. 5) and a high sulfate content in limestone flour (893 ppm) than $CaCO_3$ (37 ppm).

(4)

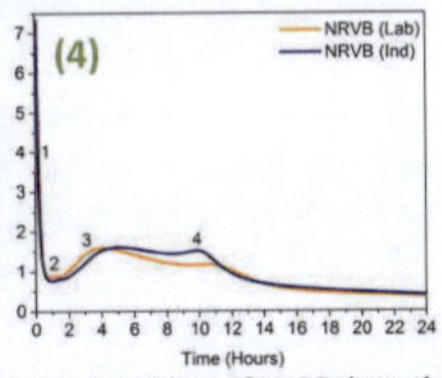

The cement paste was cured at 20 °C at relative humidity. After 28 days, the compressive strength was 8.2 ± 0.2 MPa for NRVB (Lab) and 7.15 ± 0.04 MPa for NRVB (Ind).

(5)

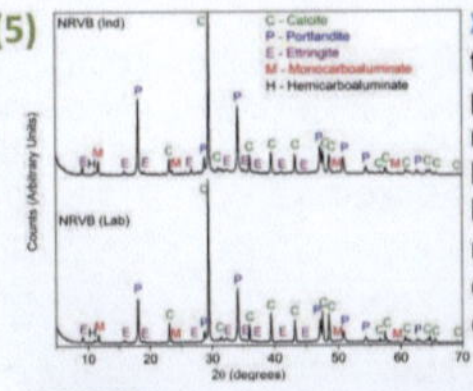

XRD analysis (Fig. 5) revealed the presence of calcite, portlandite, ettringite, monocarboaluminate, hemicarboaluminate. The higher intensity of monocarboaluminate is also observed in ^{27}Al MAS NMR data (Fig. 6).

(6)

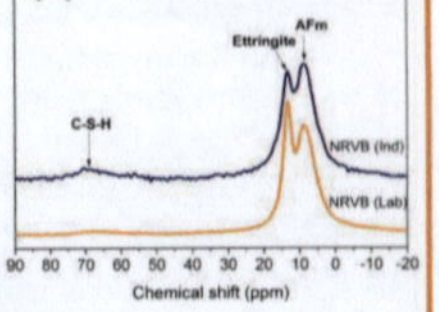

NRVB (Lab)

(7)

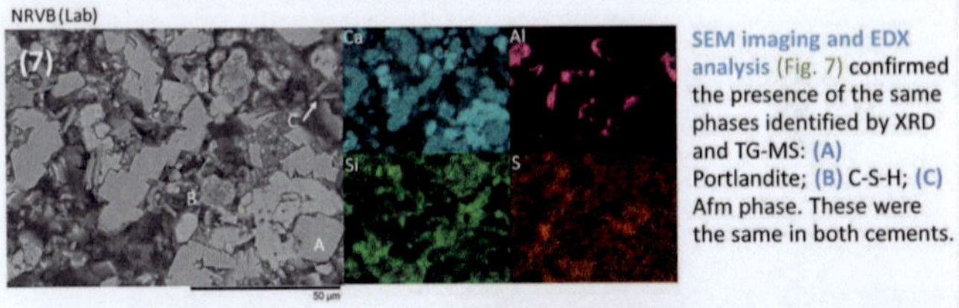

SEM imaging and EDX analysis (Fig. 7) confirmed the presence of the same phases identified by XRD and TG-MS: (A) Portlandite; (B) C-S-H; (C) Afm phase. These were the same in both cements.

The porosity of the two materials was measured using Mercury Intrusion Porosimetry. NRVB (Lab) gave a total porosity of 38 ± 1 % and NRVB (Ind) had a total porosity of 32 ± 1 %. Due to the low compressive strength of NRVB, this technique is not suitable to measure the pore size distribution (due to compression of pores at high pressures), therefore a complementary technique, using X-ray Computed Tomography was used:

- The VOI size chosen was 0.42 mm³ (250 x 250 x 250 voxels at 3 µm resolution)
- The threshold was established using the line shape of the image histograms
- Segmented porosity of NRVB (Lab) (Fig. 8) was 39 % and of NRVB (Ind) was 35 % (Fig. 9)

(8)

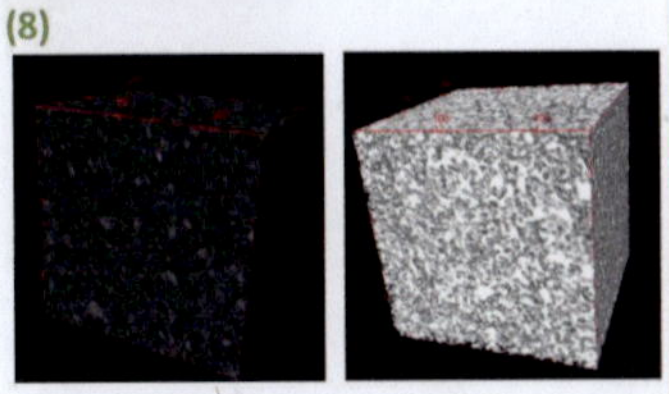

(9)

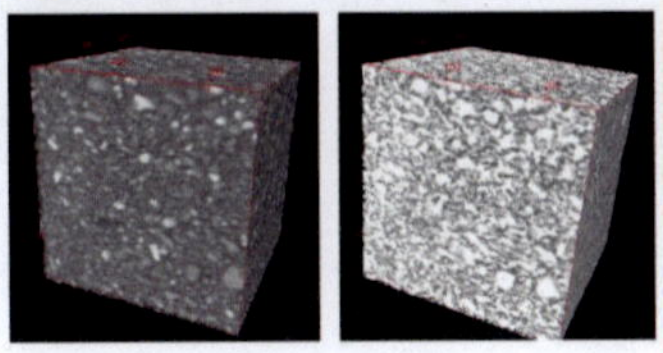

3. Preliminary Leaching Experiments

NRVB (Ind) was prepared in the same way as described above, and cured at 40 °C at relative humidity for 28 days. Monoliths were prepared, each with 15 mm height x 15 mm diameter, with the ends sealed with epoxy to allow only radial diffusion. Three different types of groundwater were prepared (Fig. 10).

(10)

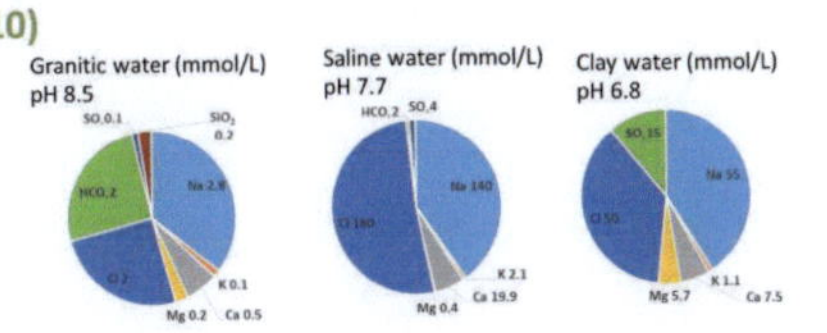

(11)

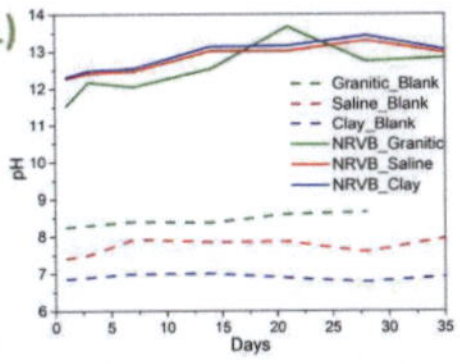

NRVB (Ind) was placed in 50 mL of groundwater, at 40 °C. Sampling was performed at specific intervals until 35 days, in controlled oxygen and nitrogen environment. pH was measured (Fig. 11) and an increase was observed in line with modelling predictions. Elemental concentrations were also measured, using ICP-OES.

Post-leaching characterisation using XRD (Fig. 12) showed some differences in the intensity of ettringite and monocarboaluminate. In terms of porosity (by MIP), control NRVB showed a porosity of 45 ± 1 %, which was very similar for NRVB leached with granitic, clay and saline groundwater, 46, 45 and 42 ± 1 %, respectively.

(12)

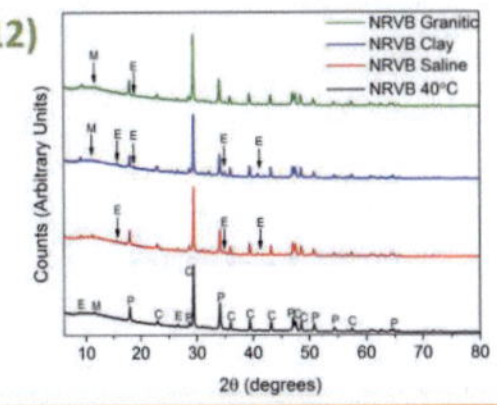

4. Future Work:

Three different leaching experiments are underway. Two of them are long-duration experiments (1 to 1.5 years), where controlled nitrogen or CO_2 environment will be applied. For the first experiment, replacement of the groundwater will take place at specific intervals, whereas for the second experiment, no groundwater will be replaced, and the cement samples only analysed at the end of the experiment. The third experiment will be a short-term experiment, where the groundwater will flow through the cement. Post-leaching characterisation will include ICP-OES, XRD, SEM, MIP and XCT. Concurrent experiments using the Cebama reference cement are also being performed.

@ISL_Sheffield
@cementsatshef

References

A.J. Francis et al (1997), NSARP Report no: S/97/014

Acknowledgements: The authors wish to acknowledge funding for this research from Radioactive Waste Management Limited and the European Commission Horizon 2020 Research and Training Programme of the European Atomic Energy Community (EURATOM) (H2020-NFRP-2014/2015) under grant agreement n° 662147 (CEBAMA). Steve Williams and Amy Shelton (RWM) is thanked for their comments and input to this work. The authors would also like to acknowledge Dr. Andrea Hamilton and Dr. Nicolas Beaudoin, from University of Strathclyde, for the ongoing assistance with XCT measurements.

TECHNOLOGY FOR BUSINESS

Bentonite interaction with saline high-pH solutions

Tiina Heikola[1] • Sirpa Kumpulainen[2] • Leena Kiviranta[2]
VTT Technical Research Centre of Finland Ltd[1],
Saanio & Riekkola Oy, Helsinki, Finland[2]

Introduction

Dissolution of cementitious materials in a repository induces high pH-leachates that may alter the mineralogical and chemical composition of the bentonite buffer and as a result jeopardize the set safety functions of the materials. Low-pH cement materials have been developed to reduce such a risk.

Objectives

To study the:

- possible alterations of bentonite in alkaline leachates,
- buffering capacity of bentonite against high pH solutions
- to facilitate modelling of bentonite/high-pH interaction

Experimental

Anaerobic glove-box

(Ar atmosphere, O_2< 0.1 ppm, low CO_2, T=25±1 °C)

- Batch tests (solid/solution 20g/200g)
- Two bentonites: - Wyoming-type **Na-bentonite**
  - Milos-type **Ca-bentonite**
- Four simulated leaching solutions (nominal pH= 8.3, 9.7, 11.3 and 12)

Results

pH-values:

- Experiments pH 11.3 and pH 12 (Fig. 1) show rather quick drop and levelling of pH-values after each leachate renewal. Quite similar values are seen for both bentonites (especially experiment pH 12).
- Experiment pH 9.7 shows rather constant pH-values well below the initial leachate value.
- Experiment pH 8.3 (the reference) the pH-values show rather constant values around the initial value.

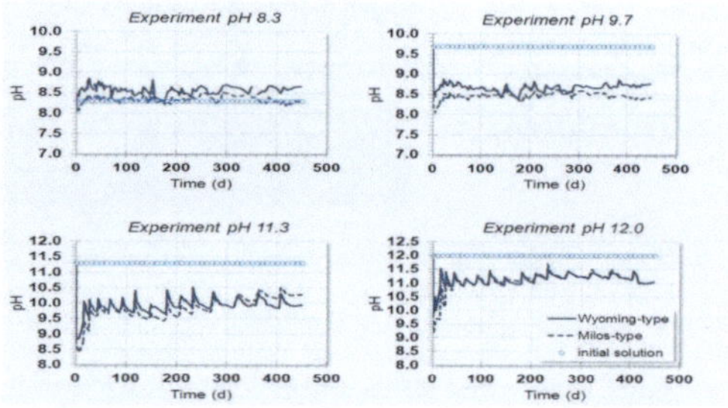

Figure 1. Measured pH values in the four leaching solutions.

Solution chemistry

- Release of silica (SiO_2) was observed at the beginning in experiments pH 11.3 and pH 12 (Fig. 2).
- Ca accumulated in all experiments, but especially in experiment pH 12 (Fig.2).
- Both bentonites in experiments pH 12 released detectable amounts of Al. In other experiments the levels were below detection limit.

Bentonite chemistry and mineralogy

Chemical analyses showed:

- decrease in SiO_2 and Al contents
- increase in Ca and C contents in experiments pH 11.3 and pH 12

Mineralogical investigations showed precipitation of:

- calcite and vaterite in experiments pH 11.3 and pH 12
- possibly zeolite (laumontite $Ca(Al_2Si_4O_{12})*4H_2O$) in Dep pH 9.7 and 12
- possibly CSH phases (Fig. 3) in experiments pH 12.

Changes in the smectite structure had occurred as there was:

- increase in clay fraction CEC (UV-VIS),
- increase in smectite layer charge (structural calculations)
- beidellitisation (structural calculations, Greene-Kelly) in experiment pH 12

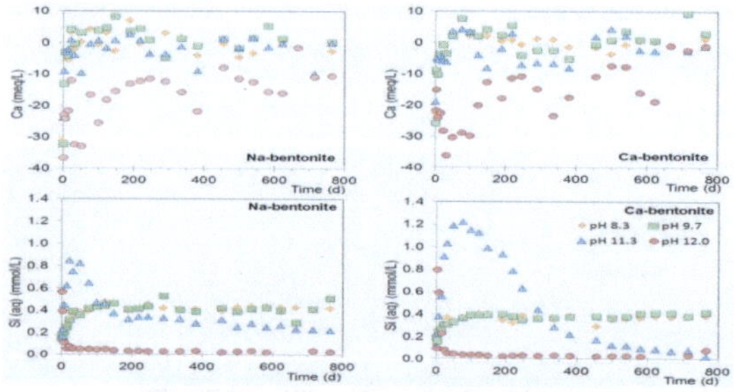

Figure 2. Ca and Si results as a function of time.

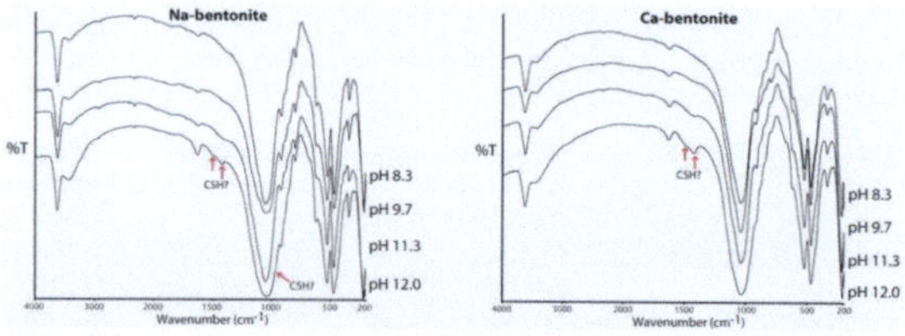

Figure 3. FTIR patterns of Wyomin-type and Milos-type materials.

Bentonite particles exposed to the alkaline leachate (pH 12) resisted to stay in dispersion, which would indicate cementation. However, their swelling pressure had increased (Fig. 4).

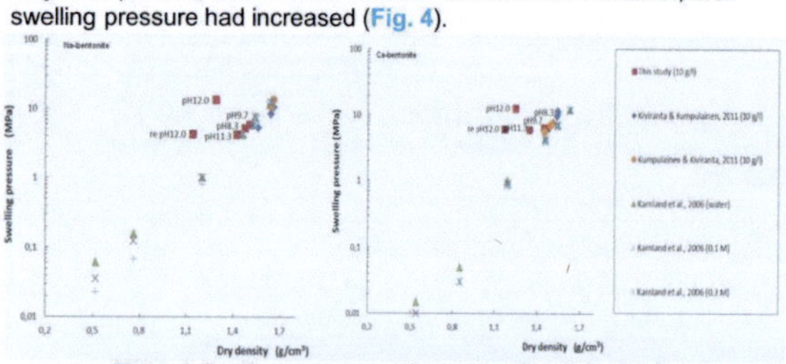

Figure 4. Swelling pressure results as a function of dry density.

Conclusions

Both bentonites showed buffering against high pH, especially at the beginning of the experiments. A slight reduction of the buffering capacity was observed in experiments pH 11.3 and pH 12. In the mineralogy and chemical composition of the bentonites only small alterations were detected, mainly in experiments pH 11.3 and pH 12. The measured swelling pressures showed increase in experiments pH 12.

Similar batch experiment at higher temperature (60 °C) underway in order to investigate the effect of temperature. The results from this experiment will available in the end of this year.

Acknowledgements This work has been funded by Posiva Oy

References: Karnland O., Olsson S., Nilsson U. 2006. Mineralogy and sealing properties of various bentonites and smectite-rich clay materials. SKB TR-06-30; Kiviranta L., Kumpulainen, S., 2011. Quality control and characterization of bentonite materials. Posiva WR 2011-84; Kumpulainen S., Kiviranta L. 2011. Mineralogical, chemical and physical study of potential buffer and backfill materials from ABM test package 1. Posiva WR 2011-41

Quantitative visualization of heterogeneous transport processes at the host rock - cement interface

J. Kulenkampff[1], M. Gründig[1], S. Gruhne[1], H. Lippold[1], J. Lippmann-Pipke[1,2], K. Jantschik[3], H.C. Moog[3]
[1]HZDR, Institute of Resource Ecology, Dresden, Germany
[2]BGR, Hannover, Germany,
[3]GRS, Braunschweig, Germany

Co-funded by the European Commission · HORIZON 2020 · Cebama · HZDR HELMHOLTZ ZENTRUM DRESDEN ROSSENDORF

Aim

Degradation at the contact between engineered materials and host rock is assumed to be a heterogeneous process, with mutual interactions between structural alterations and transport process properties in both materials. In order to elucidate this complex type of process, we apply PET (positron emission tomography) as means for direct process observation by quantitative spatiotemporal visualization of radiotracer concentration. Heterogeneous transport parameters are then derived from these observations.
Here, we successfully tested the procedure on a generic sample, which was produced of halite and salt concrete.

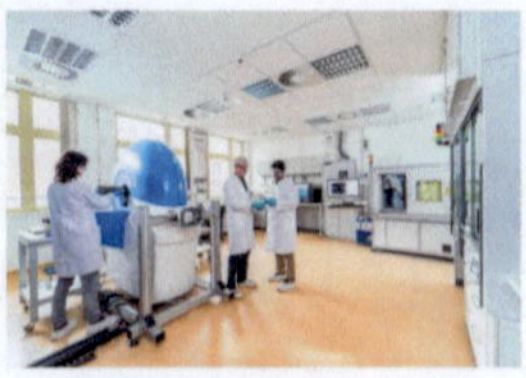

Fig. 1. Reactive transport imaging laboratory at the HZDR research site in Leipzig.
In the front: High-resolution PET-scanner (Elysia-Raytest), background: µCT-scanner (Nikon).
Photo: HZDR/André Künzelmann

Sample and Procedure

Generic material:

- Halite drill core with an aged salt concrete insert.

Procedure:

- Labelling of formation fluid with ^{22}Na (0.6 mL, 3.16 MBq).
- Injecting labelled fluid into the head space between end cap and sample surface.
- Diffusion of the tracer into the sample.
- Repeated PET-imaging of the tracer distribution over 70 days.
- Corrections and calibration.
- Parameter derivation.

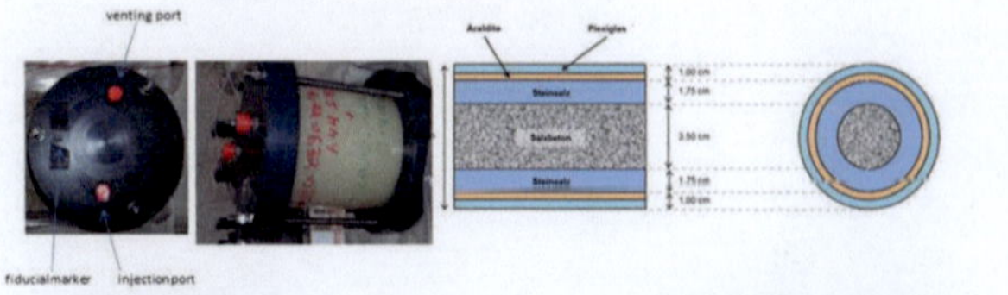

Fig. 2. Sample.
Top: Photo and cross sections.
Left: µCT-image.
Right: µCT segmented joint between host rock and cement.

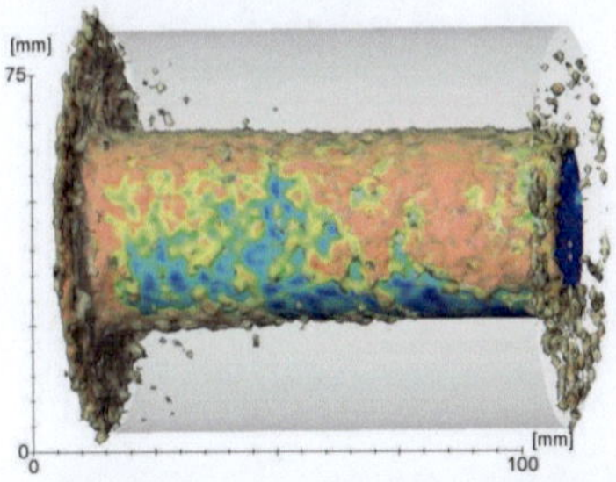

Fig. 3. Initial PET-image. Tracer distribution shortly after injection (red: high activity, blue: low activity). The tracer directly penetrated into the joint between cement and host rock. Starting from this initial patchy distribution, it then diffused predominantly into the cement (central section).

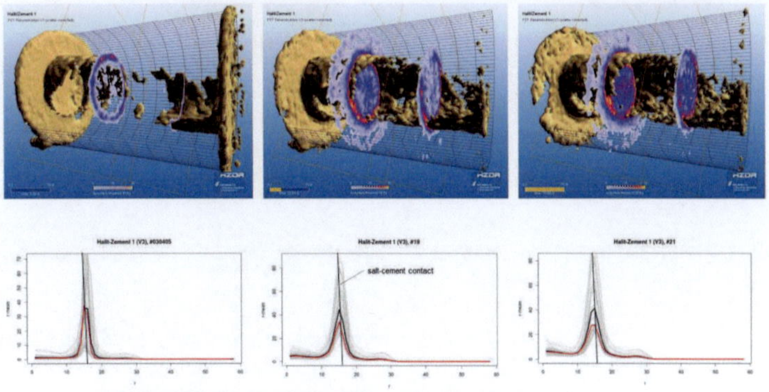

Fig. 4. Three PET time frames (directly after injection, after 22 and after 70 days).
Top: Tracer distribution iso-surface with two cross sections.
Bottom: Circular mean values vs. radius (grey: individual transaxial slices, black: mean value, red: median).
The position of the maximum value is stationary and coincides with the position of the joint.
The distribution broadens with time to the left-hand side (cement), whereas the right-hand side remains fixed.

Result

- Found open joint at material contact (µCT and PET).
- Formation fluid immediately penetrated into the joint with patchy pattern.
- Thickness of the water film is 1-10 µm – derived from activity per voxel (below µCT-resolution of 50 µm).
- Quantification of diffusion into the cement:
  - calculation of circular mean values of activity concentration yields radial tracer distribution,
  - calculation of quartiles of this distribution as measure for penetration depth,
  - cement: after 10 days significant spreading of the distribution, after 70 days propagation about 5 mm
  - salt: no significant propagation within this period (to be continued).
- Outlook: estimation of diffusion coefficients:
  - direct fit of cylindrical solution of diffusion equation,
  - inversion of FEM-simulations.

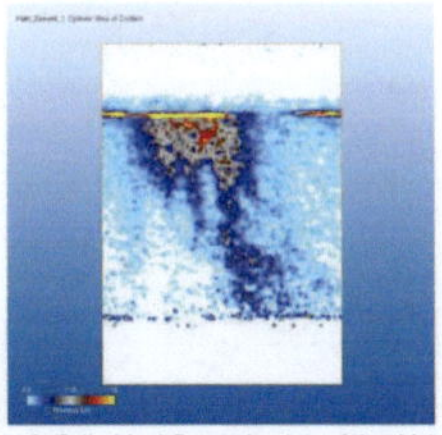

Fig. 5. Cylindrical flat projection of the thickness of the fluid film, derived from measured activity concentration referring to the initial concentration.

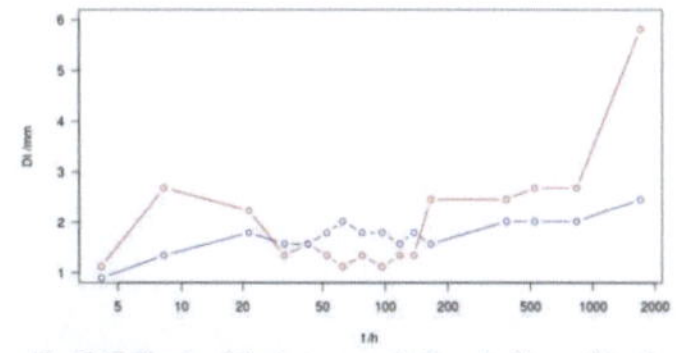

Fig. 6. Estimate of the tracer penetration depths vs. time from the spreading of the radial tracer distribution in the direction of the cement (red) and halite (blue) (Fig. 4, bottom): The interquartile ranges of the distributions (differences of the positions of the 25%-, resp. 75%-quantiles, and median position) were defined as approximate penetration depth.

Acknowledgement
The research leading to these results has received funding from the European Union's European Atomic Energy Community's (Euratom) Horizon 2020 Programme (NFRP-2014/2015) under grant agreement, 662147 – Cebama

Dr. Johannes Kulenkampff | Institute of Resource Ecology | Reactive Transport Division | j.kulenkampff@hzdr.de | www.hzdr.de

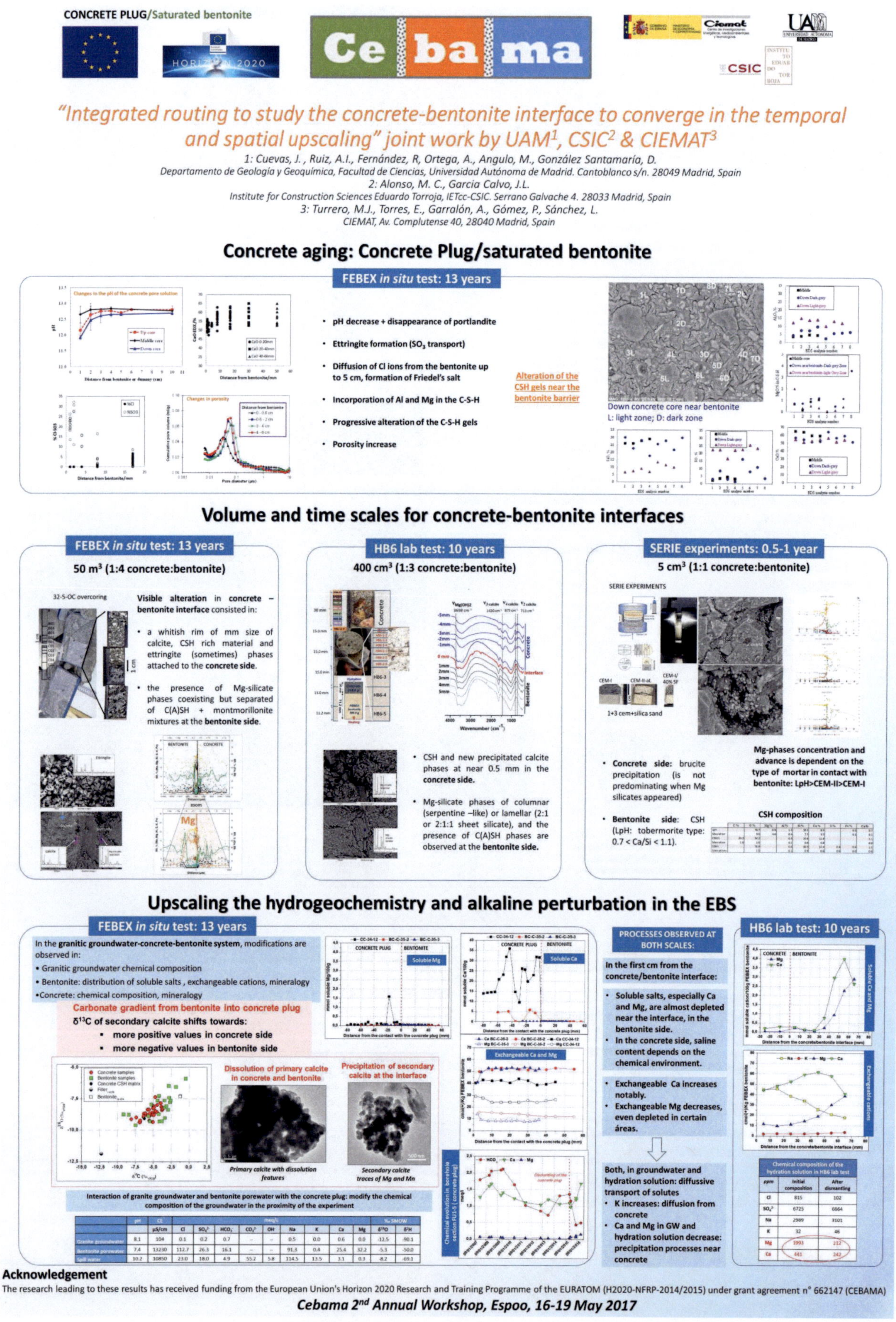

CONCRETE PLUG/Saturated bentonite

"Integrated routing to study the concrete-bentonite interface to converge in the temporal and spatial upscaling" joint work by UAM[1], CSIC[2] & CIEMAT[3]

1: Cuevas, J. , Ruiz, A.I., Fernández, R, Ortega, A., Angulo, M., González Santamaría, D.
Departamento de Geología y Geoquímica, Facultad de Ciencias, Universidad Autónoma de Madrid. Cantoblanco s/n. 28049 Madrid, Spain
2: Alonso, M. C., Garcia Calvo, J.L.
Institute for Construction Sciences Eduardo Torroja, IETcc-CSIC. Serrano Galvache 4. 28033 Madrid, Spain
3: Turrero, M.J., Torres, E., Garralón, A., Gómez, P., Sánchez, L.
CIEMAT, Av. Complutense 40, 28040 Madrid, Spain

Concrete aging: Concrete Plug/saturated bentonite

FEBEX *in situ* test: 13 years

- pH decrease + disappearance of portlandite
- Ettringite formation (SO_3 transport)
- Diffusion of Cl ions from the bentonite up to 5 cm, formation of Friedel's salt
- Incorporation of Al and Mg in the C-S-H
- Progressive alteration of the C-S-H gels
- Porosity increase

Alteration of the CSH gels near the bentonite barrier

Down concrete core near bentonite
L: light zone; D: dark zone

Volume and time scales for concrete-bentonite interfaces

FEBEX *in situ* test: 13 years

50 m³ (1:4 concrete:bentonite)

Visible alteration in concrete – **bentonite interface** consisted in:

- a whitish rim of mm size of calcite, CSH rich material and ettringite (sometimes) phases attached to the **concrete side**.
- the presence of Mg-silicate phases coexisting but separated of C(A)SH + montmorillonite mixtures at the **bentonite side**.

HB6 lab test: 10 years

400 cm³ (1:3 concrete:bentonite)

- CSH and new precipitated calcite phases at near 0.5 mm in the **concrete side.**
- Mg-silicate phases of columnar (serpentine –like) or lamellar (2:1 or 2:1:1 sheet silicate), and the presence of C(A)SH phases are observed at the **bentonite side.**

SERIE experiments: 0.5-1 year

5 cm³ (1:1 concrete:bentonite)

SERIE EXPERIMENTS

1+3 cem+silica sand

- **Concrete side:** brucite precipitation (is not predominating when Mg silicates appeared)
- **Bentonite side:** CSH (LpH: tobermorite type: 0.7 < Ca/Si < 1.1).

Mg-phases concentration and advance is dependent on the type of mortar in contact with bentonite: LpH>CEM-II>CEM-I

CSH composition

Upscaling the hydrogeochemistry and alkaline perturbation in the EBS

FEBEX *in situ* test: 13 years

In the **granitic groundwater-concrete-bentonite system**, modifications are observed in:

- Granitic groundwater chemical composition
- Bentonite: distribution of soluble salts , exchangeable cations, mineralogy
- Concrete: chemical composition, mineralogy

Carbonate gradient from bentonite into concrete plug

$\delta^{13}C$ of secondary calcite shifts towards:

- **more positive values in concrete side**
- **more negative values in bentonite side**

Dissolution of primary calcite in concrete and bentonite

Precipitation of secondary calcite at the interface

Primary calcite with dissolution features

Secondary calcite traces of Mg and Mn

Interaction of granite groundwater and bentonite porewater with the concrete plug: modify the chemical composition of the groundwater in the proximity of the experiment

	pH	CE	meq/L									‰ SMOW	
		µS/cm	Cl	SO_4^{2-}	HCO_3^-	CO_3^{2-}	OH	Na	K	Ca	Mg	$\delta^{18}O$	δ^2H
Granite groundwater	8.1	104	0.1	0.2	0.7	–	–	0.5	0.0	0.6	0.0	-12.5	-90.1
Bentonite porewater	7.4	13230	112.7	26.3	16.1	–	–	91.3	0.4	25.4	32.2	-5.3	-50.0
Spill water	10.2	10850	23.0	18.0	4.9	55.2	5.8	114.5	13.5	3.1	0.3	-8.2	-69.1

Soluble Mg

Soluble Ca

Exchangeable Ca and Mg

Chemical evolution in borehole section FU1-5 (concrete plug)

PROCESSES OBSERVED AT BOTH SCALES:

In the first cm from the concrete/bentonite interface:

- **Soluble salts, especially Ca and Mg, are almost depleted near the interface, in the bentonite side.**
- **In the concrete side, saline content depends on the chemical environment.**

- **Exchangeable Ca increases notably.**
- **Exchangeable Mg decreases, even depleted in certain áreas.**

Both, in groundwater and hydration solution: diffusive transport of solutes

- **K increases: diffusion from concrete**
- **Ca and Mg in GW and hydration solution decrease: precipitation processes near concrete**

HB6 lab test: 10 years

Soluble Ca and Mg

Exchangeable cations

Chemical composition of the hydration solution in HB6 lab test		
ppm	Initial composition	After dismantling
Cl	815	102
SO_4^{2-}	6725	6664
Na	2989	3101
K	32	46
Mg	1993	212
Ca	441	242

Acknowledgement

The research leading to these results has received funding from the European Union's Horizon 2020 Research and Training Programme of the EURATOM (H2020-NFRP-2014/2015) under grant agreement n° 662147 (CEBAMA)

Cebama 2nd Annual Workshop, Espoo, 16-19 May 2017

Rapid development of a reactive transport code with Fenics and Reaktoro

u^b UNIVERSITÄT BERN
PAUL SCHERRER INSTITUT PSI

Leonardo Hax Damiani[a,b], Philipp Krejci[a,b], Sergey V. Churakov[a,b] and Georg Kosakowski[a]
[a]LES, Paul Scherrer Institut, CH-5232, Villigen PSI, Switzerland
[b]Institute for Geological Sciences, University of Bern (CH), Switzerland

Cebama

Motivation & Background

We investigate the geochemical evolution of the natural and engineered barrier system of **deep geological repositories** for nuclear waste.

Cement and **clay** based materials are important components of the barrier system. Due to the low permeability of these materials fluid flow is suppressed and **diffusion** dominates transport.

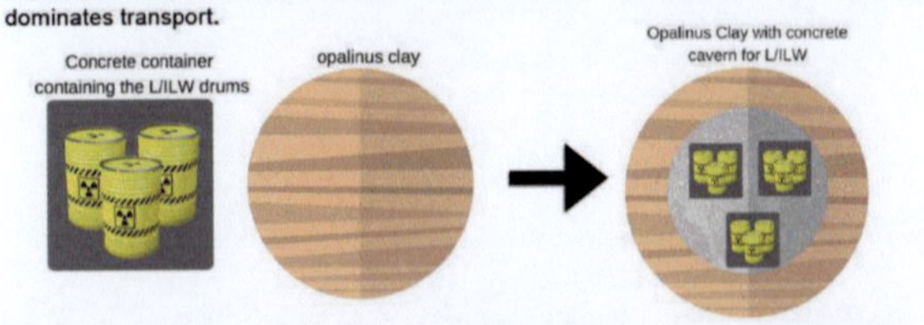

Diffusive transport in these media is affected by the presence of **minerals with charged surfaces.** These charges partition the pore space and influence transport of ions.

Cement and clay are not in thermodynamical equilibrium. Geochemical gradients across cement/clay interfaces drives transport of chemical species. Changes in the chemical composition trigger the **precipitation and dissolution of minerals** which **change locally porosity** and associated properties like permeability and diffusivity. These changes affect large-scale transport properties across the interface and might degrade the functionality of natural and engineered barriers. We develop a flexible **open-source** code that takes into account **transport of ions** in presence of minerals with charged surfaces, which is applicable to **complex chemical systems**, and provides coupling between mineral transformation and transport properties of the medium.

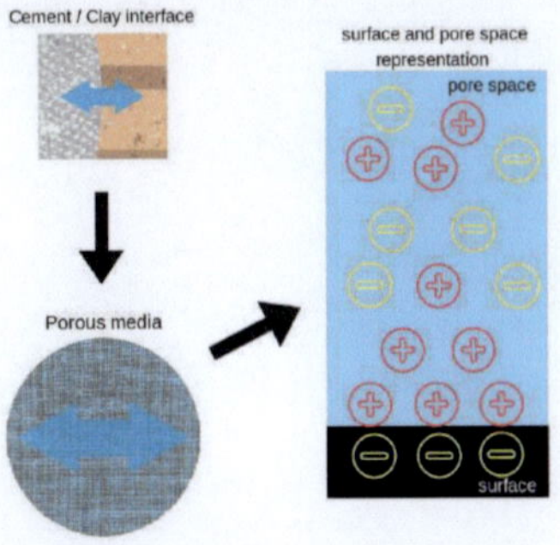

Methodology

- **Operator Splitting approach**: transport and chemical problems solved separately and iteratively.
- **Electrochemical transport**: Implemented using FEniCS, an open source finite element library for solving partial differential equations. [www.fenicsproject.org].
- **Chemical solver**: Use of Reaktoro solver for calculating the geochemical equilibrium [www.reaktoro.org].
- **Donnan Approach**: Used for partitioning the pore space according to the influence of the surface charges.

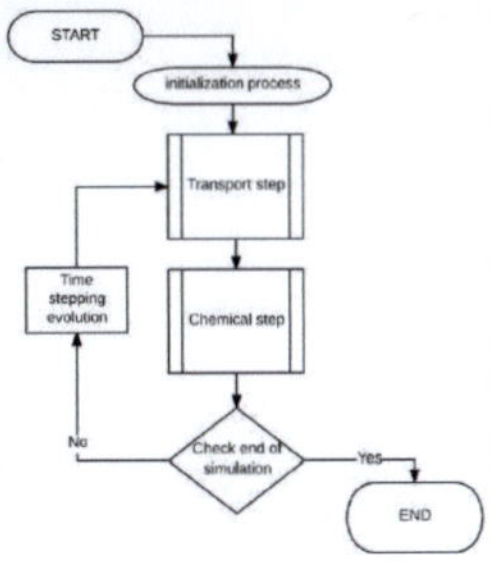

FEniCS PROJECT
NUMFOCUS OPEN CODE = BETTER SCIENCE
GEMS SELEKTOR

Donnan Approach

- The **electric field** induced by charged mineral surfaces is responsible for the electrostatic repulsion of same charged ions and attraction of opposite charged ions.
- This volume, with averaged properties, is defined as the Donnan volume (DV).
- Bulk volume and DV have **different compositions**. Transport in both volumes is described with a set of coupled transport equations.

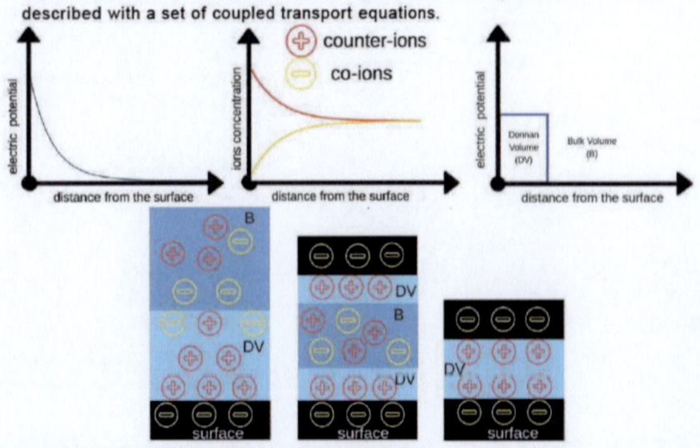

Application & Benchmarking

- Diffusion of **HNO_3** across a membrane is studied by imposing a concentration gradient in presence of a constant background concentration of **KCl**.

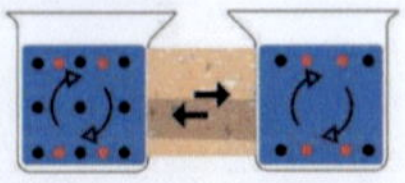

- In the experiment, a **porous cellulose acetate membrane** represents a non reactive porous medium. A conservative tracer **HTO** (tritiated water) is used to measure the transport properties of the membrane.
- KCl and HNO_3 are present in the form of ions **K^+, Cl^-, H^+** and **NO_3^-** with **different mobilities.**
- Diffusion of ions with different mobility is cross-coupled due to **electroneutrality** condition.
- The diffusion of H^+ and NO_3^- **induces** changes in concentration of K^+ and Cl^- due to cross-coupling.
- Results from our code (NPS) **agree** well with the experimental data.

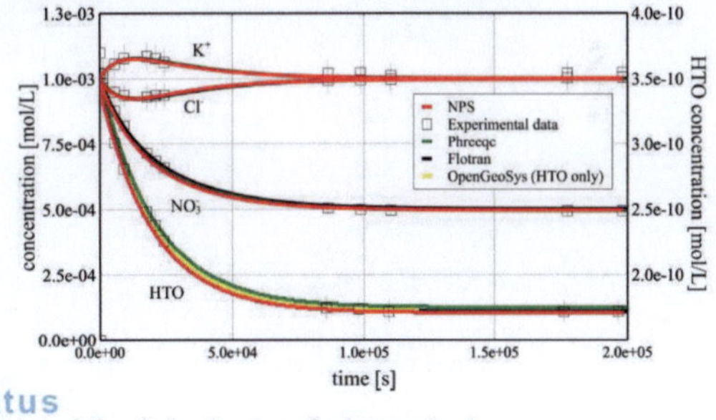

Status

- Implementation of a functional **reactive transport** code:
  - Implementation of the **Nernst-Planck** equations with standard Galerkin FE method
  - Prototype implementation with **discontinuous Galerkin** FE method
  - Coupling of transport and chemical solver **REAKTORO** via python interface, which allows to run chemical setups created with GEM-Selektor V3 and PHREEQC
- We have **successfully** verified the implementation of the Nernst-Planck equations (benchmarks and application).
- Currently ongoing: Implementation of the **Donnan Approach** into Reaktoro and GEMS.
- Current work on transport across reactive cement/clay interfaces based on data from the **Mont Terri CI experiment**.

ACKNOWLEDGMENTS

This work was supported by the Swiss State Secretariat for Education, Research and Innovation (SERI) under contract number 15.0186-2. The opinions expressed and arguments employed herein do not necessarily reflect the official views of the Swiss Government. Authors receive partial financial support from Nagra.

ACKNOWLEDGMENTS

This project is part of CEBAMA and addresses key scientific questions related to the use of cement-based materials in nuclear waste repositories including interface processes between cement-based materials and host rocks, retention of high priority radionuclides, and the development of models predicting changes in transport properties of cement-based materials.

CONTACT INFORMATION

Leonardo Hax Damiani
leonardo.hax@psi.ch

Prof. Dr. Sergey V. Churakov
sergey.churakov@psi.ch

PD Dr. Georg Kosakowski
georg.kosakowski@psi.ch

LABORATORY FOR WASTE MANAGEMENT

LES is the Swiss competence center for geochemistry and multi scale radionuclide and mass transport in argillaceous rocks and cement and their applications to deep geological systems and Swiss radioactive waste repositories.
https://www.psi.ch/les

PAUL SCHERRER INSTITUT (PSI)

2nd annual Workshop of the Cebama project

Low-pH cement carbonation rate and impact on microstructure gas transport and shrinkage

Ekoé KANGNI-FOLI
IRSN
31. avenue de la division Leclerc
92262 Fontenay-aux-roses France

Context and objectives

Low-pH concrete sheathing for Cigeo's sealing areas

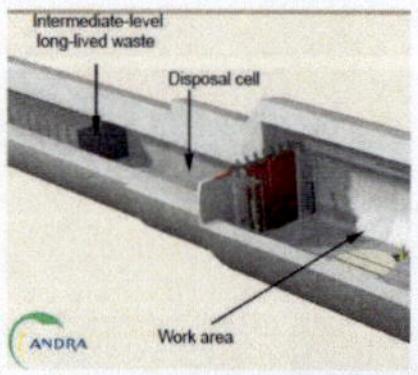

Low-pH concrete

- CO_2 diffusion within the concrete
- Formation of calcium carbonate
- Decalcification of calcium hydrates

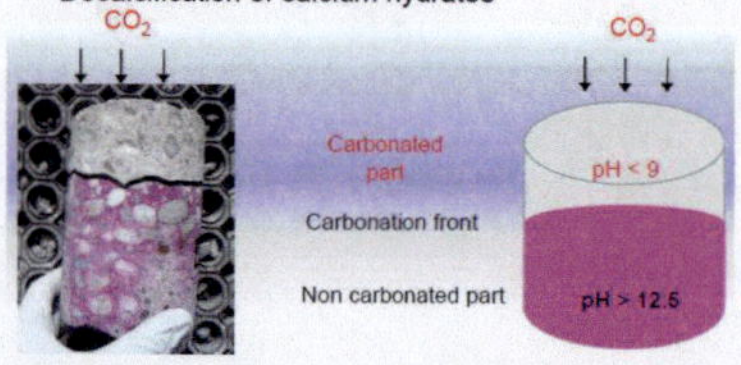

Low-pH concrete properties

- (Ca/Si) ratio decreased by adding silica
  → alkalinity. hydration heat and creep lowered
- Low-pH buffering capability < CEM I due to $Ca(OH)_2$ lower content
  → carbonation effect = cracking rate ↗
  → transport properties deteriorated

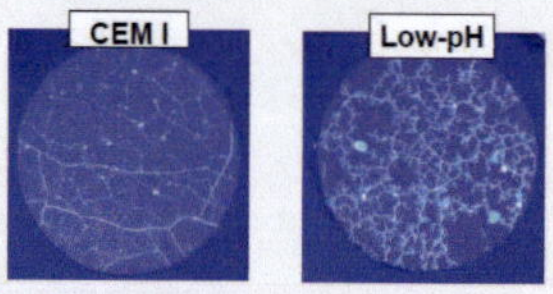

Paste impregnated with fluorescent resin under UV light
Auroy et al. (2015) Cem. Concr. Res. 74. 44-58

What about the durability of low-pH concrete?
In term of chemical, mineralogical , microstructural and macroscopical changes induced by the carbonation

Approaches

Multi scales analysis

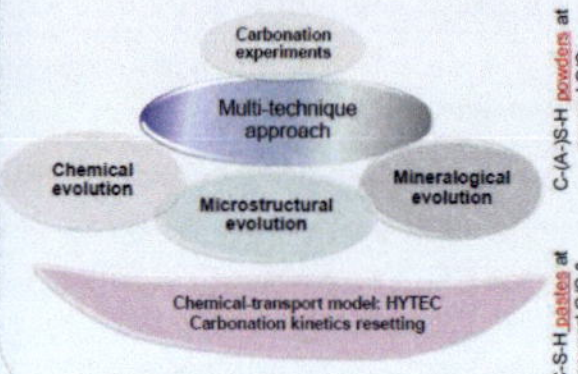

Study of the evolution of the basic unit: C-(A-)S-H

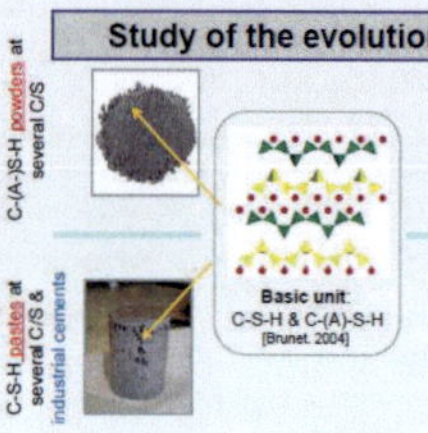

- *Description of the ultimate carbonation's state*
- Representativity of accelerated carbonation *vs* natural carbonation
- Influence of Ca/Si ratio

Changes induced by carbonation

- *Identifying the carbonation kinetics*
- Influence of Ca/Si ratio
- Representativity of accelerated carbonation *vs* natural carbonation
- Could the result obtained on low-pH model paste be extrapolated on industrial paste ?

Macroscopical's scale study: shrinkage

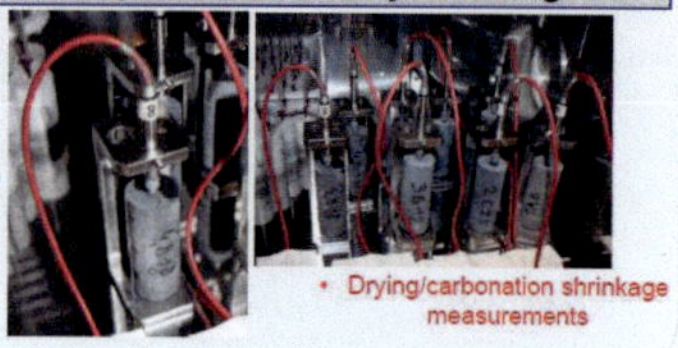

- Drying/carbonation shrinkage measurements

Auroy (2014) CEA-R-6401

Experiments

Samples

	C-S-H							C-A-S-H	low-pH industrial cements	
	C_3S	C_3S + nanosilica			C-S-H synthetic			C-A-S-H synthetic	50% CEM III/A 50% silica fume	50% CEM I 50% silica fume
	Paste				Powder				Paste	
Ca/Si	3	1.4	0.95-1	0.8	1.4	0.95-1	0.8	0.8	0.85	0.85
Al/Si	-	-	-	-	-	-	-	0.05	> 0.05	> 0.05
W/C	0.4-0.6	0.6	0.65	0.7	50				0.6	

Complexity of the mineralogy →

Accelerated carbonation device

P_{CO2} = 3%. 25° C. 55% HR. 7 days→ 1.5 years

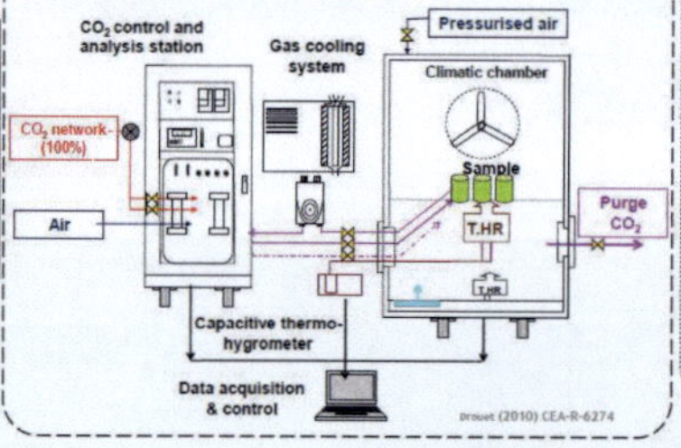

Drouet (2010) CEA-R-6274

Multi-technique analysis

- NMR ^{29}Si + ^{27}Al → adequate probe for C-S-H / C-A-S-H structure analysis
- XRD/ TGA → Phases' identification & quantification
- X Ray Tomography/Autoradiography ^{14}C → microstructure's probes
- Modeling/simulation (HYTEC) → chemical kinetics determination by inverse analysis

Chemistry and mineralogy

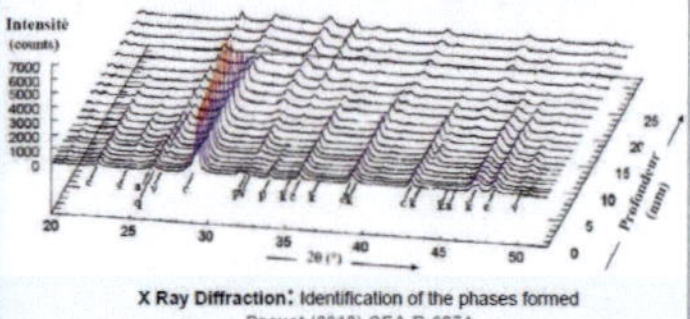

X Ray Diffraction: Identification of the phases formed
Drouet (2010) CEA-R-6274

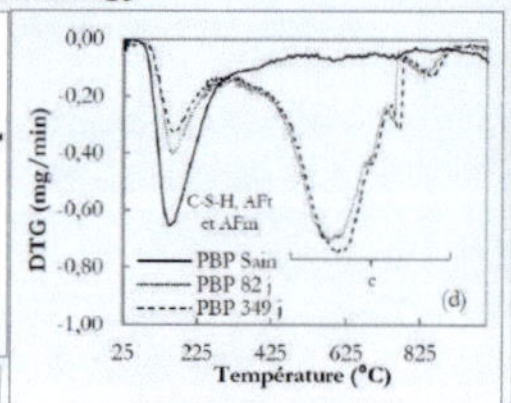

Thermo Gravimetric analysis: Quantification of the phases formed
Auroy (2014) CEA-R-6401

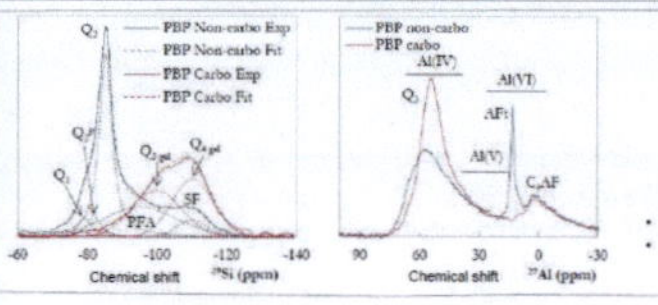

Nuclear magnetic resonance

- Identification/quantification of carbonated phases
- Suivi de dégradation des C-(A)-S-H

Auroy (2014) CEA-R-6401

Evolution of the microstructure

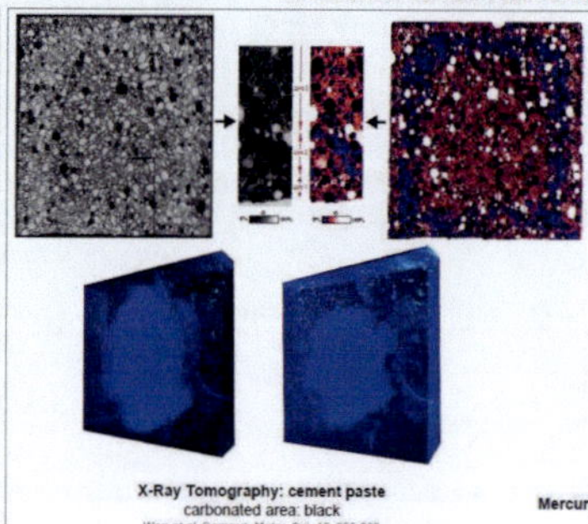

Autoradiography: mortar impregnated by ^{14}C-PMMA
White= bubbles, blue= carbonated area low porosity (≈12%), red= healthy area porosity (≈ 26%).
Black= non porous grains
Jeong et al. Comput. Mater. Sci. 69. 466-480

X-Ray Tomography: cement paste carbonated area: black
Wan et al. Comput. Mater. Sci. 45. 255-263

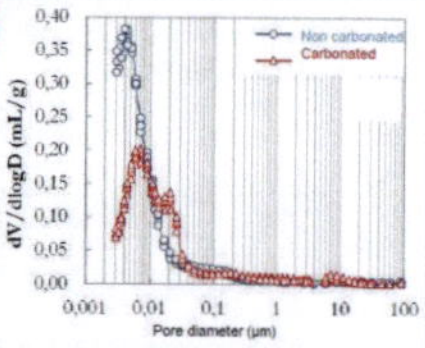

Mercury intrusion porosimetry: healthy and carbonated low-pH cement paste
Auroy (2014) CEA-R-6401

Collaborations

Jean-Baptiste d'ESPINOSE
SIMM

Alexandre DAUZERES & Émilie L'HÔPITAL
LETIS

Thibault CHARPENTIER , Stéphane POYET
Patrick LE-BESCOP
LECBA-LSDRM

Marja SIITARI-KAUPPI
Department of chemistry

UNIVERSITY OF SURREY — Cebama — NPL National Physical Laboratory

Chemical and Mineralogical Changes at the Interface between Cementitious Materials and Groundwater

E. Rastrick[1], M. Isaacs[1] and D. Read[1,2]
[1] Department of Chemistry, Faculty of Engineering and Physical Sciences, University of Surrey, Guildford, Surrey, GU2 7XH, UK
[2] National Physical Laboratory, Hampton Road, Teddington, Middlesex, TW11 0LW, UK

Introduction

The research aims to understand the chemical and mineralogical changes occurring at the interface between cementitious materials and groundwaters relevant to a radioactive waste repository. Cement and concrete can play an important role in the immobilisation of low- and intermediate-level waste. Cement is predominantly the medium that will condition inflowing groundwater to high pH, reducing radionuclide solubility as well as acting as a surface for complexation, adsorption or ion exchange with phases in the cement [1,2].

Methodology

Four cement blends have been chosen to represent the various compositions being considered for use within European radioactive waste repositories together with three synthetic groundwaters which broadly emulate granitic, clay and saline geological environments within the host rock. Radioactive tracers will be used to provide information on reaction mechanisms and timescales, notably the kinetics governing the incorporation of target elements into the cement matrix. The data gained from this research will support modelling studies allowing prediction over the spatial and temporal scales necessary for repository safety case simulations.

Cement Blend Compositions

Table 1: Cement blend compositions as set by CEBAMA project

Cement blend	OPC	GGBS	Hydrated Lime	Limestone Flour	Silica Fume
CEM 1	1	-	-	-	-
GGBS:OPC	1	9	-	-	-
NRVB	1	-	0.38	1.1	-
CEBAMA Reference mix	1	0.62	-	-	0.87

Experimental set-up

Diffusion

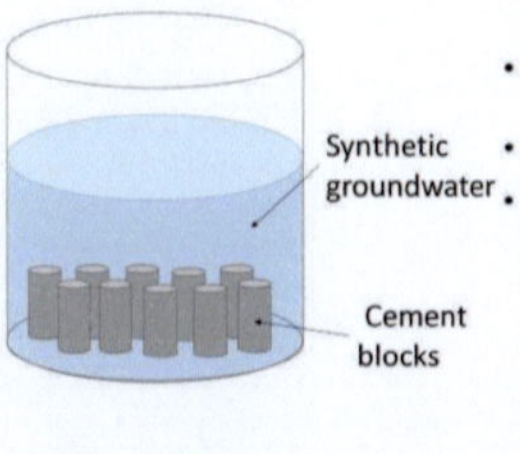

- Large number of blocks in large volume of water
- Observe uptake of groundwater ions and how this affects the phases within cement
- Periodic sampling of cement block and the surrounding groundwater
- Constant solid to liquid ratio
- Observe how the groundwater affects the cement blocks by for example, dissolution and formation of new phases, e.g. ettringite

Conditioning of groundwater

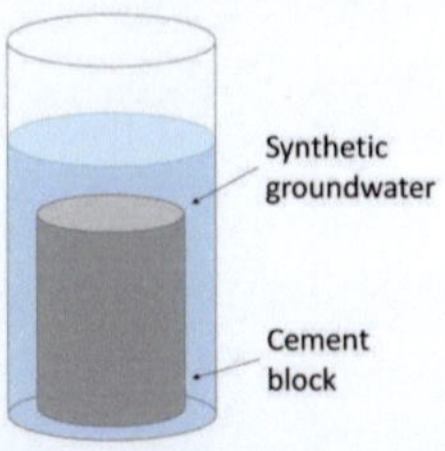

- One cement block in a smaller volume of water
- Observe how the cement conditions the surrounding groundwater
- To be analysed at the end of the two year period with no disturbances
- The experiment will allow changes in groundwater composition to be determined, for example with the dissolution of cement phases leading to an increase in pH and Ca ion concentration

Groundwater compositions

Table 2: Concentrations of groundwater ions determined by ion chromatography compared to reference concentrations

	Granitic GW reference concentration (mmol L^{-1}) [3]	Granitic GW measured concentration (mmol L^{-1})	Clay GW reference concentration (mmol L^{-1}) [4]	Clay GW measured concentration (mmol L^{-1})	Saline GW reference concentration (mmol L^{-1}) [3]	Saline GW measured concentration (mmol L^{-1})
Na^+	2.8	3.0	55.0	41.1	140.0	130.8
K^+	0.1	0.1	1.1	1.3	2.1	4.2
Ca^{2+}	0.5	0.6	7.5	6.4	19.9	24.6
Mg^{2+}	0.2	0.2	5.7	4.6	0.4	2.5
Cl^-	2.0	2.2	50.0	57.7	180.0	193.7
HCO_3^-	2.0	TBA	-	-	2.0	TBA
SO_4^{2-}	0.1	0.1	15.0	15.5	4.0	6.4
SiO_2	0.2	TBA	-	-	-	-
pH	8.2	TBA	7.4	TBA	7.7	TBA

Bulk analysis of cement

as determined by X-ray fluorescence, major constituents normalised to 100 %. LOD = limit of detection

Table 3: Major constituents within cement powders

	Units	OPC	GGBS	Hydrated lime	Lime Flour	Silica Fume
CaO	%	66.27	42.93	97.75	96.57	1.46
SiO_2	%	17.86	31.92	1.37	1.70	93.10
Al_2O_3	%	4.78	11.40	0.35	0.71	1.44
SO_3	%	3.98	4.10	0.08	0.18	0.47
MgO	%	2.75	6.76	0.20	0.29	0.88
Fe_2O_3	%	2.65	0.53	0.03	0.16	0.91
K_2O	%	1.52	1.19	0.33	0.39	1.73
TiO_2	ppm	2000	4800	61	232	50
P_2O_5	ppm	< LOD	< LOD	102	< LOD	< LOD

Table 4: Minor constituents within cement powders

	Units	OPC	GGBS	Hydrated Lime	Lime Flour	Silica Fume
Mn	ppm	363	6900	132	412	221
Sr	ppm	858	591	252	215	45
Zn	ppm	459	20	14	44	44
U	ppm	19	27	3	4	16
Th	ppm	10	28	2	3	2

Conclusions and Future Work

The synthetic groundwaters have been produced and analysed. Diffusion and conditioning experiments have initially been conducted without radioactive tracers to aid with planning of the active experiments. In the active experiments, the tracers will be introduced into the synthetic groundwaters; groundwater analysis will be conducted using liquid scintillation counting and the cement blocks analysed using digital autoradiography.

References and Acknowledgements

[1] NDA (2010) Geological Disposal: Radionuclide behaviour status report. NDA Report no. NDA/RWMD/034, 1-167
[2] Crossland I. G. and Vines S. P. (2001) Why a cementitious repository?, Nirex Report No: N/034, 1-28
[3] Gascoyne M. (2002), Influence of grout and cement on groundwater composition, Posiva Working Report, 1-44
[4] Vinsot A., Mettler S., Wechner S. (2008), In situ characterization of the Callovo-Oxfordian porewater composition, *Physics and Chemistry of the Earth*, **33**, S75-S86
We would like to thank the National Physical Laboratory for assistance with the experimental programme. The research has received funding from the European Union's Horizon 2020 Research and Training Programme of the European Atomic Energy Community (EURATOM) (H2020-NRFP-2015/2016) under grant agreement n° 662147 (CEBAMA) and from Radioactive Waste Management (RWM), UK.